D0722625

On-Chip Communication Architectures

The Morgan Kaufmann Series in Systems on Silicon

Series Editor, Wayne Wolf, Georgia Institute of Technology

The Designer's Guide to VHDL, Second Edition
Peter J. Ashenden

The System Designer's Guide to VHDL-AMS
Peter J. Ashenden, Gregory D. Peterson, and Darrell A. Teegarden

Modeling Embedded Systems and SoCs
Axel Jantsch

ASIC and FPGA Verification: A Guide to Component Modeling
Richard Munden

Multiprocessor Systems-on-Chips
Edited by Ahmed Amine Jerraya and Wayne Wolf

Functional Verification
Bruce Wile, John Goss, and Wolfgang Roesner

Customizable and Configurable Embedded Processors
Edited by Paolo Ienne and Rainer Leupers

Networks-on-Chips: Technology and Tools
Edited by Giovanni De Micheli and Luca Benini

VLSI Test Principles & Architectures
Edited by Laung-Terng Wang, Cheng-Wen Wu, and Xiaoqing Wen

Designing SoCs with Configured Processors
Steve Leibson

ESL Design and Verification
Grant Martin, Andrew Piziali, and Brian Bailey

Aspect-Oriented Programming with ***e***
David Robinson

Reconfigurable Computing: The Theory and Practice of FPGA-Based Computation
Edited by Scott Hauck and André DeHon

System-on-Chip Test Architectures
Edited by Laung-Terng Wang, Charles Stroud, and Nur Touba

Verification Techniques for System-Level Design
Masahiro Fujita, Indradeep Ghosh, and Mukul Prasad

VHDL-2008: Just the New Stuff
Peter J. Ashenden and Jim Lewis

On-Chip Communication Architectures: System on Chip Interconnect
Sudeep Pasricha and Nikil Dutt

To Come

Embedded DSP Processor Design: Application Specific Instruction Set Processors
Dake Liu

Processor Description Languages
Prabhat Mishra

On-Chip Communication Architectures

System on Chip Interconnect

Sudeep Pasricha – Nikil Dutt

ELSEVIER

AMSTERDAM • BOSTON • HEIDELBERG • LONDON
NEW YORK • OXFORD • PARIS • SAN DIEGO
SAN FRANCISCO • SINGAPORE • SYDNEY • TOKYO

Morgan Kaufmann is an imprint of Elsevier

Senior Acquisitions Editor: Charles B. Glaser
Publishing Services Manager: George Morrison
Project Manager: Mónica González de Mendoza
Assistant Editor: Greg Chalson
Cover Design: Dennis Schaefer

Morgan Kaufmann Publishers is an imprint of Elsevier.
30 Corporate Drive, Suite 400, Burlington, MA 01803, USA

This book is printed on acid-free paper.

Library of Congress Cataloging-in-Publication Data
Pasricha, Sudeep..
On-chip communication architectures: system on chip interconnect/Sudeep Pasricha, Nikil Dutt.
p. cm.
Includes bibliographical references and index.
ISBN-13: 978-0-12-373892-9 (hardback: alk. paper) 1. Systems on a chip. 2. Microcomputers—Buses 3. Computer architecture. 4. Interconnects (Integrated circuit technology) I. Dutt, Nikil. II. Title.
TK7895.E42P4 2008
621.3815—dc22

2008004691

ISBN: 978-0-12-373892-9

For information on all Morgan Kaufmann publications, visit our Web site at www.mkp.com or www.books.elsevier.com

08 09 10 11 12 13 10 9 8 7 6 5 4 3 2 1

Printed in the United States of America

Contents

Preface

Digital electronic devices such as mobile phones, video game consoles, and network routers typically contain one or more electronic (integrated circuit) chips that are composed of several components such as processors, dedicated hardware engines and memory, and are referred to as system-on-chip (SoC). These SoC designs are rapidly becoming more complex, in order to handle the ever increasing complexity of applications, fueled by the onset of the digital convergence era. Continuing improvements in process technology have allowed the integration of components previously connected at the board level onto a single chip, which further adds to the complexity.

The components on a SoC are connected together by an *on-chip communication architecture* backbone that supports all inter-component data communication, both within the chip as well as with external devices (e.g., external flash drives). These SoC communication architectures have been shown to have a significant impact on the performance, power consumption, cost, and design time of SoCs. Indeed, modern SoC design processes are increasingly becoming *communication-centric,* since reusable components (e.g., processors, memories, etc.), as well as custom hardware blocks and interfaces, need to be connected via a communication architecture fabric, with the goal of meeting various design constraints such as cost, performance, power/energy, and reliability. The move toward higher levels of abstraction have led to the notion of *electronic system level (ESL) design,* where system architects and application designers are able to capture system functionality and map desired system functionality onto a range of software and hardware configurations that exhibit differing performance, cost, power/energy, reliability, and other design metrics. A key step within an ESL design flow is the efficient use of an on-chip communication architecture fabric. Consequently, there has been a large body of work on modeling abstractions, communication protocols and standards, as well as active research on communication architecture design and exploration.

This book aims to serve as a comprehensive reference on the concepts, research, and trends in on-chip communication architecture design. We describe the basic concepts and attributes of on-chip communication architectures, to familiarize the reader with intricate details of on-chip communication architecture design and the problems facing designers. This is followed by an expansive survey of research efforts in this area, spanning the past several years, and addressing some of the major issues in on-chip communication architecture design. Finally, we present some of the trends that will shape future research in the area of on-chip communication architecture design.

AUDIENCE FOR THIS BOOK

This book is designed for a number of different audiences. Graduate and undergraduate students, as well as design engineers working in the area of SoC design should find this book useful as an introduction to the field of on-chip communication architectures. Faculty members can use this book as a textbook for a graduate or senior undergraduate course on ESL and communication-centric design. Furthermore, chip and hardware engineers who currently design at the logic or register-transfer level will find this book useful to understand the transition to higher levels of abstraction (ESL) and get insights into the key principles of communication-centric design. Finally, the book should serve as a useful reference for managers, technologists, and industry analysts who need to track trends in SoC design, ESL methodologies, and communication-centric SoC design flows.

HOW TO USE THIS BOOK

Although each chapter in this book is self-contained (with an extensive set of references at the end of each chapter), the first three chapters of this book contain the introductory material that set the stage for the design methodology of communication-centric design (Chapter 1), basic concepts of on-chip communication (Chapter 2), and contemporary/commonly used on-chip bus communication standards (Chapter 3). We recommend that these three chapters be covered for all audiences before embarking on different trajectories, based on the audience and goals. In the remainder of this section, we present two possible flows for covering the material in the book.

Figure 1 shows the suggested chapter trajectory for use of this book in a one semester/quarter graduate or upper-division undergraduate course. The first big box (Chapters 1–6 and 9) covers a snapshot of the entire communication architecture modeling, analysis, and synthesis process. After the first three introductory chapters, Chapters 4 and 5 introduce models that allow evaluation of performance and power/energy. Chapters 6 and 9 cover on-chip communication architecture synthesis and interface synthesis, respectively. Each of these chapters starts with an introductory (or generic) flow, and then surveys recent research efforts in a comprehensive manner. Next, the instructor or student has a number of choices for traversing topics, based on the focus/interest of the course and the amount of time. In conjunction with a traditional CAD course that covers physical design, logic, and register-transfer level design, the next logical step would be to cover Chapter 11. On the other hand, a course focused on synthesis and methodology issues may benefit from the chapter on encoding techniques (Chapter 7) as well as verification and security (Chapter 10). Finally, the topics of emerging architectures will round out any course by covering material on custom architectures (Chapter 8), Networks-on-Chip (Chapter 12) and emerging technologies (Chapter 13).

Figure 2 outlines a suggested flow for design engineers, practitioners, managers, technologists, industry analysts, etc. We recommend that you begin with the first three chapters to set the stage with basic terminology and concepts, before

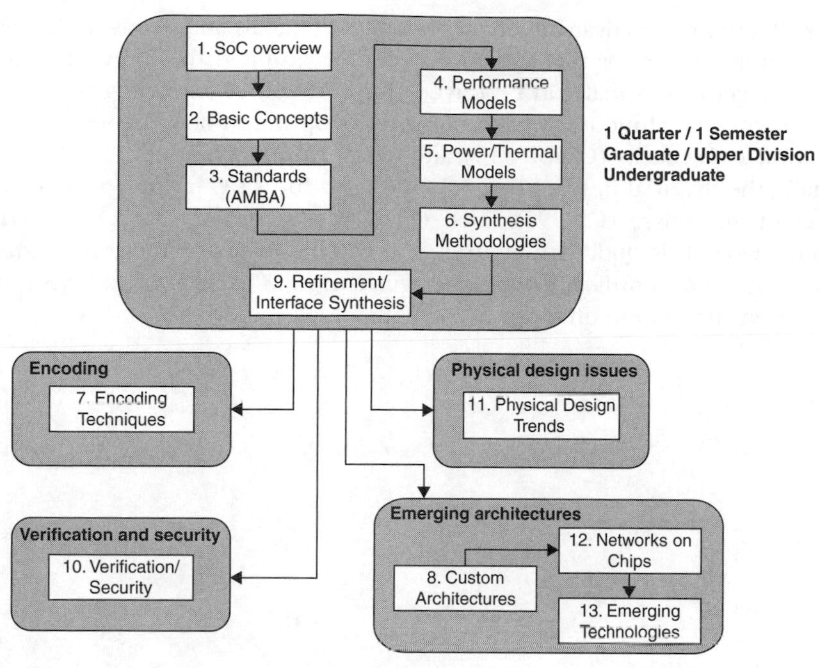

FIGURE 1

Suggested flow for a graduate/undergraduate course

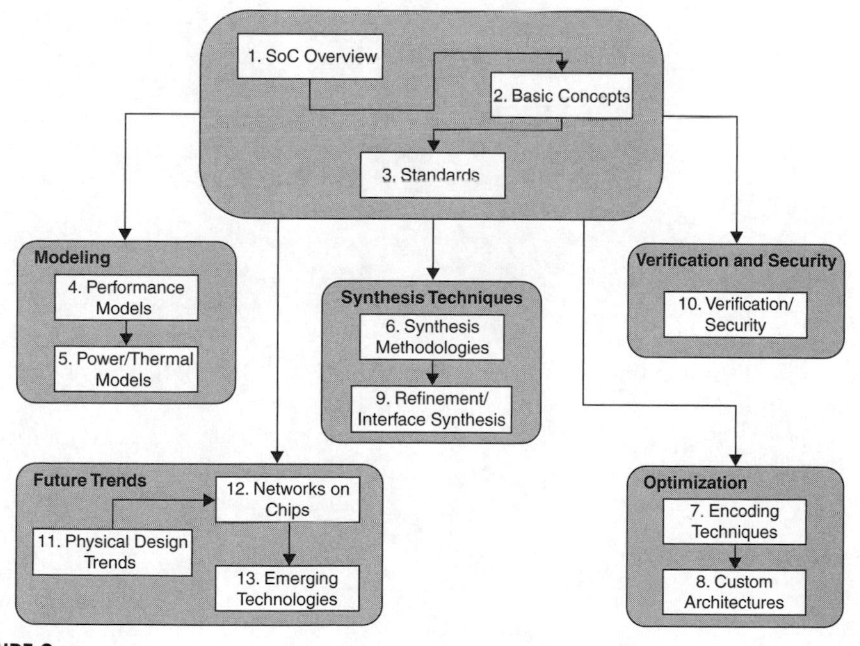

FIGURE 2

Suggested flow for a practitioners, designers, managers, technologists, etc.

embarking onto more advanced concepts. Since the goals and needs of the reader may be quite diverse for this audience, we have grouped the relevant chapters into high level topics that can be traversed in any order: modeling (Chapters 4 and 5); synthesis techniques (Chapters 6 and 9); optimization (Chapters 7 and 8); verification and security (Chapter 10); and future trends (Chapters 11–13).

Finally, the material in this book is being used for a UC Irvine Spring Quarter 2008 graduate course (CS 295) titled *On-Chip Communication Architectures*. Teaching material (including transparencies) will be posted online at *http://www.ics.uci.edu/~dutt/comm-arch-book* and will continue to be updated with material based on subsequent offerings of the course.

About the Authors

Sudeep Pasricha is an Assistant Professor at Colorado State University, with academic appointments in the CS and ECE departments, and is affiliated with the Center for Embedded Computer Systems (CECS) at UC Irvine. He received his B.E.(Hons) in Electronics and Communications Engineering from Delhi Institute of Technology, Delhi, India in 2000, an M.S. in Computer Science from the University of California, Irvine in 2005, and a Ph.D. in Computer Science from the University of California, Irvine in 2008.

Dr. Pasricha's research interests are in the areas of on-chip communication architectures, embedded systems, electronic design automation, system-level modeling languages and design methodologies, computer architecture, and VLSI CAD algorithms. He has presented several tutorials in the area of on-chip communication architecture design at leading conferences and coauthored over 25 journal and conference publications. He has received a Best Paper Award at ASPDAC 2006, a Best Paper Award nomination at DAC 2005, and several fellowships and awards for excellence in research from Delhi Institute of Technology and UC Irvine.

Nikil Dutt is a Chancellor's Professor at the University of California, Irvine, with academic appointments in the CS and EECS departments. He received a B.E.(Hons) in Mechanical Engineering from the Birla Institute of Technology and Science, Pilani, India in 1980, an M.S. in Computer Science from the Pennsylvania State University in 1983, and a Ph.D. in Computer Science from the University of Illinois at Urbana-Champaign in 1989. He is affiliated with the following Centers at UCI: Center for Embedded Computer Systems (CECS), California Institute for Telecommunications and Information Technology (Calit2), the Center for Pervasive Communications and Computing (CPCC), and the Laboratory for Ubiquitous Computing and Interaction (LUCI).

Dr. Dutt's research interests are in embedded systems, electronic design automation, computer architecture, optimizing compilers, system specification techniques, and distributed systems. He is a coauthor of six other books and over 250 conference and journal publications. His research has been recognized by Best Paper Awards at the following conferences: CHDL'89, CHDL'91, VLSI Design 2003, CODES1ISSS 2003, CNCC 2006, and ASPDAC 2006; and Best Paper Award Nominations at: WASP 2004, DAC 2005, and VLSI Design 2006. He has also received a number of departmental and campus awards for excellence in teaching at UC Irvine.

Dr. Dutt currently serves as Editor-in-Chief of ACM Transactions on Design Automation of Electronic Systems (TODAES), and as Associate Editor of ACM Transactions on Embedded Computer Systems (TECS) and of IEEE Transactions on VLSI Systems (TVLSI). He was an ACM SIGDA Distinguished Lecturer during 2001–2002, and an IEEE Computer Society Distinguished Visitor for 2003–2005.

He has served on the steering, organizing, and program committees of several premier CAD and Embedded System conferences and workshops, including ASPDAC, DATE, ICCAD, CODES1ISSS, CASES, ISLPED, and LCTES. He serves on, or has served on the advisory boards of ACM SIGBED, ACM SIGDA, and IFIP WG 10.5. He is a Fellow of the IEEE, an ACM Distinguished Scientist, and recipient of the IFIP Silver Core award.

Acknowledgments

This book would not have been possible without a great deal of help from many people. We are deeply indebted to Ioannis Savidis and Eby Friedman for contributing the chapter on physical design trends for interconnects. Their insights and expertise on the topic have greatly improved the quality of the finished book. Additionally, Andreas Gerstlauer, Ilya Issenin, Per Gunnar Kjeldsberg, Ioannis Savidis, and Sungjoo Yoo carefully reviewed drafts of this manuscript and provided invaluable comments and suggestions. Without their assistance, this book would not be in its present state.

The chapters in this book benefited greatly from the input of several researchers. We would like to extend special thanks to the following individuals for reviewing the manuscript and giving valuable feedback that led to numerous improvements: Luis Angel Bathen, Elaheh Bozorgzadeh, Arup Chakroborty, Karam S. Chatha, Siddharth Choudhuri, Jesse Dannenbring, Mohammad Ali Ghodrat, Peter Grun, Aseem Gupta, Houman Homayoun, Minyoung Kim, Kyoungwoo Lee, Gabor Madl, Mahesh Mamidipaka, Sorin Manolache, Prabhat Mishra, Jayram Moornikara, Preeti Ranjan Panda, Shinichi Shibahara, and Qiang Zhu.

We are very grateful to several people at Elsevier Publishing for their support. We would like to express our gratitude to Chuck Glaser for all his patience and support through the long and arduous process of getting the manuscript ready and published. Special thanks to Matthew Cater and Gregory Chalson for all their help with obtaining copyright permissions and for being accessible to respond to our many concerns during the writing of this book. Thanks also to Monica Mendoza for taking care of last minute details during the publishing of this book.

Last, but by no means the least, we would like to express our deep appreciation for our families and friends for providing us the assistance and encouragement to complete this book. Directly or indirectly, their guidance, friendship and support contributed immensely to the realization of this book.

Sudeep Pasricha and Nikil Dutt
Irvine, California, May 2008

Acknowledgments

List of Contributors

Ioannis Savidis and Eby G. Friedman
University of Rochester, Rochester, NY, USA

CHAPTER

Introduction

1

1.1 TRENDS IN SYSTEM-ON-CHIP DESIGN

Advances in silicon technology continue to leapfrog projections: At the time this book was being written, we were seeing numerous announcements for billion-transistor chips, whereas only a few years ago integrated circuit transistor counts were in the millions. Such single chip integrated circuits are commonly referred to as *system-on-chip* (SoC), and typically consist of several complex heterogeneous components such as programmable processors, dedicated (custom) hardware to perform specific tasks, on-chip memories, input–output interfaces, and an on-chip communication architecture that serves as the interconnection fabric for communication between these components. The dual forces of advances in technology, coupled with an insatiable demand for convergent computing devices (e.g., smart phones that include cameras, GPS devices, MP3 players) have fueled the need for complex chips that incorporate multiple processors dedicated for specific computational needs. These emerging *multiprocessor system-on-chip* (MPSoC) designs typically consist of multiple microprocessors, and tens to hundreds of additional components. Figure 1.1(a) shows an example of a small MPSoC from the multimedia domain, that incorporates two ARM9 microprocessors running embedded software, several on-chip memories, DMA (direct memory access) and LCD controllers, peripherals (e.g., timer and interrupt controller), and external interfaces (e.g., USB and Ethernet), all of which are integrated via an on-chip bus architecture consisting of multiple shared interconnected buses. Another example of a more complex MPSoC is the IBM Cell [1] (shown in Fig. 1.1(b)) used in the Sony PlayStation 3 gaming console. It consists of nine processors—eight special-purpose synergistic processing units (SPU) that perform dedicated computing tasks, and a single general purpose power processor unit that performs generalized processing, and oversees the activities on the chip. Additionally, the Cell has on-chip Level 2 (L2) cache memory, interface components to interact with external electronic systems, and a ring bus-based on-chip communication architecture that facilitates data communication between the components on the chip. Figure 1.1 clearly shows that emerging MPSoCs will use a variety of on-chip bus communication architectures that are tuned to the requirements of the application, architecture, as well as the available technology.

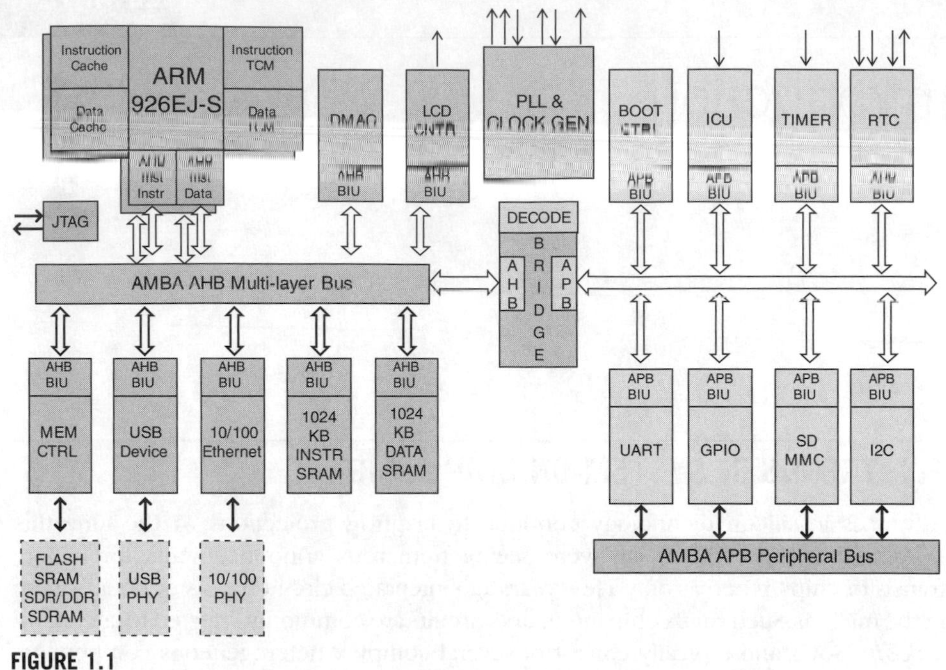

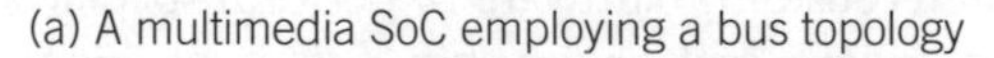

FIGURE 1.1

(a) A multimedia SoC employing a bus topology

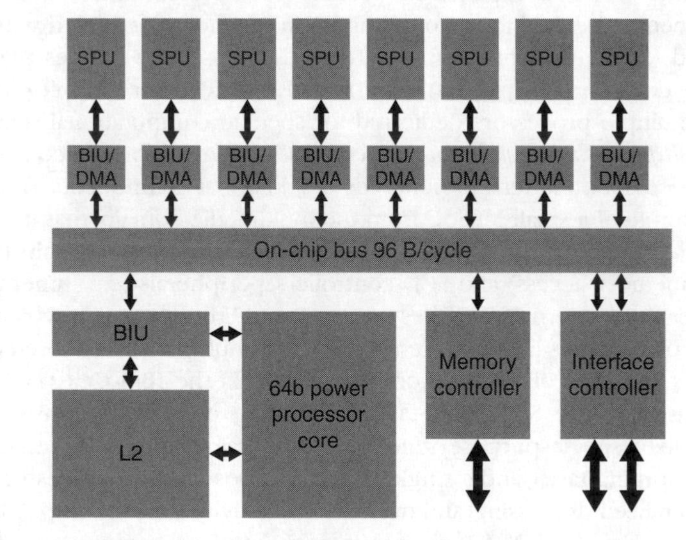

FIGURE 1.1

(b) the IBM Cell processor employing a ring topology

The availability of such a large number of devices on a chip enables new ways of realizing system functionality using a combination of software and hardware, where each MPSoC computational element can be viewed as an *intellectual property* (IP) block. These IP blocks can be a library element from a previous design (i.e., a *hard macro*), a parameterizable component to meet design requirements (i.e., a *soft macro*), or a custom-designed computational block to meet the performance, power, and other design constraints. The architectural landscape for these computational IP blocks range from off-the-shelf *general purpose processor* (GPP) cores, through *application-specific instruction-set processors* (ASIPs), *reconfigurable cores* (RCs), and custom-designed *application-specific integrated circuit* (ASIC) blocks. In going from a GPP to an ASIC block, the system designer is typically trading off flexibility (general purpose programmability) vs. performance, power efficiency, and cost. Furthermore, a plethora of domain-specific programmable processor cores have been developed—either as standalone computational engines or as computational assists to address the specific needs of an application (e.g., *digital signal processors* (DSPs), network processors, and *graphics processing units* (GPUs)).

In the context of such IP-based design methodologies, the communication architecture is both a key enabler, as well as a key differentiator for realizing complex MPSoCs. As will be described in Section 1.3, contemporary IP-based design flows are based on the premise of design reuse, with the system designer stitching together disparate IP blocks using the critical communication architecture fabric for data communication between the IP blocks. Thus the communication architecture *enables* mixing and matching of different IP blocks to create an MPSoC. Furthermore, the communication architecture provides designers the ability to explore multiple communication schemes, topologies and protocols, thereby providing product *differentiation* to meet a diverse, multi-dimensional set of design constraints, including performance, power/energy/temperature, cost, reliability, and time-to-market.

1.2 COPING WITH SoC DESIGN COMPLEXITY

The continuing improvements in silicon technology provide a great opportunity (we can integrate more and more devices on an integrated circuit), but also result in the famous "designer productivity gap," as shown in Fig. 1.2. In this figure, the *x*-axis shows progression in time, while the *y*-axis has two lines: The solid line shows the rate of growth of chip complexity (as measured by the number of logic transistors per chip), while the dotted line captures designer productivity (as measured by the average number of transistors designed by a staff engineer in a month). As can be seen in Fig. 1.2, the relentless march of progress in silicon technology is far outstripping our ability to effectively utilize these transistors for working designs in a short amount of time. As product lifetimes continue to shrink, time-to-market becomes a key design constraint.

Several strategies have been developed to cope with this widening designer productivity gap. We outline two major strategies in this chapter. The first is the elevation of the design process for architects to the *electronic system level* (ESL), where early design decisions, exploration and platform decisions are made above the traditional logic/register transfer level (RTL) (this is described further in Section 1.3). The second is aggressive exploitation of *design reuse* at the ESL. The reuse factor can be enhanced by increasing the granularity of library elements

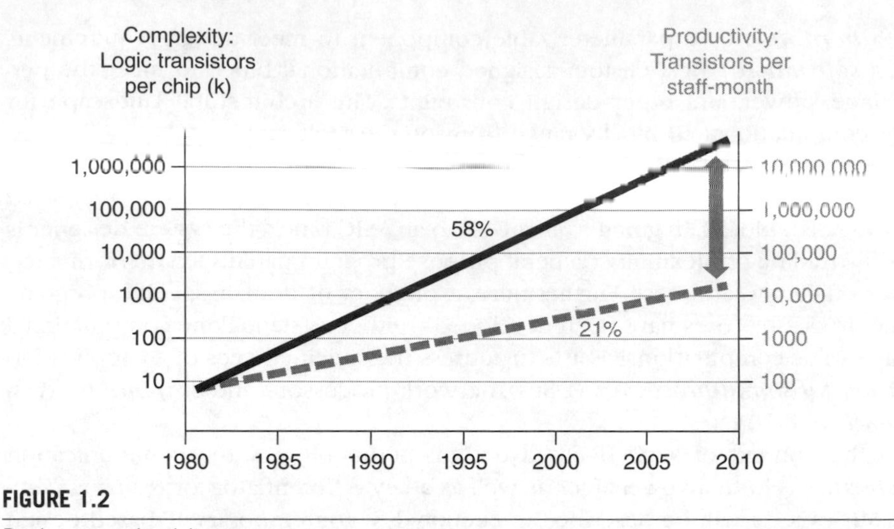

FIGURE 1.2

Designer productivity gap

from standard cells to hardware (e.g., custom accelerator) and software (e.g., processor core) IP blocks. Collections of these IP blocks can be configured into *platforms* designed for specific application domains, which further increases design reuse. Furthermore, the move toward migration of functionality to software allows reuse of software libraries that can be mapped onto existing platforms and IP blocks, which again enhances design reuse. The term *Platform-based Design* [2] is often used to represent a preconfigured silicon architecture using a number of software and hardware IPs, and is aimed at a specific application domain or even an end application (e.g., video compression, network packet processing). Such platforms typically allow for some degree of customization through architectural parameters for each IP, and often include an accompanying software toolchain to ease application development. The system architect can tweak platform parameters to tune the design for a range of performance, power, and *quality of service* (QoS) attributes.

In this regard, it is important to note that the communication architecture may be customized as well. Indeed, every instance of a silicon platform uses a specific class of communication architecture, and the platform provider may allow the architect to specify or modify some key communication parameters (e.g., bus widths and protocols) to allow for further customization of the design. As we will emphasize throughout this book, the on-chip communication architecture fabric in an MPSoC platform has a significant impact on its power, performance, cost, reliability, and time-to-market.

1.3 ESL DESIGN FLOW

Figure 1.3 shows an idealized MPSoC ESL design flow. Most designs start from customer or marketing requirement specifications that describe the overall application

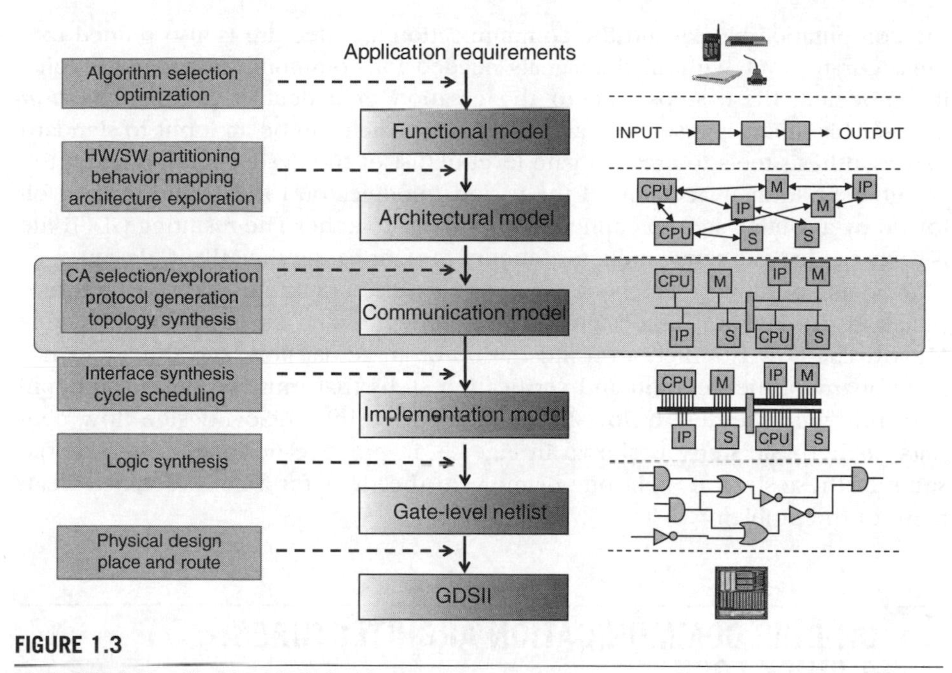

FIGURE 1.3

Ideal ESL design flow for MPSoCs (the on-chip communication architecture design phase is highlighted)

at a very abstract level. In the first step, designers select the algorithms to use, perform optimizations, and create a *functional model* of the application in a high level language such as C/C++. This is a very high level sequential-execution model of the application which captures its entire functionality. In the next step, designers perform hardware/software partitioning—mapping part of the functionality into hardware components and the remaining functionality to software. Architecture exploration is performed to select the best possible hardware or software candidates for the functional (or behavior) mapping. The result is an *architecture model* of the system where the computation entities in the system have been defined. Note, however, that these entities communicate with each other in an abstract manner, for instance using high level message passing. It is the responsibility of the subsequent *on-chip communication architecture synthesis* step to define a communication architecture for the system. This synthesis step performs design space exploration to select a particular communication architecture (e.g., hierarchical bus, ring bus, bus matrix) that best satisfies the constraints of the application. Once a communication architecture has been selected, further exploration is performed to define its topology and values for protocol parameters (e.g., arbitration scheme, bus width, and frequency). The resulting model is called a *communication model*. It is an enhanced architecture model of the system with well-defined computation and communication entities. The model is further refined in the next step, in which behaviors inside computation blocks are scheduled at cycle boundaries (cycle-accurate behavior). The interface between the computation blocks and the communication architecture is also refined to a pin-accurate level, with all the signals needed for communication being explicitly modeled. These steps lead to the creation of a detailed *implementation model*. This model is essentially an RTL model, which can be an input to standard logic synthesis tools to create a gate level netlist of the design. Subsequently, the designer performs placement of the various modules on the chip floor plan, followed by a routing step to connect the modules together. The resulting GDSII file is then handed to a semiconductor foundry for fabricating the MPSoC design.

A major portion of this book is devoted to the communication architecture phase in the MPSoC ESL design flow, as highlighted in Fig. 1.3. Note that this MPSoC ESL flow is a high level illustration of an actual flow, and thus does not show many of the iteration and verification steps that must be performed concurrently with the design flow. In addition, while the MPSoC design flow consists of the basic steps as shown in Fig. 1.3, in practice designers often merge some of these steps, or split one or more of the steps for a more detailed treatment of the problem.

1.4 ON-CHIP COMMUNICATION ARCHITECTURES: A QUICK LOOK

The components in an MPSoC design invariably need to communicate with each other during application execution. For instance, a microprocessor fetches instructions from memory components, or writes to external memories by sending data to an on-chip memory controller. It is the responsibility of the on-chip communication architecture to ensure that the multiple, co-existing data streams on the chip are correctly and reliably routed from the source components to their intended destinations. In addition to correctness, the on-chip communication architecture must provide latency or bandwidth guarantees to ensure that the application performance constraints are satisfied. A latency guarantee implies that a data unit must traverse the communication architecture and reach its destination within a finite amount of time, determined by a latency bound (e.g., 40ns from source to destination). A bandwidth guarantee implies that a group of data units must traverse a portion of the communication architecture at a certain data rate, as determined by the bandwidth requirements (e.g., 100 megabits/second from source to destination). Depending on the performance requirements of an application, various types of on-chip communication architectures can be used, as described in following subsection.

1.4.1 Types of On-Chip Communication Architectures

A basic building block of most on-chip communication architectures in MPSoC designs is the single shared bus. This is the simplest on-chip communication architecture, consisting of a set of shared, parallel wires to which various components are connected. Only one component on the bus can have control of the shared wires at any given time to perform data transfers. This limits the parallelism and achievable performance in the system, which makes it unsuitable for most MPSoC applications that can have tens to hundreds of components. Consequently, the single shared bus architecture is not scalable to meet the demands of MPSoC applications.

Figure 1.4 shows various kinds of on-chip communication architectures that are used in MPSoC designs. Many contemporary MPSoC designs mostly use shared bus-based communication architectures. Figure 1.4(a) shows a hierarchical shared bus architecture, which consists of a hierarchy of buses interconnected using bridge components. Shared buses higher up in the hierarchy are typically operated at higher clock frequencies, and are used to connect high speed, high performance components. On the other hand, shared buses lower down in the hierarchy are operated at lower frequencies to save power, and connect high latency, low performance components. Figure 1.4(b) shows a ring type bus, similar to that used in the IBM Cell MPSoC. The ring bus is actually a set of unidirectional, concentric and pipelined buses which allow high frequency operation and high bandwidth transfers between components on the bus. Figure 1.4(c) shows an ad-hoc bus architecture, where buses are operated at different frequencies and components can have point-to-point links with each other, as needed. Finally, Figure 1.4(d) shows the bus matrix (or crossbar bus) where a crossbar type architecture connects processors (and their local bus components) on the left to memories and peripherals on the right. This kind of architecture is a combination of shared bus and point-to-point interconnections.

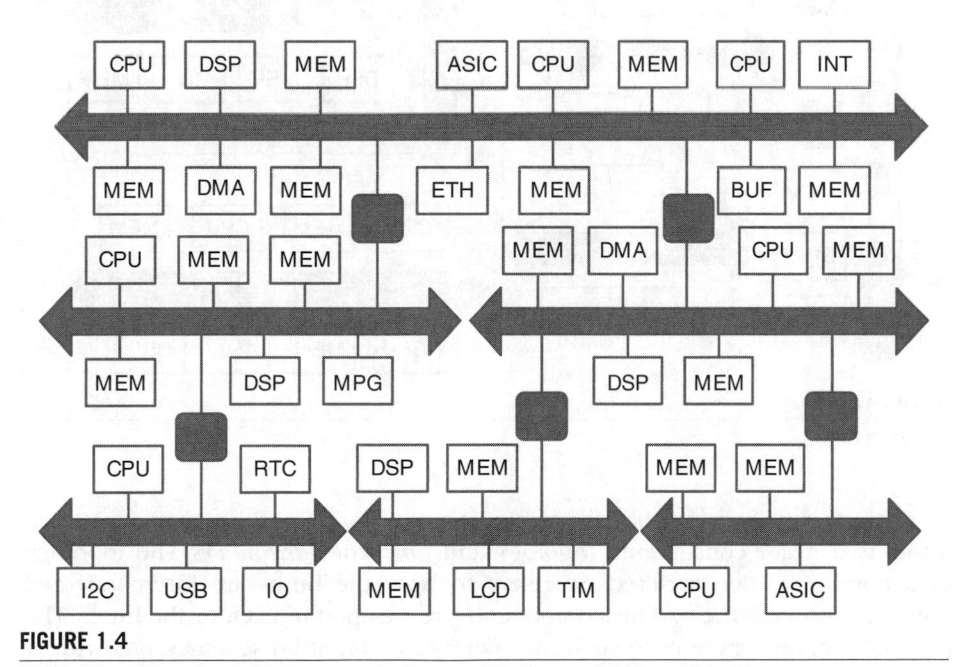

FIGURE 1.4

MPSoC bus-based on-chip communication architectures: (a) hierarchical bus

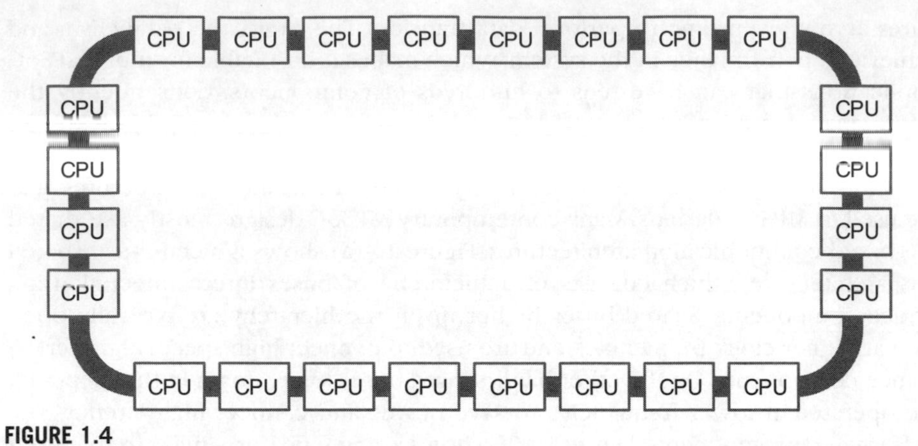

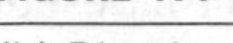

FIGURE 1.4

(b) Ring bus

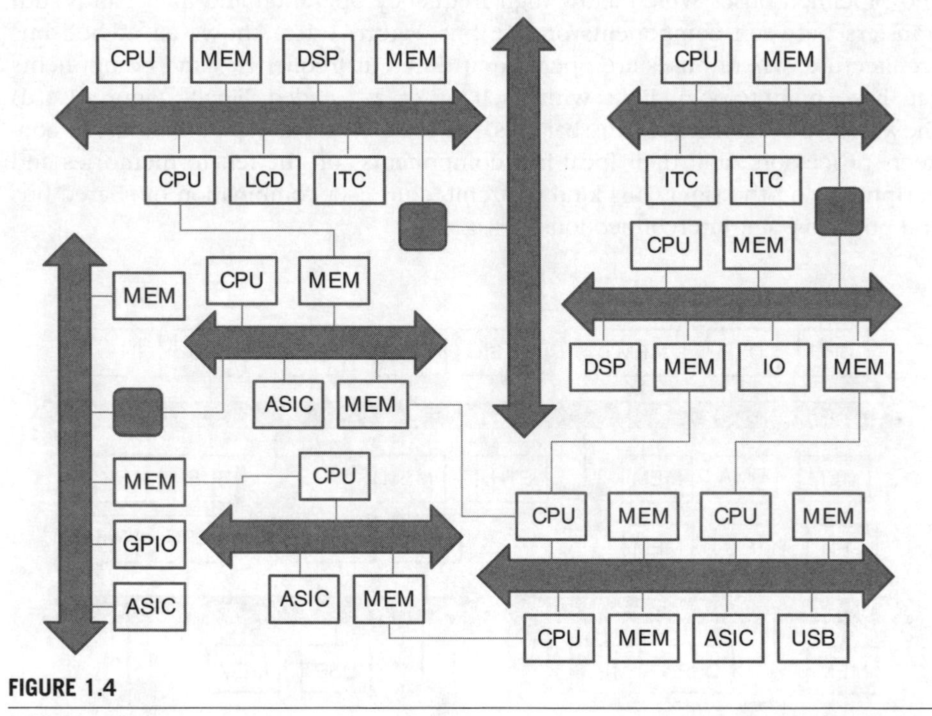

FIGURE 1.4

(c) Ad-hoc bus

Each of the above bus-based on-chip communication architectures is defined by its two major constituents: *topology* and *protocol parameters*. The topology of a communication architecture refers to how the buses are interconnected together, and how the various components are mapped to each of the buses. The protocol parameters refer to such parameters as arbitration schemes, bus widths, bus clock frequencies, buffer sizes, and burst transfer sizes, which are specific to

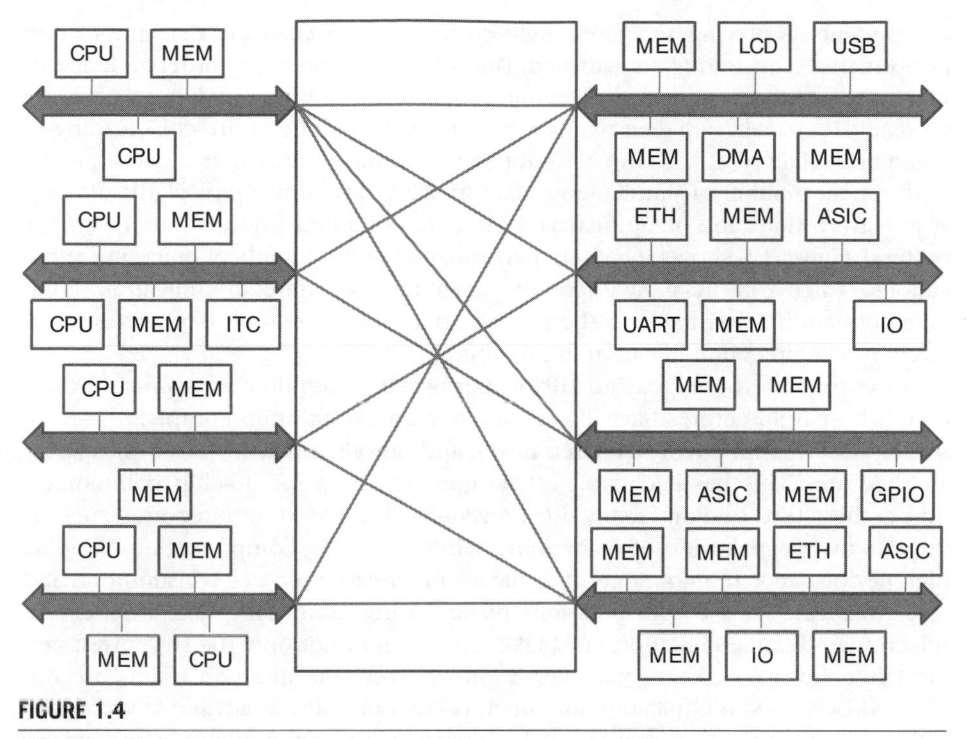

FIGURE 1.4

(d) bus matrix (or crossbar bus)

the protocol used by the communication architecture. Designing a communication architecture thus implies determining both its topology and protocol parameter values.

It has been projected that bus-based architectures cannot scale up with an increasing number of components, and also given the increasing amount of on-chip wire delays (resulting in timing unpredictability). Thus future MPSoC designs with hundreds of components will make use of network-on-chip (NoC) communication fabrics, where instead of shared buses, packet switched network fabrics with routers are used to transfer data between on-chip components. However, NoCs are still in their early phase of research and development, and concrete implementations of NoC-based MPSoCs are only now beginning to appear. The primary focus of this book is for bus-based on-chip communication architectures, although we will also introduce NoCs toward the end of this book.

Having introduced the different types of communication architectures used in MPSoC designs, we now look at how increasing application complexity and technology scaling have affected on-chip communication in recent years.

1.4.2 Impact of Increasing Application Complexity

With increasing application complexity and the rising number and variety of components being integrated into MPSoC designs, communication between on-chip components is playing an increasingly critical role in ensuring that application performance constraints are satisfied. Due to the increasing interdependence of various components on a chip, a seemingly insignificant bottleneck in transferring a single data stream between two components can stall the entire chip, leading to a functional failure of the chip. This not so uncommon scenario is a consequence of the sheer number of simultaneous data streams traversing a typical MPSoC chip at any given time, and being managed by finite communication resources (wires, buffers). Figure 1.5 shows the rising performance requirements of emerging applications, which will inevitably increase the amount of data communication traffic on a chip and further increase the probability of unforeseen bottlenecks encountered in on-chip communication in the future.

To cope with the increasing MPSoC performance requirements, on-chip communication architectures have also undergone an evolution in complexity—from shared buses, to hierarchical shared buses, and onto bus matrix (or crossbar bus) architectures. This has had two notable consequences for on-chip communication architecture design. Firstly, since advanced bus-based architectures such as the bus matrix make use of many more wires and logic components to support high performance requirements, they have a much larger power consumption and area overhead. Thus design decisions made during communication architecture selection and implementation must take into account not only the supported performance, but also ensure that overall chip power and area constraints are not violated by the communication architecture. Secondly, these advanced communication architectures have enormous design spaces that are not so easy to explore. The combination of different topologies, component mapping choices, and protocol parameter values for a communication architecture can easily create a design space with billions of possible configurations, which makes it incredibly difficult for designers to choose which configurations to explore in the finite amount of time available in an MPSoC design cycle. Thus the task of designing on-chip

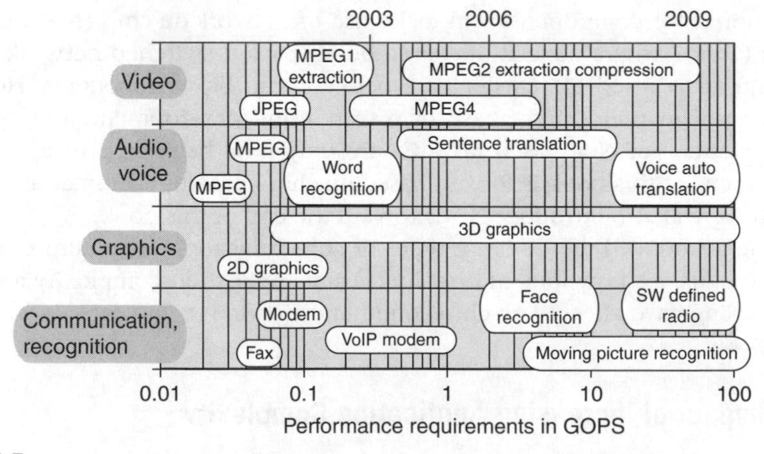

FIGURE 1.5

Increasing performance requirements for emerging applications [3,4]

communication architectures today has become a major challenge for designers, requiring a careful, time-consuming decision process to adequately balance the different constraints (power, performance, area) of the application.

1.4.3 Impact of Technology Scaling

The increasing levels of component integration would not have been possible without technology scaling that has enabled a reduction in transistor size, allowing more of them to be integrated into the same area with each passing technology generation. However, these technological advances that have ushered the industry into the *deep submicron* (DSM) era, with the ongoing commercialization of the 90 and 65nm processes, have introduced new challenges for on-chip communication architecture design. Precise control of the fabrication process in DSM technologies is almost impossible, leading to process uncertainties that cause non-uniformity of sheet resistance and an increase in coupling noise between adjacent wires in buses. In addition, decreasing wire pitch and increasing aspect ratio with technological advances further accelerates these issues. The end result of these factors is that signal propagation delay on wires (i.e., interconnect delay) is increasing with each technology generation, which puts a limit on the communication performance.

According to the *International Technology Roadmap for Semiconductors* (ITRS) 2005 predictions (Fig. 1.6) [4], the gap between interconnection delay and gate delay will increase to 9:1 at the 65nm technology. This is in sharp contrast to the 2:1 gap between interconnection delay and gate delay at the 180nm technology.

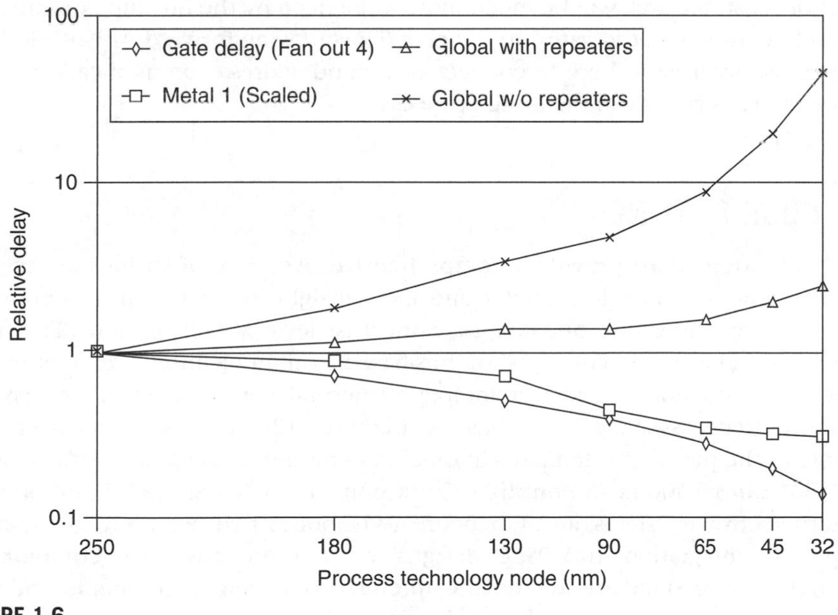

FIGURE 1.6

Relative delay comparison of wires vs. process technology [4]

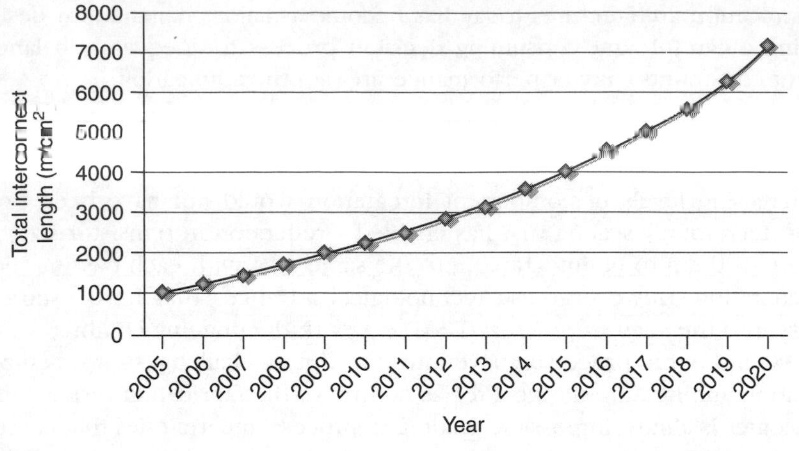

FIGURE 1.7

Trend of total interconnect length on a chip [4]

This indicates that communication, and not computation, will be the key performance bottleneck in DSM technologies. In addition, total wire length on a chip is expected to amount to 2.22km/cm^2 by the year 2010 (Fig. 1.7) [4]. Another observation is the increase of power dissipation due to the charging and discharging of interconnection wires on a chip. According to [4,5], the capacitance portion of the interconnect contributes to about 30% of the total capacitance of a chip, and soon the interconnect will consume about 50 times more power than logic circuits [6]. This means that for MPSoC designs in the DSM era, the performance, power consumption, cost, and area will be much more influenced by the on-chip communication architecture than the gates on the chip. Thus moving forward, MPSoC designs will necessarily have to be interconnect-aware and address communication architecture issues very early in the design process.

1.5 BOOK OUTLINE

This book attempts to provide a comprehensive overview of various aspects of on-chip communication in MPSoCs, and gives insight into why on-chip communication architectures are becoming a critical issue in MPSoC designs. The next chapter presents basic concepts of bus-based communication architectures—introducing commonly used terminology, structural components, wiring issues, and DSM effects associated with bus architectures. Chapter 3 gives an overview of some of the prevailing standards in on-chip communication architectures (such as AMBA 2.0/3.0, Altera Avalon, IBM CoreConnect, Sonics SMART Interconnect, STMicroelectronics STBus, and Opencores Wishbone) that are used to facilitate component integration in MPSoC designs. While these bus-based communication architecture standards define the interface between components and the bus architecture, as well as the bus architecture that implements the data transfer protocol, socket-based standards give a lot more freedom to a designer, with respect to the choice and implementation of the bus architecture, since they only provide definitions for designing component interfaces. In principle, such socket-based interface standards enable designers to reuse IPs designed for disparate communication protocols, thereby facilitating an IP-based design methodology. An overview of some of the popular socket-based interface standards (such as open core protocol (OCP), virtual component interface (VCI), and device transaction level (DTL)) is also presented in the chapter.

Having presented a background on the basic concepts of bus-based communication architectures, and prevailing standards, the next few chapters address the important problem of understanding the on-chip communication architecture design space, to aid in selecting the best communication architecture configuration for an application. Chapter 4 looks at models for the performance estimation of communication architectures. Various stochastic, simulation, and hybrid modeling abstractions are described here, that trade-off modeling complexity with estimation speed. These models capture details of the communication architecture design space and allow designers to estimate the performance of different communication architecture configurations. Chapter 5 presents various models for power and thermal estimation of communication architectures that allow systems designers to perform early estimation of the power and thermal characteristics for different communication architecture configurations. The chapter also highlights the need for PVT (process, voltage, temperature) variation-aware power estimation in future ultra DSM (UDSM) technologies. These performance and power/thermal models for communication architectures are used as part of various techniques, presented in Chapter 6, to select, configure, and design communication architectures that meet the requirements of a given application. While the emphasis here is on high level exploration based synthesis of communication architectures, novel approaches that couple physical implementation-awareness during early exploration, and co-synthesize memory and communication architectures are also presented, along with an overview of lower level physical and circuit level techniques for communication architecture design.

Once the various estimation models and design approaches have been studied, the next few chapters focus on specific research efforts in the area of communication architecture design. Chapter 7 gives an overview of a large body of work on various encoding techniques used to transform the data to be transmitted on communication architectures, in order to reduce switching power, propagation delay, crosstalk noise (both inductive and capacitive), and transmission errors. Schemes that jointly optimize for several design goals, and provide trade-offs to the designer between power, performance, and reliability, are also presented. Chapter 8 looks at custom bus architectures (such as split/segmented, serial, CDMA, asynchronous, and reconfigurable architectures) that attempt to address the shortcomings of standard on-chip communication architectures by utilizing new topologies and protocols to obtain improvements for common design goals, such as performance and power. Chapter 9 takes a step back to present various refinement methodologies that aim to take the communication architecture design process from the functional level down to the implementation level (Fig. 1.3) using a combination of novel modeling abstractions and refinement techniques. The chapter then dives into the problem of interface synthesis, which is a critical issue in such methodologies. Techniques are presented for efficiently synthesizing interfaces to correct protocol mismatches, and to ensure that interfaces using the same protocol but having different level of details correctly interact with each other. Chapter 10 tackles verification and security issues in communication architecture design. The chapter first presents techniques to verify the properties and constraints of on-chip communication protocols and communication architecture logic components, such as arbiters, and techniques for the verification of IP blocks being integrated into an SoC, to ensure that they are compliant with the on-chip communication protocol. Subsequently, the focus shifts to security issues, with a look at security features built into bus-based communication architecture standards, and an overview of research efforts that modify existing communication architecture standards to enhance overall system security.

The final part of the book presents trends in the design of on-chip communication architectures over the next several years. Chapter 11, contributed by Savidis and Friedman, presents physical design trends for on-chip communication architecture design. The focus here is on understanding the effect of DSM on interconnect design, as well as to give an overview of low power, high speed circuit design techniques that can be used in global power, and clock distribution networks. A discussion on emerging 3-D interconnects is also presented. Chapter 12 outlines networks-on-chip (NoCs), a promising new direction in on-chip communication design, and a hot area of research at the time of writing this book. The chapter describes various design aspects that characterize NoCs, such as the topology, switching schemes, routing algorithms, flow control mechanisms, clocking strategies, and QoS support. A survey of several NoC architectures that have been proposed in literature is also presented, along with a discussion on the current status and open problems in the NoC domain. Chapter 13 concludes the book by presenting three emerging technologies on the horizon that propose new paradigms for replacing traditional on-chip metallic interconnects. The chapter presents an overview and challenges associated with *optical interconnects*, that make use of light and an on-chip optical medium to transfer data; *RF/wireless interconnects*, that transfer data on a chip wirelessly using transmitting and receiving antennas integrated on the same chip; and *carbon nanotubes*, which have been shown to be much less susceptible to DSM effects and have been shown to have lower power dissipation and better performance than metallic interconnects. All three of these emerging technologies have several issues and open problems (such as the need for improvements in fabrication technology) that must be resolved before they can be adopted as part of on-chip interconnect fabrics. With the rapid advances in technology, however, it is only a matter of time before one or more of these technologies becomes feasible and advantageous to use in tomorrow's ultra large scale integrated (ULSI) designs.

REFERENCES

[1] IBM Cell Project, http://www.research.ibm.com/cell.

[2] K. Keutzer, S. Malik, A. Newton, J. Rabaey and A. Sangiovanni-Vincentelli, "System level design: Orthogonolization of concerns and platform-based design," *IEEE Transactions on Computer-Aided Design of Integrated Circuits and Systems*, Vol. 19, No. 12, 2000, pp. 1523–1543.

[3] N. Ventroux, F. Blanc, R. David and T. Collette, "Reconfigurable multiprocessor system-on-chip for embedded applications," *6th International Forum on Application-Specific Multi-Processor SoC*, Colorado, USA, 2006.

[4] Semiconductor Industry Association. The International Technology Roadmap for Semiconductors, 2005 edition. SEMATECH: Austin, TX., 2005.

[5] A. P. Chandrakasan and R. W. Brodersen, *Low Power Digital CMOS Design*, Kluwer Academic Publishers, Norwell, MA, 1995.

[6] W. J. Dally, "Computer architecture is all about interconnect," in *8th International Symposium on High-Performance Computer Architecture*, Cambridge, MA, 2002.

CHAPTER

2 Basic Concepts of Bus-Based Communication Architectures

Buses are one of the most widely used means of communicating between components in a system-on-a-chip (SoC) design. The simplicity and efficiency of transferring data on buses has ensured that they remain the preferred interconnection mechanism today. A bus connects possibly several components with a single shared channel. The shared channel can be physically implemented as a single wire (i.e., a serial bus or a set of wires) which makes up a parallel bus. This parallel bus is the typical implementation choice for a bus in almost all widely used on-chip bus-based communication architectures. Although a bus is essentially a broadcast medium, in most cases data transmitted on the bus is meant for a particular component and is ignored by the other components. Any data transmitted by a component moves from its output pins to the bus wires and is then received at the input pins of the destination component. The destination component typically sends an acknowledgement back to the transmitting component to indicate if the data was received. A *bus protocol* is used to explicitly define a communication transaction through its temporal (e.g., duration and sequence of messages exchanged) and spatial (e.g., message size) characteristics. The bus protocol also determines which component may access the shared bus if multiple requests to send (or receive) data appear on the bus at the same time. Bus-based communication architectures usually consist of one or more shared buses as well as logic components that implement the details of a particular bus protocol. In this chapter, we review the basic concepts of bus-based communication architectures. Section 2.1 presents the terminology and major components used in describing bus-based communication architectures. In Section 2.2, we discuss the organization and characteristics of buses. Section 2.3 presents different types of data transfers on buses. Section 2.4 outlines the diverse topologies resulting from bus-based communication architecture implementations. Section 2.5 briefly describes issues arising from the physical implementation of bus wires (or signals). Finally, Section 2.6 discusses deep submicron (DSM) effects that are becoming increasingly dominant for on-chip buses with complementary metal-oxide semiconductor (CMOS) technology scaling.

2.1 TERMINOLOGY

We begin by reviewing the basic terminology used to describe bus-based communication architectures, and systems deploying these architectures. Figure 2.1 shows a simple SoC design in which several (computational) components are interconnected using a bus-based communication architecture. Components which initiate and control read and write data transfers are referred to as *masters*. The *Processor* and *DSP* (*digital signal processor*) components in Fig. 2.1 are examples of master components that read/write data from/to other components in the system. Every master component is connected to the bus using a set of signals which are collectively referred to as a *master port*. The components that simply respond to data transfer requests from masters (and cannot initiate transfers themselves) are referred to as *slaves*, and have corresponding *slave ports*. The three memory blocks in Fig. 2.1 are examples of slaves that can handle requests for data read and write from other components (e.g., Processor, DSP), but cannot initiate such transfers themselves. The component ports are actually a part of its *interface* with the bus. An interface can be simple, consisting merely of the set of connecting wires to the bus (i.e., master or slave ports). Or it could be more complex, consisting of buffers, frequency converters, etc. in order to improve communication performance.

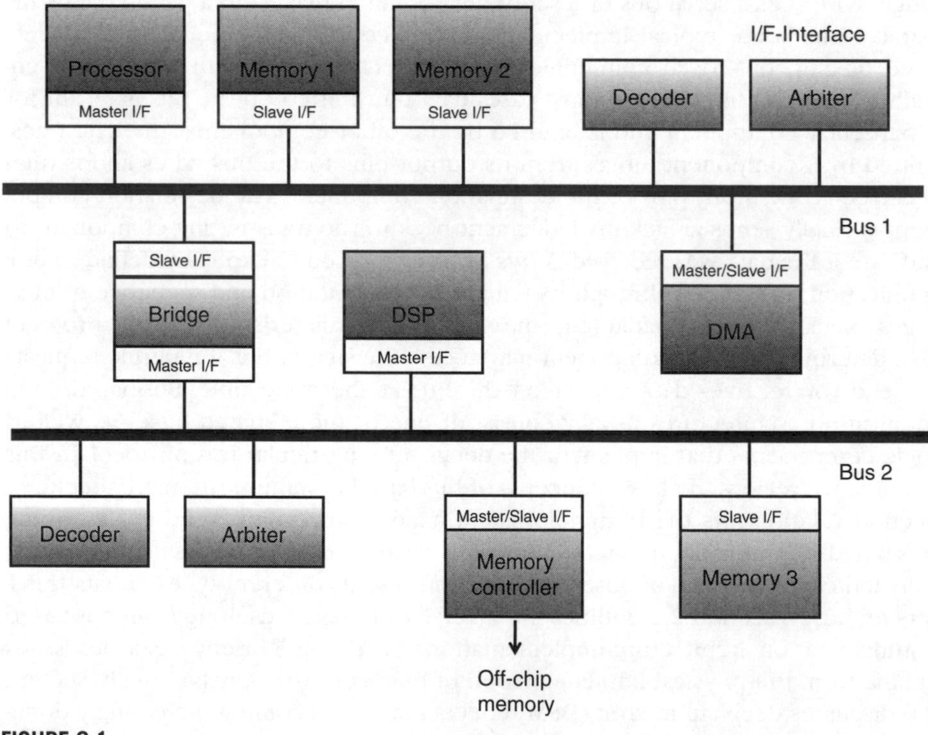

FIGURE 2.1

Example of an SoC with a bus-based communication architecture

Some components can have both master and slave ports, which means that they can act as both masters and slaves. These components are *master/slave hybrid* components. For instance, the *DMA* (*direct memory access*) component in Fig. 2.1 has a slave port that allows the *Processor* to write into (and read from) the *DMA* configuration register file, in order to initialize and configure it. Once configured, the *DMA* component uses its master port to initiate and control data transfers between memory blocks (which would otherwise have been managed by the *Processor*; as a result the *Processor* is freed up to perform other activity which typically improves system performance). Similarly, the *Memory Controller* component has a slave port which is used by the DSP component to initialize and configure its functionality. Once configured, the Memory Controller can initiate and control data transfers with external memory components connected to it, using its master port.

In addition to the wires, a bus-based communication architecture also consists of logic components such as *decoders*, *arbiters*, and *bridges*. A *decoder* is a logic component that decodes the destination address of a data transfer initiated by a master, and selects the appropriate slave to receive the data. It can either be a separate logic component, or integrated into a component interface. An *arbiter* is a logic component that determines which master to grant access to the bus, if multiple masters access the bus simultaneously. Typically, some form of a priority scheme is used, to ensure that critical data transfers in the system are not delayed. Finally, a *bridge* is a logic component that is used to connect two buses. It can have a fairly simple implementation if it connects two buses with the same protocols and clock frequencies. However, if the two buses have different protocols or clock frequencies, some form of protocol or frequency conversion is required in the bridge, which adds to its complexity. A bridge connects to a bus using a master or a slave port, just like any other component. The type of port used to connect to a bus depends on the direction of data transfers passing through it. For instance, in the example shown in Fig. 2.1, the *DMA* and *Processor* components on *Bus 1* initiate and control data transfers to *Bus 2* by sending data to the slave port of the bridge on *Bus 1*, which transfers it to its master port on *Bus 2* and sends the data to its destination. Since the *DSP* and *Memory Controller* do not initiate and control data transfers to components on *Bus 1*, a single bridge is sufficient as shown in Fig. 2.1. However, if these components needed to transfer data to *Bus 1*, another bridge with a slave port on *Bus 2* and a master port on *Bus 1* would be required.

2.2 CHARACTERISTICS OF BUS-BASED COMMUNICATION ARCHITECTURES

Bus-based communication architectures are defined by various architectural and physical characteristics that can have many different implementations. These implementation choices have trade-offs that can significantly affect the power, performance, and occupied area of the communication architecture. In this section, we describe the major characteristics of bus-based communication architectures and discuss some of their common implementation choices.

2.2.1 Bus Signal Types

Bus signals (or wires) are broadly classified into three categories, as shown in Fig. 2.2. *Address* signals are used to transmit the address of the destination for a data transfer on the bus. The number of signals used to transmit the address is typically a power of 2 (common values are 16, 32, or 64) and referred to as the address bus width. However, in some cases, this number can also be an arbitrary value (i.e., not a power of 2), depending on the number of components in a system. The address signals are collectively referred to as the *address bus*. Although most systems have a single shared address bus for both reads and writes, it is possible to have separate address buses for read and write data transfers. Having multiple address buses improves the concurrency in the system, since more data transfers can occur in parallel. However, this comes at the cost of larger number of wires which can increase area and power consumption.

Data signals are used to transmit data values to their destination addresses. The data signals are collectively referred to as the *data bus*. The typical number of signals in a data bus is 16, 32, 64, 128, 256, 512, and 1024 signals (called data bus width). However, this number can vary and have other values depending upon specific requirements of systems. The choice of data bus width is important because it determines whether any packing or unpacking of data is necessary at component interfaces. For instance, consider a case where the memory word size of a memory component is 64 bits and the data bus width is 32 bits. Then, every time a master requests data from the memory, the read data needs to be unpacked (or split) into two data items of 32 bits in width before being transmitted onto the bus. The data also needs to be packed (or merged) at the master interface before being sent to the master component. The packing and unpacking of data at the interfaces introduces an overhead in terms of power, performance, and area of the interface logic. Alternatively, if the data bus width was set to 64 bits, no such packing and unpacking of data would be required. Thus the size of the data bus is typically application specific, and in many cases depends on the memory word size of the memory components used in the system. Much like the address bus, the data buses can either be implemented as a single shared bus for both reads and writes, or separate data buses for reads and writes. Separate data buses improve concurrency and performance in the system, at the overhead of additional bus wire area and power

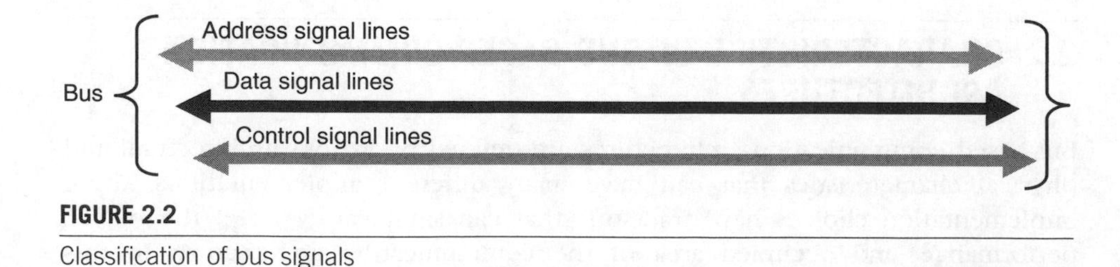

FIGURE 2.2

Classification of bus signals

consumption. It is also possible to combine the data and address buses by multiplexing them over a single set of wires. This may be useful for small, low cost SoCs where high performance communication is not as important as low area footprint, simplified wire routing, and low pin counts at component interfaces.

Control signals are used to send information about the data transfer, and are typically bus protocol specific. *Request* and *acknowledge* signals are the most common control signals, which transmit a data transfer request from a master, and the acknowledgment for a request or data received, respectively. Data size (or *byte enable*) control signals indicate the size of data being transmitted on the bus (particularly useful when the size of the data being transmitted is smaller than the data bus width). Sometimes slaves can signal an error condition to the master over special status control signals, if data cannot be read or written at the slave. Frequently, when multiple data items (called a *data burst*) need to be transmitted on the data bus by a component, there are control signals to indicate the number of data items to the destination. There are control signals that can transmit information about the source of the transmitted data such as a unique ID value identifying the transmitting component. Control signals can also transmit information about the data being transmitted to the destination, such as whether the data is cacheable, bufferable, write-through, or write-back.

2.2.2 Physical Structure

We now look at the structural implementation details of the shared bus signals. Traditionally, shared buses have been implemented using tri-state buffers that drive bidirectional lines, as shown in Fig. 2.3(a). Tri-state implementations of buses are commonly used in off-chip/backplane buses. The advantage of tri-state bidirectional buses is that they take up fewer wires and have a smaller area footprint. However, due to higher power consumption, higher delay (which can limit performance), and problems with debugging tri-state buffers, their use is restricted in modern bus-based on-chip communication architectures. Other efficient (and

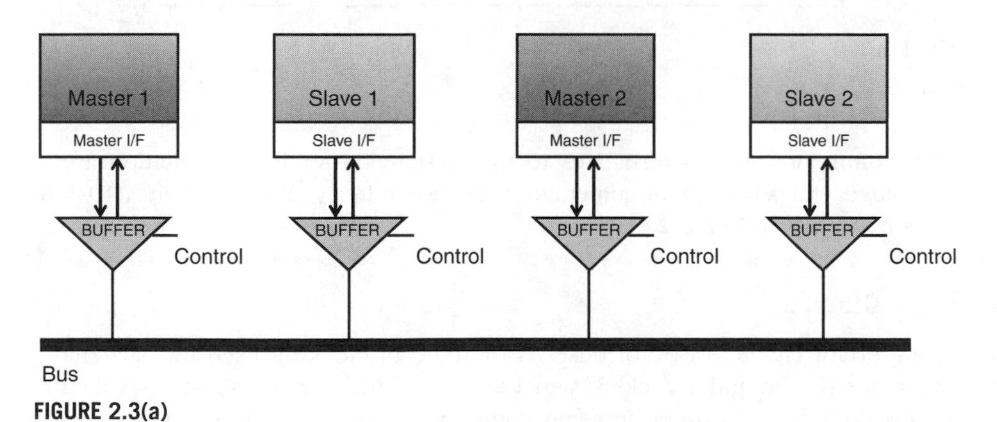

FIGURE 2.3(a)

Shared bus implementation alternatives: tri-state buffer based bidirectional signals

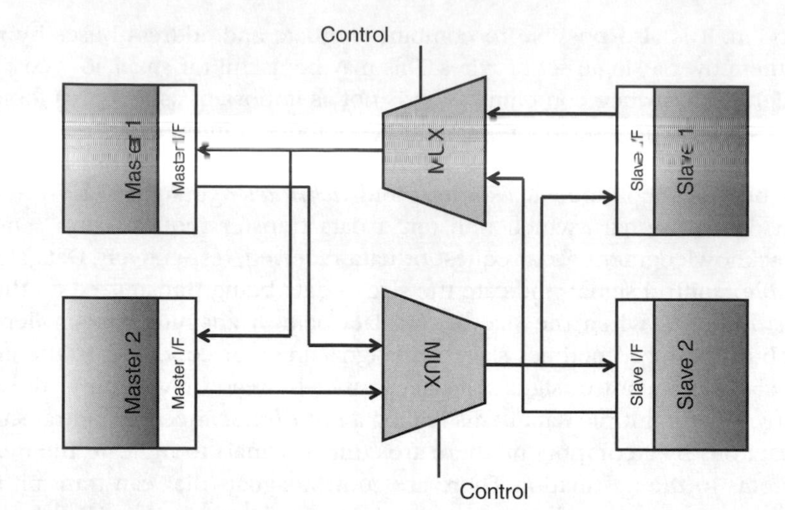

FIGURE 2.3(b)

MUX-based

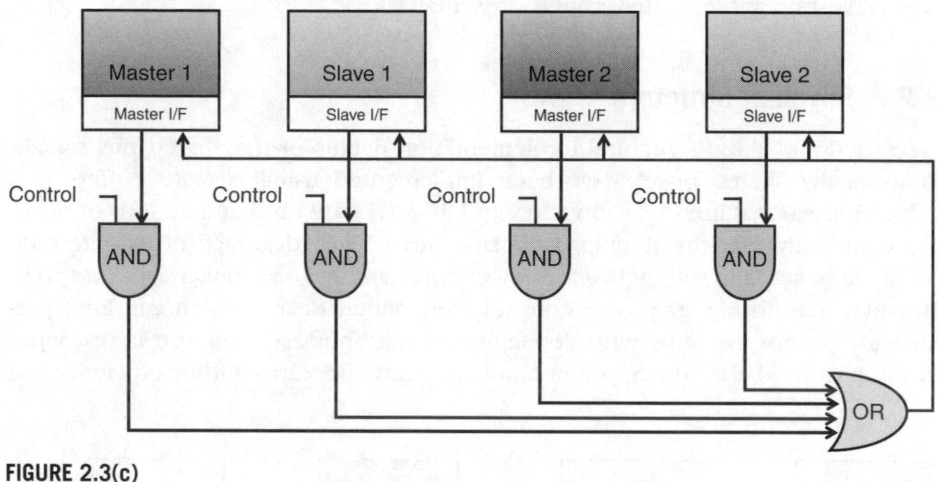

FIGURE 2.3(c)

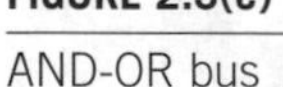

AND-OR bus

more commonly used) alternatives to the tri-state buffer implementation are a multiplexer (MUX) based implementation, as shown in Fig. 2.3(b) and an AND–OR structure as shown in Fig. 2.3(c).

2.2.3 Clocking

An important characteristic of buses is the type of clocking used for data transfers. A bus that includes a clock signal as one of the control signals is called a *synchronous* bus. Transfers between components usually require some form of handshaking. An example of a synchronous bus is shown in Fig. 2.4(a), where a

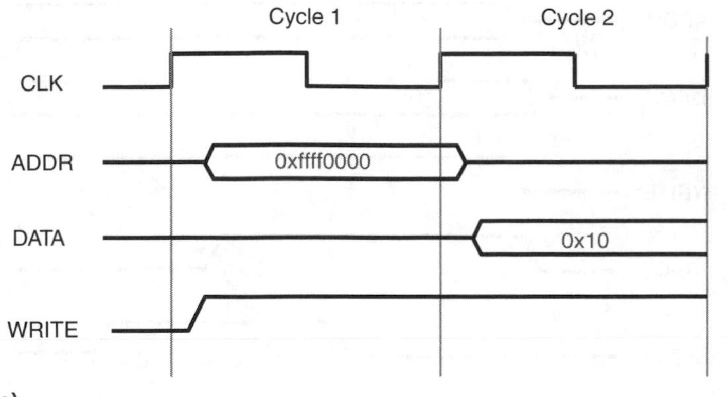

FIGURE 2.4(a)

Clocking strategies for buses: synchronous bus

data item is written by a master and is received by its destination in a total of two clock cycles. The master sends the address (*ADDR*) in the first clock cycle, and asserts the *WRITE* control signal which indicates to the destination that the master wants to write data to it. The data to be written is sent at the beginning of the second clock cycle, and is sampled off the bus by the destination slave at the rising edge of the third clock cycle. The slave can then optionally assert a response control signal (*RESP*) sending the bit value that corresponds to an *OK*, which indicates to the master that there was no error and the write proceeded as intended. The clock signal is essential for synchronization purposes. Synchronous buses enable fast data transfers, but can require frequency converters at component interfaces, since not all components connected to the bus can be expected to run at the same clock frequency as the bus. Indeed, in modern bus-based SoC designs, processors typically run at a clock frequency that is two to four times the frequency of the bus clock. It might also be required to pipeline the synchronous bus by inserting register slices (or buffers) [1] on it, because the signal delay for long wire lengths can actually exceed the bus clock cycle time [2]. In such cases, the register slices allow buffering of data to ensure that the destination can still sample the data as expected at the end of a clock cycle (although not at the end of the same cycle in which the data was transmitted). As an example, consider a synchronous bus clocked at a frequency $f = 100\,\text{MHz}$. A signal on the bus has $1/f = 10\,\text{ns}$ to travel from source to destination. If the signal delay on this bus is 20 ns, then a register slice inserted in the middle of the bus allows the signal to be buffered at the end of the first cycle, and then reach the destination at the end of the next cycle. This is discussed in more detail in Section 2.5. Most standard bus-based communication architectures [1, 3–5] (described in more detail in Chapter 3) use synchronous buses, since they provide high data throughput performance. Synchronous buses are almost always used in the critical path of an SoC design, such as the processor–memory interconnection.

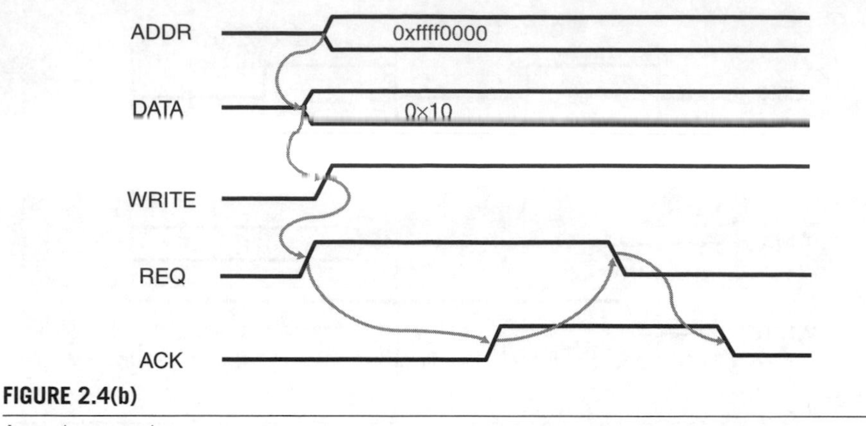

FIGURE 2.4(b)

Asynchronous bus

In an *asynchronous bus*, shown in Fig. 2.4(b), no clock signal is present in the control signals of the bus. In this case, bus synchronization occurs with the help of a handshake protocol that uses request-acknowledgement signals to ensure that the data transfer was completed successfully. For instance, for the example in Fig. 2.4(b) in which a master must write a data to a slave, the address (*ADDR*) and data (*DATA*) for the write data transfer is driven onto the bus by the master, along with a control signal to indicate a write transaction (*WRITE*). The master also asserts the request (*REQ*) control signal. When the destination slave sees the request signal, it samples the data off the bus, and then sends an acknowledgement (*ACK*) signal to inform the master that the data was received. The transmitting master then lowers the request (*REQ*) signal. The handshaking between the source and destination components is usually more extensive in asynchronous buses than in synchronous buses. An example of an asynchronous bus is the MARBLE bus, proposed by Bainbridge and Furber [6]. Asynchronous buses are typically slower than synchronous buses because of the additional overhead of the handshaking protocol, used for synchronization. They also require additional synchronization signals. However, they do not need additional frequency converters like synchronous buses and thus consume less area than synchronous buses. Asynchronous buses also do not suffer from *clock skew* (a phenomenon in which the clock signal from the clock arrives at different components at different times, causing potential timing errors) or the overhead of clock power dissipation, unlike synchronous buses.

2.2.4 Decoding

Whenever data needs to be transferred on a shared bus, the source component transmits the address of the destination component and the data. Each component in an SoC design is typically assigned an address map (i.e., a range of addresses). It is the job of a *decoder* to decode the address and select the appropriate destination component to receive the data. Decoding can be implemented either in a centralized or a distributed manner. Figure 2.5(a) shows a centralized implementation of a decoder. The decoder takes the address of a data transfer issued by a

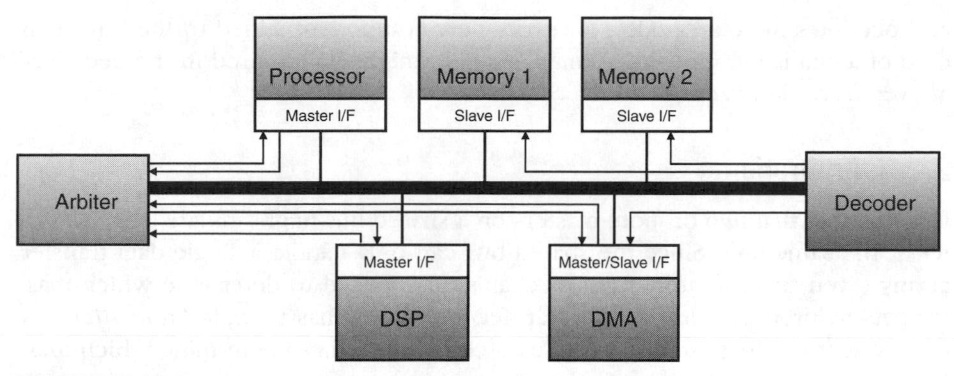

FIGURE 2.5(a)

Different implementation strategies for decoder and arbiter: centralized

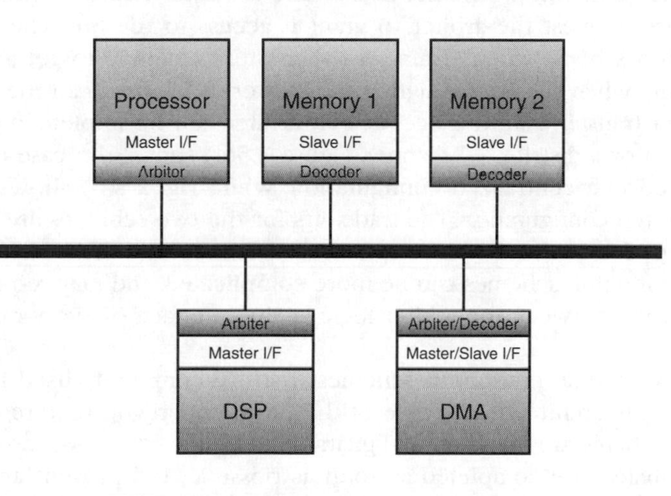

FIGURE 2.5(b)

Distributed

master as its input and then sends a select signal to the appropriate slave component to indicate that data needs to be read or written to the slave. The centralized implementation has the advantage that minimal change is required if new components are added to the system, which makes this scheme easily extensible. Another implementation scheme for the decoder is the distributed decoding approach, shown in Fig. 2.5(b). All the slaves have their own separate decoders in this scheme. When the address is transmitted on the shared bus by the master, the decoders at every slave interface decode the address to determine if the transfer is intended for them. Such a distributed scheme has the advantage of utilizing fewer signals compared to the centralized approach, which needs extra signals to connect the centralized decoder to every slave on the bus. However, there is more hardware duplication in the distributed case, because every slave on the bus now decodes the address, as opposed to a single decoder decoding the address in the centralized scheme. Thus, distributed decoding usually requires more logic and occupies more area. Also, for every new component added to the bus or in case of a change in the address map, changes might be required in the decoders at every slave interface.

2.2.5 Arbitration

It is possible that two or more masters on a shared bus might initiate a data transfer at the same time. Since the shared bus can only handle a single data transfer at any given time, an arbitration mechanism is needed to determine which master gets to proceed with its data transfer, and which has to wait. An ***arbiter*** is a component on the shared bus that uses certain criteria to determine which master gains access to the bus, if more than one master request access to it simultaneously. The criteria used to determine which master gains access to the bus is called the *arbitration scheme*. Every time a master needs to transfer data, it needs to first request the arbiter to grant it access to the bus. The arbiter uses its arbitration scheme to determine if the master is eligible to get access to the bus, and only when the arbiter signals the master to proceed can the master initiate the data transfer. Like the decoder, an arbiter can be implemented in either a centralized or a distributed manner. Figure 2.5(a) shows the case of an arbiter implemented in a centralized configuration, while Fig. 2.5(b) shows the arbiter in a distributed configuration. The trade-offs for the two schemes are the same as for the decoder. It should be noted that distributed arbitration for the more complex arbitration schemes can be more complicated, and may require synchronization signals between the arbiter logic at the interface of every master on the shared bus.

There are several arbitration schemes that are commonly used in bus-based communication architectures. Some of the basic underlying requirements for an arbitration scheme are that it should guarantee fairness of access, ensure that critical data transfers are completed as soon as possible, and prevent any starvation scenarios (for instance, due to a request from a master never getting access to the bus, which stalls the master execution). One of the most commonly used arbitration schemes is the *static priority* (SP) scheme, in which masters on a bus are assigned fixed priority values. The master having the highest priority always gets access to the bus. The SP scheme can be implemented in a pre-emptive or a non-pre-emptive manner. In a pre-emptive implementation, an ongoing lower priority data transfer is terminated immediately without being completed if a request for bus access is received from a higher priority master. In a non-pre-emptive implementation, the ongoing lower priority data transfer is allowed to complete before the bus is granted to a higher priority master. The SP scheme is simple to implement and can provide high performance by ensuring that critical data transfers, such as between processor and memory, always get higher priority. However, this scheme must be implemented carefully as it can lead to starvation of lower priority masters, which might never be able to get access to the bus if there are frequent bus accesses by higher priority masters. The *round-robin* (RR) arbitration scheme can ensure that there is no starvation in the system. In this scheme, access to the bus is granted in a circular (round-robin) manner, to every master on the bus, which guarantees that every master will eventually get access to the bus. A master relinquishes control over the bus when it no longer has any data to send (or has held the bus for the maximum allowed time) and passes the ownership to the next master in line. The RR scheme is also simple to implement, and can ensure equitable bandwidth distribution on a bus, but suffers from a drawback compared to the SP scheme, in that critical data transfers may have to wait a long time before they can proceed. The *time division multiple access* (TDMA) arbitration scheme is another popular scheme that can guarantee a fixed, higher bus bandwidth to masters with higher data transfer requirements, while also ensuring that lower priority masters do not starve. In this scheme, each master is assigned time slots (or time frames) of varying lengths, depending on the bandwidth requirements of the master. The choice of number of time slots to assign to each master is extremely important. The length of the allocated time frame should be long enough to complete at least a single data transfer, but not so long that other critical data transfers have to wait for a long time to get access.

While the schemes discussed above are essentially single level schemes, more complex arbitration schemes have also been proposed. For instance, it is possible to combine two arbitration schemes to create a two level arbitration scheme: Sonics SMART Interconnect [7] (described in Chapter 3) makes use of a *two level TDMA/RR* arbitration scheme. In this scheme, a TDMA arbitration scheme allocates time slots to various masters. If a master does not have any data to transfer during its time slot, a second level RR scheme selects another master to grant bus access to. Such a scheme thus enables better utilization of the bus, compared to the TDMA scheme, at the cost of a more complex implementation requiring more logic and occupying more area. Another complex, but highly efficient arbitration scheme is the *dynamic priority* (DP) scheme that can dynamically vary the priorities of the masters at runtime (i.e., while the system is executing). Unlike the SP scheme, additional logic is used to analyze data traffic at runtime, and the priorities are dynamically adapted to the changing traffic profiles of an application. Such a scheme can ensure better performance since it can efficiently track changing traffic profiles and ensure that masters that need to send larger amounts of data get higher priority. However, the implementation cost of such a scheme can be high, requiring several registers to keep track of priorities and data traffic profiles at various points during execution. A simpler variant of the DP scheme is the *programmable priority* (PP) scheme, which allows the application to write into the arbiter's programmable registers and set the priority for masters on the bus dynamically.

Since arbiters are invoked for every transfer on the bus, they are considered to be in the critical path of a bus-based communication architecture and must be designed with great care. An arbiter with a complex arbitration scheme implementation, that takes more than one cycle to make a decision, can severely reduce performance. While it might make sense to use a complex multi-cycle arbitration scheme for some applications, in other cases better performance can be achieved by using a simpler, single cycle arbitration scheme. Sometimes pipelining a complex multi-cycle arbiter implementation can also improve performance. These scenarios motivate the need to profile the application early in the design flow, to explore and select an appropriate arbiter implementation. Models for such performance exploration of bus-based communication architectures are presented in more detail in Chapter 4.

2.3 DATA TRANSFER MODES

Typically, data can be transferred over a bus using one of possibly several transfer modes. While some of the basic transfer modes are supported by all standard bus-based communication architectures, other modes are more specialized and specific to certain standard bus protocols. In this section, we review some of the main data transfer modes used in bus-based communication architectures.

2.3.1 Single Non-pipelined Transfer

The simplest form of data transfer on a bus is the single non-pipelined data transfer mode. In this mode, the master first requests access to the bus from the arbiter, and when it is granted access, sends out the address in the next cycle, and then writes data in the subsequent cycle (for a write data transfer) or waits for the slave to send the read data in the subsequent cycle(s). Figure 2.6 shows an example of a master performing two single read data transfers in a sequential manner. The master requests access to the bus from the arbiter at the beginning of the first cycle by asserting the *BUSREQ* control signal. The arbiter grants access to the master in the second cycle by asserting the *GRANT* control signal. Once the master sees that it has been granted access to the bus, it sends out the address (*A1*) of the slave to read data from at the beginning of the third cycle. The slave samples the read request at the beginning of the fourth cycle and then sends back the requested data (*D_A1*) in the same cycle. The read data is sampled off the bus by the master at the beginning of the fifth cycle. To read another data from the slave, the master again requests access to the bus from the arbiter, at the beginning

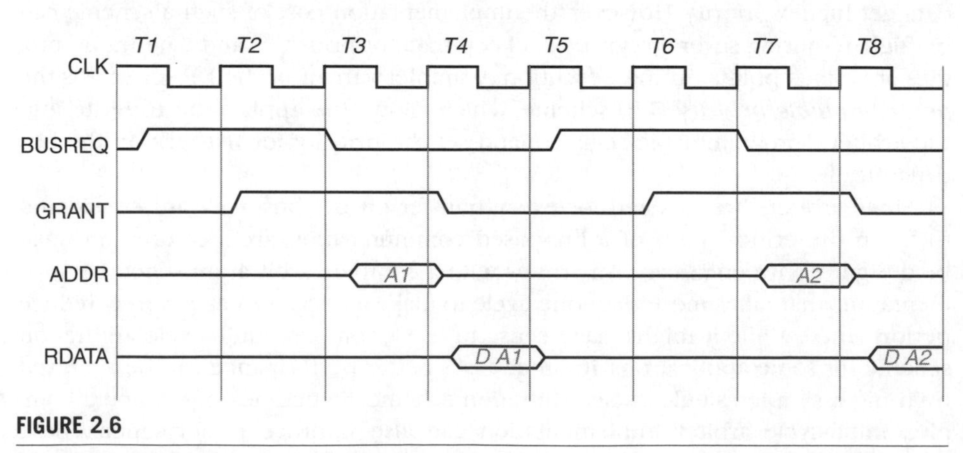

FIGURE 2.6

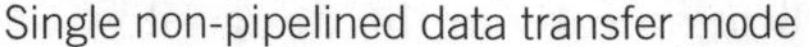

Single non-pipelined data transfer mode

of the fifth cycle. The sequence of events that follow are similar to that of the first data transfer, and the master samples the data from the slave at the beginning of the ninth clock cycle. Note that unlike in this example, slaves typically take multiple cycles to return data and in some cases even to write data. Also note that in the case where only one master is connected to the bus, arbitration is not needed since there is no possibility of simultaneous bus transfers. In such a case, the first two cycles (bus request and grant) are absent and the data transfer will take only two cycles. Conversely, it is possible that for the case when arbitration is required (i.e., when there are multiple masters connected to the bus), the arbiter takes multiple cycles to decide which master to grant bus access to, as discussed earlier. Such a scenario is possible when the arbiter makes use of a complex arbitration scheme, such as the DP-based one; or for the case when the bus clock frequency is so high that it takes multiple clock cycles for the arbiter to get a response from its chosen arbitration scheme. For the example shown in Fig. 2.6, the single non-pipelined data transfer mode takes as many as four cycles to complete a single read data transfer, under the assumption that arbitration is needed and takes a single cycle. Single non-pipelined transfers, as described in this section, typically occur in bus-based communication architectures with multiplexed address and data buses. We now look at transfer modes that allow us to reduce this number of cycles.

2.3.2 Pipelined Transfer

The pipelined data transfer mode overlaps the address and data phases of multiple data transfers to improve bus performance (i.e., bus throughput). Figure 2.7 shows an example of a pipelined data transfer for two write data transfers initiated by separate masters. At the beginning of the first cycle, both masters (*M1*, *M2*) request access to the bus. The arbiter grants master *M1* access to the bus in the same cycle. Master *M1* then sends the address of its destination slave (*A1*) in the second cycle and the data to write (*D_A1*) in the third cycle. The arbiter grants access to the bus to the second master *M2*, even before its write transfer is finished. This allows *M2* to send the address of its destination slave (*A2*) in the third cycle. Notice that the address phase of the transfer by master *M2* overlaps with the data phase of master *M1*. Finally, master *M2* sends the write data in the fourth cycle, to complete the transfer. Such an overlapped transfer improves bus utilization and reduces the time for a data transfer. Pipelined transfers typically require a more complex arbiter implementation that can perform pipelined (or overlapped) arbitration. Additionally, pipelined transfers are only possible in bus implementations with separate address and data buses (i.e., with no multiplexing of address and data signals).

2.3.3 Burst Transfer

We saw in Fig. 2.6 that multiple data transfers from the same master required arbitration for every individual data transfer. The burst transfer mode improves bus performance by requesting arbitration only once for multiple data transfers.

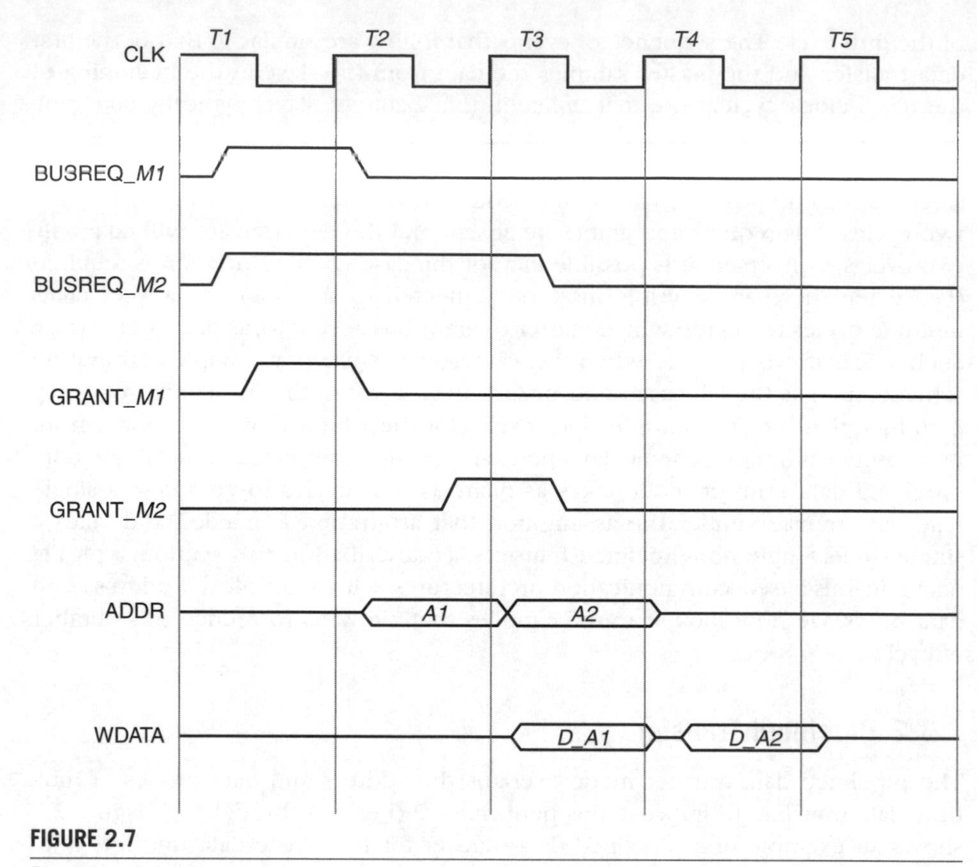

FIGURE 2.7

Pipelined data transfer mode

Figure 2.8(a) shows an example of a non-pipelined, burst data transfer by a master. The scenario depicted has a master needing to write four data items to a slave on the bus. At the beginning of the first cycle, a master requests access to the bus for a "burst" of four data items, and is granted the access by the arbiter at the beginning of the second cycle. Typically, control signals (not shown in the figure) from the master inform the arbiter of the length of the burst (four in this case). The master then proceeds to send the address of the first data (*A1*) item in the third cycle, and then the data to write to the slave (*D_A1*) in the fourth cycle. Since the arbiter has already granted bus access to the master for a burst of four data items, re-arbitration at this point is not required, and the master simply proceeds to send the address of the next data item (*A2*) in the burst at the beginning of the fifth cycle. The data transfer continues till all four data items have been sent to the slave. As can be seen from Fig. 2.8(a), the overhead of arbitration for each data item sent by the master is avoided in a burst transfer, which significantly reduces data transfer time, compared to the single transfer mode shown in Fig. 2.6. Performance can be improved even further if pipelining is allowed within the burst transfer. Figure 2.8(b) shows the same case as Fig. 2.8(a) where a master

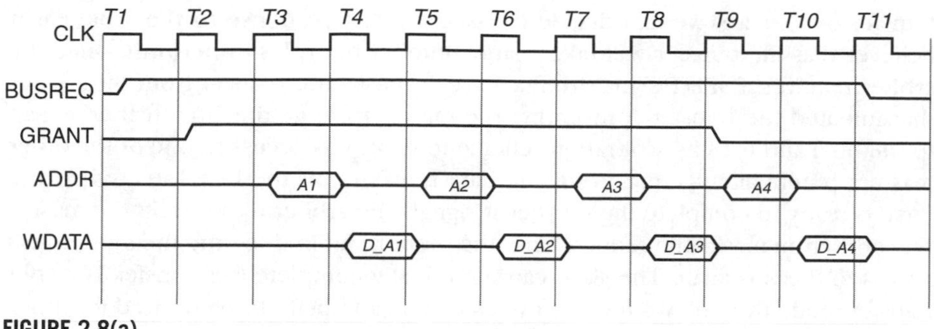

FIGURE 2.8(a)

Example of master writing four data items in burst transfer mode: non-pipelined burst transfer mode

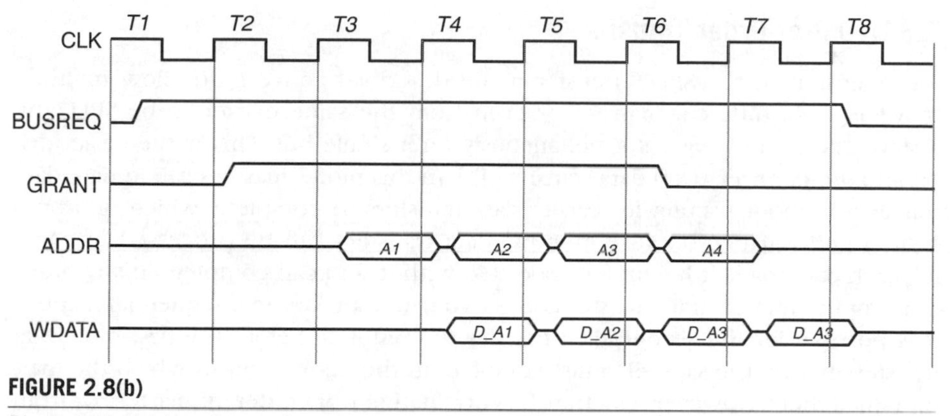

FIGURE 2.8(b)

Pipelined burst transfer mode

sends four data items to a slave, but this time the address and data phases of the data transfers within the burst are overlapped. This pipelined burst mode reduces the data transfer time compared to the non-pipelined burst mode in Fig. 2.8(a), and thus considerably improves bus utilization and performance.

2.3.4 Split Transfer

During a data transfer on a bus, it is possible that a slave can take multiple cycles to return the requested data or write the data. Since the bus in such a case is typically held by a master and no other master can gain access to it till the transfer is completed, the bus remains idle for multiple cycles till the slave completes the transfer. Such a scenario results in under-utilization of the bus and reduces performance. A *split transfer* [4] is a special type of transfer mode that can improve bus utilization in such cases by "splitting" the data transfer and allowing the idle cycles that would otherwise be spent waiting for the slave to be utilized for data transfers by other masters. A split transfer mode typically works as follows. When a data transfer occurs, a slave can decide to issue a SPLIT response to the arbiter if it believes that the transfer will take a large number of cycles to perform. Once the arbiter receives a SPLIT signal from a slave, it masks the request from the master that initiated the transfer (preventing the master from getting any further access to the bus) and uses its arbitration scheme to grant bus access to one of the other masters potentially waiting to initiate data transfers on the bus. Later, when the slave is ready to complete the transfer, it signals the arbiter to "un-split" the master. The arbiter un-masks the request from the master, and in due time the master gets access to the bus again. The slave can then finally complete the transfer. The split transfer mode thus allows the idle cycles in a data transfer to be utilized for other data transfers, and is therefore an effective mechanism for improving the communication performance in bus-based systems. Of course a prerequisite for using this transfer mode is the presence of split capable slaves and arbiters.

2.3.5 Out-of-Order Transfer

An extension of the SPLIT transfer mode described above is to allow multiple transfers from different masters, or even from the same master, to be SPLIT by a slave and be in progress simultaneously on a single bus. This is the basic idea behind *out-of-order (OO)* data transfers [1]. In this mode, masters can initiate data transfers without waiting for earlier data transfers to complete, which improves system performance because multiple data transfers can be processed in parallel. Each data transfer has an ID associated with it, and can complete in any order. This implies that even if a master issues two data transfers, in a sequential manner, it is possible for the second data transfer to complete before the first one. Data transfers having the same ID must complete in the same order in which the master issued them. However, data transfers originating from different masters or from the same master but having different IDs have no ordering restrictions and can complete in any order. This ability to complete data transfers out of order means that data transfers to faster memory blocks can be completed without waiting for earlier transfers to slower memory blocks. As a result, bus utilization and overall system performance are improved significantly.

Predictably, there is overhead involved in the implementation of such an advanced and complex data transfer scheme. Firstly, additional signals are needed to transmit IDs for every data transfer in the system. Secondly, master interfaces need to be extended to handle data transfer IDs and be able to reorder received data. Thirdly, slaves require additional logic at their interface to decode and process IDs, and ensure that the proper data transfer ordering is maintained. The *read (or write) data reordering depth* is a parameter that specifies the maximum number of read (or write) data transfers pending in the slave that can be reordered. Larger reordering depths can significantly improve performance, but also require more logic and increase system cost. Therefore, a designer must be careful in deciding a value for this parameter. There is typically a maximum value of reordering depth beyond which the performance does not improve for a given application [8]. This threshold corresponds to the maximum level of data traffic parallelism in the application, and can be obtained after performance profiling (described in more detail in Chapter 4). Finally, additional bus logic must also be added (to the arbiter module, or separately) to ensure that data transfer ordering is maintained, for transfers originating from multiple masters.

2.3.6 Broadcast Transfer

Typically, data transfers on the bus involve just two components—a master and a slave. Appropriate select signals ensure that only the source and destination components sample data onto and off the bus. However, it is possible for the data on the bus to be "visible" to other components on the bus, besides the two that are involved in the transfer. This is because every time a data item is transmitted over a bus, it is physically broadcast to every component on the bus. A *broadcast transfer* is one that involves a source component transmitting data on the bus, and multiple components sampling the data off the bus. One of the uses of this transfer mode is for snooping and cache coherence protocols. When several components on the bus have a private cache which is fed from a single memory module, a problem arises when the memory is updated (for instance, when a cache line is written to memory by a component). In such a case it is essential that the private caches of the components on the bus invalidate (or update) their cache entries to prevent reading incorrect values. Broadcasting allows the address of the memory location (or cache line) being updated to be transmitted to all the components on the bus, so that they can invalidate (or update) their local copies.

2.4 BUS TOPOLOGY TYPES

Bus-based communication architectures can have several different types of bus arrangements or topology structures which affect the cost, complexity, power, and performance profiles of the communication architecture. Figure 2.9 shows the major bus topology types that are used in SoC designs. The simplest scheme for component interconnection is the *single bus* topology shown in Fig. 2.9(a). All the components in the system are connected to a single shared bus. An example of a commercial SoC with a shared bus is the DaVinci family of digital video processing SoCs from Texas Instruments [25]. While such a configuration is sufficient for very small SoCs having only a few components, it does not scale well to handle larger systems. This is because a single bus allows only a single data transfer at a time. A more efficient topology that allows multiple data transfers in parallel is the *hierarchical bus* topology shown in Fig. 2.9(b). In this topology, the components are connected to multiple buses that interface with each other using a bridge component. Concurrent data transfers are possible on each bus, provided that the components are allocated to the buses in such a manner that there is minimum interaction between components on different buses. Since buses can have different clock frequencies, the bridge component can be quite complex, to handle interbus transactions, data buffering, frequency conversion, etc. There are several commercial SoCs today that make use of the hierarchical bus topology, such as the customizable multiprocessor ARM PrimeXsys SoCs [27] that are widely used

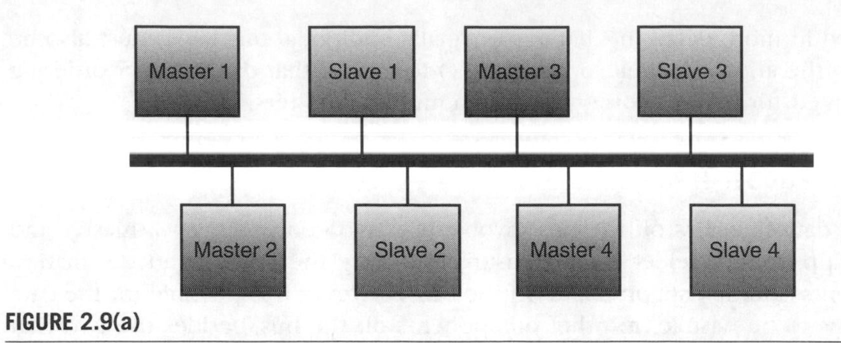

FIGURE 2.9(a)

Different bus-based communication architecture topology structures: single bus

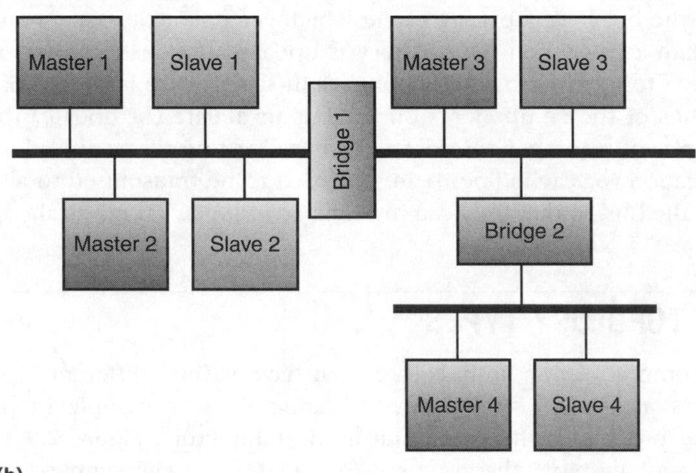

FIGURE 2.9(b)

Hierarchical bus

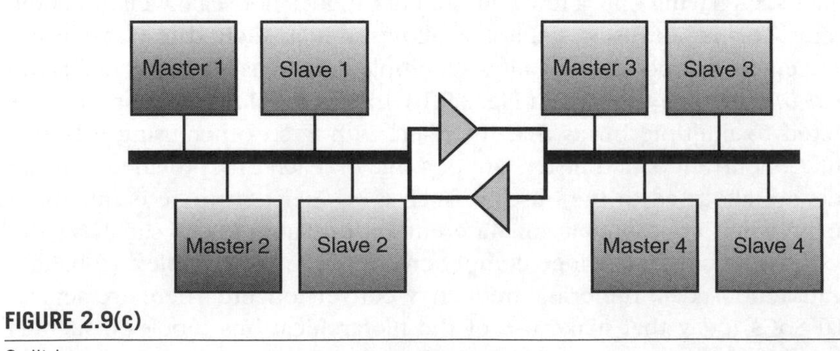

FIGURE 2.9(c)

Split bus

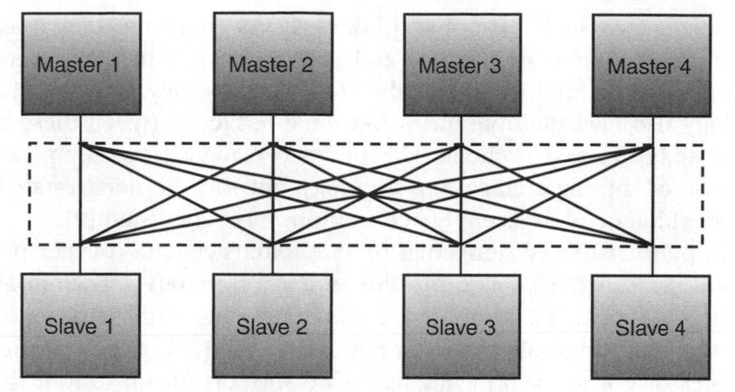

FIGURE 2.9(d)

Full bus crossbar (or point-to-point bus)

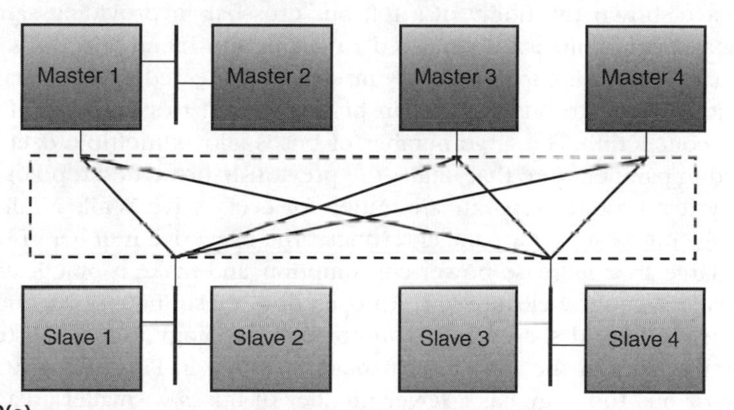

FIGURE 2.9(e)

Partial bus crossbar

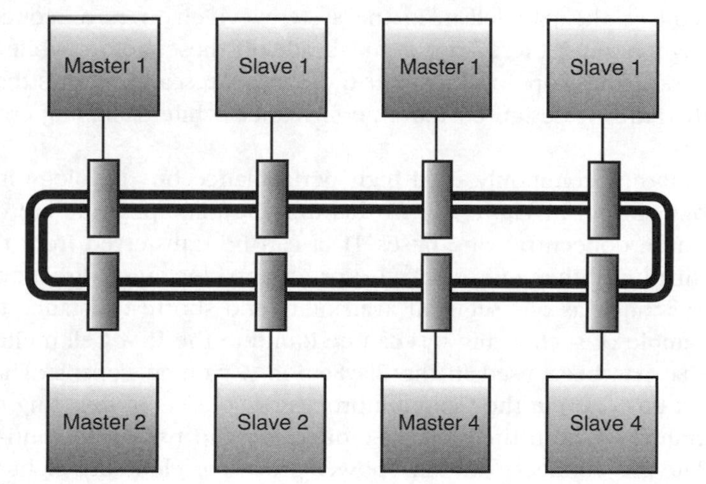

FIGURE 2.9(f)

Ring bus

in handheld devices such as mobile phones, PDAs (personal digital assistants), GPS (global positioning system) units, and PMPs (personal media players). A simpler variant of the hierarchical bus is the *split bus* topology shown in Fig. 2.9(c). This topology also uses multiple buses, but the interface between these buses is a simple tri-state buffer-based scheme. This prevents using a more complex protocol as in the case of the hierarchical bus topology, but the simpler tri-state interface can be more efficient as far as energy consumption is concerned [9].

For high performance systems that require extensive data transfer parallelism, the *full bus crossbar* (also called *full bus matrix*) topology shown in Fig. 2.9(d) is a suitable choice. An example of a commercial SoC with a full bus crossbar topology is the Niagara multiprocessor SoC from SUN [26], which connects eight SPARC processor cores (each having hardware support for up to four threads) to four L2-cache banks, an I/O bridge, and an FPU (floating point unit). While this solution might be excessive for smaller systems [10], several research efforts [11–13] have shown the utility of a full bus crossbar in providing significantly higher data throughput rates compared to single and hierarchical bus architecture alternatives. In this topology, every master is connected to every slave in the system with a separate bus, which can be considered to be a kind of point-to-point interconnection. The large number of buses allows multiple data transfers to proceed in parallel. Note that unlike the previously presented topologies, a full crossbar system requires separate arbitration for every slave. While a full crossbar bus topology offers superior parallel response, the excessive number of buses can take up a large area, increase power consumption, and make it practically impossible to achieve routing closure [14]. To overcome these limitations, one alternative is to use a hybrid shared bus/point-to-point topology, which clusters some of the components in the full crossbar bus, as shown in Fig. 2.9(e). Such a *partial crossbar* bus topology has a fewer number of buses, a smaller area, reduced power consumption, and less wire congestion than a full crossbar bus topology [14–16]. However, the clustering of components in the partial crossbar bus topology also reduces the parallelism in the system, which in turn reduces performance. Designers must therefore carefully trade-off these factors while designing a partial crossbar bus topology. Chapter 6 describes research efforts that attempt to optimally trade-off design cost and performance while designing crossbar bus architectures.

Finally, another commonly used high performance bus topology is the *ring bus* topology, shown in Fig. 2.9(f). In this topology, components are connected to one or more concentric ring buses. Data can be transferred from the source to the destination either in a clockwise or an anti-clockwise direction, depending on factors such as bus segment availability and shortest distance to destination. An example of such a ring bus can be found in the IBM Cell multiprocessor SoC [28] that has been used in the PlayStation 3 gaming console. The element interconnect bus (EIB) in the Cell multiprocessor consists of four ring buses, two of which transfer data in the clockwise direction and two in the anti-clockwise direction. The EIB connects the PPE (power processor element), eight SPEs (synergistic processor elements), a MIC (memory interface controller), and an external BIC (bus interface controller). The EIB ring bus was chosen over a full crossbar bus primarily because of its lower area footprint, while still offering an acceptable bandwidth.

2.5 PHYSICAL IMPLEMENTATION OF BUS WIRES

With the rising complexity and ever increasing levels of component integration in SoC designs, the volume of data transfers between components has also increased. To meet performance requirements, bus clock frequencies have been steadily increasing, since data throughput is a function of bus clock frequency, as given by the relation:

$$throughput_{bus} = width_{bus} \times clock\_frequency_{bus}$$

where the *throughput* is in terms of megabits per second if the *width* is specified in terms of bits and the *frequency* in terms of megahertz (MHz). Now, a rise in bus clock frequency implies a shorter bus clock cycle period. For instance, a bus with a clock frequency of 100 MHz has a bus clock cycle duration of 10 ns, whereas a bus with a higher clock frequency of 500 MHz has a bus clock cycle duration of only 2 ns. This has major implications as CMOS process technology continues to shrink. Bus wires are implemented as long metal lines on a silicon wafer, and transmit data using electromagnetic waves which cannot travel faster than a finite speed limit. With shrinking process technology, logic components such as gates have also correspondingly decreased in size. However, the wire lengths have not shrunk accordingly, resulting in relatively longer communication path lengths between logic components in newer technologies. Worse, due to increasing bus clock frequencies, the time allowed for a signal on the bus to travel from the source to its destination in a single bus clock cycle has reduced considerably, as discussed above. Another way of stating this fact is that the distance that can be covered by a signal on the bus in a single clock cycle has been reduced with increasing clock frequencies and shrinking CMOS process technology. Consequently, it can take multiple cycles to send a signal across a chip. For instance, it has been estimated that in the 50 nm process technology node, the signal propagation delay will be as high as 6–10 bus clock cycles [17] to send a signal from one end of the SoC to the other. This increase and unpredictability in signal propagation time can have serious consequences for the performance and correct functioning of the SoC design.

Several ways of tackling this problem have been proposed. Hierarchical [4, 5] or split bus [9] communication architectures partition otherwise long bus lines into shorter ones, separated by bridges, or tri-state buffer structures, respectively. This makes it possible for signals to traverse a bus segment in a single clock cycle. Hierarchical bus architectures such as AMBA 2.0 [4] allow different buses to operate at different bus clock frequencies. Utilizing multiple clock domains separated by bridge logic components allows better signal propagation management, since signals need to traverse smaller wire lengths. Another commonly used technique makes use of register slices [1] or buffers to pipeline long bus wires. Such a scheme enables a signal to be in flight for several cycles, taking a single clock cycle to move between successive pipeline stages, before finally reaching its destination. Carloni and Sangiovanni-Vincentelli [18] described one such approach to pipeline long wires having a latency of more than one cycle. Pipeline logic elements (called relay stations) were inserted into long wires after the physical layout phase, to ensure latency insensitive design. Yet another technique is to make use of asynchronous buses which discard the clock signal altogether, in favor of more elaborate handshaking based synchronization mechanism. The MARBLE bus architecture [6] is one example of an asynchronous communication architecture. It is also possible to make use of globally asynchronous, locally synchronous (GALS) techniques for communication, which use asynchronous handshaking synchronization for long wires that interconnect smaller synchronous regions, consisting of components connected via a synchronous bus. Finally, there are several low level techniques that are commonly used to reduce signal propagation delay on any wire, such as by inserting repeaters, or varying the dimensions of the wires (wire sizing) [19].

2.6 DISCUSSION: BUSES IN THE DSM ERA

With the scaling of CMOS technology below 90 nm, SoC designs have entered the DSM era, characterized by high levels of component integration, high clock frequencies, and low signal voltages. In addition to an increase in signal propagation delay, which requires making architectural changes to buses (as discussed in Section 2.5), DSM effects will create severe signal integrity problems that will make it harder to guarantee error-free data communication on buses. The signal integrity problem can be defined as the scenario where the received signal at the destination is different from the transmitted signal at the source driver, for a bus wire. This happens because of signal degradation caused by various DSM effects that create noise (i.e., a deviation of a signal from its intended or ideal value). Some of the important DSM effects that can cause noise on buses include crosstalk, external electromagnetic interference, transmission line effects, and soft errors. These effects are described below.

Crosstalk is the phenomenon of noise being caused on a signal *A* due to the coupling with another signal *B*. Due to the close proximity of bus wires, near-field electromagnetic coupling causes inductive and capacitive crosstalk on the bus signals. Even when wires are far apart, crosstalk can still be present between signals due to coupling facilitated by the common substrate, a shared power supply or ground, or a shared signal return path. As wires become narrower (with technology scaling) and clock frequencies increase, fringing field effects and inductance effects become larger for wires, leading to higher inductive and capacitive crosstalk. *Electromagnetic interference* (EMI) from large external electric and magnetic fields can couple into circuits and create unwanted noise. As highly integrated, portable wireless communication SoCs increasingly consist of analog, RF, and digital circuits, EMI due to external and internal coupling will increase. Long on-chip buses in particular will be the sources and receptors of EMI noise. *Transmission line effects* will arise due to discontinuities in wires that are modeled as transmission lines. In DSM technologies, when a wire is longer than 1/10 of the wavelength of the signal frequency component that is transmitted, the wave nature of the propagated signal must be modeled, otherwise significant errors may result. Wires will thus have to be modeled as transmission lines to avoid errors during signal analysis. Discontinuities in these transmission lines (due to various factors such as capacitive loads, vias, wire bends, package pins, crossover wires, and non-ideal receivers) can result in impendence mismatches. Such mismatches will create noise as a result of signal reflections at the discontinuities. Finally, signal integrity will also be influenced by *soft errors* that are caused by a collision of thermal neutrons (produced by the decay of cosmic ray showers) and/or alpha particles (produced by impurities in the substrate). Highly integrated SoCs will be particularly susceptible to soft errors that will create spurious pulses and interfere with signals on buses.

As a result of all the DSM effects described above, it will become harder to guarantee error-free data transfers on buses. Reduced signal swings in DSM technologies will result in a further reduction of voltage noise margins, increasing the probability of transmission errors in the presence of even the smallest sources of noise. Many other factors such as increasing wire resistance due to skin effect at high frequencies, increasing number of metal layers that increase cross-layer coupling, and timing errors due to jitters will cause new challenges in DSM technologies. These problems have been well summarized in several books [20–24]. It is very important that emerging tools and methodologies for on-chip communication architecture design be able to handle not only the increased number of wires, but also allow designers to predict and address DSM issues as early in the design flow as possible, to reduce design iterations (instead of finding and fixing the problems in post-layout). We will revisit DSM-aware methodologies, techniques, and architectures throughout this book.

2.7 SUMMARY

In this chapter, we presented some of the basic concepts of bus-based communication architectures. We first introduced the components and terminology used to describe these communication architectures and then covered some of their major characteristics such as bus signal types, physical structure, clocking, decoding, and arbitration. We presented an overview of some of the basic data transfer modes that are used during data transfers, and then described some of the more advanced transfer modes intended to improve bus utilization and throughput performance. Some commonly used bus topology structures were described, and finally we discussed some of the issues in the physical implementation of bus wires. In the next chapter, we will look at examples of some standard bus-based communication architectures that are widely used in SoC designs.

REFERENCES

[1] ARM AMBA AXI Specification www.arm.com/armtech/AXI.

[2] S. Pasricha, N. Dutt, E. Bozorgzadeh and M. Ben-Romdhane, "FABSYN: Floorplan-aware bus architecture synthesis," *IEEE Transactions on Very Large Scale Integration Systems (TVLSI)*, Vol. 14, No. 3, March 2006, pp. 241–253.

[3] "STBus Communication System: Concepts and Definitions," *Reference Guide*, STMicroelectronics, May 2003.

[4] ARM AMBA Specification and Multi layer AHB Specification (rev2.0), http://www.arm.com, 2001.

[5] IBMCoreConnect Specification, http://www.ibm.com/chips/techlib/techlib.nsf/productfamilies/CoreConnect_Bus_Architecture.

[6] W. J. Bainbridge and S. B. Furber, "Asynchronous macrocell interconnect using MARBLE," in *Proceedings of Fourth International Symposium on Advanced Research in Asynchronous Circuits and Systems*, 1998, pp. 122–132.

[7] Sonics, "Sonics μNetworks technical overview," Sonics Inc, June 2000.

[8] S. Pasricha, N. Dutt and M. Ben-Romdhane, "Extending the transaction level modeling approach for fast communication architecture exploration," *Design and Automation Conference (DAC 2004)*, San Diego, CA, June 2004, pp. 113–118.

[9] Cheng-Ta Hsieh and M. Pedram, "Architectural energy optimization by bus splitting," in *Proceedings of IEEE Transactions on Computer-Aided Design of Integrated Circuits and Systems (IEEE TCAD)*, Vol. 21, No. 4, April 2002, pp. 408–414.

[10] S. Brini, D. Benjelloun and F. Castanier, "A flexible virtual platform for computational and communication architecture exploration of DMT VDSL modems," in *Proceedings of DATE*, 2003, pp. 164–169.

[11] V. Lahtinen, E. Salminen, K. Kuusilinna and T. Hamalainen, "Comparison of synthesized bus and crossbar interconnection architectures," in *Proceedings of the 2003 International Symposium on Circuits and Systems*, 2003, pp. 433–436.

[12] Y. Zhang and M. J. Irwin, "Power and performance comparison of crossbars and buses as on-chip interconnect structures," in *Conference Record of the Thirty-Third Asilomar Conference on Signals, Systems, and Computers*, 1999, pp. 378–383.

[13] M. Loghi, F. Angiolini, D. Bertozzi, L. Benini and R. Zafalon, "Analyzing on-chip communication in a MPSoC environment," in *Proceedings of Design, Automation and Test in Europe Conference and Exhibition*, 2004, pp. 752–757.

[14] S. Pasricha, N. Dutt and M. Ben-Romdhane, "Constraint-driven bus matrix synthesis for MPSoC," *Asia and South Pacific Design Automation Conference (ASPDAC 2006)*, Yokohama, Japan, January 2006, pp. 30–35.

[15] S. Murali and G. De Micheli, "An application-specific design methodology for STbus crossbar generation," in *Proceedings of Design, Automation and Test in Europe* (DATE), 2005, pp. 1176–1181.

[16] S. Pasricha, Y. Park, F. Kurdahi and N. Dutt, "System-level power-performance trade-offs in bus matrix communication architecture synthesis," *International Conference on Hardware/Software Codesign and System Synthesis (CODES+ISSS 2006)*, Seoul, Korea, October 2006.

[17] R. Ho, K. W. Mai and M. A. Horowitz, "The future of wires," *Proceedings of the IEEE*, Vol. 89, April 2001, pp. 490–504.

[18] L. P. Carloni and A. L. Sangiovanni-Vincentelli, "Coping with latency in SoC design," *IEEE Micro*, Vol. 22, No. 5, September/October 2002, pp. 24–35.

[19] A. B. Kahng, S. Muddu and E. Sarto, "Interconnect optimization strategies for high-performance VLSI designs," in *Proceedings of Twelfth International Conference on VLSI Design*, 1999, pp. 464–469.

[20] H. B. Bakoglu, *Circuits, Interconnections, and Packaging for VLSI*, Addison-Wesley, Reading, MA, 1990.

[21] W. J. Dally and J. H. Poulton, *Digital Systems Engineering*, Cambridge University Press, Cambridge, UK, 1998.

[22] C.-K. Cheng, J. Lillis, S. Lin and N. Chang, *Interconnect Analysis and Synthesis*, Wiley-Interscience, 1999.

[23] Q. K. Zhu, *Interconnect RC and Layout Extraction for VLSI*, Trafford, 2002.

[24] M. Nakhla and R. Achar, *Introduction to High-Speed Circuit and Interconnect Analysis*, Omniz Global Knowledge Corporation, 2002.

[25] D. Talla, "An innovative HD video and digital image processor for low-cost digital entertainment products," *Proceedings, HotChips*, 2007.

[26] S. Phillips, "VictoriaFalls: Scaling highly-threaded processor cores," *Proceedings, HotChips*, 2007.

[27] ARM PrimeXsys Platform, http://www.arm.com.

[28] IBM Cell Project, http://www.research.ibm.com/cell.

CHAPTER

3 On-Chip Communication Architecture Standards

System-on-chip (SoC) designs typically have several different types of components such as processors, memories, custom hardware, peripherals, and external interface IP (intellectual property) blocks that need to communicate with each other. In SoC design houses, some of these components might be designed from scratch, while others are reused from previous designs or procured from external IP vendors. Each of these components has an interface to the outside world consisting of a set of pins that are responsible for sending/receiving addresses, data, and control information to/from other components. The choice of pins at the interface is governed by the particular bus protocol of the communication architecture. In order to seamlessly integrate all these components into an SoC design, it is necessary to have some kind of a standard interface definition for the components. Without a standard interface definition, the component interfaces will not be compatible with the bus architecture implementation, and consequently will not function correctly. In such a scenario, the components will require the design of logic wrappers at their interfaces (more details in Chapter 9) to correctly interface with the bus architecture being used. These logic wrappers, however, require additional area on the chip and can be time consuming to design and verify.

To speed up SoC integration and promote IP reuse over several designs, several bus-based communication architecture standards have emerged over the past several years. A communication architecture standard defines a specific data transfer protocol, which in turn decides the number and functionality of the pins at the interface of the components. Usually, bus-based communication architecture standards define the interface between components and the bus architecture, as well as the bus architecture that implements the data transfer protocol. Many of the bus architecture standards give designers a certain amount of freedom to implement the bus architecture in one of many ways. For instance, most bus-based communication architecture standards give designers freedom to select arbitration policies that are suitable for the specific SoC being designed. Socket-based bus interface standards on the other hand give absolute freedom to a designer, with respect to the choice and implementation of the bus architecture, since they only provide definitions for designing component interfaces. Any proprietary or standard bus architecture implementation can be selected, and components with socket-based interface definitions can be connected to it using logic adapters. This allows a great deal of flexibility for IP reuse. In this chapter, we present various standards used in SoCs with bus-based communication architectures. Section 3.1 details some commonly used bus-based communication architecture standards. Section 3.2 describes popular socket-based bus interface standards. Finally, we briefly discuss some of the standards for off-chip interconnects in Section 3.3.

3.1 STANDARD ON-CHIP BUS-BASED COMMUNICATION ARCHITECTURES

Since the early 1990s, several on-chip bus-based communication architecture standards have been proposed to handle the communication needs of emerging SoC designs. Some of the popular standards include ARM Microcontroller Bus Architecture (AMBA) versions 2.0 [1] and 3.0 [2], IBM CoreConnect [3], STMicroelectronics STBus [4], Sonics SMART Interconnect [5], OpenCores Wishbone [6], and Altera Avalon [7]. The next few sections describe these bus-based communication architecture standards in more detail. Since these standards are constantly evolving, the descriptions of these standards presented here are meant to serve as exemplars, to highlight the capabilities and features required from on-chip communication architectures for supporting diverse SoC application requirements.

3.1.1 AMBA 2.0

AMBA version 2.0 [1] is one of the most widely used on-chip communication standards today. The goal of this standard is to provide a flexible high performance bus architecture specification, that is technology independent, takes up minimal silicon area, and encourages IP reuse across designs. AMBA 2.0 defines three distinct bus standards:

1. *Advanced high performance bus (AHB)*, which is a high performance bus meant to connect high bandwidth, high clock frequency components such as microprocessors, DMA (Direct Memory Access) controllers, off-chip memory interfaces, and high bandwidth on-chip memory blocks.
2. *Advanced system bus (ASB)*, which is a light-weight alternative to the AHB bus, meant to connect high clock frequency components that do not need the advanced protocol features of AHB.
3. *Advanced peripheral bus (APB)*, is a low complexity bus optimized for low power operation, and meant for high latency, low bandwidth peripheral components such as timers, UARTs (universal asynchronous receivers/transmitters), user interface (e.g., keyboard) controllers, etc.

Figure 3.1 shows an example of a typical AMBA-based system, with the buses arranged in a simple hierarchical bus topology (see Chapter 2). The AMBA AHB (or ASB) bus typically acts as a backbone bus that provides a high bandwidth interface between the components involved in a majority of the transfers. The bridge component on the high performance bus is used to interface to the lower bandwidth APB bus, to which most of the low bandwidth peripherals are

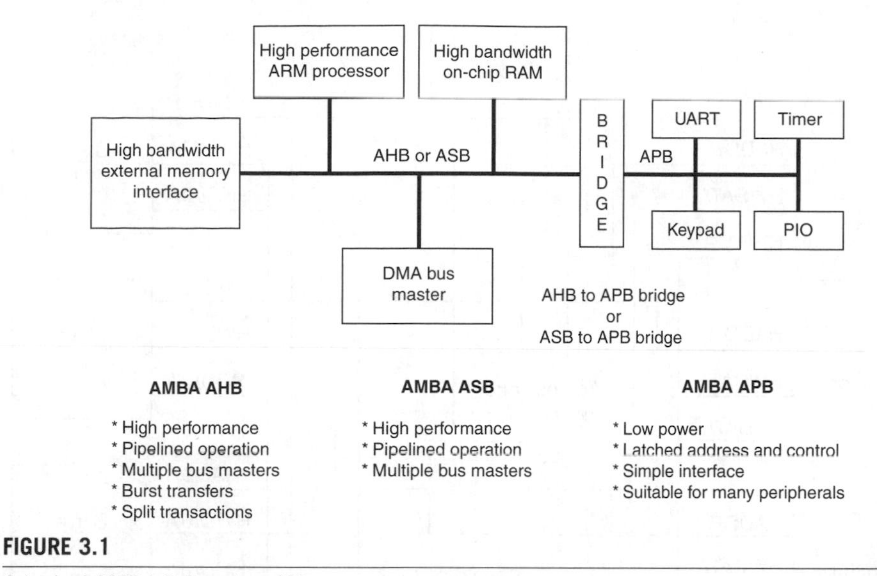

FIGURE 3.1

A typical AMBA 2.0 system [1]
Source: ARM Inc.

connected. Note that Fig. 3.1 shows just one of the topology arrangements possible with the AMBA 2.0 buses. Other topology configurations such as single shared bus, multi-layer AHB (or bus matrix) are also possible and will be discussed later in this section. Also note that the AMBA specification [1] is described at the architectural level granularity, and does not provide any information about the electrical characteristics of the bus, which are dependent on the selected manufacturing process technology. Even the timing specifications are only provided in terms of signal behavior at the cycle level—the exact timing requirements depend on the selected complementary metal-oxide semiconductor (CMOS) process technology and frequency of operation. The AMBA specification [1] defines all the signals, transfer modes, structural configuration, and other bus protocol details for the AHB, ASB, and APB buses. Since the specification recommends using the AHB over the ASB for all new designs, and the features of the ASB are simply a subset of the AHB features, we will not present details of the ASB bus. We now describe the specifications of the AHB and APB buses.

3.1.1.1 *Advanced High Performance Bus*

The AHB bus standard describes a high performance bus that supports advanced features for high bandwidth, low latency data transfers. AHB can be used to connect multiple master components, and supports high performance data transfer features such as pipelined operations, burst mode transfers, and split transactions. A single 32-bit address bus is used by the masters to transmit the addresses of the slaves required to complete the read or write data transfer requests. Since the AMBA specification restricts all its on-chip buses to a non-tri-state implementation, the AHB has separate data buses for reads and writes. The data buses have a minimum recommended width of 32 bits, but can have any values ranging through 8, 16, 32, 64, 128, 256, 512, or 1024 bits, depending on application bandwidth requirements,

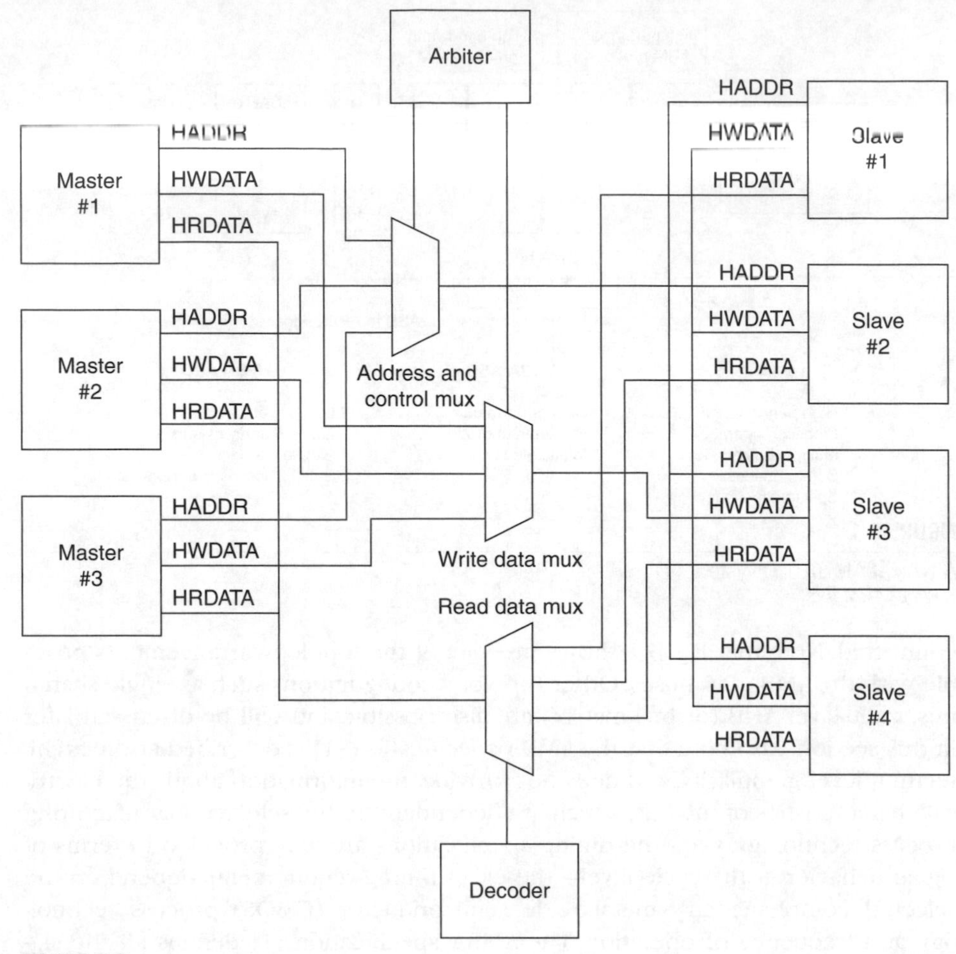

FIGURE 3.2

AHB multiplexer interconnection scheme [1]
Source: ARM Inc.

component interface pin constraints, and the bit width of words accessed from memory modules (e.g., embedded DRAM). Figure 3.2 shows the structure of a simple AHB shared bus with 3 masters and 4 slaves. As can be seen, the AHB makes use of a central multiplexer-based interconnection scheme, along with centralized arbiter and decoder modules to manage data transfers. The masters on the bus drive their address signals (*HADDR*) and control signals (not shown in the figure) whenever they want to perform a data transfer, and the arbiter determines which of the masters will have its address, control, and possibly write data (or *HWDATA*) signals routed (broadcast) to all the slaves on the bus. The decoder is used to decode the destination slave address, select the appropriate slave to receive the data transfer request, and route back response (and possible read data or *HRDATA*) signals to the masters.

Figure 3.3 shows a basic data transfer on the AHB bus. An AHB transfer consists of an address phase that lasts for a single cycle, and a data phase, that can require

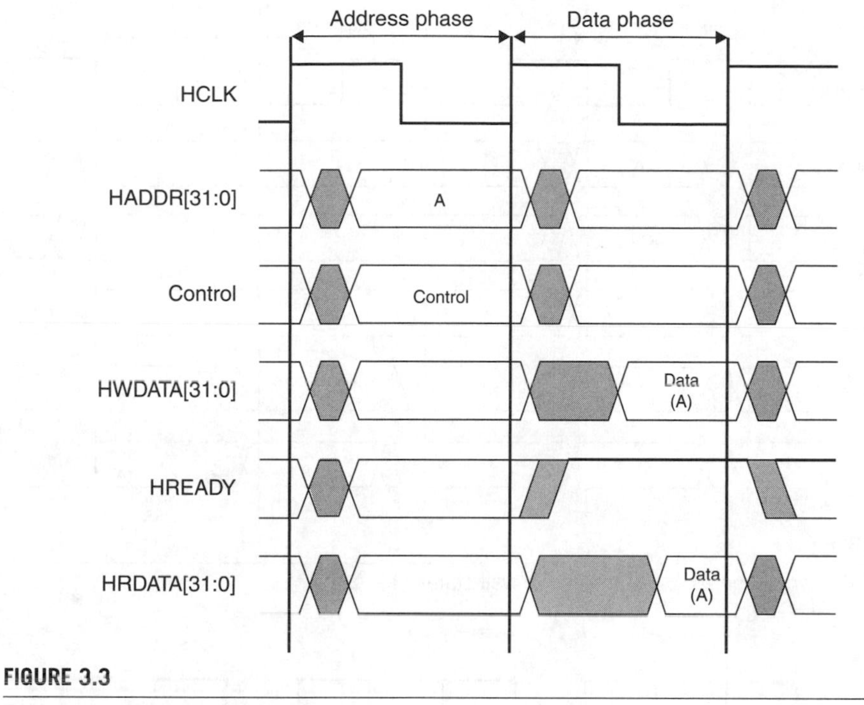

FIGURE 3.3

Basic data transfer on AHB bus [1]
Source. ARM Inc.

one or more cycles. In the figure, the master drives the address and control signals on the bus after the rising edge of the clock (*HCLK*) in the address phase. The slave samples (i.e., reads) the address and control signals at the next rising edge of the clock, which marks the beginning of the data phase. In the case of a read, the slave drives the appropriate data onto the read data bus (*HRDATA*) in this data phase cycle, followed by the master sampling the data off the bus on the third rising edge of the clock. Otherwise, in the case of a write, the master drives the data onto the write bus (*HWDATA*) in the data phase, followed by the slave sampling the data off the bus on the third rising edge of the clock. Note that Fig. 3.3 shows data on both the read and write buses for illustration purposes only (corresponding to a read or a write, respectively). In practice, the read data bus will be idle on a write, and the write data bus will be idle during a read.

It is possible that the slave can require more than one cycle to provide a response to the master, either because it needs one or more cycles to read the requested data, or to get in a state to write data. In such cases, the slave can introduce wait cycles by lowering the *HREADY* signal to indicate to the master that additional time is required to complete the transaction. Figure 3.4 shows the case for a read (or write) operation, where the slave inserts two wait cycles by lowering the *HREADY* signal for two cycles, before driving the read data onto the read data bus (or sampling the write data from the write data bus). Since the address and data phases of a transfer occur in separate cycles, it is possible to overlap the address phase of one transfer with the data phase of another transfer, in the same clock period. Such

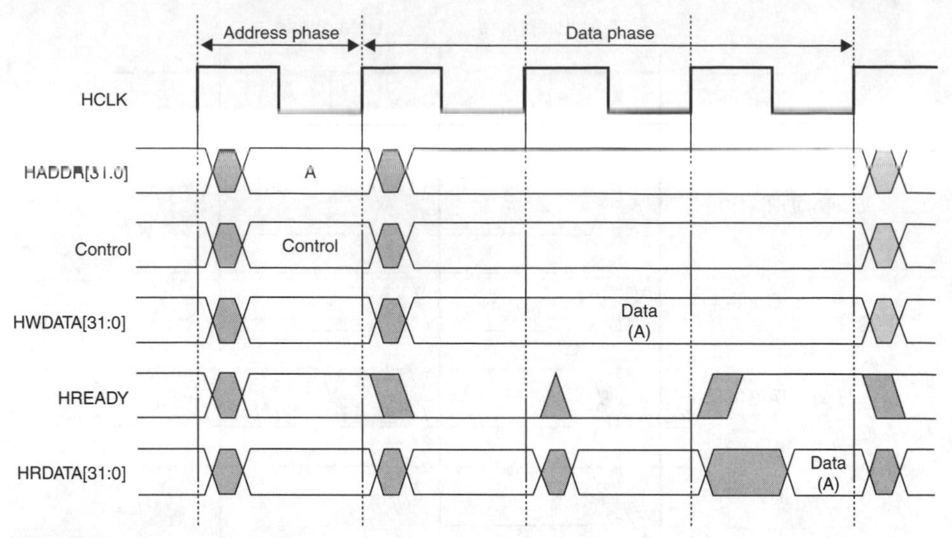

FIGURE 3.4

Basic data transfer on AHB bus with slave wait states [1]
Source: ARM Inc.

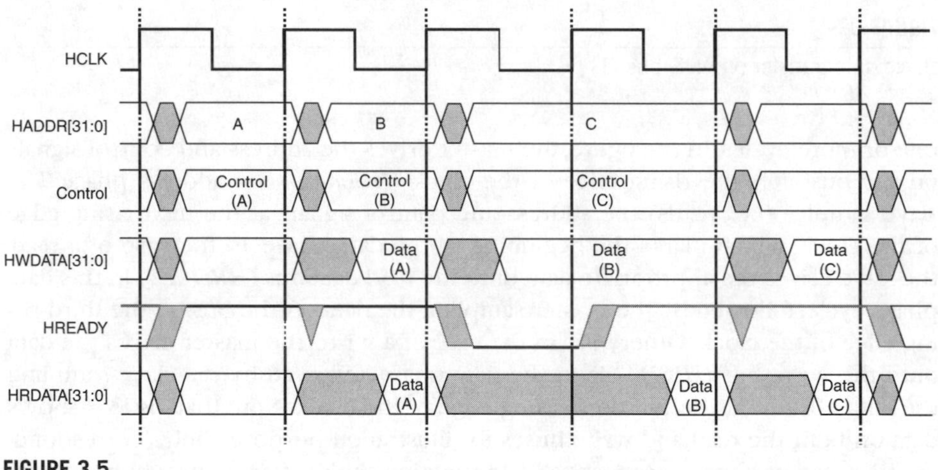

FIGURE 3.5

Pipelined data transfers on AHB bus [1]
Source: ARM Inc.

a *pipelined* operation of the AHB bus, depicted in Fig. 3.5, allows for high performance operation.

In addition to the address and data phases, an arbitration phase is necessary for all data transfers if there is more than one master connected to a bus. The arbitration phase ensures that only one master gains access to the bus and proceeds with its data transfer at any given point of time. Figure 3.6 shows the centralized arbitration scheme on an AHB bus to which three master components are connected. Whenever a master needs to initiate a read or write data transfer,

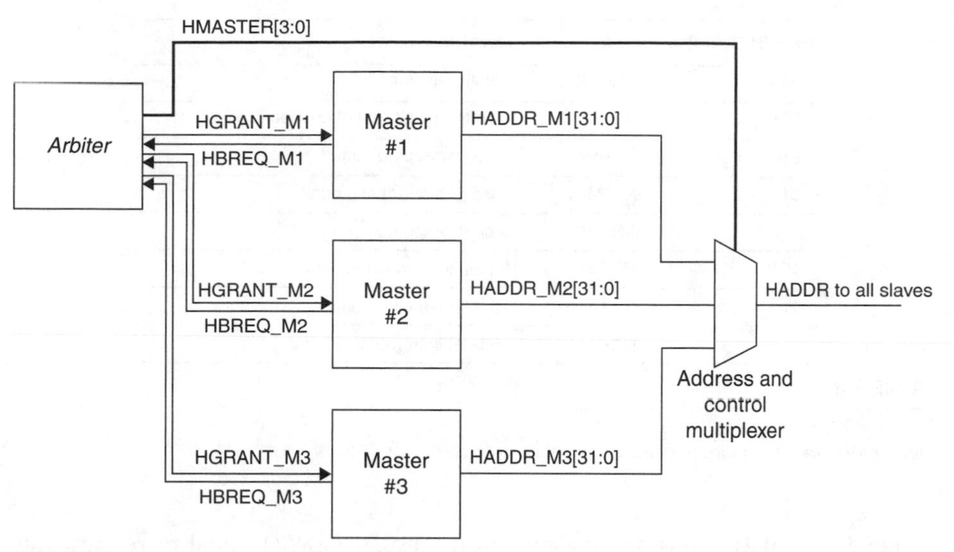

FIGURE 3.6

Arbitration on AHB bus [1]
Source: ARM Inc.

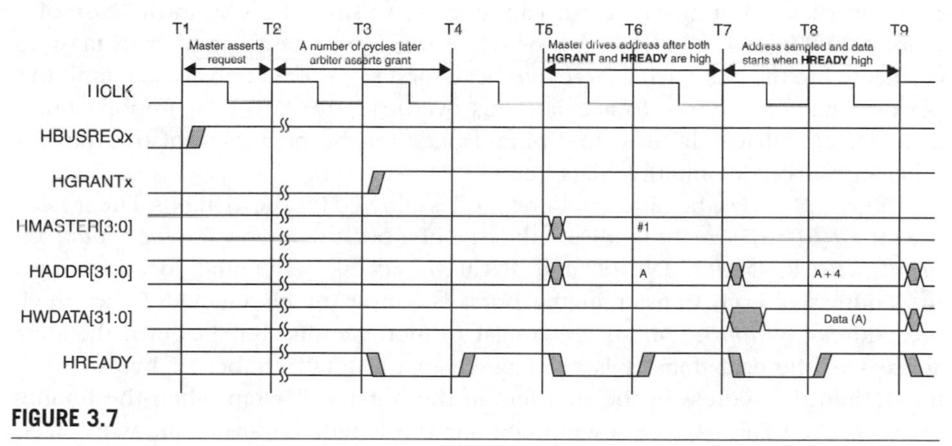

FIGURE 3.7

Cost of arbitration on AHB bus [1]
Source: ARM Inc.

it drives the *HBUSREQ* signal to the arbiter, requesting it for access to the AHB bus. The arbiter samples the *HBUSREQ* signals from all the masters and uses its arbitration policy to decide which master gets granted access to the bus access. The AHB specification does not identify a particular arbitration scheme—instead, a designer is given the freedom to implement any suitable scheme depending on the target application latency and bandwidth requirements. Once the arbitration scheme selects a master, the arbiter drives the *HGRANT* signal to the selected master, indicating it has been granted access to the bus and can proceed with its transaction. All the other masters must wait until re-arbitration, after completion of the current data transfer. Figure 3.7 shows an example of how the arbitration

HBURST[2:0]	Type	Description
000	SINGLE	Single transfer
001	INCR	Incrementing burst of unspecified length
010	WRAP4	4-beat wrapping burst
011	INCR4	4-beat incrementing burst
100	WRAP8	8-beat wrapping burst
101	INCR8	8-beat incrementing burst
110	WRAP16	16-beat wrapping burst
111	INCR16	16-beat incrementing burst

FIGURE 3.8

Different burst modes on the AHB bus [1]
Source: ARM Inc.

proceeds on an AHB bus: the master asserts the *HBUSREQ* signal to request bus access from the arbiter, which samples the requests at the next rising clock edge, uses its internal arbitration scheme to select a master, and then grants it access to the bus by asserting the *HGRANT* signal. Note that there is typically at least a one cycle overhead for arbitration that can increase to several cycles in the case of a more complex arbitration scheme, or when there are a large number of masters connected to the bus. Such a large overhead for a single data transfer can limit the performance on the bus. To alleviate this overhead, the AHB bus supports burst data transfers that only need to arbitrate once (at the beginning of the transaction) for transferring multiple data items.

Figure 3.8 shows the different burst modes allowed for the AHB bus. The master uses the *HBURST* signal to indicate the size of a burst data transfer. *Incrementing bursts* (INCR, INCR4, INCR8, and INCR16) access sequential locations, and the address of each transfer in the burst is simply an increment of the previous address. *Wrapping bursts* are similar to incrementing bursts, but if the start address of the data transfer is not aligned to the total number of bytes in the burst, then the address of the transfers in the burst will wrap when the boundary is reached. For instance, a wrapping burst of length 4, transferring word-sized (4 byte) data items will wrap at 16 byte boundaries. So if the start address of a transfer is 0x64 h, then the four addresses in the burst will be 0x64 h, 0x68 h, 0x6Ch, and 0x60 h. Figure 3.9 presents an example of a wrapping burst of length 4 (WRAP4). Notice the wait state inserted by the slave (by lowering the *HREADY* signal) for the first data transfer. Since the burst of word transfers will wrap at 16 byte boundaries, the transfer to address 0x3C is followed by a transfer to address 0x30 h. An incrementing burst of length 4 (INCR4) on the other hand would have continued beyond the 16 byte boundary, and would access the following sequence of addresses: 0x38 h, 0x3Ch, 0x40 h, and 0x44 h.

It is possible for an AHB burst to be interrupted by the arbiter, in case a higher priority master needs to transfer data on the bus. If a master loses access to the bus in the middle of a burst, it must reassert its *HBUSREQ* signal to again arbitrate

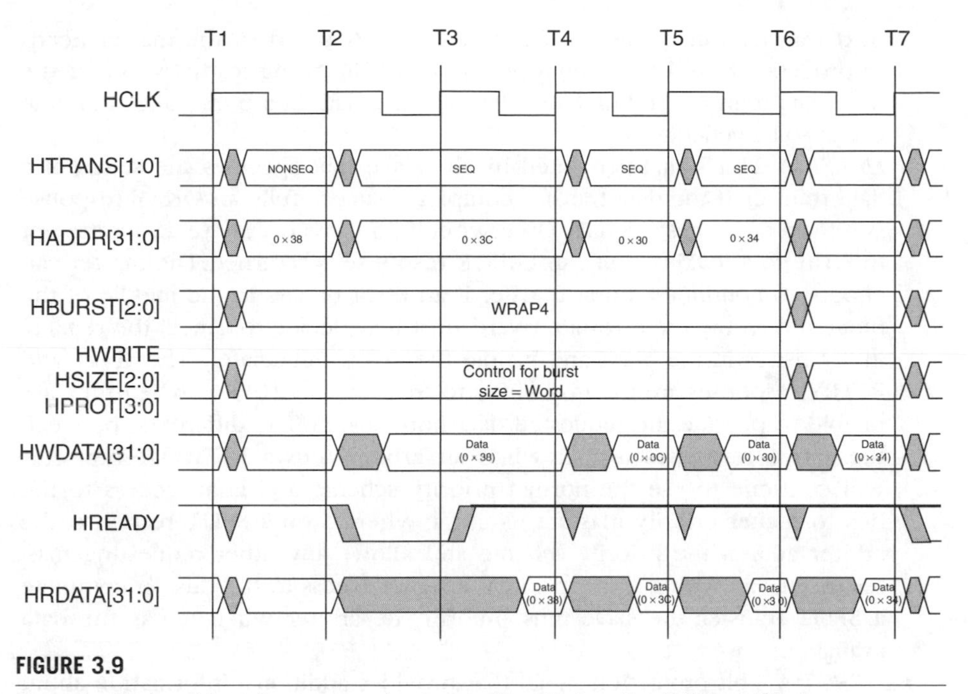

FIGURE 3.9

Example of a burst transfer—a wrapping burst of length 4 on the AHB bus [1]
Source: ARM Inc.

for access to the bus. Master and slave module AHB interfaces must be designed to correctly handle early burst termination. If a master requires that its burst transfer not be interrupted, it must assert the *HLOCK* signal when requesting the bus from the arbiter, to indicate that the bus needs to be *locked* away from other masters, for the duration of the burst transfer. When the arbiter sees the *HLOCK* signal for a master request, it must ensure that no other master is granted access to the bus once this master is granted access to the bus and its first data transfer has commenced. In addition to *HLOCK*, there are several other control signals used during data transfers, such as:

- *HWRITE*: A 1-bit signal generated by the master that indicates the transfer direction—a write when it is high, or a read when it is low.
- *HSIZE*: A 3-bit signal generated by the master that indicates the size of the data transfer. There are 8 possible allowed values for data transfer size ranging from 8 bits (000) to 1024 bits (111). This signal is particularly useful if the size of data being transmitted on the bus is smaller than the bus width.
- *HTRANS*: A 2-bit signal generated by a master to indicate the type of a transaction. There are four types of transfers possible in AHB: (i) *NONSEQ*, which is usually a single transfer or the first transfer in a burst, (ii) *SEQ*, which specifies the remaining transfers in a burst, (iii) *IDLE*, which indicates that no data transfer is required and is generally used when a master is granted access to the bus, but does not need to transfer any data, and (iv) *BUSY*, which indicates that the master is continuing with a burst transfer, but the

next transfer cannot take place immediately (e.g., when the master needs to process a read data for multiple cycles before being ready to receive the next data item in the burst; or when a write data in a burst takes multiple cycles to be generated).

- *HRESP*: A 2-bit signal generated by the slave that specifies the status of a data transfer. If the data transfer completes successfully, an OKAY response is returned on these signals. Otherwise, if an error occurs (e.g., an attempt to write to a ROM region), an ERROR response is returned. The master can choose to continue a burst transfer if an error occurs in the middle of the burst and rectify the error afterward, or it may decide to cancel the remaining transfers in the burst and handle the error immediately. The SPLIT and RETRY responses are used by slaves to free up the AHB bus when they are unable to provide the requested data immediately. The difference between these two responses is that when an arbiter sees a RETRY response, it will continue to use the normal priority scheme and grant access to the bus to higher priority masters, as usual; whereas on a SPLIT response, the arbiter adjusts the priority scheme and allows any other requesting master (even one with a lower priority) to gain access to the bus. To complete a SPLIT transfer, the slave must inform the arbiter when it has the data available.
- *HPROT*: A 4-bit protection signal that provides additional information about a data transfer. It is typically used by a component requiring some level of protection control, and can be used to specify, for instance, if the transfer is an opcode (*OPC*) fetch or a data access. Other possible uses include indications for a privileged mode access or a user mode access, and specifying if a transfer is bufferable or cacheable.

As mentioned above, a SPLIT transfer enables a slave to free up access to a bus, if it believes that the data requested from it (by the master) will not be available for several cycles. In such a scenario, the slave records the ID of the master from which the request initiated (to restart the transfer again at a later time) and asserts an *HSPLIT* signal to the arbiter. The arbiter then masks (i.e., ignores) requests from the master that was SPLIT, and grants other masters access to the bus. This process is shown in Fig. 3.10, where a SPLIT on the second cycle causes the arbiter to grant bus access to another master. When the slave is ready to complete the transfer, it signals the arbiter and sends the ID of the master that was involved in the split transfer. The arbiter unmasks the master, and eventually grants the master access to the bus, to complete the data transfer. The entire process is transparent to the masters making the request. Thus SPLIT transfers allow the time that would have otherwise been spent waiting for the data from the slave, to be utilized in completing another transfer, which enables better utilization of the bus. Note that an AHB master can only have a single outstanding transaction at any given time. If more than one outstanding transaction needs to be handled by a master component, it requires an additional set of request and grant signals for each such outstanding transaction. Both SPLIT and RETRY transfers can cause bus deadlocks, and therefore care must be taken while implementing them.

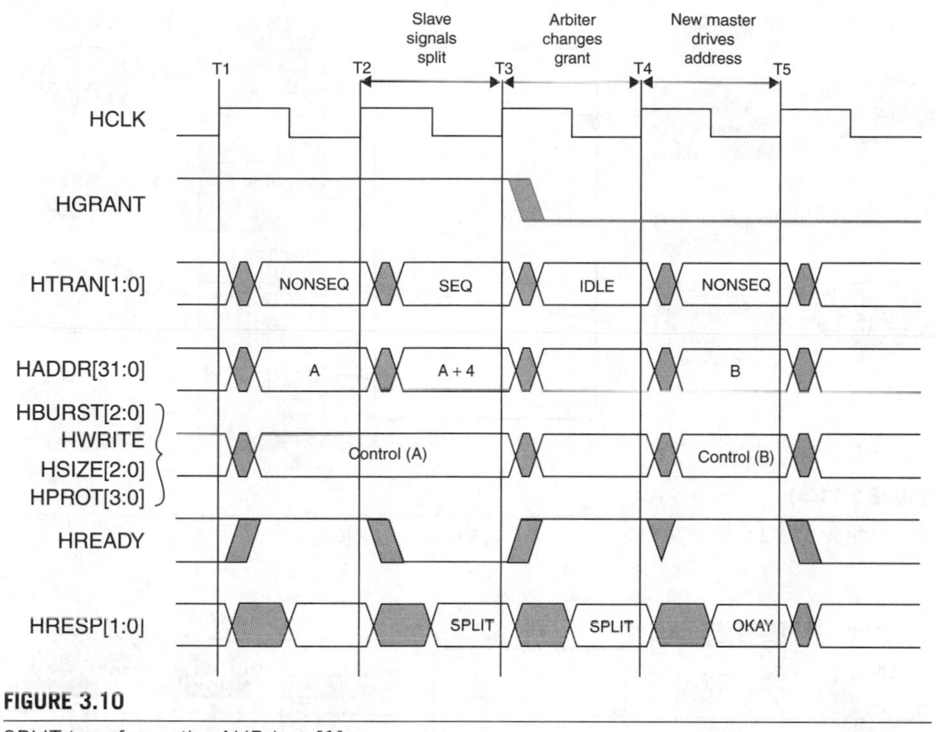

FIGURE 3.10

SPLIT transfer on the AHB bus [1]
Source: ARM Inc.

AHB Bus Matrix Topology

In addition to the basic hierarchical bus topology, where an AHB bus uses a bridge to interface with an APB bus, AHB bus-based communication architectures can have other topologies as well, such as a hierarchical bus topology with multiple AHB (and APB) buses interfacing with each other through bridges. For SoC designs that require very high bandwidths and require multiple concurrent data transfers, the hierarchical bus architecture may be insufficient. For such designs, an AHB multi-layer bus matrix [8] topology offers a more suitable communication infrastructure. Figure 3.11(a) shows an example of a 2 master, 4 slave AHB full bus matrix topology, that has multiple buses in parallel, to support concurrent data transfers and high bandwidths. The *Input Stage* is used to handle interrupted bursts, and to register and hold incoming transfers from masters if the destination slaves cannot accept them immediately. The *Decoder* generates select signals for slaves, and also selects which control and read data inputs received from slaves are to be sent to the master. The *Output Stage* selects the address, control and write data to send to a slave. It calls the *Arbiter* component, which uses an arbitration scheme to select the master that gets to access a slave, if there are simultaneous requests from several masters. Unlike in traditional hierarchical shared bus architectures, arbitration in a bus matrix is not centralized, but distributed so that every slave has its own arbitration. One drawback of the full bus matrix scheme is

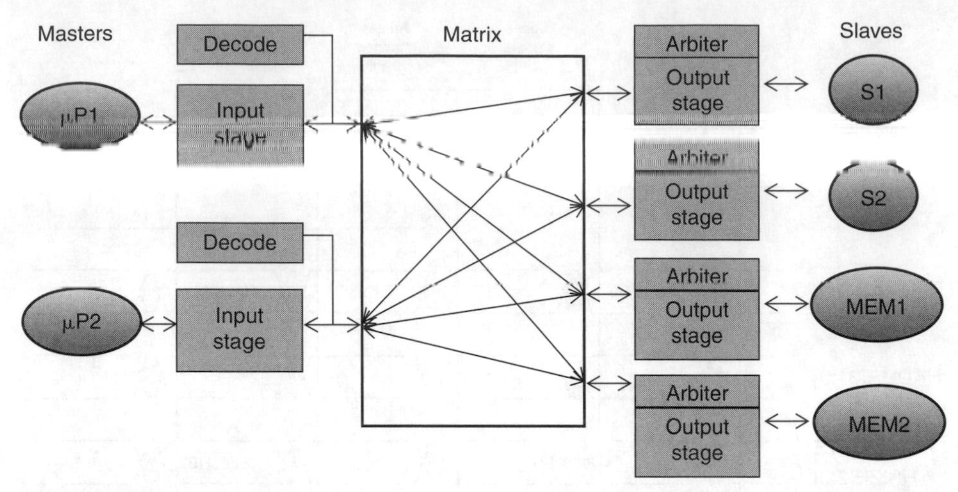

FIGURE 3.11(a)

An example of a 2 master, 4 slave AHB: full bus matrix topology

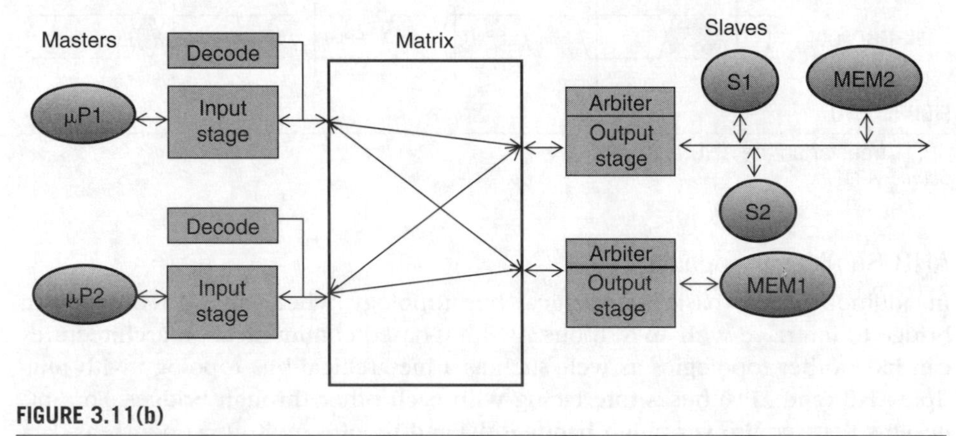

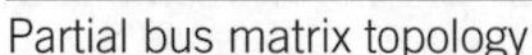
FIGURE 3.11(b)

Partial bus matrix topology

that it connects every master to every slave in the system, resulting in a very large number of buses (and consequently wires and bus logic components). Such a configuration, therefore, achieves high performance at the cost of high power consumption and a larger area footprint. For systems that have less stringent performance requirements, a partial AHB bus matrix [8, 9] topology can be used. Figure 3.11(b) shows a partial AHB bus matrix configuration that clusters components onto shared buses, to reduce the number of buses in the matrix. This partial matrix configuration offers less potential bandwidth due to the likelihood of data traffic conflicts on the shared buses, when compared to the full bus matrix. However, the partial matrix configuration consumes less power and takes up a smaller chip area, which is a desirable characteristic for communication architectures. Pasricha et al. [10] showed how a full AHB bus matrix can be reduced to a

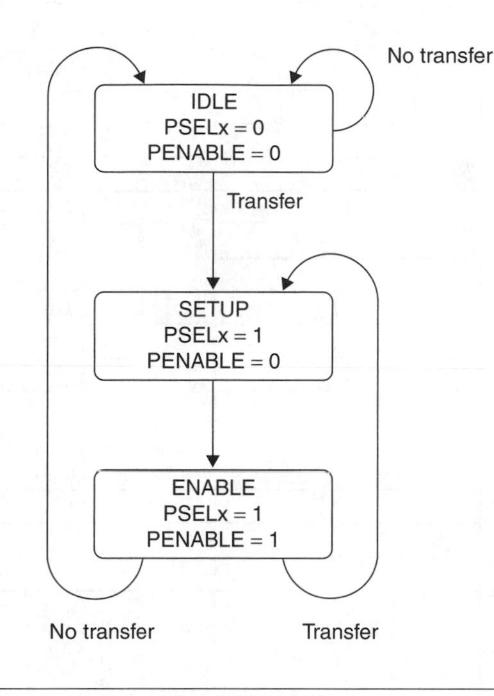

FIGURE 3.12

State diagram representing activity of the APB bus [1]
Source: ARM Inc.

partial bus matrix topology for a multiprocessor system-on-chip (MPSoC) design, by optimally reducing the number of buses while still meeting all application performance constraints. This methodology is described in more detail in Chapter 6.

3.1.1.2 *Advanced Peripheral Bus*

The APB bus standard defines a bus that is optimized for reduced interface complexity and low power consumption. This bus is meant to interface with the AHB (via a bridge), connecting low bandwidth (or high latency) peripheral components that do not require the advanced features of high performance buses such as the AHB. The APB allows only non-pipelined data transfers, and has only a single master—the bridge that connects the AHB bus to the APB bus. Typically, the APB bus operates at a much lower clock frequency (which helps reduce power consumption) and has a smaller bus width, compared to the AHB bus. Figure 3.12 depicts a state diagram that represents the activity on an APB bus. The *IDLE* state is the default state, in which the APB bus remains when there are no transfer requests from the AHB bus. When a transfer request arrives from the AHB bus via the AHB–APB bridge, the APB bus moves to the *SETUP* state and asserts the appropriate slave select signal to select the slave on the APB bus that is required to participate in the transfer. The APB bus remains in the *SETUP* state for one cycle, and this time is spent in decoding the address of the destination peripheral component. The APB bus moves to the *ENABLE* state on the next rising edge of the clock and asserts the *PENABLE* signal to indicate that the transfer is ready to be performed. This state also typically lasts for one cycle, after which it can go back

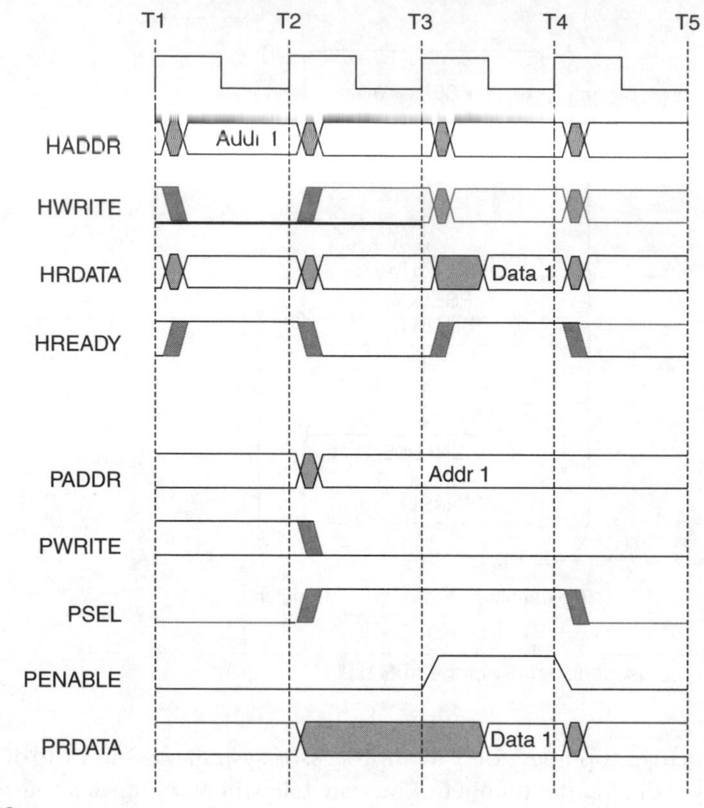

FIGURE 3.13

Read data request from the AHB to the APB bus [1]
Source: ARM Inc.

to the *SETUP* stage if another transfer follows the current transfer or to the *IDLE* state if no further transfers are required.

Figure 3.13 illustrates a read request as it propagates from the AHB to the APB bus. The top four signals belong to the AHB bus while the four signals at the bottom belong to the APB bus. The master component on the AHB bus drives the address and control signals onto the AHB bus in the first cycle. These are sampled by the AHB-APB bridge component on the rising edge of the next clock cycle (T2). Since the transfer is intended for the APB bus, the appropriate select signal (*PSEL*) is asserted and the APB bus transitions from the *IDLE* state to the *SETUP* state in cycle T2. This is followed by the *ENABLE* state in cycle T3, in which the slave receives the request and returns the read data. The returned data can usually be directly routed back to the master on the AHB bus, where it will be sampled off the bus at the next rising clock edge (T4). However, for high performance systems, the returned data can be first registered at the bridge, and then driven to the appropriate master in the following cycle. While this approach requires an extra cycle, it can allow the AHB bus to operate at a much higher clock frequency, which allows an overall improvement in system performance. Figure 3.14

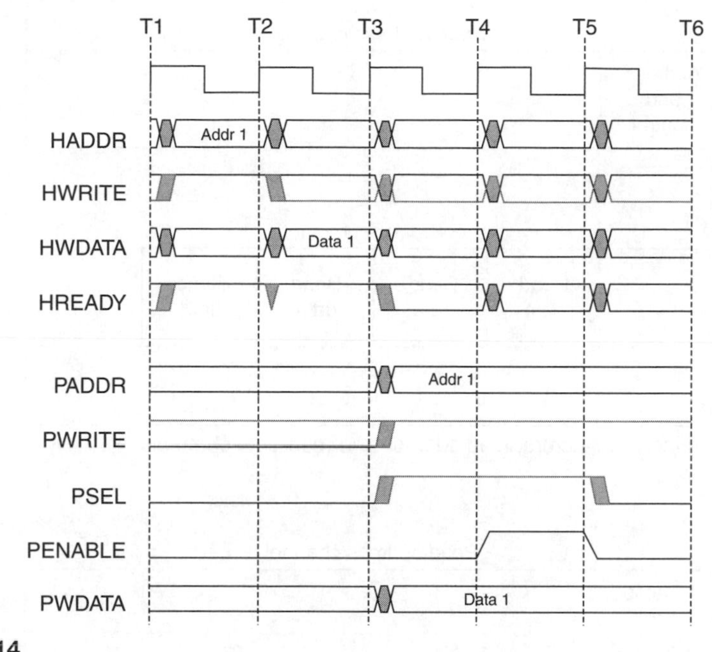

FIGURE 3.14

Write data request from the AHB to the APB bus [1]
Source: ARM Inc.

shows a similar scenario, for a write transfer on the APB bus. The bridge samples the address, control, and data signals from the master, holding these values for the duration of the write transfer on the APB bus (as it switches through the *SETUP* and *ENABLE* states).

3.1.2 AMBA 3.0

The AMBA 3.0 bus architecture specification [2] introduces the Advanced eXensible Interface (AXI) bus that extends the AHB bus with advanced features to support the next generation of high performance MPSoC designs. The goals of the AXI bus protocol include supporting high frequency operation without using complex bridges, flexibility in meeting the interface, and performance requirements of a diverse set of components, and backward compatibility with AMBA 2.0 AHB and APB interfaces. We now look at the AXI specification in more detail (Figure 3.15).

3.1.2.1 *Advanced Exensible Interface*

The AXI bus standard proposes a burst-based, pipelined data transfer bus, similar to the AHB bus, but with additional advanced features and enhancements. The main features of AXI and its differences with the AHB standard are presented in Table 3.1, and elaborated in more detail below.

The AXI specification describes a high level channel-based architecture for communicating between masters and slaves on a bus. Five separate channels are defined: *read address*, *read data*, *write address*, *write data*, and *write response*.

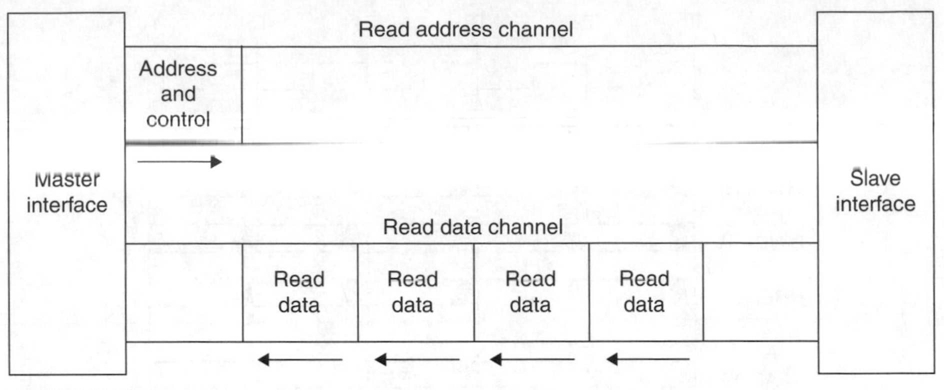

FIGURE 3.15(a)

AMBA AXI channel architecture: read address and read data channels

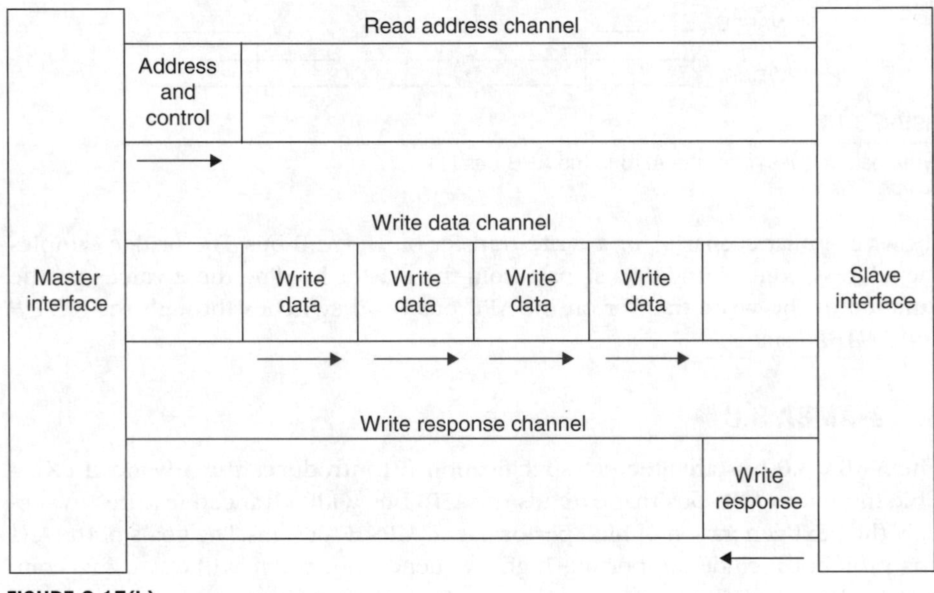

FIGURE 3.15(b)

Write address, write data, and write response channels [2]
Source: ARM Inc.

Just like for the data bus in AHB, the data channel width in AXI can range from 8 to 1024 bits. The read channels are shown in Fig. 3.15(a). The address and control information for a read transfer is sent by the master on the read channel, while the read data and response information from the slave is received on the read data channel. Figure 3.15(b) shows the write data channels. The address and control information for a write transfer is sent on the write address channel, while the write data is transmitted on the write data channel. A one byte strobe signal is

Table 3.1 Contrasting features of AXI and AHB

AMBA 3.0 AXI	AMBA 2.0 AHB
Channel-based specification, with five separate channels for read address, read data, write address, write data, and write response enabling flexibility in implementation.	Explicit bus-based specification, with single shared address bus and separate read and write data buses.
Burst mode requires transmitting address of only first data item on the bus.	Requires transmitting address of every data item transmitted on the bus.
OO transaction completion provides native support for multiple, outstanding transactions.	Simpler SPLIT transaction scheme provides limited and rudimentary outstanding transaction completion.
Fixed burst mode for memory mapped I/O peripherals.	No fixed burst mode.
Exclusive data access (semaphore operation) support.	No exclusive access support.
Advanced security and cache hint support.	Simple protection and cache hint support.
Register slice support for timing isolation.	No inherent support for timing isolation.
Native low-power clock control interface.	No low-power interface.
Default bus matrix topology support.	Default hierarchical bus topology support.

included for every 8 bits of write data, to indicate which bytes of the data bus are valid. This is useful for cases where there is a mismatch between the size of the data being transferred and the data bus width. A separate write response channel provides the slave a means to respond to write transactions. A write completion signal occurs once for every burst (and not for every data transfer) to indicate the status of the write at the slave. The five separate channels provide implementation flexibility to a designer and can be implemented in any one of three ways:

1. *Shared address bus and shared data buses (SASD)*: A single shared address bus is coupled with a bidirectional data bus that handles both reads and writes. Such a configuration is typically useful for smaller, low complexity embedded systems.
2. *Shared address bus and multiple data buses (SAMD)*: A single shared address bus is coupled with separate, unidirectional read and write data buses. Since the address bus bandwidth is typically less than that of the data buses (as only one address needs to be sent for a burst data transfer), interconnect complexity can be reduced while still maintaining performance, by using a shared address bus.

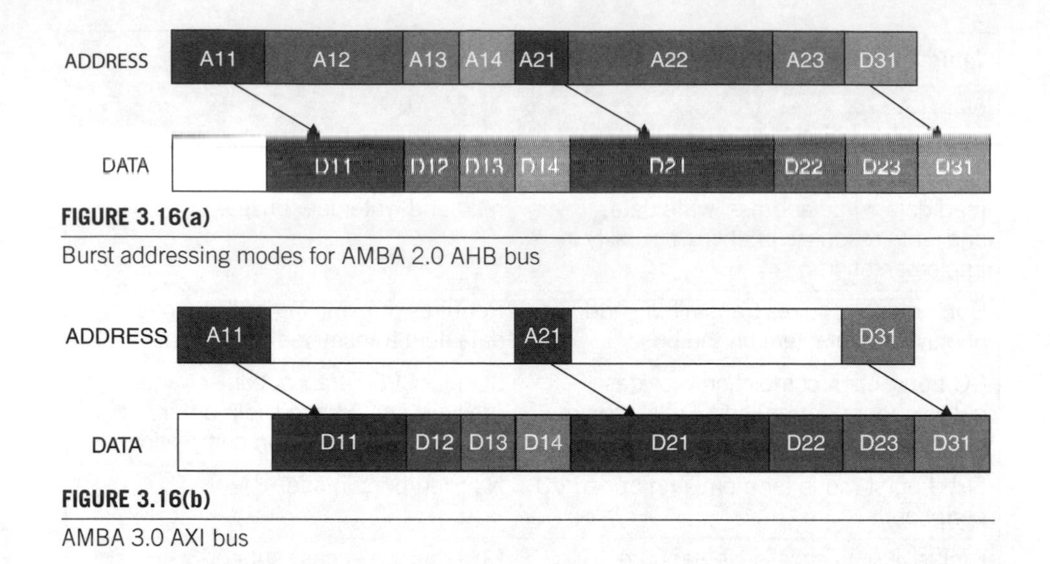

FIGURE 3.16(a)

Burst addressing modes for AMBA 2.0 AHB bus

FIGURE 3.16(b)

AMBA 3.0 AXI bus

3. ***Multiple address buses, multiple data buses (MAMD)***: A separate address bus for reads and writes is coupled with separate read and write data buses. This configuration has the largest interconnect complexity, but also the best performance of the three implementation alternatives.

The *MAMD* mode in particular allows multiple concurrent read and write transactions to occur independent of each other, which can be very useful for high performance SoC designs. In contrast, AHB only explicitly supports the *SAMD* implementation mode.

One significant difference between AXI and AHB is the way addressing is handled during burst data transfers. In AHB, every data transfer in a burst requires an address to be transmitted on the address bus. In contrast, AXI requires the address of only the first data item in the burst to be transmitted. Figure 3.16(a) shows burst data transfers on an AHB bus that require an address for every data item transmitted. Contrast this with the same scenario on an AXI bus, shown in Fig. 3.16(b), where only the address of the first data item in a burst is transmitted. It is the responsibility of the slave to calculate the address of the subsequent transfers in the burst. Because only a single address is transmitted per burst, the address buses in AXI are freed up to handle other transactions. Figure 3.17 shows how read and write transactions can occur simultaneously in an *SAMD* implementation of AXI, because the address bus is freed up during a burst. The AHB in contrast must wait for a burst to complete before initiating another transfer, which results in under-utilization of its data buses. Another enhancement in AXI, compared to AHB, comes in the form of support for an additional burst type. AXI not only supports all the incrementing and wrapping burst types present in AHB, but also an additional *fixed burst mode*. In this mode, the address of every data item transferred in a burst remains the same. This burst type is very useful for repeated accesses to the same location, such as for data transfers with an I/O peripheral FIFO (first-in-first-out).

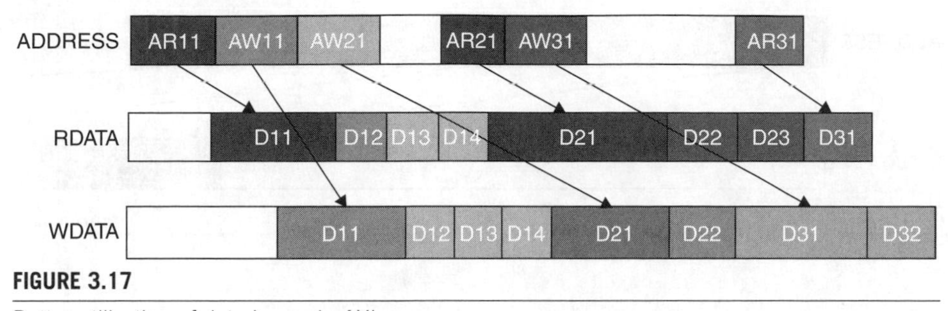

FIGURE 3.17

Better utilization of data buses in AXI

Another important AXI feature is its support for *out-of-order (OO) transaction completion* which is an advanced feature that maximizes data throughput and improves system efficiency. AXI masters have the ability to issue multiple outstanding addresses, which means that transaction addresses can be issued without waiting for earlier transactions to complete. This is accomplished by assigned IDs to read and write transactions issued by the masters. The AXI specification lays down certain guidelines to govern the ordering of transactions. Transactions from different masters have no ordering restrictions and can complete in any order. Transactions from the same master but with different ID values can complete in any order. However, a sequence of write transactions with the same ID value must complete in the same order in which the master issued them. For a sequence of read transactions with the same ID value, two possible scenarios exist: (i) reads with the same ID value are from the same slave, in which case it is the responsibility of the slave to ensure that the read data returns in the same order in which the addresses are received; and (ii) reads with the same ID value are from different slaves, in which case the AXI bus logic must ensure that the read data returns in the same order that the master issued the addresses in.

The ability to finish transactions OO allows completion of transactions to faster regions of a memory (or peripheral), without waiting for earlier transactions to slower regions. This feature reduces the effect of transaction latency and improves system performance. Figure 3.18 shows a comparison between AHB and AXI for a scenario where a master must access a slow (i.e., high latency) slave. Figure 3.18(a) shows how an access to a slow slave in AHB holds up the master and the bus till the slave is ready to return the data. Note that using SPLIT transactions on AHB can free up the bus, but not the master, which is still stalled. In contrast, in the AXI case, as shown in Fig. 3.18(b), the master need not wait to get the data back from the slave before issuing other transactions. Both the bus and the master are freed up in AXI, which allows better performance, higher efficiency, and greater bus utilization. Although theoretically any number of transactions can be reordered on the AXI bus, a practical limit is placed by the read/write data *reordering depth* at the slave interfaces. The read or write data reordering depth of a slave is the number of addresses pending in the slave that can be reordered. A slave that processes all transactions in order is said to have a data reordering depth of 1. This reordering depth must be specified by the designer for both reads and writes, and involves a trade-off between hardware complexity and parallelism in the system—a larger reordering

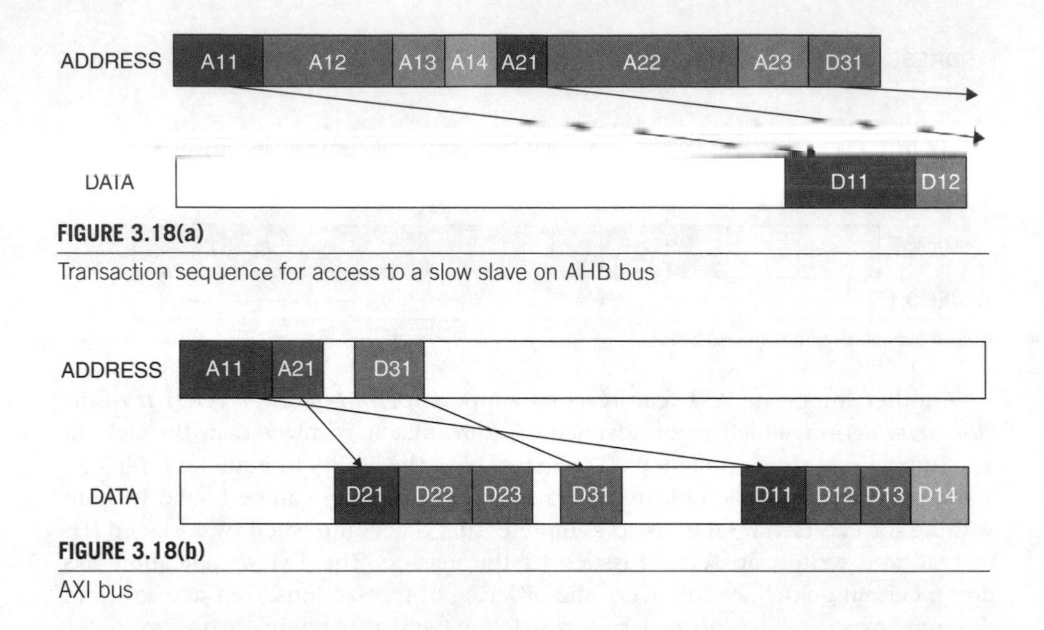

FIGURE 3.18(a)

Transaction sequence for access to a slow slave on AHB bus

FIGURE 3.18(b)

AXI bus

depth requires greater hardware complexity, but also improves parallelism and possibly overall system performance. However, there is a limit beyond which increasing the reordering depth for a slave does not improve performance because there is typically a limit to the maximum number of concurrent transfers possible in an application, as shown by Pasricha et al. [11].

Other areas where AXI improves on and differs from the AHB feature set include:

- *Semaphore operations*: AXI provides support for semaphore type operations using an exclusive access mechanism that does not require the bus to remain locked to a particular master for the duration of the transaction. The support for semaphore type operations in AXI therefore does not affect maximum achievable bandwidth or data access latency. In contrast, AHB does not provide any support for semaphore type operations. An exclusive access on AXI is initiated when a master performs an exclusive read from a slave address location. At some later time, the master attempts to complete the exclusive access by attempting to write to the same location. The exclusive write access is signaled by the slave as successful if no other master has written to the location between the read and write accesses, or as a failure if another master has written to the location between the read and write accesses. A slave is required to have additional logic (such as a monitor unit for each exclusive-capable master ID that can access it) if it supports exclusive accesses.
- *Cache support*: AXI provides support for system level caches and other performance enhancing components with the help of two 4-bit cache hint signals, one for each of the read and write channels. These cache hint signals provide additional information about how the transaction can be processed. While AHB provides support for basic cache hints (with a 2-bit signal) such as if data is bufferable or cacheable, AXI extends this by providing additional

signals to specify write-through and write-back cache allocation strategies for reads and writes, as well as providing the designer the option of customizing the hint signals for other purposes such as flushing cache and page tables.

- *Protection support*: AXI utilizes two 3-bit signals, one each for the read and write data channels, to provide protection against illegal transactions. There are three levels of protection possible, each represented by a single bit of the protection signals: (i) *normal or privileged access*, used by certain masters to indicate their processing mode and to obtain access to special resources in a system, since privileged accesses typically provide greater access within a system, (ii) *secure or non-secure accesses*, used in systems where a greater degree of differentiation between processing modes is required, and (iii) *instruction or data accesses*, to indicate if the transaction is an instruction or a data access. In comparison, AHB provides support for normal/privileged accesses and instruction/data accesses, but not for secure/non-secure accesses.
- *Low power support*: AXI supports an optional set of signals for low power operation. These signals target two classes of peripherals. The first consists of peripherals that require a power-down sequence, and can have their clocks turned off (to save power) only after they enter a low power state. These peripherals require a signal from the system clock controller to determine when to begin the power-down sequence. The second type of peripherals are those that do not require a power-down sequence, and can assert a signal to indicate when it is acceptable to turn off their clock. AXI provides support for both types of signals. In contrast, AHB does not include any signals for low power peripheral operation support.
- *Recommended topology*: The AXI specification assumes a default bus matrix topology implementation, without any requirement for complex bridges. Such an implementation is in keeping with the advanced feature set of AXI, and is suitable for contemporary high performance designs. However, this does not limit AXI in any way from being used in a hierarchical bus topology, if required. The AHB specification, on the other hand, assumes a default hierarchical bus topology arrangement, involving an AHB bus interfacing with an APB bus via a bridge. This is in keeping with the comparatively less advanced AHB feature set, but does not limit it from being implemented in a bus matrix topology.
- *Register slice support*: The clock frequency on a bus puts a limit on the length of its bus wires during physical layout. This is because a signal can only travel a finite distance on the chip in a single clock cycle [12]. As the clock frequency of the bus is increased, the clock cycle period (which is the inverse of the clock frequency) is reduced, and the distance that can be traveled by a signal shrinks. In fact, it can take multiple cycles for a signal to travel between the ends of a chip [13] in Deep Submicron (DSM) process technologies. For high performance SoC designs, high clock frequencies on the bus are essential to meet performance constraints. To sustain these high bus clock frequencies and ensure correct operation on long interconnects, AXI proposes using one or more register slices on a bus. These register slices latch signal information coming from the source and then retransmit it toward the destination.

The advantage here is that the signal need only cover the distance between latches in a single clock cycle. Thus, by inserting register slices, timing closure (i.e., ensuring timing requirements of a design are met) becomes relatively easier. Introducing the register slices will of course increase the latency (in terms of number of cycles required) for communicating on the bus. However, the bus can now be operated at a much higher frequency (compared to a bus with no register slices inserted) which can improve overall system performance. AHB, in contrast, does not provide any such means of alleviating the problem of meeting timing for high performance, high frequency systems.

AXI Bus Matrix Topology

Just like the AHB protocol, AXI can be connected in a bus matrix topology. ARM distributes an AXI configurable interconnect (ACI) IP at the register transfer level (RTL) called *PL300* [30] (recently superseded by the next version—*PL301* [31]) that allows designers to connect several masters and slaves together with a configurable AXI bus matrix fabric. The structure of the AXI bus matrix is somewhat similar to that of the AHB bus matrix described earlier, but differs in its support for additional features of the AXI protocol such as independent control for decoupled read, write, and response channels, and OO transaction completion. Components are connected to the AXI bus matrix using interface routers. Each master is connected to a slave interface router, and each slave is connected to a master interface router. The interface routers are part of the AXI bus matrix fabric. These routers essentially consist of multiplexing and de-multiplexing elements to ensure appropriate connectivity with other components connected to the bus matrix. The select signal for each router is generated from a control block that is unique for each channel and interface. The control blocks store the routing information necessary to enforce the ordering constraints within the AXI protocol, and consist of arbiters, decoders, content addressable buffers (CABs), and FIFO elements. The bus matrix fabric does not buffer addresses and data—slaves supporting outstanding transactions must provide the required storage locally. Much like in the case of the AHB bus matrix, a full AXI bus matrix (that connects all the masters to all the slaves in a system) supports high bandwidth but can be prohibitively expensive as it requires a very large number of wires and bus logic components. In a lot of cases where a somewhat lower performance is acceptable, a partial AXI bus matrix that clusters components onto shared buses to reduce the number of wires and bus logic components in the matrix may be more suitable. Pasricha et al. [9] showed how a partial AXI bus matrix can be automatically synthesized from a full AXI bus matrix for MPSoC designs. This methodology, which reduces the number of buses in the matrix while satisfying all application performance constraints, is described in more detail in Chapter 6.

3.1.3 IBM CoreConnect

The IBM CoreConnect [3] on-chip communication architecture standard is another popular synchronous bus-based standard that shares many similarities with the AMBA standard. It defines three types of buses: (i) *processor local bus (PLB)*, which is a high performance bus used to connect high speed processor and memory components, (ii) *on-chip peripheral bus (OPB)*, which is used to connect lower performance peripheral components, and (iii) *device control register (DCR)* bus, which is a simple, high latency bus used to access the status and control registers of the PLB and OPB masters. The CoreConnect standard targets a hierarchical bus topology implementation, similar to AMBA 2.0, with the OPB and PLB buses interfacing with each other using a bridge as shown in Fig. 3.19. We now look at the specifications of the PLB, OPB, and DCR buses in more detail.

3.1.3.1 *Processor Local Bus*

The PLB is a synchronous, high performance bus, similar in many aspects to the AMBA 2.0 AHB, and used to interconnect high performance processor, ASIC, and memory cores. The main features of PLB are summarized below:

- Shared address, separate read and write data buses (*SAMD*).
- Decoupled address, read data, write data buses.
- Support for 32-bit address, 16, 32, 64, and 128-bit data bus widths.
- Dynamic bus sizing—byte, half-word, word, and double-word transfers.
- Up to 16 masters and any number of slaves.
- AND–OR implementation structure.
- Pipelined transfers.
- Variable or fixed length burst transfers.
- Support for 16–64 byte bursts.
- SPLIT transfer support

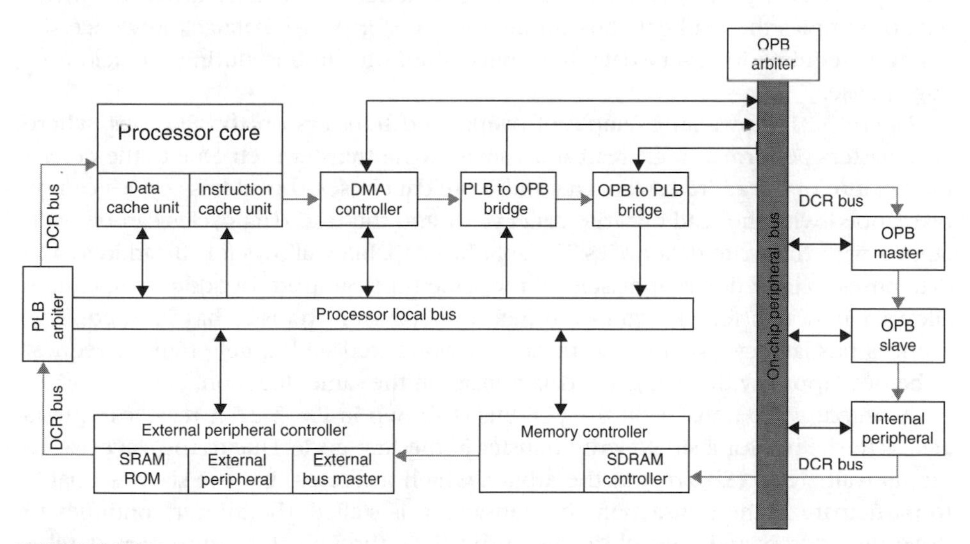

FIGURE 3.19

An example of CoreConnect-based SoC design [3]

Reprint Courtesy of International Business Machines Corporation copyright (2001) © International Business Machines Corporation

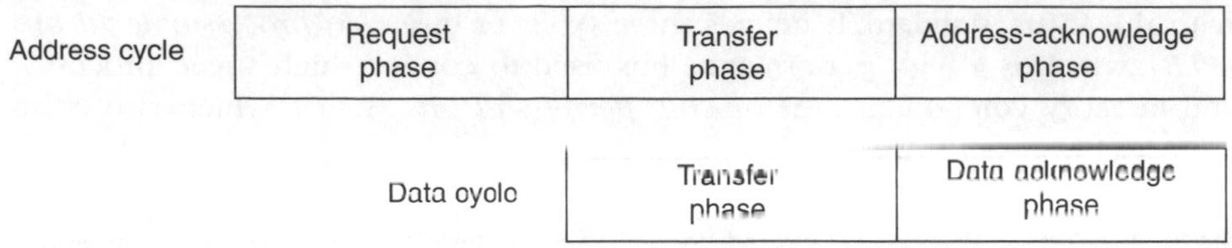

Address cycle	Request phase	Transfer phase	Address-acknowledge phase
Data cycle		Transfer phase	Data acknowledge phase

FIGURE 3.20

PLB address and data cycles [3]

Reprint Courtesy of International Business Machines Corporation copyright (2001) © International Business Machines Corporation

- Overlapped read and write transfers (up to 2 transfers per cycle).
- Centralized arbiter.
- Four levels of request priority for each master, programmable secondary arbitration.
- Locked transfer support for atomic accesses.
- Latency timer (to limit a master's tenure on PLB during bursts).

A PLB transaction consists of two cycles: address and data, as shown in Fig. 3.20. The address cycle has three phases: request, transfer, and address acknowledge. A PLB bus transaction is initiated when a master drives its address and control signals, and sends a bus access request to the arbiter, during the request phase. Once the arbiter grants the master access to the bus, the address and control information is sent to the slave in the transfer phase. The address cycle is terminated by the slave latching the address and control information during the acknowledge phase. The data cycle has two phases: transfer and acknowledge. The master drives the write data or samples the read data bus, during the transfer phase. Data acknowledge signaling is required for every data item transmitted on the bus, during the acknowledge phase.

Figure 3.21 shows an example of overlapped transfers on the PLB bus, where two masters perform a single read and a single write transfer each. Due to the decoupled nature of the address, read data, and write data buses, the address cycles can be overlapped with the read or write data cycles, and the read data cycles can be overlapped with the write data cycles. The split bus capability allows for the address and data buses to have different masters at the same time. Support for address pipelining allows a new transfer to begin even before the current transfer has finished. This reduces bus latency by allowing the latency associated with a new transfer request to be overlapped with an ongoing data transfer in the same direction.

A simple write transfer on the PLB bus is shown in Fig. 3.22. A master requests access to the bus for a single write transfer in the first cycle. Due to the slave asserting its wait signal (*Sl_wait*) to the arbiter, which indicates that the slave is unable to participate in the transaction, the transaction is stalled. The arbiter continues to drive the address and control signals to the slave through this entire period, till it receives the slave address acknowledge (*Sl_AddrAck*) signal in the fourth cycle. The slave then asserts the write data acknowledge (*Sl_wrDAck*) and write transfer complete (*Sl_wrComp*) signals to indicate the end of the transaction.

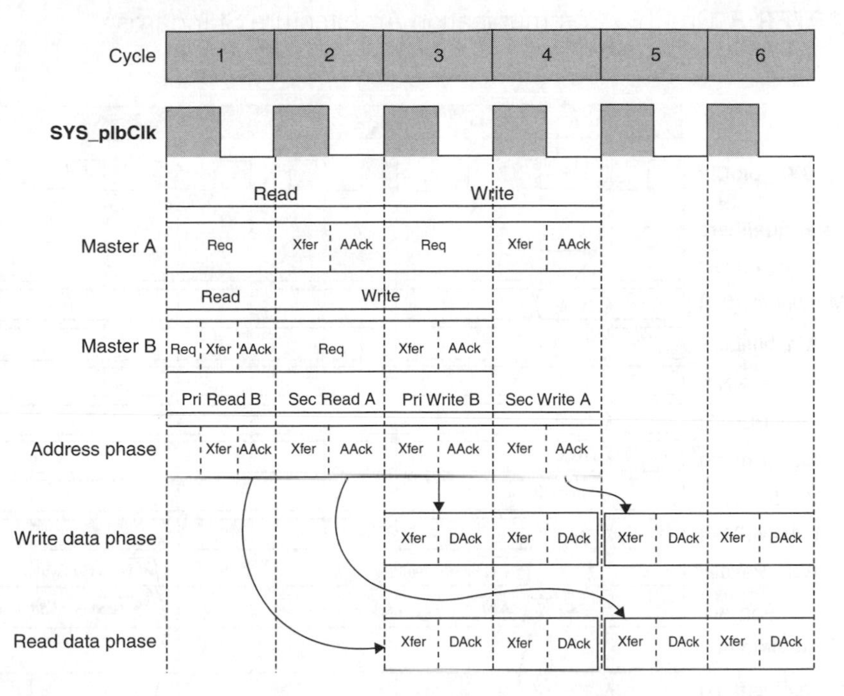

FIGURE 3.21

Example of overlapped PLB transfers [3]

Reprint Courtesy of International Business Machines Corporation copyright (2001) © International Business Machines Corporation

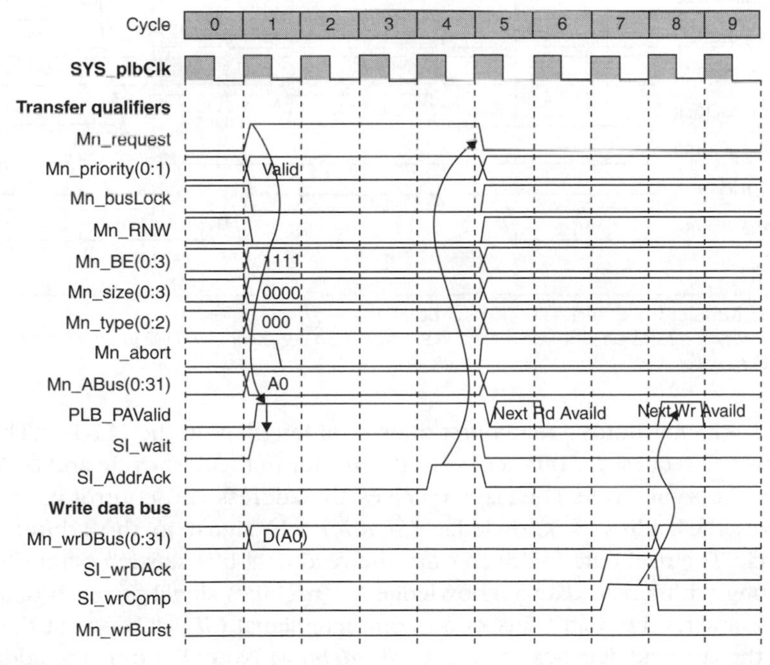

FIGURE 3.22

Example of a single write transfer on a PLB bus [3]

Reprint Courtesy of International Business Machines Corporation copyright (2001) © International Business Machines Corporation

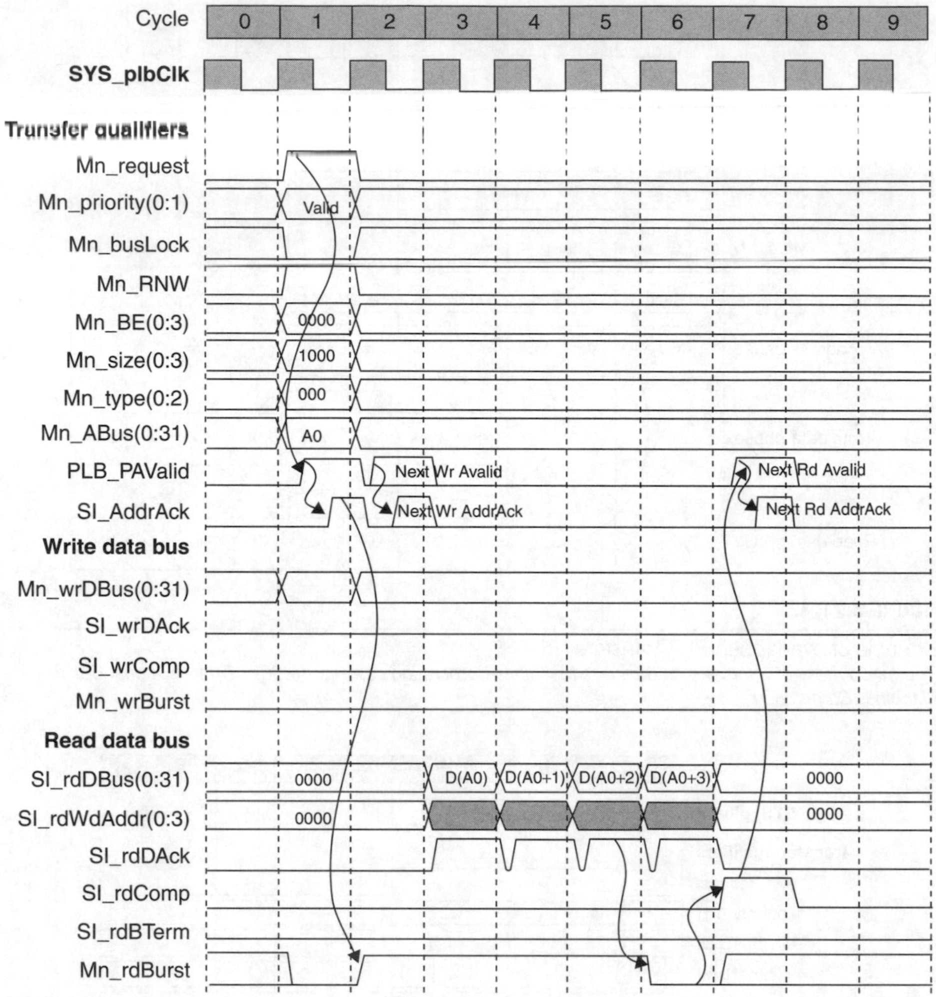

FIGURE 3.23

Read burst transfer (of length 4) on a PLB bus [3]

Reprint Courtesy of International Business Machines Corporation copyright (2001) © International Business Machines Corporation

Figure 3.23 illustrates a read burst transfer of length 4 on the PLB bus. The master asserts the request for bus access to the arbiter in the first cycle and is granted access in the same cycle. The slave receives the address and control information and drives the address acknowledge (*Sl_addrAck*) signal to the arbiter in the same cycle. The read data is driven onto the read data bus starting from the third cycle, along with a read data acknowledge (*Sl_rdDAck*) signal for each data item. The slave asserts the read transaction complete signal (*Sl_rdComp*) in the cycle prior to the last read data acknowledge (*Sl_rdDAck*). Note that only the address of the first burst data needs to be transmitted by the master and it is the responsibility of the slave to internally increment the addresses sequentially for each transfer (just like in AMBA 3.0 AXI).

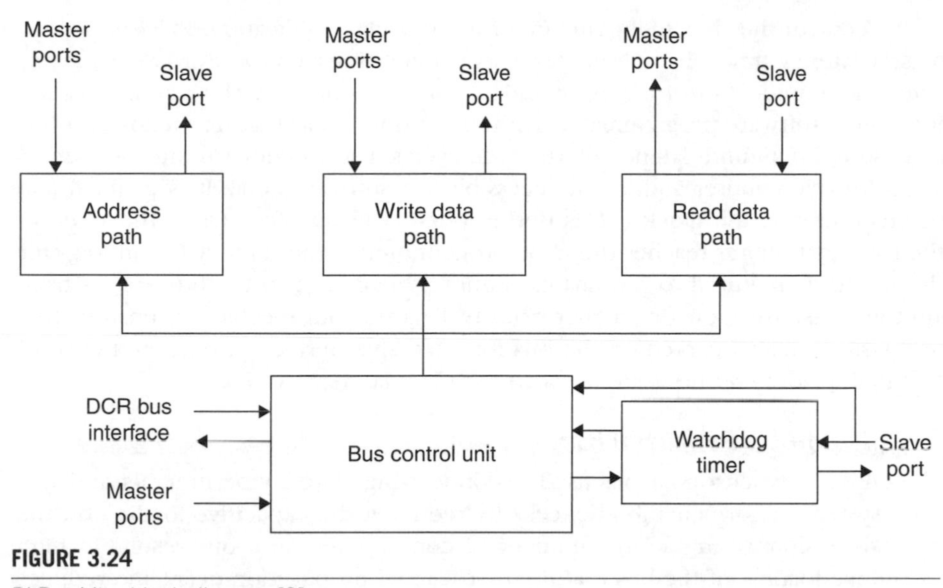

FIGURE 3.24

PLB arbiter block diagram [3]

Reprint Courtesy of International Business Machines Corporation copyright (2001) © International Business Machines Corporation

Figure 3.24 shows a block diagram of the PLB arbiter that handles arbitration for up to eight masters. It consists of several components that are described below:

- *Bus control unit (BCU)*: This supports arbitration for masters on the PLB bus. Each master typically drives a 2-bit signal that encodes four priority levels (highest, high, low, lowest), indicating the priority of the request to the arbiter. The arbiter uses this information, in conjunction with the requests from other masters to decide which master to grant the bus access to. In case of a tie, the arbiter provides a choice of using either a fixed (static) priority scheme or a fair, round-robin (RR) priority scheme. The BCU consists of four 32-bit DCRs to control and report its status: (i) *PLB arbiter control register (PACR)*, which is used to program the choice of static or RR priority schemes to be used in case of an arbitration tie, (ii) *PLB error address register (PEAR)*, which contains the address of the access where a bus time-out error occurred, (iii) *PLB error status register (PESR)*, which contains bits to identify time-out errors on PLB bus transfers, the master initiating the transfer, and the type of transfer (read or write), and (iv) *PLB revision ID register (PREV)*, that contains the revision ID of the PLB arbiter core.
- *Address Path Unit:* It contains necessary MUXes to select master address that will be driven to the slaves.
- *Read Data Path Unit:* It contains necessary steering logic for the read data bus.
- *Write Data Path Unit:* It contains necessary steering logic for the write data bus.
- *Watchdog Timer:* It provides the necessary handshake to complete a transfer, if a master's request times out on the PLB.

To control the maximum latency of a particular application, PLB supports a master latency timer in each master. This latency timer consists of two 8-bit registers: a *Latency Count Register* and a *Latency Counter*. The Latency Count Register is software programmable, with the option of hardwiring the lower 4 bits to ensure a minimum latency of 16 clock cycles. The Latency Counter is used as a clock cycle counter and is not accessible via software code. It is enabled and begins counting the clock cycles during burst data transfers. Once the value of the Latency Counter reaches the value programmed in the Latency Count Register, the master is required to terminate its burst if a request to the bus arrives from another master of equal or higher priority. This timeout mechanism ensures that no master remains parked on the bus for excessive periods of time, and ensures that high priority requests are serviced with low latency.

3.1.3.2 *On-Chip Peripheral Bus*

The OPB is a synchronous bus used to connect high latency peripherals, and alleviate system performance bottlenecks by reducing the capacitive loading on the PLB (since connecting a large number of components on a bus results in high capacitive loading of the bus that increases signal propagation delay and reduces performance). A bridge module is used to interface the OPB bus with a PLB bus. The OPB is more advanced than the simple APB peripheral bus used in AMBA, as indicated by its main features, summarized below:

- Shared address bus, multiple data buses (*SAMD*).
- Up to a 64-bit address bus width.
- 32- or 64-bit read, write data bus width support.
- Support for multiple masters.
- Bus parking (or locking) for reduced transfer latency.
- Sequential address transfers (burst mode).
- Dynamic bus sizing—byte, half-word, word, and double-word transfers.
- MUX-based (or AND-OR) structural implementation.
- Single cycle data transfer between OPB masters and slaves.
- 16 cycle fixed timeout provided by arbiter, to reduce transfer latency (can be disabled by a slave).
- Slave retry support, to break possible arbitration deadlocks.

A basic data transfer on an OPB bus is shown in Fig. 3.25. A master requests access to the bus for a read data transfer in the second cycle. The arbiter grants access to the bus, based on its arbitration scheme, and the master subsequently assumes ownership of the bus by asserting the select (*M1_select*) signal. The master then drives the address and control signals, which is sampled off the bus by the slave. The slave drives the read data onto the data bus and asserts the transfer acknowledge (*Sl2_xferAck*) signal in the next cycle. The master latches the data off the bus in the following cycle and de-asserts select to end the transaction.

For reducing arbitration latency during multiple transfers, a master can park or lock itself on the OPB bus. The parked master is allowed to access the bus without any delay due to an arbitration cycle, as long as no request is asserted by another master. Like in the case of the PLB, the OPB masters capable of long parked/locked

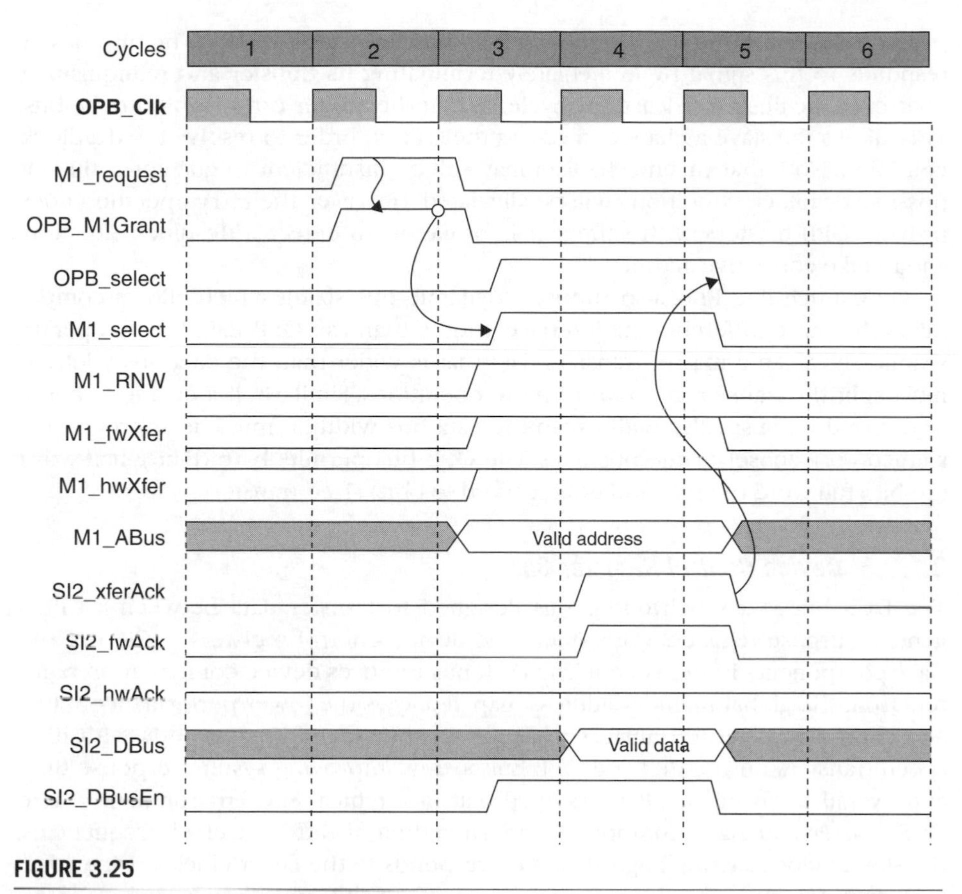

FIGURE 3.25

Basic read data transfer on OPB [3]

Reprint Courtesy of International Business Machines Corporation copyright (2001) © International Business Machines Corporation

access have Latency Counters to insure a low latency response for requests from other masters. Multiple data transfers to sequential addresses can take advantage of the sequential transfer mode in OPB, which is similar to a burst transfer. Unlike the PLB, a master on the OPB bus must drive addresses for each data item transferred on the bus. OPB also allows the overlapping of a new arbitration request with the final cycle of an ongoing data transfer, to avoid wasting a cycle for bus request and improve performance. The arbiter module is simpler than the PLB arbiter, with only two registers: one for programmable priority and another for controlling bus parking. There is flexibility in choosing either a fixed (static) scheme or a fair, RR scheme. These options are dynamically programmable, allowing for adjustments in priority based on varying traffic profiles or operation modes. The watchdog timer module in the arbiter implements a timeout feature if a slave does not respond to a master request within 16 clock cycles. If a slave must take longer to complete the transfer, it is allowed to inhibit the timer counter in the watchdog.

To alleviate possible deadlock scenarios on the bus, OPB supports the slave retry operation. This allows an OPB slave to assert a retry signal if the slave cannot perform the required bus operation at a particular instant of time. The bus master responds to this signal by immediately terminating its transfer and relinquishing control of the bus for at least one cycle, so that the arbiter can re-arbitrate the bus. This allows the slave a chance to access the OPB in order to resolve the deadlock condition. Note that this mechanism may still be insufficient to guarantee that all possible deadlock conditions will be alleviated. However, the retry operation does provide OPB masters with sufficient information to detect a deadlock situation, and to take corrective action.

OPB, much like PLB, also supports dynamic bus sizing which allows components that have different data interface widths than the OPB data bus to operate seamlessly. When a master transfers data that is wider than the data bus width, it must split the transfer into two or more operations. Similarly, if a data item being transferred has a smaller width than the data bus width, it must be appropriately aligned on a subset of the bus lines. The OPB bus permits byte (8-bit), half-word (16-bit), full-word (32-bit), and double-word (64-bit) sized transfers.

3.1.3.3 *Device Control Register Bus*

The DCR bus is a synchronous bus designed to transfer data between a CPU's general-purpose registers (GPRs) and the device control registers (DCRs) of the slave components in the system. The DCR bus removes device configuration registers from the global memory address map. It allows the lower performance status and control read/write transfers to occur separately, and concurrently with high speed transfers on the PLB and OPB buses, thus improving system response time and overall performance. It is assumed that in a typical SoC environment where DCR master and slave components are operating at different clock frequencies, the slower clock's rising edge always corresponds to the faster clock's rising edge. DCR transactions control the configuration of on-chip peripherals such as interrupt controllers, timers, arbiters, bridges, etc. The main features of the DCR bus are summarized below:

- 10-bit, up to 32-bit address bus.
- 32-bit read and write data buses.
- 4-cycle minimum read or write transfers (extendable by slave or master).
- Slave bus timeout inhibit capability.
- Multi-master arbitration.
- Privileged and non-privileged transfers.
- Daisy-chain (serial) or distributed-OR (parallel) bus topologies.

The DCR bus consists of the address, read and write data buses, and the DCR read, DCR write, master ID, privilege level, timeout wait, and acknowledge signals. Slaves can have privileged registers that can only be accessed by privileged transactions from a master. Any non-privileged transaction meant for a privileged DCR is ignored by the slave and results in a timeout. A typical transfer on the DCR bus is initiated by a master asserting the DCR read or write command signals, and driving the address and appropriate control signals. Slaves decode the command, address, privilege level, and master ID to determine whether to claim the transfer or not.

A slave can claim a transfer by asserting the timeout wait or acknowledge signals. Since requests can time out if a response is not received by the master, a slave that takes longer to complete the transfer must assert the timeout wait signal to prevent a timeout. Asserting the acknowledge signal implies that a write operation is complete, or that read data has been driven onto the read data bus. If no slave responds to the transfer request, a timeout occurs and the master terminates the command.

The DCR bus can be implemented by daisy-chaining the slave components or by creating a distributed-OR structure out of the slave devices. The daisy-chain approach allows for easier chip level wiring while the distributed-OR approach allows for easier chip level timing closure. For the case of the daisy-chain configuration, data moves along the ring-like network connecting all the slave components, and each slave component either passes along the unmodified data input, or puts its data onto its data bus output. In the distributed-OR implementation, each slave directly receives data from the master, and places its data output onto the system OR logic. It is possible for multiple masters to be connected to the DCR bus. In such a case, an arbiter is required to negotiate access to the DCR bus.

3.1.4 STMicroelectronics STBus

STMicroelectronics' STBus [4] on-chip communication architecture is an evolutionary on-chip interconnection standard developed for microcontroller consumer applications such as set-top boxes, ATM networks, digital cameras, etc. It is closely related to the VSIA (virtual sockets interface alliance; described in Section 3.2.2) industry interface standard [14] to ease compatibility and integration with third party IP blocks. The STBus standard defines three types of synchronous buses (or bus protocols) having varying levels of complexity in terms of performance and implementation: (i) *Type 1*, which is the simplest bus protocol intended for peripheral register access, (ii) *Type 2*, which is a more complex bus protocol that supports pipelined operations and SPLIT transactions, and (iii) *Type 3*, which is the most advanced bus that implements OO transactions, compound operations, and transaction labeling/hints. These buses are implemented with a MUX-based structure, and can be arranged in either a shared bus, partial crossbar (partial bus matrix), or a full crossbar (full bus matrix) topology. We describe each of these bus protocols in more detail below.

3.1.4.1 *Type 1*

The Type 1 or *peripheral* STBus standard is the simplest protocol in the STBus family that is meant to interconnect components such as general-purpose input/output (GPIO), UARTs, and simple controllers that require medium data-rate communication with the rest of the system. A simple handshaking mechanism is used to ensure correct transmission. Operations on the buses are defined in terms of *OPCs* which define, for instance, whether an operation is a read or a write. The Type 1 bus supports the LOAD (read) and STORE (write) data operations, with an address bus size of 32 bits, and a possible data size of a byte (8-bit), half-word (16-bit), word (32-bit), and double-word (64-bit). Figure 3.26 shows a simple data transfer on the Type 1 bus. The initiator (master) sends a data transfer request (*REQ*) to

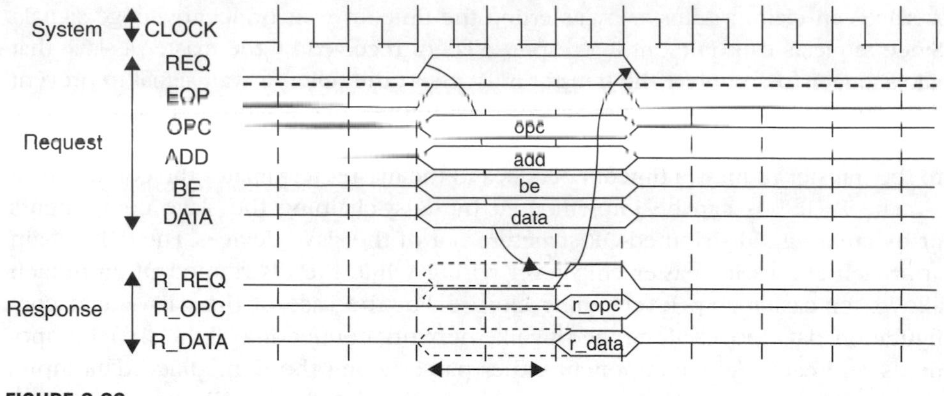

FIGURE 3.26

Basic data transfer on STBus Type 1 bus [4]
Source: STMicroelectronics

the target (slave) by sending the *OPC* for the transfer (either read or write), the transfer address (*ADD*), and the byte enable (*BE*) to specify which bytes in the bus are significant, based on the width of the data involved in the transfer (1–8 bytes). The write data bus (*DATA*) to send the write data (for a write data transfer). The slave indicates it has received the transfer request by asserting the handshake signal (*R_REQ*), and then proceeds to return the read data on the read data bus (*R_DATA*) or write data in its address space for a write data transfer. The slave returns an optional response opcode (*R_OPC*) to indicate any errors during the transaction.

The Type 1 bus is similar to the IBM CoreConnect DCR bus, since it is also used to program the internal configuration registers of components connected to the STBus communication architecture. A register decoder block is responsible for performing address decoding, data transfer routing, and arbitration (if more than one master is connected to the bus) for transfers on the Type 1 bus.

3.1.4.2 *Type 2*

The Type 2 or the *basic* STBus standard supports all the Type 1 functionality, and additionally provides support for pipelined operations, SPLIT transactions, compound operations, source labeling, and some priority and labeling/hint information. This bus protocol is targeted at high performance components. The use of SPLIT transactions and pipelined operation improves bus efficiency and performance. The Type 2 standard supports the basic LOAD (read) and STORE (write) operations, with an address bus size of 32 bits and allowed data bus sizes of 8, 16, 32, 64, 128, or 256 bits. Additionally, the Type 2 standard also supports compound operations which are built from one or more primitives. The supported standard compound operations include:

- *READMODWRITE*: An atomic operation that transfers read data from the slave to the master, but leaves the slave *locked* until a write transfer from the same master completes, replacing the information at the specified address in the slave.

- *SWAP*: An atomic operation that exchanges a data value from the master with the data held in a specified location in a slave.
- *FLUSH*: An operation used to ensure the coherence of main memory while allowing local copies associated with a slave to remain coherent. The operation returns a response when any copies of the data associated with a physical address (which are held by a slave module) are coherent with the actual data at the physical address. The slave may retain a copy of the data.
- *PURGE*: An operation used to ensure the coherence of main memory while ensuring that stale local copies are destroyed. The operation returns a response when any copies of the data associated with a physical address (which are held by a slave module) are coherent with the actual data at the physical address, while removing any copies of the data held by the slave.
- *USER*: This is reserved for user defined operations that can implement useful operations specific to particular applications.

3.1.4.3 *Type 3*

The Type 3 or *advanced* STBus standard supports all Type 1 and Type 2 functionality, but additionally supports packet shaping and OO transaction completion. These features make this bus protocol suitable for very high performance components. The Type 3 supports the same basic and compound operations as the Type 2 standard. Packet shaping allows optimum bandwidth allocation in Type 3 buses, with only the minimum number of clock cycles required to carry out a transaction being used. In Type 1 and Type 2 buses, every request requires a response, which wastes bandwidth. In contrast, a Type 3 requires only a single response from a slave for multiple write data operations by a master, or a single read request from a master for multiple data reads from a slave. Due to this asymmetry between request and response phases, the bandwidth allocation in the Type 3 protocol is optimized compared to the Type 1 and Type 2 protocols. The use of OO transaction completion further reduces latency on the bus since a master waiting for a response from a slave no longer blocks access to other slaves. Transaction IDs associated with data transfers allow the components to have up to 16 transactions in progress.

3.1.4.4 *STBus Components*

The STBus node shown in Fig. 3.27 is the main component of the STBus on-chip communication architecture. It consists of two main blocks: the *control logic* and the *data path*. The control logic is responsible for the arbitration, decoding, and contains other bus logic required for implementing advanced protocol features (such as OO transaction completion), while the data path represents the topology of the communication architecture, which can be a shared bus, a full crossbar, or a partial crossbar configuration. Master components connect to the initiator interfaces, while slaves connect to the target interfaces of this node architecture.

The control logic is responsible for arbitration, and supports several arbitration schemes which are described below:

- *Fixed priority*: The priorities of the masters are static or hardwired.
- *Variable priority*: The priorities of the masters can be dynamically changed during system operation by writing to special programmable priority registers

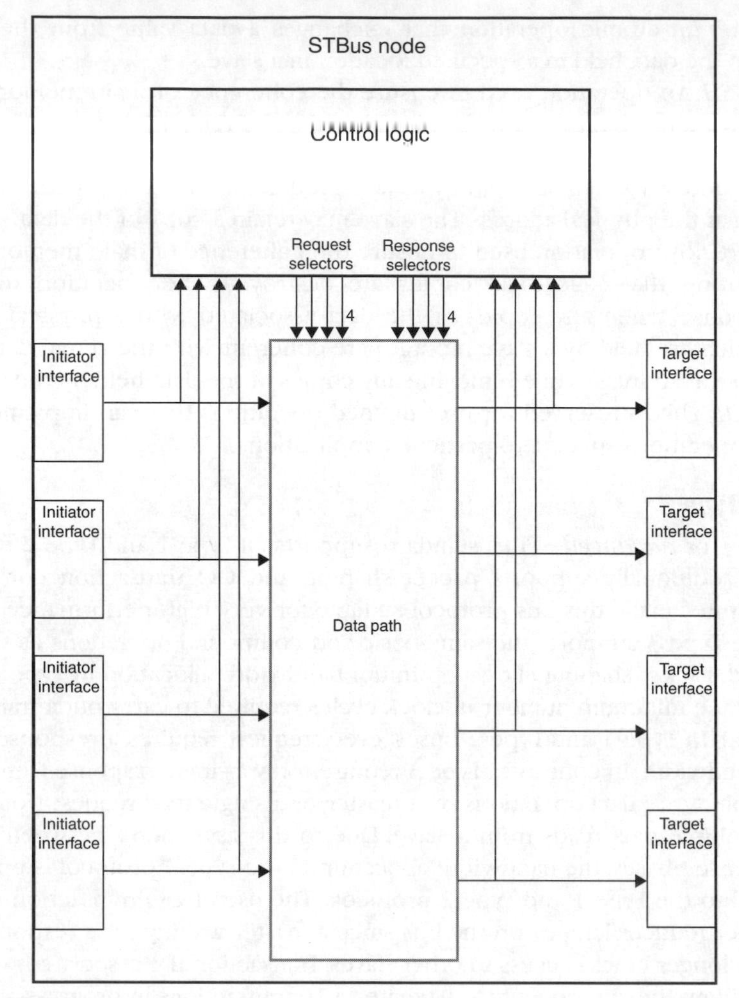

FIGURE 3.27

Block diagram of STBus node [4]
Source: STMicroelectronics

in the node. If two or more masters have the same priority value stored in their registers, the master with the higher fixed (hardwired) priority gets preference.

- *Least recently used (LRU)*: Masters are granted access to the bus in the order of the longest time since the last grant. If two or more masters have been waiting for the same amount of time, the master with the higher fixed (hardwired) priority gets preference.
- *Latency-based*: Each master has a register associated with it containing the maximum allowed latency in clock cycles. If the value is 0, then it needs to have zero cycle latency when a request is received, and such a master must be granted bus access as soon as possible. Each master also has a counter, which is loaded with its maximum latency value each time the master

makes a request. At every subsequent cycle, the counter is decremented. The arbiter grants bus access to the master having the lowest counter value. If two or more masters have the same counter values, the grant will be given to the master with the higher fixed (hardwired) priority.

- *Bandwidth-based*: Each master has a register associated with it containing its bandwidth, expressed in terms of clock cycles per fixed period. Each master also has a counter, loaded with a starting value obtained from its bandwidth register. At the beginning of an operation, the counter for the first master (selected based on the hardwired priority) starts decrementing for the entire duration of the time slot allocated to this master. If during this period the first master makes any requests, they are granted. If during this period no master makes any requests, the first master eventually loses its slot, and the next master based on hardwired priority is selected. However, if during this period the first master makes no requests but requests are received from other masters, its counter is stopped and the bus is granted to the next master in order of hardwired priority (whose counter subsequently begins to decrement since it is now using bandwidth). If a master consumes its bandwidth (i.e., its bandwidth counter has reached 0), it cannot be granted bus access again till the end of the fixed time period. This process continues till all masters have consumed their allocated bandwidth, at which point the arbitration process starts again. This scheme is similar to a TDMA/static-priority two level arbitration scheme.
- *STB*: This is a hybrid of the latency-based and variable priority schemes. In its normal state, arbitration proceeds just like in a variable priority scheme. As in the latency-based scheme, masters also have an associated maximum latency register and counter. Each master also has an additional *latency-counter-enable* bit. If this latency-counter-enable bit is set, and the counter value is 0, then a master is said to be in *panic state*. In the case when one or more masters are in a panic state, the normal variable priority scheme is overridden, and the masters in panic state are granted bus access, in the order of highest priority.
- *Message-based*: This is a fixed priority scheme which allows masters having a higher priority and a priority flag set, to interrupt the message transfer on the bus. Note that in the normal fixed priority scheme, a message transfer cannot be interrupted while it is in progress.

From the description of the supported arbitration schemes in STBus, it can be seen that the control unit requires several registers to hold the latency, bandwidth, and/or priority values for the masters in the system. In addition to the node, other components used in the STBus communication architecture include:

- *Size converters*: Components used to allow communication between two STBus IP blocks having different data bus widths.
- *Type converters*: Components used to allow communication between two STBus IPs following different STBus protocol types (e.g., between Type 1 and Type 3).
- *Buffer*: A FIFO-based component that is used as a retiming stage between two IPs following the Type 2 or Type 3 protocol. A buffer is useful to break

critical paths between components that are far apart from each other on the system floorplan. In this aspect, it is similar to the register slice proposed in AMBA 3.0 AXI.

3.1.5 Sonics SMART Interconnect

The Sonics SMART Interconnect [5] is another on-chip communication architecture standard designed to ease component interoperability and provide high performance for a wide range of applications. The standard comprises of three synchronous bus-based interconnect specifications which differ in their level of complexity and performance: (i) *SonicsMX*, which is a high performance interconnect fabric with advanced features; (ii) *SonicsLX*, which is also a high performance interconnect fabric, but with less advanced features compared to SonicsMX; and (iii) *S3220*, which is a peripheral interconnect, designed to connect slower peripheral components. Both SonicsMX and SonicsLX natively support the open core protocol (OCP) version 2.0 [15] wrapper-based interface standard (described in Section 3.2.1). Additionally, components with AMBA 2.0 AHB and AMBA 3.0 AXI natives interfaces can also be plugged into these interconnection fabrics using pre-designed interface bridge logic components. These features are crucial in maximizing reuse of IP cores. Sonics SMART Interconnect provides a highly configurable communication architecture solution for contemporary SoC designs. It is supported by the *SonicsStudio* development environment [5] that allows automated configuration, data analysis, and performance verification for the communication architecture. We now examine the specifications of the three bus-based interconnect fabrics that make up the Sonics SMART Interconnect, in more detail.

3.1.5.1 *SonicsMX*

SonicsMX is the third generation of *socket-based* (described in Section 3.2) synchronous interconnect fabrics from Sonics, targeted at high performance, sophisticated SoC applications such as WCDMA/3G wireless headsets, video game consoles, and portable multimedia players. Its main features are summarized below:

- Pipelined, non-blocking, and multi-threaded communication support.
- Split/outstanding transactions for high performance.
- Configurable data bus width: 32, 64, or 128 bits.
- Multiple topology support—shared bus, full crossbar, partial crossbar.
- Socket-based connection support, using native OCP 2.0 [15] interface between components and interconnect.
- Bandwidth and latency-based arbitration schemes to obtain desired quality of service (QoS) for components threads.
- Register points (RPs) for pipelining long interconnects and providing timing isolation.
- Protection mode support.
- Advanced error handling support.
- Fine-grained power management support.

SonicsMX supports the full crossbar, partial crossbar, and shared bus interconnection topology. A crossbar configuration can have a maximum size of 8 × 8, with a maximum of 8 masters connected to 8 slaves. A shared bus configuration can connect up to 32 components with up to 16 masters connected to 16 slaves. *Protocol points (PPs)* are used to interconnect different topology configurations, and consist of logic for frequency and data width conversion. Up to 4 crossbar and shared buses can be joined in a single instance of SonicsMX, to support up to 64 cores. A single *register target (RT)* component is used as interface to the internal configuration registers of a SonicsMX instance, for any dynamic reconfiguration. Multiple SonicsMX instances (interconnected via bridges) can be used in an SoC design. SonicsMX makes use of *RPs* to pipeline long interconnects in order to break long combinatorial timing paths and achieve the desired bus clock frequency of operation. *RPs* are essentially small FIFOs with a depth that is configurable, to provide queuing support if needed. This is similar in concept to register slices in AMBA 3.0 AXI and buffers in STMicroelectronics' STBus.

Initiator (master) and *target* (slave) components are connected to the interconnect fabric (which natively support the OCP 2.0 interface) through *initiator* and *target agents (TAs)*, respectively. An *agent* contains bus logic such as a bridge to connect a component with a mismatched OCP version 1.0 [16], AMBA AHB [1] or AXI [2] interface, *data width converters* to handle mismatched component and interconnect data widths, *flip-flops* to adapt component timing to the interconnect clock frequency, and *RPs*, which provide FIFO-based transaction buffering if required. Figure 3.28 shows an example of a SonicsMX instance that consists of a *crossbar topology (XB)* and a *shared link/bus (SL)* topology interconnected using *PP* connectors. The initiators connect to the architecture via *initiator agents (IAs)* while the targets connect to it via the *TA*. Components with a

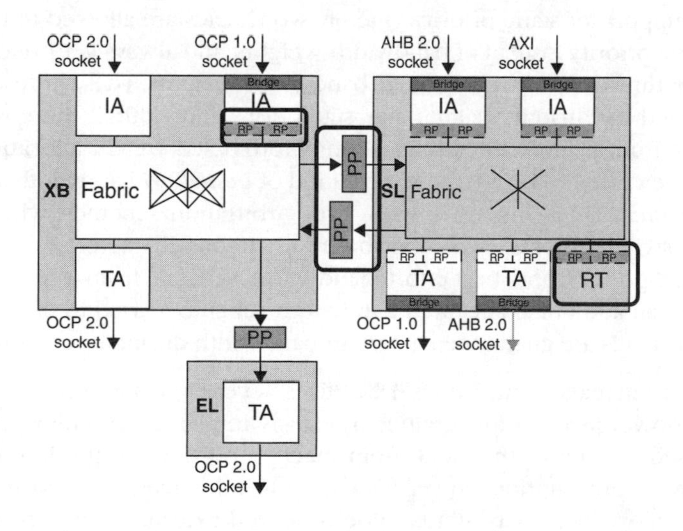

FIGURE 3.28

Example of SonicsMX system with crossbar (*XB*) and shared bus (*SL*) topologies interconnected using *PP* connectors [5]
Source: Sonics Inc.

non-native (i.e., non-OCP 2.0) interface require a bridge at their respective agents. *RPs* provide FIFO buffering/queuing between the components and the interconnect. An *extender link (EL)* is used to connect a target (or initiator) that lies far away from the interconnect fabric on the system floorplan. *ELs* are optimized to span large distances, and use *PP* connectors to interface with the interconnection fabric. An *RT* component is used for interfacing with the internal configuration registers of the SonicsMX instance. Note that all the buffering components in the SonicsMX interconnect fabric can increase the transaction latency, and must be used judiciously. For instance, there is typically at least a one cycle penalty for using flip-flops, *RPs*, and *PPs*.

For multi-threaded components in a system, SonicsMX supports defining mappings between initiator threads and target threads. Resources within the interconnect fabric such as *RPs* and *PPs* are allocated for each of the thread mappings. For cases where independent resources and flow control is not required for every thread (e.g., for multi-threaded initiator components with limited concurrency), thread collapsing at the initiator socket is supported. This can reduce the overhead of allocating unnecessary resources in the interconnect fabric, without degrading performance.

The arbitration schemes used in SonicsMX guarantee QoS requirements for an application. Three QoS levels are defined for use, each characterized by a different arbitration policy:

- *Weighted QoS*: This mode uses a bandwidth-based arbitration scheme, where the available bandwidth is distributed among initiators, based on the ratio of bandwidth weights configured at each initiator agent.
- *Priority QoS*: This mode uses two arbitration schemes. It extends the bandwidth-based arbitration scheme from the weighted QoS mode, by adding support for static priority. One or two threads are allowed to be assigned a static priority instead of bandwidth weights, and always get preference over other threads that are allocated bandwidth weights. These threads that are assigned bandwidth weights may starve for bandwidth, if there is excessive traffic from priority threads. The bandwidth-based weight allocation scheme can therefore be considered to be a kind of *best-effort* bandwidth scheme.
- *Controlled QoS*: This mode uses three arbitration schemes, which dynamically switch among each other based on traffic characteristics. In addition to the priority and best-effort bandwidth scheme from the priority QoS mode, an additional allocated bandwidth scheme is used to ensure that certain threads are guaranteed a certain bandwidth during operation.

For power management, SonicsMX utilizes several mechanisms to reduce idle and active power levels. A fine grained internally implemented clock gating mechanism is used to remove the clock from inactive portions of the interconnect to reduce power consumption. Power management interfaces (signals) at each interface socket allow its activity status to be observed externally, and enable activity-dependent power management (e.g., waking up powered-down targets). Coarse grain power management is supported by using power control logic external to the SonicsMX instance. The external power control logic manages power by removing clock or supply voltage from the entire interconnect. Every SonicsMX instance provides a single power management interface for the entire interconnect, which allows the external power control logic to know when the clock or supply voltage can be restored or removed without disrupting communications.

An optional access protection mechanism is implemented in SonicsMX to designate protected regions within the address spaces of certain targets. This mechanism can dynamically specify protected region sizes and locations at runtime. The mechanism can define access permissions as a function of which initiator can access a protected region, the type of transaction (read or write) being requested, or what state the target is currently in. Each target is allowed to have up to eight protection regions.

3.1.5.2 *SonicsLX*

SonicsLX is a third generation of socket-based synchronous interconnect fabrics from Sonics, targeted at mid-range SoC designs. It supports pipelined, multithreaded, and non-blocking communication on its buses. It also has support for SPLIT transactions to improve bus utilization. SonicsLX can be arranged in a full or partial crossbar topology and supports the weighted and priority QoS modes, as described above for SonicsMX. The SonicsLX features are a subset of the SonicsMX feature set. Table 3.2 summarizes the main differences between SonicsMX and SonicsLX.

3.1.5.3 *Sonics Synapse 3220*

The Sonics Synapse 3220 synchronous interconnect fabric is targeted at low bandwidth, physically dispersed peripheral target (slave) cores. The main features of the 3220 interconnect are a subset of the SonicsMX and SonicsLX interconnect fabrics. Its main characteristics are summarized below:

- Up to 4 OCP-compliant initiators, and 63 OCP-compliant targets.
- Up to 24-bit configurable address bus.
- Configurable data bus widths—8, 16, 32 bits.
- Fair arbitration scheme, with high priority allowed for a single initiator thread.

Table 3.2 Comparison of SonicsMX and SonicsLX feature set

Features	SonicsMX	SonicsLX
Data Width Conversion	Full	Full
Quality of Service Management	Multi-level QoS	Two level QoS
Advanced Power Management	Configurable	Fixed
Advanced Security Management	Full	Reduced
Interrupt and Error Management	Full	Reduced
Side Band Signaling Management	Full	None

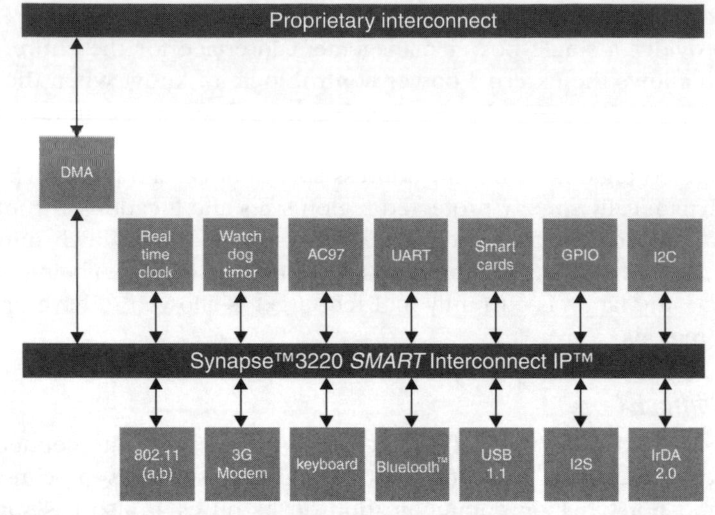

FIGURE 3.29

Example of Synapse 3220 interconnect fabric interfacing with the rest of the system [5]
Source: Sonics Inc.

- Power management interface.
- Exclusive (semaphore) access support.
- Error detection and recovery—watchdog timer to identify unresponsive peripherals.
- Protection mode support.

Figure 3.29 shows some of the typical peripheral targets that are connected to the Synapse 3220 interconnect fabric. The fabric interfaces with the main interconnect fabric (SonicsMX or another proprietary interconnect) using a DMA type block that acts like a bridge.

3.1.6 OpenCores Wishbone

The Wishbone bus-based on-chip communication architecture standard [6] is an open-source standard that proposes a single, high speed synchronous bus specification to connect all the components in an SoC design. Since it is open source, designers can download synthesizable Wishbone RTL components available for free from the OpenCores website [6]. However, due to lack of default support for advanced features (e.g., OO transaction completion, SPLIT transactions, power management, etc.), its scope is limited to small- and mid-range embedded systems. The main features of the Wishbone high speed bus standard are summarized below:

- Multiple master support.
- Up to 64-bit address bus width.
- Configurable data bus width 8- to 64-bit (expandable).
- Supports single or block read/write operations.
- Read-modify-write (RMW) support for semaphore type operations.

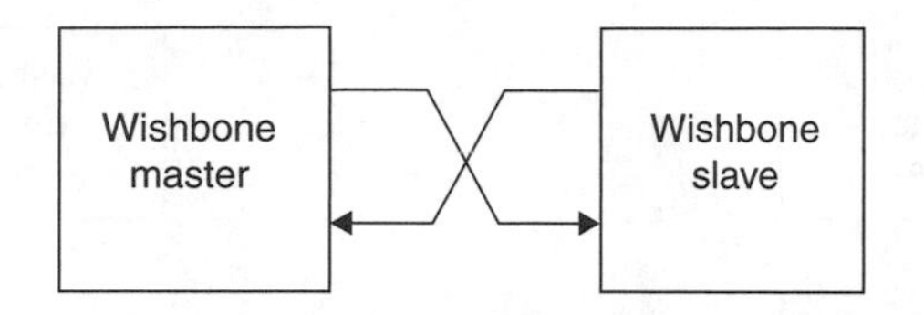

FIGURE 3.30(a)

Different topologies of the Wishbone bus architecture: point-to-point

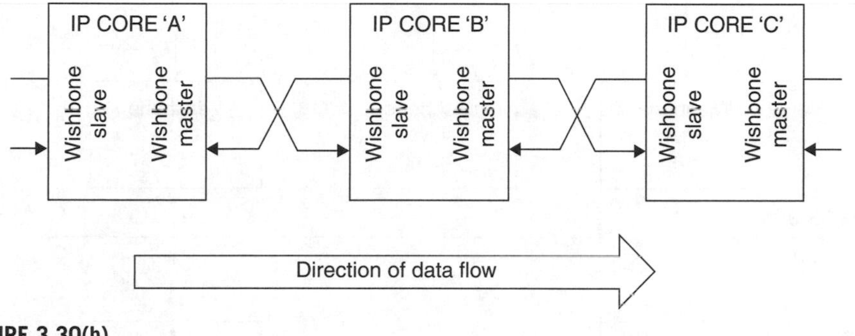

FIGURE 3.30(b)

Data flow

- Supports point-to-point, data flow, shared bus, or crossbar topology.
- Retry mode support.
- User defined tags, for error signaling, cache hints, etc.
- Flexible arbitration scheme (fixed priority, RR, etc.).

The Wishbone interface is highly configurable, and allows a user to customize tags or signals to support specific application requirements. Thus, designers can create their own customized version of the Wishbone standard to suit a particular applications need. While this customizability is desirable, it can make developing generic components (e.g., bridges to interface with other standards such as AMBA) more difficult. Figure 3.30 shows the different possible topologies in which the Wishbone on-chip communication architecture can be structured. For very small systems, the master and slaves can be directly connected using *point-to-point* bus links, as shown in Fig. 3.30(a). For systems with a sequential data flow, where data flows from component to components, the *data flow* topology is recommended, as shown in Fig. 3.30(b). Data flows in a pipelined manner, exploiting parallelism and thus speeding up execution. To interconnect several masters and slaves effectively and with the fewest resources, a *shared bus* topology can be used, as shown in Fig. 3.30(c). The arbitration scheme of the arbiter (not shown in the figure) is left up to the system designer. The shared bus can have either a tri-state-based, or a MUX-based implementation. For higher performance systems, a *crossbar* topology can be used, as shown in Fig. 3.30(d). While this topology offers better performance due to higher parallelism, it requires more logic and

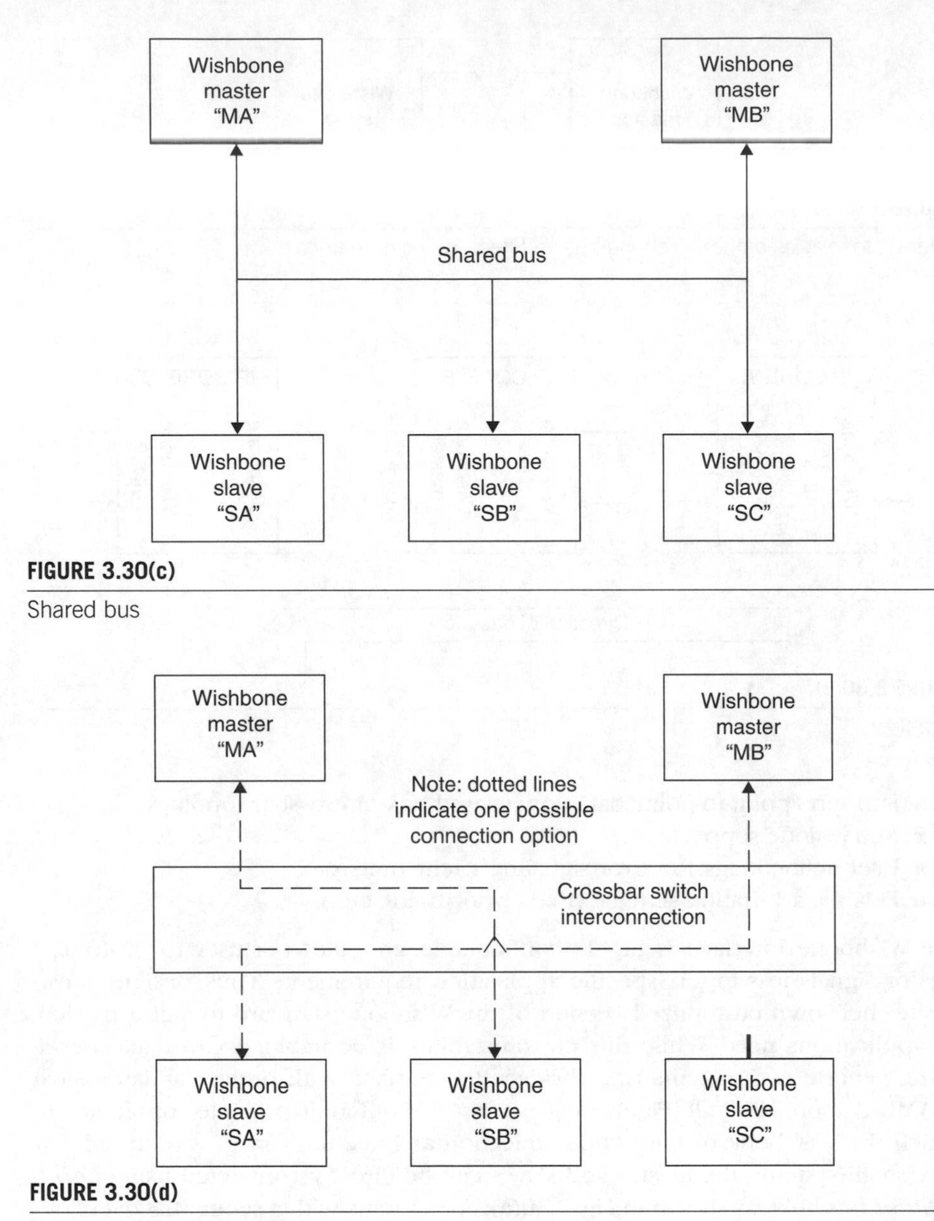

FIGURE 3.30(c)

Shared bus

FIGURE 3.30(d)

Full or partial crossbar [6]

routing resources. As a rule of thumb, a crossbar switch with two masters and two slaves takes up twice the interconnect logic as a similar shared bus system with the two masters and two slaves.

Wishbone supports the basic single read/write data transfers with handshaking, and block (burst) read/write data transfers, like the other standards. Semaphore type operations, that allow multiple components to share common resources, are supported via the RMW operation. This type of operation is commonly used in disk controllers, serial ports, and memories. In this operation, once an arbiter grants the bus to a master, no other master is allowed to access the bus till the selected master has read, modified, and written back data to the slave. Such a *locking* of the bus can be inefficient, especially if the slave takes multiple cycles to respond. More efficient semaphore type operation support was discussed earlier in the context of other bus standards such as AMBA 3.0 AXI and STBus Type 3, where the bus need not remain locked for the entire duration of the semaphore operation.

3.1.7 Altera Avalon

The Altera Avalon [7] synchronous bus-based communication architecture is targeted at system-on-programmable-chip (SoPC) designs, and is comprised of two standards: Avalon memory mapped (Avalon-MM) [32] and Avalon streaming (Avalon-ST) [33].

3.1.7.1 *Avalon-MM*

The Avalon-MM standard defines an interface to connect memory-mapped master and slave peripherals such as microprocessors, UARTS, memory, timers, etc. Configurability is an important attribute of the Avalon-MM interface, and components can choose to use a small set of signals if they only support simple transfers (e.g., a simple ROM interface requiring only address, read data, and control signals). Components requiring more complex transfer types will support a larger set of signals (e.g., high speed memory controller that supports pipelined bursts). The Avalon-MM signals are a superset of several other bus standards. For example, most Wishbone interface signals can be mapped to Avalon-MM signals types, making it easy to include Wishbone components into Avalon-MM systems.

Avalon-MM is implemented as a synchronous bus crossbar, as shown in Fig. 3.31. The crossbar has an integrated interrupt controller, and supports optional logic to transfer data across multiple clock domains, and across multiple interface widths. Pipeline registers can be added at any point on a crossbar bus to increase the value of the maximum allowed clock frequency on the bus. Separate read and write data buses can have widths of up to 128 bits. The crossbar buses support the burst transfer mode to hide arbitration latency for individual transfers, as well as fixed latency and variable latency pipelined reads. For the fixed latency case, with a specified pipeline latency of *N*, the slave must present valid read data on the *N*th rising clock edge after the end of the address phase. Figure 3.32(a) shows an example of pipelined read data transfers with a fixed latency of 2 cycles. In the variable latency case, the slave can take an arbitrary number of cycles after the address phase to put the read data on the bus, as shown in Fig. 3.32(b). An additional signal is used by the slave in this case to signal to the switch fabric when valid data has been put on the read data bus.

Several arbitration schemes are allowed in the Avalon-MM crossbar, such as:

- *Fairness-based shares*: This scheme assigns each master port with an integer value of transfer shares with respect to a slave port. One share signifies permission to perform one transfer. As an example, if *Master 1* is assigned

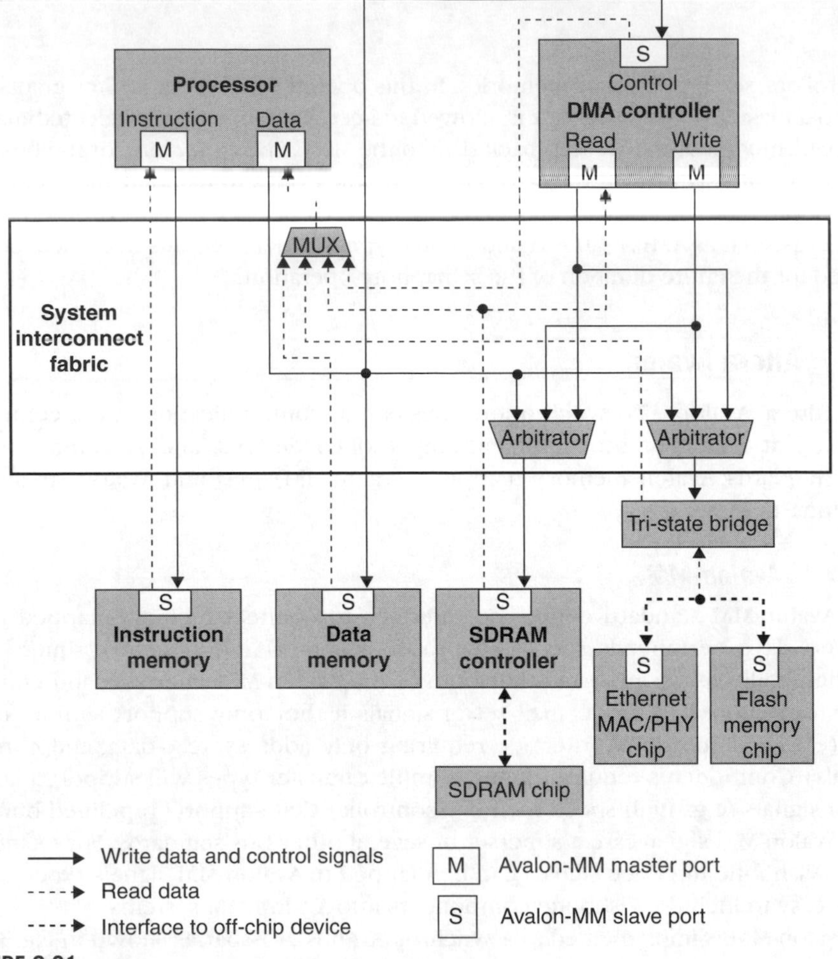

FIGURE 3.31

Avalon-MM crossbar switch implementation example [7]

Altera is a trademark and service mark of Altera Corporation in the United States and other countries. Altera products are the intellectual property of Altera Corporation and are protected by copyright laws and one or more U.S. and foreign patents and patent applications

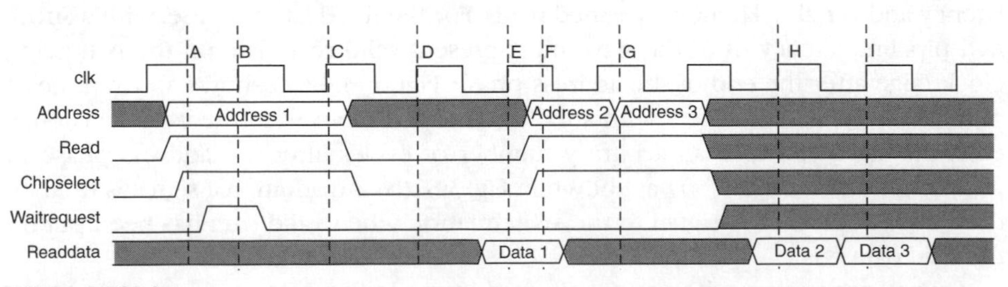

FIGURE 3.32(a)

Pipelined read data transfer in Avalon-MM, with fixed latency = 2 cycles

Altera is a trademark and service mark of Altera Corporation in the United States and other countries. Altera products are the intellectual property of Altera Corporation and are protected by copyright laws and one or more U.S. and foreign patents and patent applications

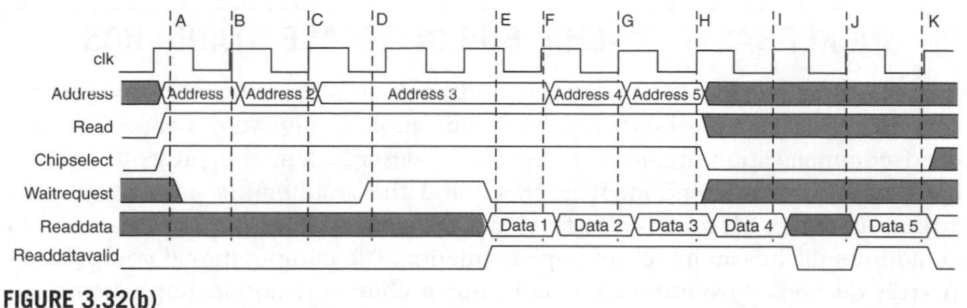

FIGURE 3.32(b)

Variable latency [7]

Altera is a trademark and service mark of Altera Corporation in the United States and other countries. Altera products are the intellectual property of Altera Corporation and are protected by copyright laws and one or more U.S. and foreign patents and patent applications

3 shares and *Master 2* is assigned 4 shares, then the arbiter grants *Master 1* access for 3 transfers, followed by a grant to *Master 2* for 4 transfers. The arbiter cycles through this process indefinitely. If a master stops requesting transfers before it exhausts its shares, it forfeits all of its remaining shares, and the arbiter grants access to another requesting master.

- *RR*: Masters are granted access to a slave in a cyclic, RR manner, ensuring a fair bus access distribution among requesting masters.
- ***RR with minimum share values***: In this scheme, a slave port can define a value for the minimum number of shares in each RR cycle. This results in the arbiter granting at least *N* shares to any master port when it begins a sequence of transfers. By declaring a minimum share value *N*, a slave indicates that it is more efficient at handling continuous sequential transfers of length *N*. Since burst transfers provide even higher performance for continuous transfers to sequential addresses, the minimum share value does not apply for slave ports that support burst transfers—the burst length takes precedence over minimum share value.

3.1.7.2 *Avalon-ST*

The Avalon-ST standard defines an interface optimized for the unidirectional flow of data, with support for multiplexed streams, packets, and DSP (digital signal processor) data. Avalon-ST is implemented as a synchronous point-to-point communication bus. The interface signals can be used to describe traditional streaming interfaces that consist of a single stream of data without knowledge of channels or packet boundaries. The interface can also support more complex protocols such as burst and packet transfers with packets interleaved across multiple channels. Packet transfer between the source and destination components is supported by using three interface signals: *startofpacket*, which is used by the source to indicate the cycle with the start of packet; *endofpacket*, which indicates the cycle containing the end of the packet; and the optional *empty* signal, which indicates the number of symbols that are empty during the cycles that mark the end of a packet. At the time of publishing of this book, Avalon-ST is a relatively new standard and its specification document is sparse on advanced features and supported modes.

3.2 SOCKET-BASED ON-CHIP BUS INTERFACE STANDARDS

A socket-based on-chip bus interface standard defines the interface of a component that connects to a bus-based communication architecture. Unlike the bus-based communication architecture standards described in the previous section that define the bus–component interface and the architectural implementation of the bus, socket-based bus interface standards only define the interface and do not address the bus architecture implementation. The computational components are truly decoupled from the communication architecture and its implementation in this scenario. Figure 3.33 illustrates an example of a system utilizing a socket-based bus interface standard. The standard interface definitions allow components to be designed with a standard interface, without committing to a particular communication architecture implementation. This improves IP reuse flexibility, since components can now be connected to any of the wide array of standard bus-based communication architectures described previously. Figure 3.33 shows how designers are free to choose any standard or proprietary bus architecture (e.g., AMBA, CoreConnect, STBus) to implement the actual communication primitives, when using a socket-based interface (I/F) standard. The only requirement for seamless component and bus architecture integration in this case is *adapter* logic components that can map the component interface to the bus architecture fabric protocol. Such adapter or translation logic is not required if the bus architecture implementation natively supports the socket-based interface definition. As an example, since the Sonics SMART Interconnect [5] bus architecture fabric natively supports the signals in the OCP 2.0 [15] socket-based interface standard, no translation logic is required between the bus and components that have an OCP 2.0 socket interface.

Socket-based interface standards must be generic, comprehensive, and configurable to capture the basic functionality and advanced features supported by a

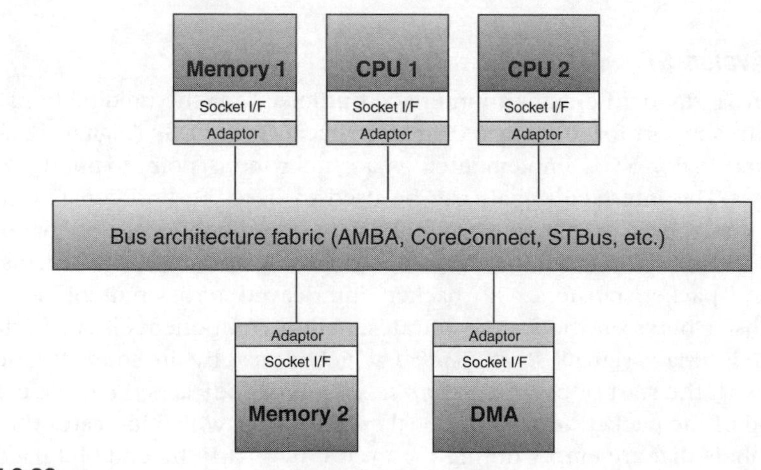

FIGURE 3.33

Example of system implemented with socket-based interface standards wide array of standard bus-based communication architectures. This ensures that there are no possible incompatibilities when a component with a socket-based interface is connected to the bus-based communication architecture that actually implements the communication primitives. Socket-based interface standards have the disadvantage of requiring adapter or translation logic at the bus–component interface. This results in additional design time to create the adapter logic, an increase in chip area, and possible performance penalties due to every data transfer propagating through the adapter logic that has an intrinsic delay associated with it. However, the adapter logic only needs to be created once and can be subsequently reused in several designs. The benefits that accrue from using socket-based bus interface standards such as improved IP reusability across designs and greater flexibility to explore (or change) diverse bus architecture implementations should not be underestimated.

Several socket-based bus interface standards have been proposed over the years. Some of the popular interface standards include OCP [15], virtual component interface (VCI) [14], and device transaction level (DTL) [17]. Because of the implementation flexibility it offers, AMBA 3.0 AXI [2] can also be considered to be a type of bus interface standard. OCP, however, is by far the most popular industry standard as far as socket-based bus interface specifications go. Because the VCI interface standard is a subset of the OCP, and DTL is a proprietary standard with very little publicly available information, we will only briefly review these two interface standards. We now present a comprehensive overview of the OCP socket based interface standard.

3.2.1 Open Core Protocol

The OCP version 2.0 [15] socket-based bus interface standard defines a high performance, synchronous, bus architecture-independent interface between IP cores. It promotes IP design reuse by allowing IP cores to be independent of the architecture and design of the systems in which they are used. It is also highly configurable and can be optimized to use only the necessary features required for communicating between two components, which saves chip area. OCP essentially defines a point-to-point interface between two components, one of which must be a master and the other a slave. Its key features include:

- Point-to-point synchronous interface.
- Bus architecture independent.
- Configurable data flow (address, data, control) signals for area-efficient implementation.
- Configurable sideband signals to support additional communication re-quirements.
- Pipelined transfer support.
- Burst transfer support.
- OO transaction completion support.
- Multiple threads.

Figure 3.34 shows an example of a simple SoC system consisting of three IP cores with OCP interfaces connected to an on-chip bus. The on-chip bus can

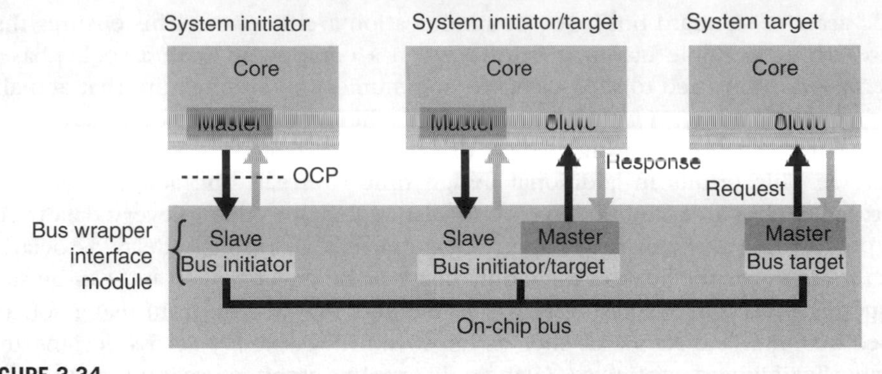

FIGURE 3.34

Example of system implemented with OCP socket-based interface standard [15]
Source: OCP-IP

belong to any one of the standard bus-based on chip communication architectures such as AMBA 2.0/3.0, CoreConnect or STBus, described in Section 3.1. A bus wrapper interface module is required to translate and map the OCP interface signals of the IP cores, to the signals of the on-chip bus. The wrapper interface module must act as the complementary side of the point-to-point OCP connection for each IP core port that is connected to the bus that is, for the case of a master port on the IP core, the connected wrapper module must act as a slave, and for a slave port on the IP core, the connected wrapper module must act as a master (as shown in Fig. 3.34). A data transfer in such a system proceeds as follows: the master (initiator) sends address, data and control information via the OCP interface signals to its corresponding slave (target) bus wrapper interface module. The interface module converts the OCP request to an on-chip bus request, which is then transmitted to the destination. The request is received by the wrapper interface module at the destination, and converted from an on-chip bus request to an OCP request. This OCP request is then transferred from the (master) wrapper interface module to the (slave) destination, which takes the appropriate action.

3.2.1.1 *OCP Signals*

The OCP interface is synchronous, with a single clock signal. Thus all its signals are driven with respect to, and sampled by the rising edge of the clock. The OCP interface signals are divided into three categories: data flow, sideband, and test signals. A small subset of the data flow signals are required to be supported by all OCP interfaces. The remaining data flow signals, as well as all the sideband and test signals are optional. With the exception of the clock, all OCP signals are unidirectional and point-to-point.

Data flow signals consist of a set of signals, some of which are used for data transfers, while others are configured to support any additional communication requirements between the master and slave components. Data flow signals can be divided into the following categories:

- *Basic signals*: These include the clock, address (separate), read and write data, transfer type, and handshaking/response signals between the master and the slave. Only the clock and transfer type signals are mandatory for an OCP interface, the remaining signals being optional. The widths of the address, read data, and write data are configurable, and not limited to being multiples of eight. The transfer type indicates the type of data transfer operation issued by a thread running on a master, and can be any one of the following:

— *Read*: Reads data from the addressed location in a slave.
— *Write*: Writes data to the addressed location in a slave.
— *Idle*: No operation is required to be performed.
— *Broadcast*: Writes data to the addressed location, which may be mapped to more than one slave.
— *Exclusive read*: Reads from a location in a slave and locks it, preventing other masters from writing to the location (exclusive access). The location is unlocked after a write to it from the original master that caused the lock to be set.
— *Linked read*: Reads data from the addressed location in a slave, and sets a reservation in a monitor for the corresponding thread, for the addressed location. Read or write requests from other masters to the reserved location are not blocked from proceeding, but may clear the reservation.
— *Non-posted write*: Writes data to the addressed location in a slave, unlocking the location if it was locked by an exclusive read, and clearing any reservations set by other threads.
— *Conditional write*: Only writes to the addressed location in a slave if a reservation is set for the corresponding thread. Also clears all reservations on the location. If no reservation is present for the corresponding thread, no write is performed, no reservations are cleared, and a FAIL response is returned.

The handshaking signals are used by the master and the slave to synchronize data transfers, and the response signals are used by the slave to signal whether a request is valid or if an error occurred.

- *Simple extensions*: These include signals to indicate the address region (e.g., register or memory), BEs for partial transfers and core-specific configurable signals that send additional information with the transfer request, read data, write data, and the response from the slave. The configurable signals can transmit information about data byte parity, error correction code values, FIFO full or empty status, and cacheable storage attributes.
- *Burst extensions*: These signals are used to support burst transfers. They specify details about the data burst, such as
  — whether it is a precise length or unknown length burst;
  — burst length (for a precise length burst);
  — wrapping, incrementing, exclusive-OR (used by some processors for critical-word first cache line fill from wide and slow memory systems) or streaming (fixed address) burst mode;
  — packing or non-packing mode - for the scenario where data is transferred between OCP interfaces having different widths, the packing mode aggregates data when translating from a narrow to wide OCP interface, while the non-packing mode performs stripping when translating from a wide to narrow OCP interface;

  - — minimum number of transfers to be kept together as an atomic unit when interleaving requests from different masters onto a single thread at the slave;
  - — whether a single request is sufficient for multiple data transfers, or if a request needs to be sent for every data transfer in the burst;
  - — the last request, write data, or response in a burst.
- *Tag extensions*: These signals are used to assign tags (or IDs) to OCP transfers to enable OO responses and to indicate which transfers should be processed in order.
- *Thread extensions*: These signals are used to assign IDs to threads in the master and slave, and for a component to indicate which threads are busy and unable to accept any new requests or responses.

Sideband signals are optional OCP signals that are not part of the data flow phases, and can change independent of the request/response flow (but are still synchronous to the rising edge of the clock). These signals are used to transmit control information such as interrupts, resets, errors, and other component-specific information. They are also used to exchange status and control information between a component and the rest of the system. Finally, the OCP *Test* signals are also a set of optional signals, and are responsible for supporting scan, clock control, and IEEE 1149.1 (JTAG), for testing purposes.

Some of the OCP signals can be grouped together because they must be active at the same time. The OCP data flow signals can be combined into three *groups* of request, response, and data handshake signals. These groups in turn map one-on-one onto their respective protocol *phases*—request, response, and data handshake phases. An OCP *transfer* consists of several phases, and different types of transfers are made up of different combinations of phases. Every transfer must have a request phase. Read type requests always have a response phase, but write type transfers can be configured to with or without the response and data handshake phases. Generally, in an OCP read or write transfer, the request phase must precede the data handshake phase which in turn must precede the response phase. Burst *transactions* are comprised of a set of transfers linked together, and having a defined address sequence and number of transfers. This hierarchy of elements that are part of OCP is summarized in Fig. 3.35.

3.2.1.2 *OCP Profiles*

As mentioned earlier, the OCP interface can be configured to meet the requirements of the communicating components. The OCP 2.0 specification [15] introduces several pre-defined *profiles* that define a configuration of an OCP interface. These profiles consist of OCP interface signals, specific protocol features, and application guidelines. Two sets of profiles are provided:

1. *Native OCP profiles*: These profiles are meant for new components implementing native OCP interfaces. There are three profiles defined in this set:
   - *Block data flow profile*: Master type (read-only, read-write, or write-only) interface for components that require exchanging data blocks with memory.

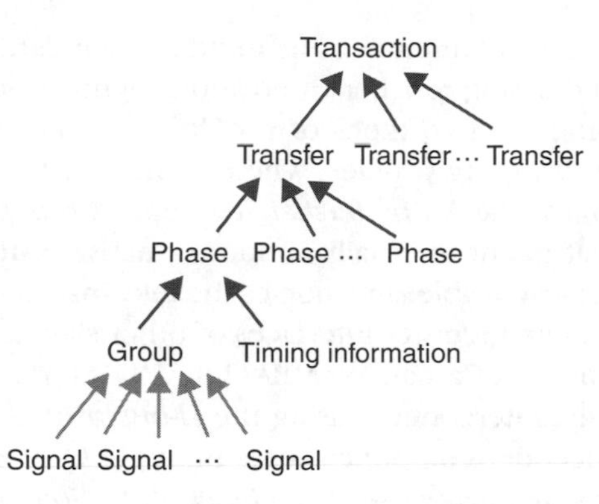

FIGURE 3.35

Hierarchy of elements that compose the OCP [15]
Source: OCP-IP

This profile is useful for managing pipelined access of defined-length traffic (e.g., MPEG macro-blocks) to and from memory.

- *Sequential undefined length data flow profile*: Master type (read-only, read–write, or write-only) interface for cores that communicate data streams with memory.
- *Register access profile*: Supports programmable register interfaces across a wide range of IP cores, such as simple peripherals, DMA, or register-controlled processing engines. Offers a control processor the ability to program the functionality of an attached component.

2. *Bridging profiles*: These profiles are meant to simplify or automate the creation of bridges between OCP and other bus protocol interfaces. There are three profiles in this set:
   - *Simple H-bus profile*: Intended to provide a connection through an external bridge to a CPU with a non-OCP interface (e.g., ARM9 processor [34] with AMBA 2.0 AHB native interface). This profile thus allows creation of OCP master wrappers to native interfaces of simple CPU type masters with multiple-request/multiple-data, read and write transactions.
   - *X-bus packet write profile*: Supports cacheable and non-cacheable instruction and data write traffic between a CPU and the memories and register interfaces of other slaves. Allows creation of OCP master wrappers to native interfaces of CPU type masters (e.g., ARM11 CPU master with AMBA 3.0 AXI native interface) with single-request/multiple-data, write-only transactions.
   - *X-bus packet read profile*: Supports cacheable and non-cacheable instruction and data read traffic between a CPU and the memories and register interfaces of other slaves. Allows creation of OCP master wrappers for native interfaces of CPU type masters (e.g., ARM11 CPU [35] master with AMBA 3.0 AXI native interface) with single-request multiple-data, read-only transactions.

These profiles are useful in several ways such as simplifying the task of integrating OCP components from different vendors, reducing the learning curve when applying OCP for standard purposes, simplifying logic needed to bridge an OCP component with another communication interface standard, improving component maintenance, and easing test bench creation. Figure 3.36 shows an example of an SoC design using the two types of profiles: the CPU and the CPU subsystem make use of the *bridging* profiles, whereas the rest of the components use the *native* OCP profiles. The *X bus packet read* and *X-bus packet write* profiles used by the CPU (that might internally support a native AMBA 3.0 AXI interface, for instance) support cacheable and non-cacheable instruction and data traffic between the memories and register interfaces of other slaves. The CPU bus subsystem (which might consist of a native AMBA 2.0 AHB interface, for instance) connects to the OCP-based interconnect using the *H-bus profile*, through an external bridge. The MPEG2 decoder component uses multiple OCP interfaces. It has two OCP master interfaces that make use of the *block data flow profile* that is suitable for managing pipelined access of defined-length traffic (e.g., MPEG macro-blocks) to and from memory. The reason for using two master interfaces is to improve parallelism and achieve higher performance. The decoder also has an OCP slave interface that uses the *register access profile*, to allow the CPU to program its operation. The DMA and media controllers also have OCP slave interfaces that use the *register access profile* for the same purpose. The DMA controller has an OCP master interface that can use either a *block data flow profile* or an *OO system interface profile* (TBD—to be defined in future revisions [15]) depending on the amount of parallelism required. The media controller has an OCP master interface that uses the *sequential undefined length data flow profile* which is a good fit for the controller because it needs to communicate a data stream with a memory-based buffer. Finally, the shared synchronous dynamic random access memory (SDRAM) controller optimizes bank and page accesses to SDRAM and can maximize performance (and minimize latency) by reordering requests. Therefore its slave OCP interface uses the *OO memory interface profile* (TBD—to be defined in future revisions [15]).

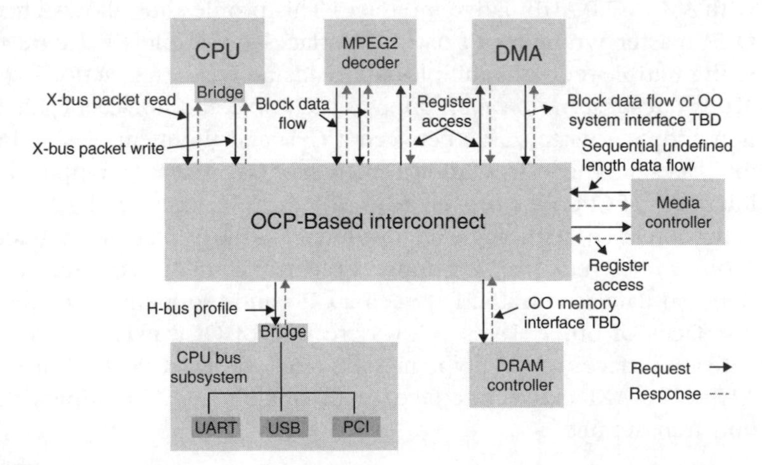

FIGURE 3.36

Example of SoC using several OCP profiles [15]
Source: OCP-IP

3.2.2 VSIA Virtual Component Interface

The virtual socket interface alliance (VSIA) VCI [14] is another point-to-point, synchronous, socket-based bus interface standard. It defines three types of interfaces having varying levels of complexity:

1. *Peripheral VCI (PVCI)*: Defines a simple handshake interface for data transfers, with support for burst transfers, address bus widths up to 64 bits, and data bus widths up to 32 bits.
2. *Basic VCI (BVCI)*: Superset of PVCI; adds support for SPLIT transactions, additional burst modes (e.g., wrapped, fixed), additional data transfer modes (e.g., locked/exclusive read), and data bus widths up to 128 bits.
3. *Advanced VCI (AVCI)*: Superset of BVCI; adds additional data transfer modes (e.g., new wrap, defined transfer modes), and support for advanced features such as OO transaction completion and multiple threads.

As can be seen, each of the interfaces described above is an enhanced and enriched version of the previous one. The interfaces proposed by the AMBA APB, AHB, and AXI bus architecture standards can be considered as somewhat analogous to the Peripheral, Basic, and AVCI interfaces, respectively. VCI actually shares many parallels with the OCP 2.0 socket-based interface standard. Unlike OCP 2.0, however, the VCI interface only contains data flow signals and does not address issues pertaining to test and control. Since the data flow signals in the VCI interface are quite similar to the OCP 2.0 interface signals (although AVCI has some additional features not found in OCP 2.0; however these have yet to be tested in silicon or verified in any form), OCP 2.0 is generally considered a functional superset of VCI.

3.2.3 Philips Device Transaction Level Protocol

The Philips DTL standard [17] defines another point-to-point, synchronous data transfer protocol. The DTL interface supports all the basic signals for single and block (or burst) data transfers, error signaling, and subword operations. Additionally, an extended DTL interface protocol specification defines optional application (or domain) specific extensions to the basic protocol. These extensions include signals for:

- *Addressing modes*: Wrapped, fixed, or decrementing addresses for block transfers.
- *2-D block operations*: Useful when operating on data stored in memory that represents a large 2-D area such as a frame buffer, and smaller 2-D accesses (such as a JPEG 8×8 block) are required.
- *Secure operations*: To indicate if a particular transaction is secure or not.
- *Buffer management*: To allow a component to request a flush of a write buffer, or to request notification when a certain data element reaches its destination.

Four major applications or OCP-like profiles related to traffic types are supported by DTL. Each of these four profiles has requirements for signals that must be implemented at the component interfaces. These profiles are presented below:

1. *Memory-mapped input/output (MMIO)*: For low bandwidth and latency critical control traffic.
2. *Memory-mapped block data (MMBD)*: For moving a block of data between a CPU (or any other component) and memory. Examples include cache line fills or cache line write-back on a CPU. This type of traffic may be both bandwidth and latency critical.
3. *Memory-mapped streaming data (MMSD)*: For moving a sequence of data items between components and memory. Such traffic is usually bandwidth critical, while latency may be less important.
4. *Peer-to-peer streaming data (PPSD)*: For moving a sequence of data items between two components. Like with the MMSD case, such traffic is usually bandwidth critical, while latency may be less important.

DTL is a proprietary interface standard (unlike OCP 2.0 and VCI which are open standards) developed by Philips, and has been used in the popular Philips Nexperia platform [29]. Although currently support for some of the more advanced features such as SPLIT or OO transaction completion, semaphore type operations, cache coherency, and read buffer management is not present, it is planned to be incorporated in future revisions of the DTL specification [17].

3.3 DISCUSSION: OFF-CHIP BUS ARCHITECTURE STANDARDS

While the focus of this book is on on-chip bus-based communication architectures, we include a brief discussion of off-chip bus architecture standards here for the sake of completeness. Off-chip buses are used to connect an SoC with external components, the most common of which are off-chip DRAM memory components such as synchronous dynamic random access memory (SDRAM), dual data rate DRAM (DDR DRAM), and Rambus DRAM (RDRAM). While on-chip embedded DRAM is beginning to become more ubiquitous in SoCs, it is more expensive. Applications with large memory requirements still rely on cheaper off-chip memories to meet storage requirements. An important motivation for the design of off-chip interconnects to connect an SoC with off-chip memory is the need to reduce pin counts, because a large number of pins can significantly increase packaging costs and system complexity. Therefore, unlike on-chip buses that make use of unidirectional multiplexed or AND–OR implementations, off-chip buses prefer bidirectional tri-state implementations to reduce pin counts.

Several off-chip, shared bus standards such as S-100 [18], PC-AT [19], Multi-Bus (II) [20], VME [20], PCI [21], and PCI-X (PCI Extended) [22] have been proposed and used in designs. PCI has undoubtedly been the most popular standard in this category, with almost the entire software infrastructure of the computer industry tied to the PCI interconnect model. However, these shared, parallel off-chip interconnects have inherent limitations such as crosstalk, excessive circuit capacitive loading, delays due to capacitive loads, high power dissipation, signal skew effects due to large distances covered, and reliability issues. As performance requirements of applications increase, higher bus clock speeds and shrinking process technology make crosstalk and capacitive delays more significant, and limit the maximum bus clock speed achievable. To alleviate the problems faced by shared parallel interconnects, switched parallel interconnect standards such as HyperTransport [23] and RapidIO [24] have been proposed, which use narrow, point-to-point connections. HyperTransport, for instance, supports narrower widths between 2 and 32 bits, and higher clock frequencies up to 800 MHz. These switched, point-to-point parallel interconnects solve the electrical loading, speed, and reliability issues of shared parallel interconnects. However, crosstalk and signal skew are still a problem. Switched serial interconnect standards such as PCIe (PCI Express) [25] and Infiniband [26] use a single signal for transmission and can achieve very high speeds, without suffering from any crosstalk effects. PCIe is fast becoming one of the most dominant off-chip standards in system design, especially because of its support for legacy PCI infrastructure. The emerging trend of using optical interconnects instead of copper-cabling for chip-to-chip interconnection is also well suited to a serial communication approach. The interested reader is directed to surveys of off-chip communication architectures by Mayhew and Krishnan [27] and Sassone [28] for related discussions on this topic.

3.4 SUMMARY

In this chapter, we presented the prevailing standards for on-chip communication architectures. Standards are essential in order to promote IP reuse and reduce the design time of the increasingly complex SoC designs today. On-chip bus-based communication architecture standards define the interface signals for components, as well as bus logic components such as arbiters, decoders, and bridges that are needed to implement the features of the proposed standard. We looked at some of the popular on-chip bus architecture standards such as ARM's AMBA 2.0 and 3.0, IBM's CoreConnect, STMicroelectronics' STBus, Sonics' SMART Interconnect, OpenCores' Wishbone, and Altera's Avalon. Another set of standards focuses on defining the component interface, but not the architecture implementation (which is left to the designer). These are the socket-based bus interface standards. Since these standards only define the component interface, the designer is free to use either a proprietary, custom bus architecture implementation, or any one of the bus architecture standards described above, such as AMBA 2.0/3.0 or CoreConnect. Socket-based bus interface standards require additional adapter logic to interface components to non-native bus architectures, and this can increase area, cost, and delay. However, the benefits of improved IP reusability across designs and greater flexibility to explore (or change) diverse bus architecture implementations is also substantial. We described some of the popular socket-based bus interface standards such as OCP, VCI, and DTL in this chapter. Finally, we briefly covered popular off-chip buses and standards which are used to connect SoCs to external DRAM memory blocks and other SoCs.

With a background on bus-based communication architectures in the last chapter (Chapter 2), and a description of prevalent communication architecture standards in this chapter, we now proceed to address the important problem of understanding the on-chip communication architecture design space, to aid in selecting the best communication architecture configuration for an application. The next chapter (Chapter 4) presents models for the performance estimation of communication architectures. These models capture details of the communication architecture design space and allow designers to estimate the performance of different communication architecture configurations. The subsequent chapter (Chapter 5) presents models for power estimation of communication architectures that allow designers to know more about the power characteristics of different communication architecture configurations. These performance and power models for communication architectures are used as part of various techniques (presented in Chapter 6) to select, configure and design a communication architecture that meets the requirements of a given application.

Brief Discussion: Evolution of On-Chip Communication Protocols

As the trend for SoCs moves toward multiple processors on a chip, on-chip communication protocols are continuously evolving. Most of the popular communication architecture (e.g., AMBA) and socket-based standards (e.g., OCP) have evolved over the last few years to accommodate the need for high performance and customizable on-chip data communication. Going forward, these standards will likely continue to evolve further, to handle the many needs of multiprocessor SoC (MPSoC) designs. Support for cache coherence mechanisms will be important in the next generation communication protocols, as multiple processors will frequently access shared memories both on and off the chip. The excessive power consumption of complex communication architectures will also necessitate more explicit support for dynamic power management, to switch off parts of the communication architecture fabric when not in use via power/clock gating. Finally, the number of on-chip communication standards has been growing over the past few years, and will possibly continue to grow in the coming years, requiring more emphasis on techniques to handle interface mismatches. Some recent research in the area of handling interface mismatches between different protocols is presented in Chapter 9.

REFERENCES

[1] ARM AMBA Specification and Multi layer AHB Specification (rev2.0), http://www.arm.com, 2001.

[2] ARM AMBA 3.0 AXI Specification, www.arm.com/armtech/AXI.

[3] IBM CoreConnect Specification, http://www.ibm.com/chips/techlib/techlib.nsf/product families/CoreConnect_Bus_Architecture.

[4] "STBus Communication System: Concepts and Definitions," *Reference Guide*, STMicro electronics, May 2003.

[5] Sonics SMART Interconnect, http://www.sonicsinc.com.

[6] Wishbone Specification, http://www.opencores.org/wishbone.

[7] Altera Avalon Interface Specification, April 2006, http://www.altera.com/.

[8] AMBA AHB Interconnection Matrix, www.synopsys.com/products/designware/amba_solutions.html.

[9] S. Pasricha, N. Dutt and M. Ben-Romdhane, "Constraint-driven bus matrix synthesis for MPSoC," *Asia and South Pacific Design Automation Conference (ASPDAC 2006)*, Yokohama, Japan, January 2006, pp. 30–35.

[10] S. Pasricha, Y. Park, F. Kurdahi and N. Dutt, "System-level power-performance trade-offs in bus matrix communication architecture synthesis," *International Conference on Hardware/Software Codesign and System Synthesis (CODES + ISSS 2006)*, Seoul, Korea, October 2006 pp. 300–305..

[11] S. Pasricha, N. Dutt and M. Ben-Romdhane, "Extending the transaction level modeling approach for fast communication architecture exploration," *Design and Automation Conference (DAC 2004)*, San Diego, CA, June 2004, pp. 113–118.

[12] S. Pasricha, N. Dutt, E. Bozorgzadeh and M. Ben-Romdhane, "FABSYN: Floorplan-aware bus architecture synthesis," *IEEE Transactions on Very Large Scale Integration Systems (TVLSI)*, Vol. 14, No. 2, March 2006, pp. 241–253.

[13] R. Ho, K. W. Mai and M. A. Horowitz, "The future of wires," *Proceedings of the IEEE*, Vol. 89, 2001.

[14] VSI Alliance™ On-Chip Bus Development Working Group, Virtual Component Interface Standard Version 2 (OCB 2 2.0), April 2001.

[15] Open Core Protocol International Partnership (OCP-IP). OCP Datasheet, Release Version 2.0/2.1, http://www.ocpip.org.

[16] Open Core Protocol International Partnership (OCP-IP). OCP Datasheet, Release Version 1.0, http://www.ocpip.org.

[17] Philips Semiconductors. Device Transaction Level (DTL) Protocol Specification, Version 2.4, February 2005.

[18] "IEEE Standard 696 Interface Devices," IEEE Computer Society, June 1982.

[19] "PC/104 Specification," PC/104 Embedded Consortium, August 2001.

[20] J. Zalewski, *Advanced Multimicroprocessor Bus Architectures*, IEEE Computer Society Press, 1995.

[21] PCI Special Interest Group, "PCI Local Bus Specification, Revision 2.2," December 1998.

[22] PCI Special Interest Group, "PCI-X 2.0 Protocol Specification Revision 2.0," July 2003.

[23] HyperTransport Consortium, "HyperTransport Technology: Simplifying System Design," October 2002, http://www.hypertransport.org.

[24] RapidIO Trade Association, "RapidIO Technical Whitepaper Rev 3," http://www.rapidio.org.

[25] PCI Special Interest Group, "PCI Express Base Specification Revision 1.0a," April 2003.

[26] Infiniband Trade Association, "Infiniband Architecture Specification, Release 1.0," October 2000. http://www.infinibandta.org.

[27] D. Mayhew and V. Krishnan, "PCI express and advanced switching: evolutionary path to building next generation interconnects," in *Proceedings of 11th Symposium on High Performance Interconnects*, 2003, pp. 21–29.

[28] P. Sassone, "Commercial trends in off-chip communication," Technical Report, Georgia Institute of Technology, May 2003.

[29] J.A. de Oliveira and H. van Antwerpen, "The Philips Nexperia digital video platform," *Winning the SoC Revolution*, G. Martin and H. Chang (Eds.), Kluwer Academic Publishers, 2003.

[30] PrimeCell® AXI Configurable Interconnect (PL300) Technical Reference Manual, http://www.arm.com/products/solutions/AXISystemComponents.html.

[31] PrimeCell® AXI Configurable Interconnect (PL301) Technical Reference Manual, http://www.arm.com/products/solutions/AXISystemComponents.html.

[32] Avalon Memory Mapped Interface Specification, May 2007, http://www.altera.com/.

[33] Avalon Streaming Interface Specification, June 2007, http://www.altera.com/.

[34] ARM9 Processor Family, http://www.arm.com/products/CPUs/families/ARM9Family.html.

[35] ARM11 Processor Family, http://www.arm.com/products/CPUs/families/ARM11Family.html.

CHAPTER

Models for Performance Exploration

4

On-chip communication architectures have numerous sources of delay due to signal propagation along the wires, synchronization (e.g., handshaking), transfer modes (e.g., pipeline access, burst transfer, etc.), arbitration mechanisms for congestion management, cross-bridge transfers, and data packing/unpacking at the interfaces. These communication delays can significantly influence the performance of system-on-chip (SoC) applications, and are in fact, a major cause of bottlenecks in many designs. It is therefore important to consider these delays when exploring SoC applications, in order to get an accurate estimation of the system performance. Figure 4.1 shows where communication architecture performance exploration is typically performed in a typical electronic system level (ESL) design flow (described in Chapter 1)—early on at the system level, before the implementation (or register transfer level, RTL) model is created. Performing

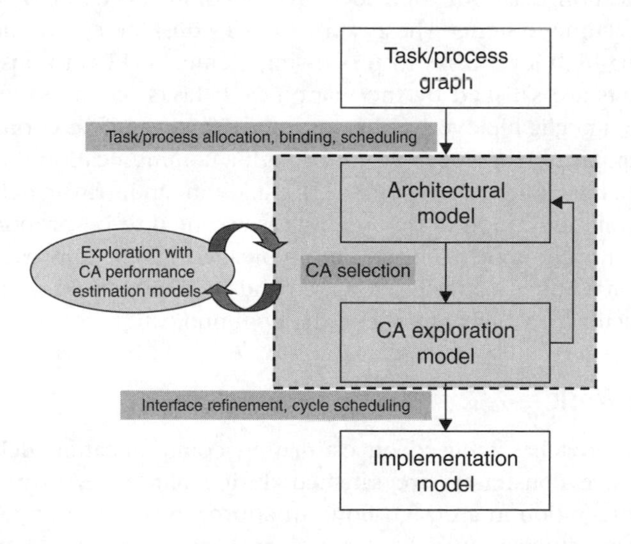

FIGURE 4.1

Communication architecture (CA) performance estimation in a typical ESL design flow communication architecture exploration early in the design flow is crucial for SoC designs, because it allows designers to manage an extremely large design space more efficiently and make design decisions that significantly affect system performance in a much shorter span of time than at lower (e.g., RTL) levels, where the amount and complexity of detail makes such an activity prohibitively time consuming. In this chapter, we review research efforts that propose performance estimation models for communication architecture exploration at the system level. Note that there is also a large body of work dealing with modeling interconnects at the circuit level to estimate interconnect delay (i.e., performance). The focus of this chapter is not on these circuit level models (Chapter 11 covers these models in detail), but rather on performance estimation models for communication architecture exploration created earlier in the design flow, at the system level. In Section 4.1, we discuss static, estimation-based models for communication architecture (CA) performance exploration. Section 4.2 presents dynamic, simulation-based exploration models. Finally, Section 4.3 describes hybrid models that incorporate static estimation and dynamic simulation to explore communication architecture performance.

4.1 STATIC PERFORMANCE ESTIMATION MODELS

Static performance estimation methods attempt to determine the performance of a system through analysis, typically yielding closed form expressions that capture the system performance as a function of certain system parameters. The key challenge here is to determine the right set of system parameters and their interactions. A number of approaches have looked at statically estimating the communication delay in applications. Very early work [1–3] focused on estimating communication delay for high level synthesis, in the context of distributed embedded real-time systems. These approaches consider communication delay while mapping high level tasks to processing elements (PEs) to ensure that real-time constraints are satisfied by the mappings. If tasks on the same PE need to communicate, a negligible overhead is assumed. However, if tasks on separate PEs need to communicate, they need to go through a communication link that can be either a shared bus or a point-to-point (P2P) link with an intrinsic delay associated with it. This communication delay on a link is assumed to be proportional to the amount of data traffic traversing the link. Typically in these approaches, the total data volume on a link is statically estimated and then multiplied by a communication scaling factor (or weight) to obtain the communication cost in the system.

4.1.1 Early Work

Subsequent approaches focused on estimating communication delay to ensure that performance constraints are satisfied during hardware/software (HW/SW) component integration in a design flow. An approach proposed by Gogniat et al. [4] captures communication delay for a given protocol using a simple equation that includes the number of data to be transferred, the width of the data elements in relation to the width of the bus, and the retrieval/communication time of a single data element. This approach is extended by Knudsen et al. [5–7] by creating a more detailed communication estimation model to be integrated with HW/SW partitioning to trade off the costs associated with different partitioning choices. Like the approach by Gogniat et al. [4], only P2P communication channels are considered by the approach. Figure 4.2(a) shows the communication model proposed by Knudsen and Madsen [5–7] that is divided into three parts: a *transmitting* model, a *channel* model, and a *receiving* model. The approach considers the possible performance degradation imposed by the HW/SW drivers, in addition to the characteristics of the communication channel. The time overhead of establishing a connection (e.g., due to arbitration) is not modeled, however, because it is more relevant to a shared communication channel than a P2P communication link.

Figure 4.2(b) shows the driver transmission parameters used to create the driver transmission delay model. The driver receives n_t words for transmission from the PE and produces n_c channel words. This may require packing or splitting of the driver input words of bit width w_t to generate channel words having bit width w_c. If the clock frequency of the transmitting processor is f_t, the number of cycles required to call the driver for transmission (transfer arguments, transfer execution flow, etc.) is c_{tc}, and the number of transmission processing cycles per driver input word is c_{tp}, then the total driver transmission delay is calculated as:

$$t_{td} = (c_{tc} + c_{tp}n_t)/f_t$$

The packing granularity parameter w_g trades off the time spent in packing the input driver words with the number of created channel words, and this directly influences the number of transmission processing cycles required per driver input

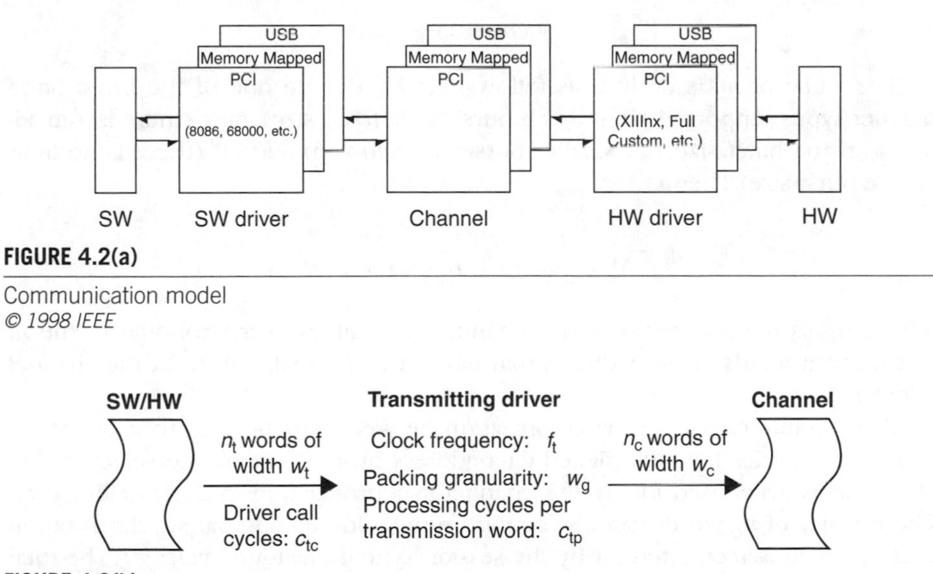

FIGURE 4.2(a)

Communication model

FIGURE 4.2(b)

Driver transmission delay parameters

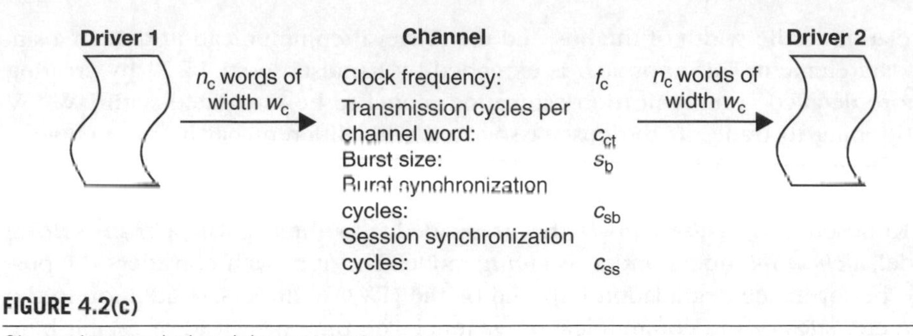

FIGURE 4.2(c)

Channel transmission delay parameters
© 1998 IEEE

word c_{tp}. For instance, one packing strategy is to pack the received words in a bit-wise manner into the channel words. This results in more processing time in the driver, but also fewer words to be transmitted on the channel. A faster packing approach would be to pack the input words on a per-input word basis, by packing as many whole input words that can fit into a channel word. An even faster approach would be to pack each input word into a single channel word. However, these faster approaches would increase the amount of words on the channel, which would incur additional delay. Therefore, a careful trade-off analysis is essential.

Figure 4.2(c) shows the channel transmission parameters that are used in the channel delay model. The n_c channel words are transmitted as a burst of $n_b - 1$ words of size s_b, and a remainder burst of size s_r, $0 \le s_r \le s_b$:

$$n_c = (n_b - 1)s_b + s_r$$

The value of n_b is derived as follows. Let b_m denote one of the three burst transfer types supported: *fixed* (each burst has a fixed size), max (there is a maximum on the burst size, but smaller bursts are allowed) and inf (there is no limit on the burst size). Then,

$$n_b = \begin{cases} 1 & \text{if } b_m = \text{inf} \\ \lceil n_{cd} s_b \rceil & \text{if } b_m = \text{fixed, max} \end{cases}$$

where n_{cd} is the number of actual channel data values corresponding to the n_t driver input words of bit width w_t that have been packed/split to fit the channel width w_c.

Every transfer requires synchronization between bursts (e.g., to account for turnaround cycles for multiplexed data/address buses, slave device select cycles, etc.) that is accounted for by the channel synchronization cycles per burst c_{sb}. The transfer of n_c words can also require some additional initial synchronization cycles, which is accounted for by the session synchronization cycles c_{ss}. The total synchronization cycles c_{cs} for the transfer of n_c words is given by:

$$c_{cs} = \lceil n_b c_{sb} \rceil + c_{ss}$$

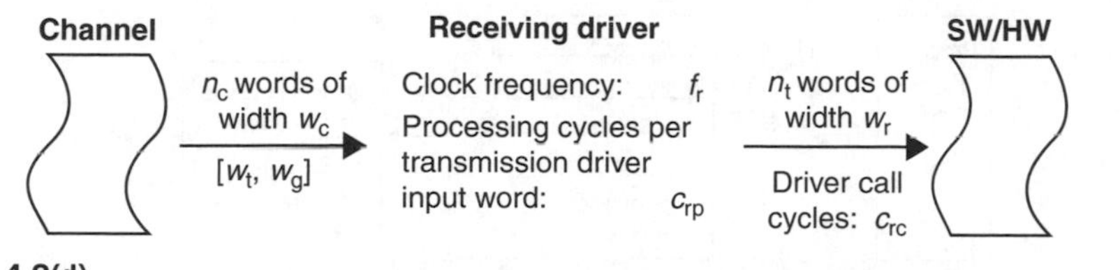

FIGURE 4.2(d)

Driver reception delay parameters for the approach proposed by Knudsen et al. [5–7] *© 1998 IEEE*

Then, for the clock frequency of the channel f_c, the number of transmission cycles per channel word c_{ct}, and the number of transmitted channel words n_c, the channel transmission delay is given as:

$$t_{cd} = (c_{cs} + c_{ct}n_c)/f_c$$

This delay assumes that the connection has already been set up, and ignores any arbitration and channel setup delays.

Figure 4.2(d) shows the driver reception parameters used in the driver reception delay model. One underlying assumption here is that the receiver receives (or has knowledge of) the parameters w_t and w_g, so it knows how data was packed at the transmitting driver. It is also assumed that $w_r \geq w_t$ so that each unpacked/unsplit word of size w_t is put on a single output word of bit width w_r at the receiver. Then, if the clock frequency of the receiving PE is f_r, the number of driver call cycles for reception is c_{rc}, and the number of receiver processing (unpacking/unsplitting) cycles per transmission driver input word is c_{rp}, the driver reception delay can be calculated as:

$$t_{rd} = (c_{rc} + c_{rp}n_t)/f_r$$

Assuming that the driver production, channel transmission and driver reception of channel words occur in a parallel, pipelined manner, the slowest part determines the total transmission delay. The maximum delay can then be estimated as:

$$t_m = \max(t_{td}, t_{cd}, t_{rd})$$

and the total transmission delay can be given by:

$$t_t = t_m + 2t_m/n_t$$

where the last term can be considered as an approximation of the pipeline startup/completion delay. Since the number of channel words may differ from the number of transmission/reception words, the pipeline startup/completion delay is not modeled accurately by the given term which is a worst case estimate.[1] The authors demonstrated the utility of these communication delay estimation models for simple examples of communication between two processors, where it was

[1]Static estimation techniques often use conservative estimates when run-time/execution data is not available.

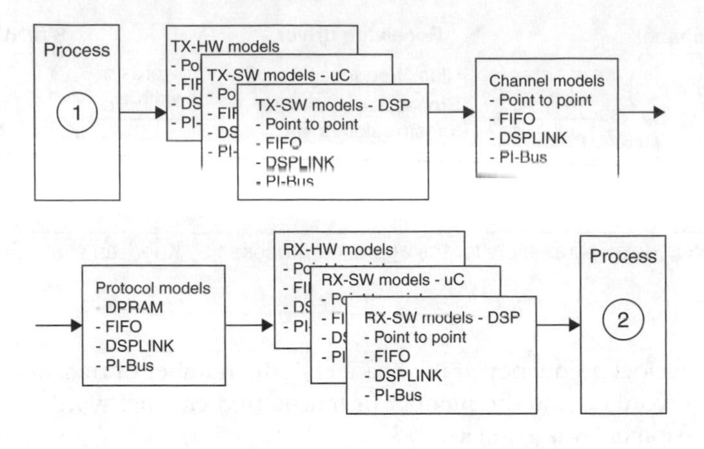

FIGURE 4.3

Communication performance model proposed by Renner et al. [8, 9]

shown that appropriately setting values for parameters such as packing granularity (w_g) and frequency values for the transmitter (f_t), receiver (f_r), and channel (f_c) can reduce overall communication delay.

4.1.2 Protocol Delay-Aware Approach

A derivative approach was proposed by Renner et al. [8, 9], and the corresponding communication performance model is shown in Fig. 4.3. The communication model in this approach considers not only the transmitter, channel, and receiver delays, but also delays arising due to the specific protocol chosen. Additionally, a library of performance parameters is required for the different implementation models of the transmitting/receiving drivers, channels, and protocols.

The transmitting driver delay model [8] is given by:

$$t_{\mathrm{p}}^{\mathrm{T}} = \frac{n_{\mathrm{c}}^{\mathrm{HW}}}{f_{\mathrm{c}}^{\mathrm{HW}}} + \frac{n_{\mathrm{c}}^{\mathrm{SW}}}{f_{\mathrm{c}}^{\mathrm{SW}}}$$

where n_c is the number of cycles required to put data onto the channel, while f_c is the clock frequency of the transmitting driver implementation. Since the driver can have only one implementation, either in hardware or software, the other term in the equation will be zero. Renner et al. [9] appended an additional term to the above equation, to compensate for overhead due to any splitting/packing of data. Table 4.1 shows the transmission parameters and delay for three possible driver implementations (*SW—DSP*, *SW—microcontroller*, *HW—FPGA*), connected to different types of channels (links). The receiving driver delay model is similar to the transmitter driver delay model.

The channel delay model for a one bit link is based on the reconfigurable switching interconnect prototype [8], that includes delays due to Field Programmable Gate Arrays (FPGAs) and dual-port RAM (DPRAM) that might be

Table 4.1 Transmitter model parameters [8]

Host	Transmission link	Cycles	t^{T}_{p}	f_c
DSP	PTP	2	40 ns	50 MHz
-SW-	DPRAM	2	40 ns	
	FIFO	2	40 ns	
	DSPLINK	4	80 ns	
μC	PTP	2	100 ns	20 MHz
-SW-	DPRAM	2	100 ns	
	CAN	4–8	200–400 ns	
FPGA	PTP	2	400 ns	5MHz
-HW-	DPRAM	2	400 ns	
	FIFO	2	400 ns	
	DSPLINK	4	800 ns	
	CAN	4–7	800 ns–1,4 μs	

used to implement required features of the selected protocols such as memory and glue logic synchronization circuits. The channel delay model is given by:

$$t_p^C = k \cdot t_{SW} + \sum_i t_{FPGA,i} + m \cdot t_{DPR} + \sum_i t_{WIRE,i}$$

where k is the number of switches used, t_{SW} is the switch delay, t_{FPGA} is the FPGA delay, m is the number of DPRAM accesses required, t_{DPR} is the DPRAM access time, and t_{WIRE} is the wire delay. Depending on the type of transmission link used, one or more terms in the equation will be eliminated. Table 4.2 shows the channel delay parameter values for different types of transmission links considered in this work.

The delay due to the protocol, for different transmission protocols, is shown in Table 4.3. While the overhead for simple P2P protocols is considered negligible, the delay due to arbitration and checksum bits in the *DSPLINK* and *CAN* bus protocols can add substantial delay. Finally, the total communication delay is calculated as:

$$t_p = t_p^T + t_p^C + t_p^P + t_p^R$$

whereas the delay when transmission occurs in a pipelined fashion is given by:

$$t_p = \max(t_p^T(t),\ t_p^C(t),\ t_p^P(t),\ t_p^R(t))$$

These performance models were subsequently used by the authors to guide their automated communication architecture synthesis, to generate communication structures for real-time embedded systems [9].

Table 4.2 Channel model parameters [8]

Transmission link	k	$\sum t_{FPGA}$	mm	$\sum t_{WIRE}$	t_p^C
K1: PTP	3	0	0	3.8ns	33.8ns
K2: DPR(1)	1	0	1	1.3ns	36.3ns
DPR(2)	2	0	1	3.2ns	48.2ns
K3: FIFO(1)	1	52.0ns	0	1.3ns	63.3ns
FIFO(2)	0	97.6ns	0	0	97.6ns
FIFO(3)	1	33.5ns	1	1.3ns	69.8ns
FIFO(4)	3	52.1ns	0	3.2ns	85.3ns
K4: CAN	3	0	0	3.8ns	33.8ns
K5: DLK	3	0	0	3.8ns	33.8ns

Table 4.3 Protocol model parameters [8]

Transmission link	Protocol cycles	t_p^P	f_c
PTP	0	0ns	–
DPRAM	0	0ns	–
FIFO	0	0ns	–
DSPLINK	12–21	116–594ns	10MHz
CAN	47–130	188–520μs	250kHz

4.1.3 AMBA 2.0 AHB Estimation

A delay model for the AMBA 2.0 AHB [10] single shared bus and hierarchical shared bus architectures was proposed by Cho et al. [11]. For a single shared bus, assuming that the number of data items to be transferred to be N_d, the number of masters on the bus N_m, and a fixed burst size of B, the latency of the bus (L_{shared_bus}) is given as:

$$L_s = N_m(3 - 2U)N_d\,S$$

$$L_b = \lceil N_d(1 - S)/B \rceil + N_d(1 - S)$$

$$L_{shared\_bus} = L_s + L_b$$

where L_s refers to single data transfer mode, that typically take three cycles (a cycle each for request, address, and data phases), and L_b represents burst transfer mode. The underlying assumption here is that the slave does not introduce any wait states, and that the request and address phases occur in the same cycle in the burst transfer mode. The parameter S ($1 \geq S \geq 0$) refers to the probability of

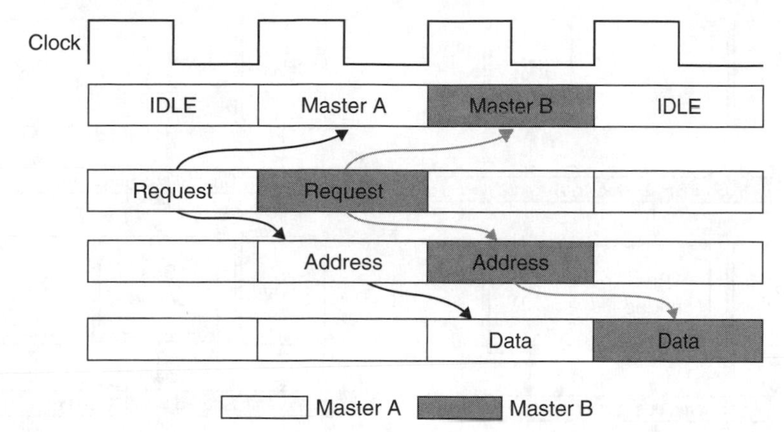

FIGURE 4.4

Effect of pipelining in hiding latency of single data transfers [11]
© 2006 ACM Press

single mode transfers that occur in the application which is implemented on a shared bus. The parameter U $(1 \geq U \geq 0)$ refers to the usage of the bus, and is a probability of continuing single transfers, in a pipelined manner. As the bus usage increases, more and more single data transfers can be completed in a pipelined manner, as shown in Fig. 4.4, which reduces the latency of single data transfers.

For the hierarchical shared bus case, where multiple shared buses are connected together via bridges, the bus latency is given as:

$$L_{\text{hierarchical\_bus}} = (N_{\text{m}}/N_{\text{l}})\, L_{\text{shared\_bus}}(1 - A) + \alpha A$$

where N_{l} is the number of layers (or buses) in the hierarchical shared bus architecture, the parameter A $(1 \geq A \geq 0)$ is the probability of the path of the data transfer passing through a bridge, and α is the bridge factor that represents the latency overhead caused by using the bridge. The first term in the equation represents the latency of data transfers occurring on a single shared bus, that do not require passing through a bridge, as can be seen in Fig. 4.5(a). As the number of layers (or shared buses) N_{l} increase in the system, the system-wide latency of such data transfers is reduced. Note that the inherent latency of such a data transfer remains unchanged as the number of buses is increased, but the overall system latency reduces because each shared bus can be executing a transfer in parallel. The second term in the equation represents the latency of data transfers that activate one or more bridges, as is shown in Fig. 4.5(b). The value of bridge factor α is given as:

$$\alpha = \sum_{i=1}^{N_{\text{l}}-1} \left(\frac{\beta}{\chi} \frac{N_{\text{m}}}{N_{\text{l}} - i} L_{\text{shared\_bus}} \right)$$

where β is the number of data transfer paths which use the same number of bridges

$$\beta = C_i^{N_{\text{l}}-1}$$

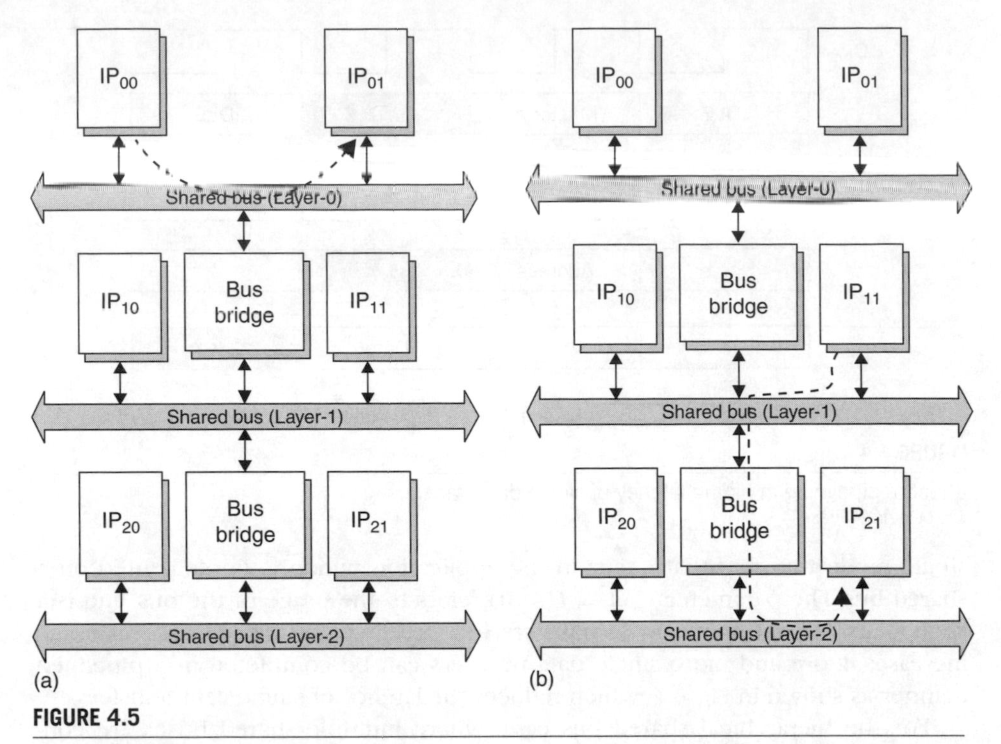

FIGURE 4.5

(a) Data transfer on a bus that does not activate a bridge, (b) data transfer on a bus that activates a bridge [11]

and χ is the number of total data transfer paths using bridges

$$\chi = \sum_{j=1}^{N_i - 1} C_j^{N_1 - 1}$$

Note that in this above-mentioned approach [11], the parameters *S*, *U*, and *A* are probabilistic and need to be determined before the bus delay/latency numbers can be obtained. Under assumptions of suitable values for these parameters, the approach was shown to have an accuracy of 96% and 85% compared to a simulation-based approach, for the single shared bus and a hierarchical, multiple-layer shared bus AMBA (ARM microcontroller bus architecture) AHB (advanced high performance bus) architecture, respectively [11].

4.1.4 Discussion: Limitations of Static Performance Estimation Methods

As can be seen from the examples of the approaches described above, static estimation requires several assumptions (e.g., probabilistic values for parameters [11], single cycle arbitration for all transfers, etc.) that are dependent on the application functionality and thus not easy to model statically. Static estimation approaches are unable to account for non-deterministic traffic generation by the components on the buses, and consequently cannot predict dynamic component (e.g., memory access) delays. These static approaches also cannot accurately account for other sources of dynamic delays such as those arising due to complex arbitration and traffic congestion, cache misses, burst interruptions, interface buffer overflows, and the effect of advanced bus features such as SPLIT transactions [10] and out-of-order (OO) transaction completion [12]. Often, static techniques tend to assume that computation and communication in a system can be statically scheduled, and are either overly optimistic when they ignore dynamic bus contention effects (e.g., Knudsen and Madsen [6]) or overly pessimistic when they assume a worst case scenario for bus contention (e.g., Yen and Wolf [1]). As a result, these static approaches have limited applicability for most medium- to large-scale SoCs, and are typically used to obtain worst case performance bounds, or provide high level (conservative) performance estimates very early in the design flow. These static models are much more useful for simpler SoC applications that can be statically scheduled and have a predictable behavior, such as non-pre-emptive periodic task-based applications. The models can also be used in larger and more complex systems to guide a dynamic scheduler and create a tight real-time schedule by providing estimates for the actual protocols that will be used in the system.

A more accurate approach to communication architecture performance estimation requires creating a model of the application that can be simulated. This allows a more accurate estimation of the data traffic behavior on the bus and the corresponding delays can be more reliably assessed. However, there are several ways in which simulation models can be created, that differ based upon the level or kind of details that are captured (or omitted) by the model. The next section looks at some of these simulation-based performance estimation modeling approaches.

4.2 DYNAMIC (SIMULATION-BASED) PERFORMANCE ESTIMATION MODELS

A lot of work has been done in the area of creating simulation models for performance estimation of communication architectures. Several approaches have been proposed, that capture the communication architecture at different levels of detail or abstraction. Typically, if a model captures more detail, it is more accurate in estimating the performance, but also slower to simulate. Simulation speed is clearly important, especially with ever increasing complexity of modern SoC designs (that results in more and more components being added to an SoC) and a correspondingly larger amount of detail that must be captured and simulated for performance estimation. There is thus an inherent trade-off between the accuracy of results and simulation speed of the model used to obtain these results.

Figure 4.6 outlines the typical modeling abstractions used to develop performance estimation models. The amount of detail captured increases at each step when going from top to bottom in the figure. Correspondingly, the simulation speed of the models also decreases as we lower the abstraction level, since the captured detail increases the time it takes to simulate the system. All of the

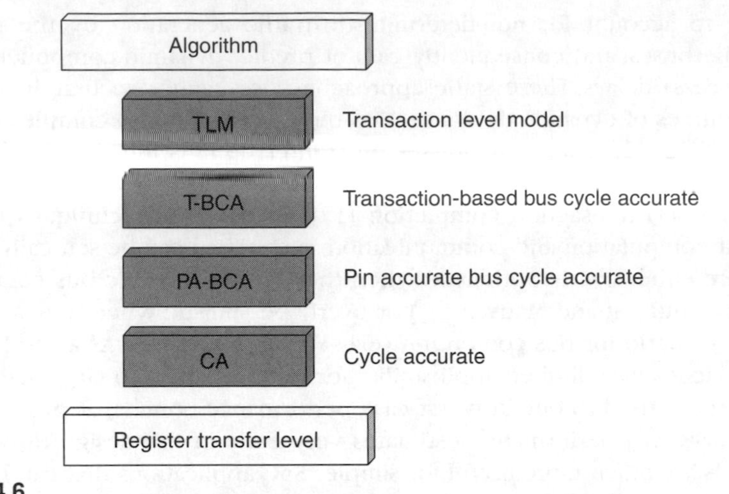

FIGURE 4.6

Modeling abstractions for communication architecture performance exploration

modeling approaches presented here use high level languages such as *C/C++* or *SystemC* [13–15], that are at least an order of magnitude faster than the RTL modeling abstraction that uses *Verilog* or *VHDL* to capture a design. The use of high level languages enables early and faster exploration of the communication architecture design space in the SoC development cycle. We now examine the classification of existing modeling abstractions in more detail.

4.2.1 Cycle Accurate Models

Cycle accurate (CA) models are extremely accurate models that capture the complete functionality, structure, communication, and timing of an SoC design. The communication between components is modeled in a pin-accurate, cycle accurate manner, that is all the hardware signals that connect a component to a bus are explicitly modeled, and all the timing and protocol induced delays are captured accurately for the bus architecture. The components are modeled in a cycle accurate manner as well—the computation inside a component is scheduled at cycle boundaries. Figure 4.7(a) shows an example of a cycle accurate model. The behavior inside the master and slave components consists of computation and communication directives that are scheduled in a cycle accurate manner. The interconnection between the components is also modeled in a pin accurate, cycle accurate manner. Such cycle accurate models are fairly close in structure to RTL level models used for synthesis. Due to the excessive level of detail that is captured at this modeling abstraction, the simulation speed is quite slow, but still anywhere between 10 and 100 times faster than RTL simulation speed (because these cycle accurate models are captured and simulated using high level languages), depending on the complexity of the design. The time to create these models is also excessive—almost comparable to the time taken for writing detailed RTL code. These models are therefore quite prohibitive, both in their modeling time

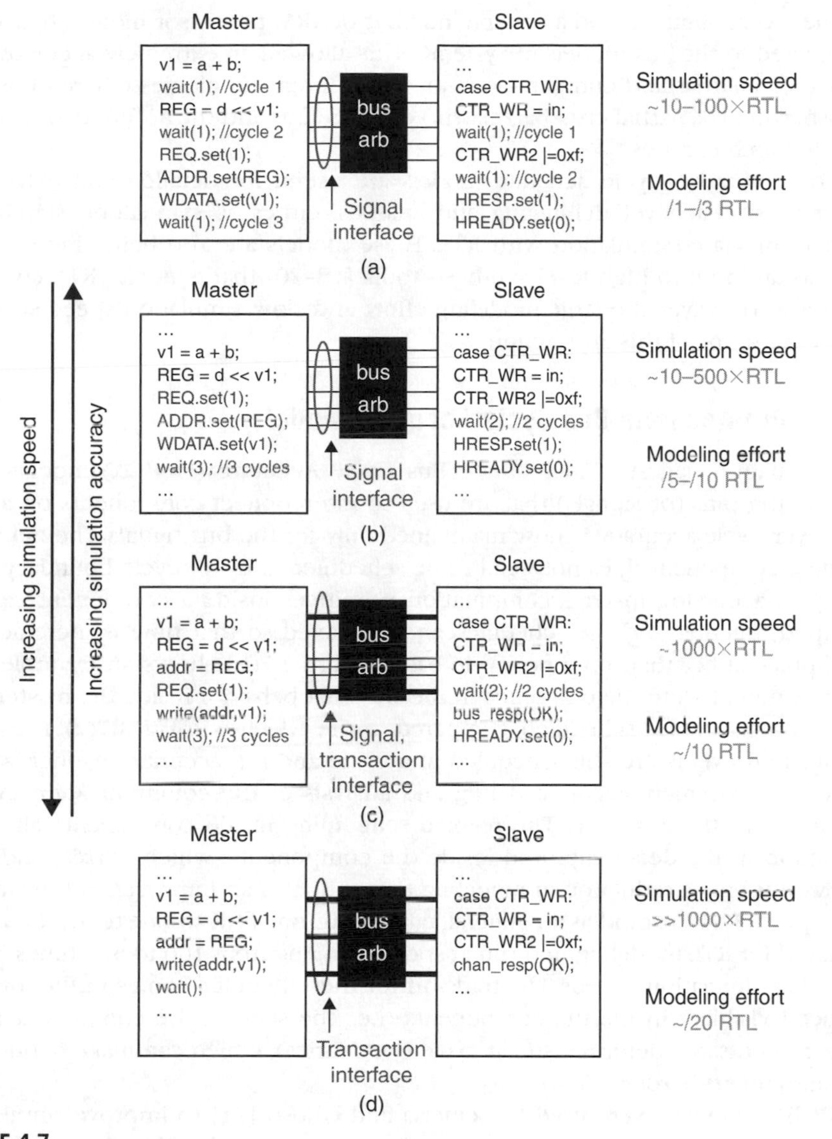

FIGURE 4.7

Trade-offs between different modeling abstractions: (a) CA, (b) PA-BCA, (c) T-BCA, and (d) TLM

and slow simulation speed to be useful for high level communication architecture exploration of SoC design today.

Regardless of these drawbacks, there have been a few approaches that have used cycle accurate models for communication architecture exploration. For instance, Loghi et al. [16] used cycle accurate models, captured using SystemC, to explore and contrast the AMBA AHB shared bus [10] and the STBus [17] communication architectures. Different classes of applications having varying data traffic profiles were simulated on a scalable number of ARM processor models that were connected to the bus models in SystemC. This allowed an extremely accurate performance analysis and comparison study for different topologies (shared bus, full crossbar/matrix, partial crossbar/matrix) of the STBus and the AMBA AHB communication architectures.

To summarize, cycle accurate models are useful for detailed and extremely accurate system level debugging and analysis, either as standalone simulation models, or via co-simulation with RTL. These models are also being increasingly used as an input to high level synthesis tools [18–20] that generate RTL code for synthesis. However, the large modeling effort and slow simulation speed severely limits the scope of this abstraction.

4.2.2 Pin-Accurate Bus Cycle Accurate Models

As the name suggests, Pin-Accurate Bus Cycle Accurate (PA-BCA) models capture all the pins (or signals) that are used to interconnect components on a bus. However, cycle accuracy is now maintained only for the bus signals. The behavior inside a component does not need to be scheduled at every cycle boundary (i.e., be cycle accurate). Instead, computation primitives inside a component can be grouped together and the schedule can be relaxed so that time is incremented in chunks, rather than on a per-cycle basis. Figure 4.7(b) shows an example of a PA-BCA model. Note the difference in the way that behavior inside the master and slave components is scheduled, compared to the CA case in Fig. 4.7(a). The communication events are still scheduled in a cycle and pin accurate manner, which allows an extremely accurate debug and analysis of the communication events occurring in these models. The relaxed scheduling inside components allows a reduction in the detail captured inside the components, which correspondingly allows not only a reduction in modeling time, but also an improvement in simulation speed. PA-BCA models, in general, take about one-fifth to one-tenth the effort required for RTL modeling, and run at speeds ranging from 100 to 500 times faster than RTL simulation speeds. The trade-off for these benefits comes in the form of reduced visibility inside the components (i.e., the state of the component is no longer necessarily deterministic at cycle boundaries), which can make debugging of components harder.

PA-BCA models were used by Séméria and Ghosh [21] to improve simulation speed over CA models, for the purpose of HW/SW co-verification. Abstract models of components were interconnected using bus functional models (BFM, which is another name for PA-BCA) for use in verifying the timing and interfaces between different components that are either implemented in hardware, or as software running on processor instruction set simulator (ISS) models. SystemC was used to capture the simulation model. Experiments showed that reducing the detail of the processor ISS model (which is initially modeled at a cycle accurate granularity and is a major simulation bottleneck) while keeping the pin accurate and cycle accurate bus model enabled a significant 20 times speedup [21]. BFM models were also found to be suitable for highly accurate communication architecture exploration of the AMBA AHB bus by Schirner and Domer [22]. Their work compared the accuracy of variants of transaction level models (TLM, described in Section 4.2.4) with BFM for exploration. An interesting approach to communication architecture performance exploration using PA-BCA models was presented by Kalla et al. [24], in which traces of component behavior were executed on a PA-BCA simulator. This abstraction of computation inside components as a high level trace allowed for as much as a 94% speedup over a CA simulation model.

To summarize, PA-BCA models abstract away cycle accuracy inside a component, while still maintaining cycle and pin accuracy for the bus, similar to the accuracy of the bus in the CA case. This raising of the abstraction level by relaxing the cycle accurate scheduling constraint compared to the CA models, allows for a speedup, not only in terms of modeling time but also as far as simulation speed is concerned. The drawback for this model is that the state of a component during simulation cannot be accurately determined at each cycle boundary (compared to a more detail CA/RTL model) which can make debugging the component harder. Whereas PA-BCA models abstract up the computation, in the next section we examine a model that abstracts up the computation as well as the communication, to further improve modeling time and increase simulation speedup.

4.2.3 Transaction-Based Bus Cycle Accurate Models

While the PA-BCA models raise the abstraction of the computation inside the SoC components, Transaction-based bus cycle accurate (T-BCA) models raise the abstraction of not only the computation, but also the communication between the components. Figure 4.7(c) shows an example of a T-BCA model. The granularity of the communication interface between components was pin accurate and cycle accurate for both the CA and PA-BCA models. In contrast, the communication interface abstraction is raised to the transaction level [25, 26] instead of the pin accurate level, while still maintaining cycle accuracy in the T-BCA model. The main idea in the T-BCA modeling abstraction is to avoid modeling all the pins at the communication interface of each component, as well as the signals that make up the buses, and instead use the notion of a *channel* and *transactions* on the channel. A *transaction* on a channel refers to a data read or write data communication event initiated by a master [25]. Buses, which are modeled as a collection of signals in the CA and PA-BCA models, are modeled as a *channel* in T-BCA. These channels present a simple *read()/write()* function interface to the components, that is called each time the component needs to read or write data to another component. In the CA and PA-BCA models, all the signals that connect a component to the bus must be modeled and made to transition correctly, not only for each of the components in the SoC design, but also for all the bus logic components such as the arbiters, decoders, and bridges. Since modern bus-based communication architecture standards such as AMBA [10], CoreConnect [23], and STBus [17] have hundreds of signals, modeling and verifying the correctness of a read or write transaction can cause hundreds of signals to trigger across several components in a CA or PA-BCA model. The development of these models thus becomes a cumbersome, time-consuming and error-prone task. The T-BCA modeling abstraction discards the notion of signals and instead makes use of faster *read()* and

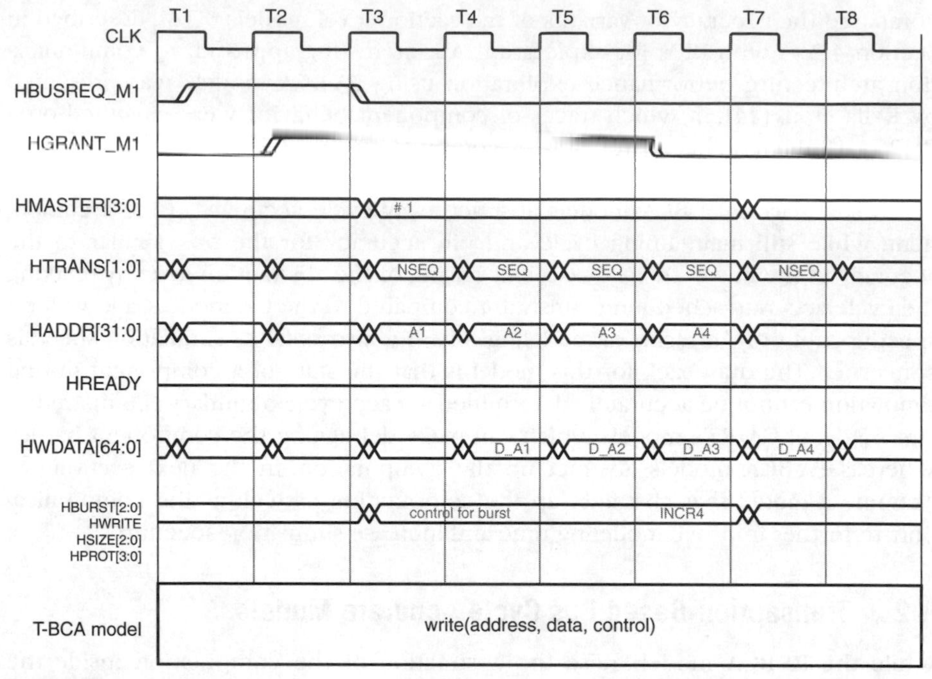

FIGURE 4.8

Comparison of CA/PA-BCA and T-BCA communication interface granularity

write( ) function calls with appropriate arguments to specify data, address, and control information.

Figure 4.8 shows a timing diagram of a write burst of length four in an SoC design using the AMBA AHB bus architecture. Only the most relevant signals have been shown for clarity. A master *M1* requests access to the bus in cycle 1 by asserting the *HBUSREQ_M1* signal. The arbiter receives the signal, and since there is no contention, grants the master access to the bus by asserting the *HGRANT_M1* signal in cycle 2. The master then proceeds to set the appropriate value for each of the control signals to indicate a write burst of four data items (which requires toggling the values of *HTRANS*, *HBURST*, *HWRITE*, *HSIZE*, etc.), starts sending the address and data in a pipelined fashion at every subsequent cycle till all data items have been sent. The signals cause the interface signals of the destination slave to toggle when it receives the request for a write data, and following which it needs to consider responding back to the master with an acknowledgement by possibly toggling the *HREADY* signal, driving it low if it requires the master to wait before sending the next data in the burst (in this case, however, this does not happen). As can be seen, a simple write transaction results in large amounts of activity in the CA and PA-BCA pin-accurate models.

A T-BCA model handles a write transaction differently, in a much simpler manner, by calling a function:

write(address, data, control)

Essentially, this function aggregates the behavior of all the signals shown in Fig. 4.8. The *address* field specifies the destination address where the *data* must be written, and *control* specifies control information, such as the burst size. Thus, instead of signals, T-BCA models make use of function semantics. The master, slave, and bus logic components all support this read/write function call interface, which not only reduces the modeling effort for T-BCA models considerably, but also speeds up simulation, since there is no need to toggle signal values at every cycle anymore. T-BCA models are up to 1000 times faster to simulate than RTL models, take approximately one-tenth the time and modeling effort and manage to maintain the state of the bus architecture accurately at a cycle accurate granularity. The trade-off in T-BCA models is the lack of visibility of signal transitions on the bus and component interfaces at every cycle, which makes the design somewhat harder to debug.

4.2.3.1 *Communication Architecture Exploration with T-BCA Models*

There have been a few approaches that have proposed using T-BCA models for communication architecture exploration. Caldari et al. [27] proposed T-BCA models for the exploration of the AMBA AHB and APB (advanced peripheral bus) architectures, using SystemC 2.0 [13–15]. The AHB and APB were captured using a program state machine (PSM) model [28], which is an instance of a heterogeneous model that integrates a hierarchical concurrent finite state machine (HCFSM) with a programming language paradigm (finite state machine (FSM) and other formal representations of communication architectures are discussed in more detail in Chapter 10). In addition to *read*() and *write*() transaction functions, certain signals such as *HREADY* and *HRESP* were also captured, in order to maintain cycle accuracy on the bus [27]. A comparison study between a PA-BCA model of the STBus and a T-BCA model of the AMBA AHB and APB buses showed a speedup of between 3× and 7× for the T-BCA model, for different traffic profiles on a small SoC testbench. A speedup of around 100× was also shown for a case study that compared their T-BCA model with a CA model for the AMBA AHB bus (both models being captured in SystemC 2.0). Another approach by Ogawa et al. [29] created a T-BCA model of the AMBA AHB bus architecture using *C* as the modeling language. The model made use of request and response signaling, in addition to using read/write transactions for communication. This T-BCA model was then used to perform design space exploration of an AMBA AHB based graphics display SoC subsystem, and enabled fast simulation performance, with a speedup of about 30× compared to CA models of the same system. A similar approach for modeling the AMBA AHB bus architecture was proposed by Kim et al. [30]. Another approach to creating T-BCA models is that used in the Synopsys Cycle Accurate SystemC models for AMBA AHB and APB [32]. These models capture signals as function calls, which enables simulation speedup while still maintaining bus cycle accuracy. These models were used by Jang et al. [31] to perform high level exploration on a network SoC design. A T-BCA model for the STBus communication architecture using SystemC 2.0 as the modeling language was also proposed by Boussctta et al. [33].

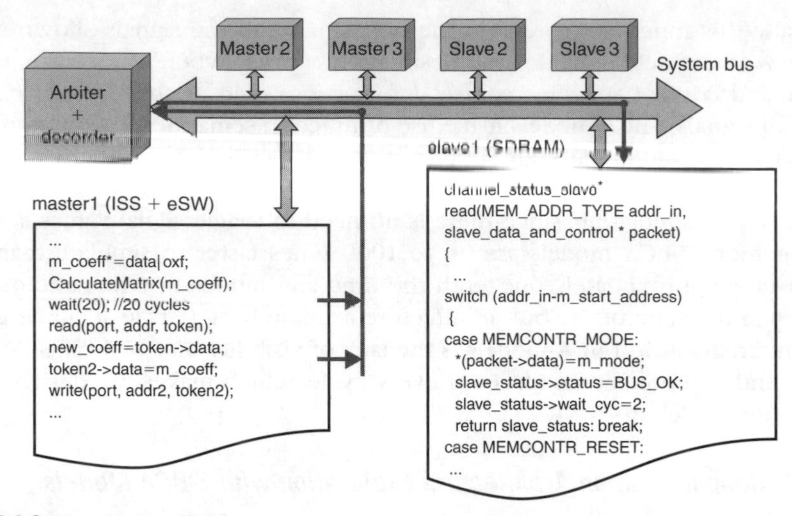

FIGURE 4.9

CCATB transaction example [35]
© 2004 IEEE

4.2.3.2 *The CCATB Approach*

An approach to further improve T-BCA modeling time and simulation speed was introduced by Pasricha et al. [34, 35]. The proposed modeling abstraction, called CCATB (cycle count accurate at transaction boundaries), is *cycle accurate* when viewed at *transaction boundaries*. A transaction in this context refers to a read or write (either a single data item or a burst of multiple data) issued by a master to a slave. Transactions in the CCATB model are similar to transactions at the TLM abstraction level [25, 26] (described in Section 4.2.4), except that additional bus protocol-specific control and timing information is passed along. Figure 4.9 shows an example of read and write transactions issued by a master (embedded software running on a processor ISS) to a slave (synchronous dynamic random access memory, SDRAM memory). The *port* parameter in the transaction specifies the port to send the read/write request on (since a master may be connected to multiple buses), *addr* is the address of the slave to send the transaction to and *token* is a structure that contains pointers to data and control information. Table 4.4 shows some of the main fields in this *token* data structure (which can be customized based on the bus protocol being used) passed by the master and received by the arbiter. The bus model in CCATB integrates arbiter and decoder modules, as well as bus protocol-specific timing details.

To speed up modeling and simulation time compared to traditional T-BCA approaches, the CCATB approach raises the abstraction to the transaction boundary level. This means that instead of maintaining cycle accuracy at every cycle boundary, the approach maintains cycle count accuracy at transaction boundaries that is, the number of cycles that elapse from the beginning till the end of a transaction is the same when compared to cycles elapsed in a detailed PA-BCA model. CCATB maintains overall cycle count accuracy, needed to gather statistics for accurate communication space exploration, while optimizing the models for

Table 4.4 Fields in token structure of a CCATB transaction [34]

Request field	Description
m_data	Pointer to an array of data
m_burst_length	Length of transaction burst
m_burst_type	Type of burst (incr, fixed, wrapping, etc.)
m_byte_enable	Byte enable strobe for unaligned transfers
m_read	Indicates whether transaction is read/write
m_lock	Lock bus during transaction
m_cachc	Cache/buffer hints
m_prot	Protection modes
m_transID	Transaction ID (needed for OO access)
m_busy_idle	Schedule of busy/idle cycles from master
m_ID	ID for identifying the master
status	Status of transaction (returned by slave)

faster modeling and simulation. The approach essentially trades off intra-transaction visibility to gain simulation and modeling speedup.

To illustrate how CCATB maintains cycle count accuracy at transaction boundaries, a comparison is presented between the CCATB model and a more detail PA-BCA model, on examples of transactions in an AMBA 2.0-based SoC design. A brief description of the AMBA 2.0 signals used in these examples is presented next, followed by details of the examples shown in Figures 4.10(a)-(c).

In AMBA 2.0, when a master needs to send or receive data, it requests the arbiter for access to the bus by raising the *HBUSREQx* signal. The arbiter, in turn, responds to the master via the *HGRANTx* signal. After invoking its arbitration scheme and selecting a winner, the arbiter drives the appropriate *HMASTERx* signals to indicate which master has gained access to the bus (this information is used by certain slaves). When a slave is ready to be accessed by a master, it drives the *HREADYx* signal high. Only when a master has received a bus grant from the arbiter via *HGRANTx* and detects a high *HREADYx* signal from the destination slave, will it initiate the transaction. The transaction consists of the master driving the *HTRANSx* signal, that describes the type of transaction (sequential or non-sequential), the *HADDRx* signals which are used to specify the slave addresses, and the *HWDATAx* signal if there is write data to be sent to the slave. Any data to be read from the slave appears on the *HRDATAx* signal lines. The master also drives control information about the data transaction on other signal lines—*HSIZEx* (size of the data item being sent), *HBURSTx* (number of data items being transferred in a burst transaction), *HWRITE* (whether the transfer is a read or a write), and *HPROTx* (containing protection information for slaves that might require it).

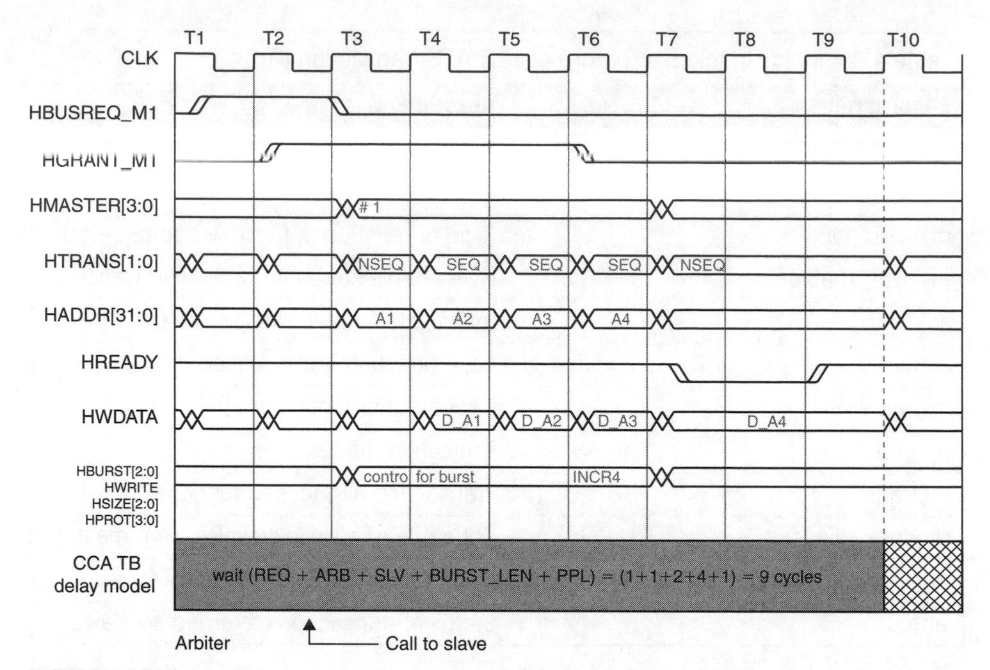

FIGURE 4.10(a)

Comparing CCATB with PA-BCA for AMBA 2.0: example transaction sequence 1

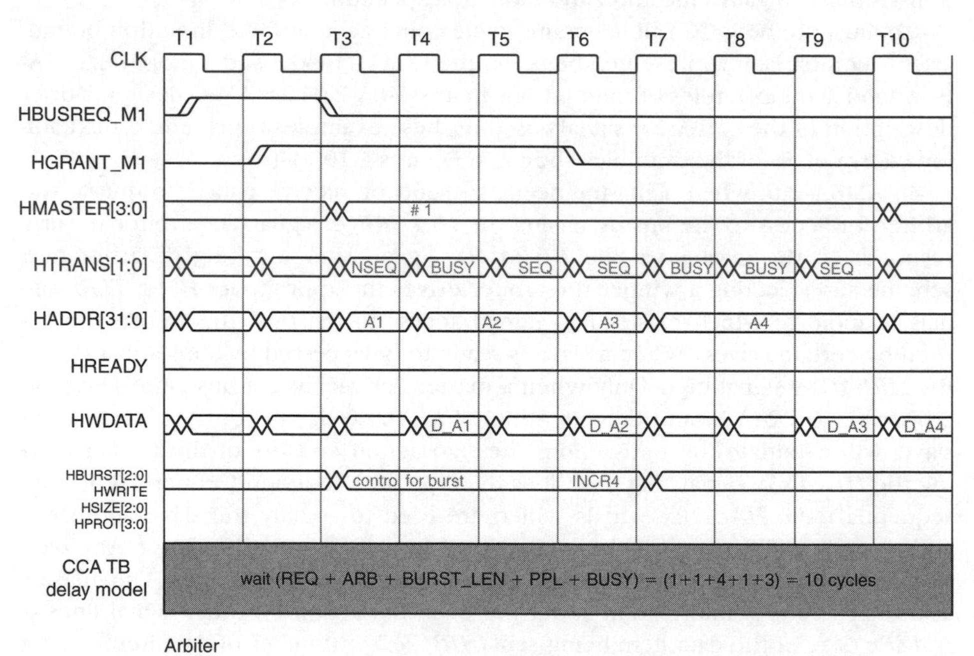

FIGURE 4.10(b)

Comparing CCATB with PA-BCA for AMBA 2.0: example transaction sequence 2

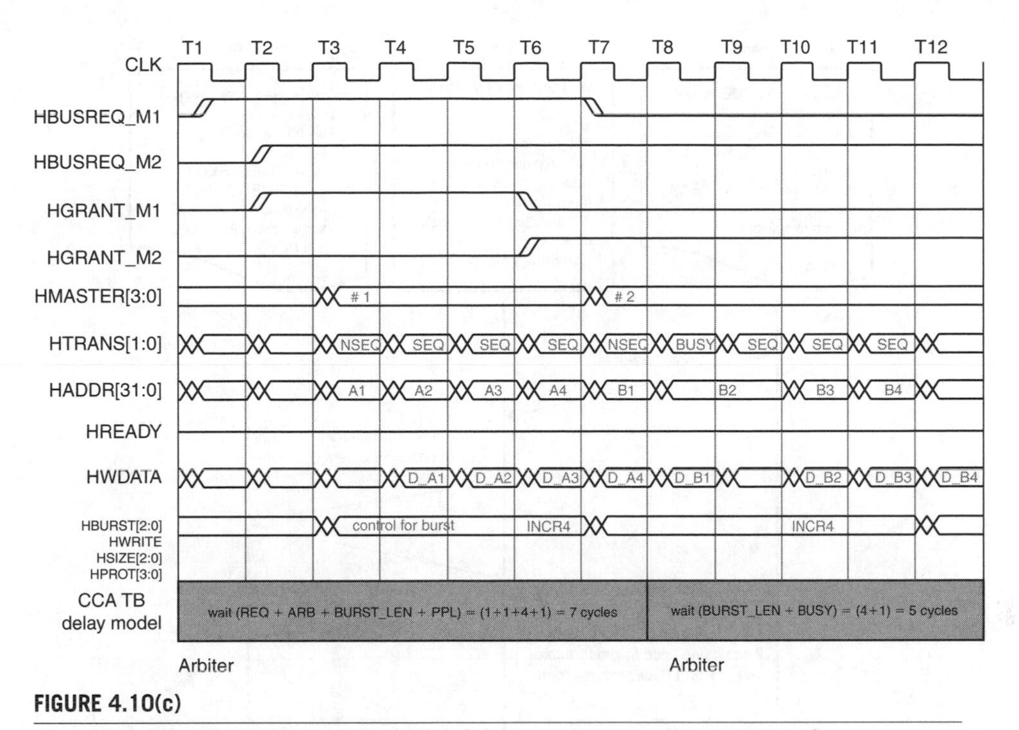

FIGURE 4.10(c)

Comparing CCATB with PA-BCA for AMBA 2.0: example transaction sequence 3

The examples shown in Fig. 4.10 are now described in more detail. In the first example in Fig. 4.10(a), a master requests an incremental write burst of four data packets and the arbiter immediately grants it access to the bus in the next cycle. The transaction is initiated and data sent to the slave, but before it can process the final data packet in the sequence, the slave needs to perform the computation with the previously written data, which consumes up to two cycles. For this duration, the slave drives the *HREADY* signal low to indicate to the master that it is not ready yet to receive the final data packet in the burst. The burst transaction resumes once the slave drives the *HREADY* signal high. The sequence of actions in the CCATB model is shown in Fig. 4.11. The arbiter accounts for the request (*REQ*) and arbitration (*ARB*) delays for the write request before invoking the slave to complete the transaction. The slave performs the write and returns a token structure that contains the status of the write and an indication to the arbiter that 2 wait states need to be inserted. The arbiter then increments simulation time with the slave delay (*SLV*), burst length (*BURST_LEN*), and pipeline startup (*PPL*) delays. The arbiter then returns the status of the writes at the end of the transaction to the master.

Figure 4.10(b) illustrates a similar scenario, but this time there is delay in generating the data at the master end instead of a processing delay at the slave end. After the write burst initiates, the master indicates that it requires extra cycles to generate write data for the slave by sending a *BUSY* status on the *HTRANS* [1:0] lines. In the CCATB model, the arbiter gets a schedule of busy cycles (*BUSY*) from

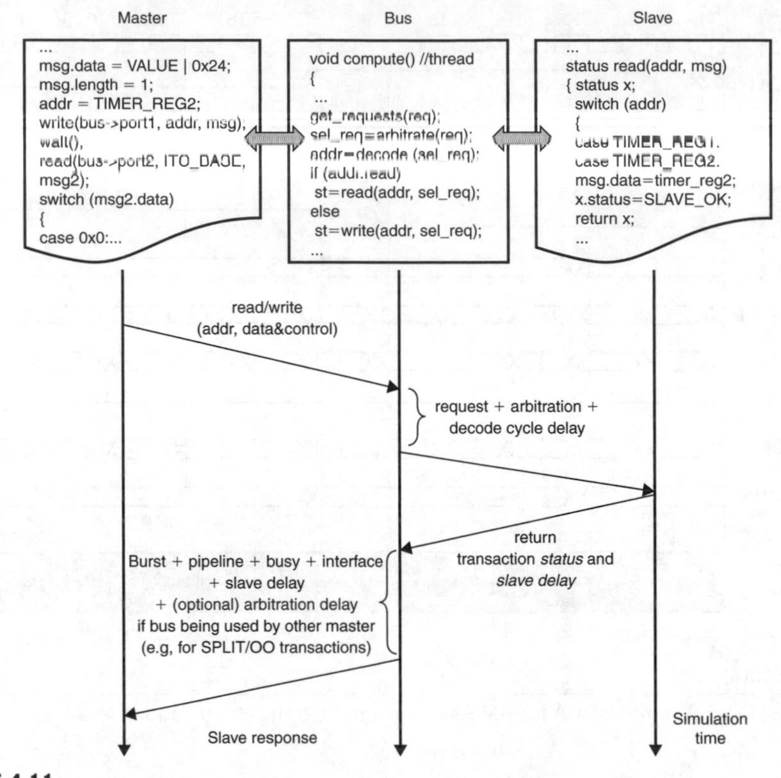

FIGURE 4.11

CCATB transaction execution sequence [34]
© 2004 IEEE

the master when it receives the transaction request, and thus it accounts for the *BUSY* cycle delay in the transaction, along with the other delays discussed above. There is no delay at the slave and consequently no increment in simulation time due to slave delay in this case.

In Fig. 4.10(c), after a master requests access to the bus for a write burst, another master requests the bus for a write burst. While there is no delay at the master or the slave end for the first write burst, there is delay in generating the data at the master end for master *M2*, which is indicated by the *BUSY* status on the *HTRANS* [1:0] lines. In the CCATB model, the arbiter accounts for the *REQ*, *ARB*, *BURST_LEN*, and *PPL* delays and increments simulation time. For the subsequent transaction by master *M2*, the request has already been registered at the arbiter and no arbitration is required, so there is no *REQ* or *ARB* delay. Since transfers are pipelined, there is also no pipeline startup delay like in the case of master *M1*. Thus there is no *PPL* delay. There is, however, delay that is dependent on the burst length (*BURST_LEN*) and the busy cycles (*BUSY*), which is accounted for by the arbiter. Like in the previous scenario, the slave does not delay either of the burst transactions, so there is no simulation time increment due to slave delay.

Since cycle accuracy is not maintained within a transaction in the CCATB model, the behavior of the bus logic components can be captured at a higher

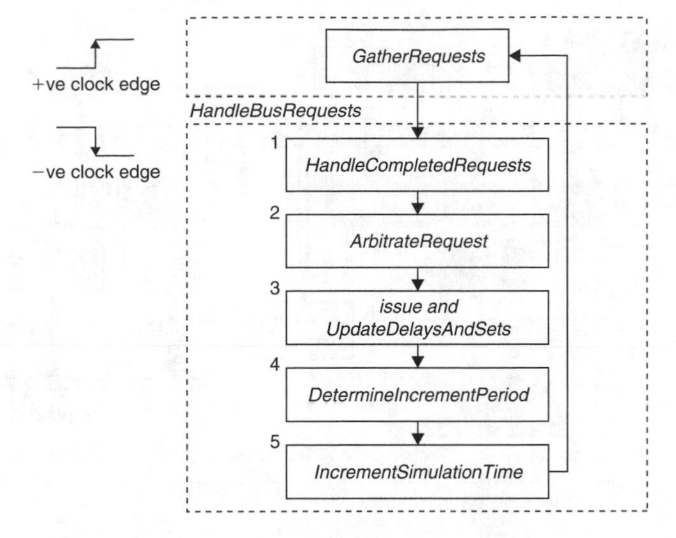

FIGURE 4.12

CCATB simulation flow

granularity and this allows a reduction in modeling time (as will be seen in the experimental results later). To explain how simulation speedup is obtained by raising the abstraction in CCATB, a brief description of the simulation model used by the approach is presented. Figure 4.12 shows a high level overview of the simulation flow in CCATB, enumerating the different procedures invoked during execution. There are two main phases in the model—the first is triggered on a positive edge of the system clock while the second is triggered on the negative edge of the system clock. In the first phase, on the positive edge of the clock, all the read and write requests in the system are gathered (*GatherRequests*). In the second phase, on the negative edge of the clock, these requests are processed by calling the *HandleBusRequests* procedure, which in turn calls various functions and subprocedures to perform different tasks. In the first step, inside the *HandleBusRequests* procedure, completed requests are handled and masters notified about any completed transactions (*HandleCompletedRequests*). In the second step, arbitration is performed on the buses in the system to select the requests that have the highest priority. The winning requests are subsequently granted access to their respective buses (*ArbitrateRequest*). In the third step, the selected read or write requests are issued to the slaves (*issue*). Additionally, responses from the slaves are received and request tokens updated with appropriate delay cycles to wait before notifying the master (*UpdateDelaysAndSets*). In the fourth step, the number of cycles to increment the simulation time are determined (*DetermineIncrementPeriod*) before finally incrementing the simulation time (*IncrementSimulationTime*) in the fifth step. This completes one iteration cycle of the simulation (which could represent one or more simulation cycles based on the amount by which simulation time is incremented) that is repeated till the simulation ends. Speedup is obtained in CCATB because delays are aggregated inside the bus module, which enables simulation time to be incremented in blocks of multiple cycles by

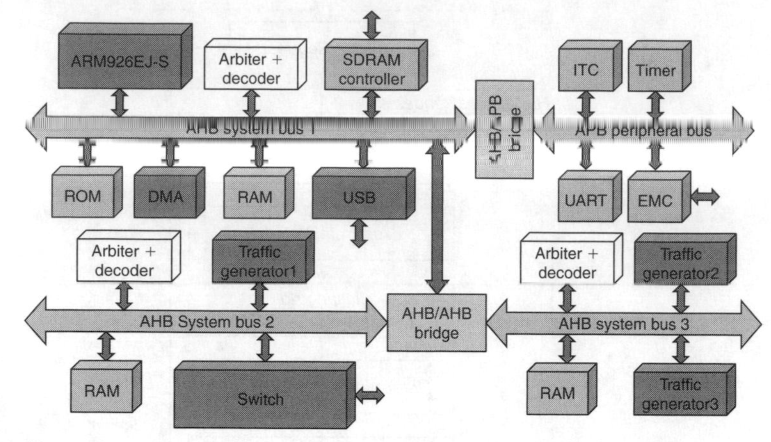

FIGURE 4.13

Example of SoC platform for comparison study [35]
© 2004 IEEE

IncrementSimulationTime, thus preventing unnecessary iterations of the simulation flow as well as unnecessary invocations of simulation components at every cycle, which can drastically speed up simulation. The interested reader is referred to [35] for a detailed treatment of the simulation model and details of each of the procedures shown in Fig. 4.12.

An experimental study was performed to compare the simulation speed of CCATB with the PA-BCA and T-BCA modeling approaches. A generic SoC platform modeled in SystemC and based around the AMBA 2.0 hierarchical shared bus communication architecture was selected, and is shown in Fig. 4.13. The SoC platform has an ARM926EJ-S processor ISS model (modeled in *C++* and encapsulated by a SystemC wrapper) with a test program running on it that initializes different components and then regulates data flow to and from the external interfaces such as USB, switch, external memory controller (EMC), and the SDRAM controller. The goal was to compare not only the simulation speeds but also to determine how the speed changed with system complexity. First the speedup was compared for a *lightweight* system comprising of just 2 traffic generator masters along with peripherals used by these masters, such as the RAM and the EMC. Gradually, the system complexity was increased by adding more masters and their slave peripherals. Figure 4.14 shows the simulation speed comparison with increasing design complexity.

Note the steep drop in simulation speed when the third master is added—this is due to the detailed non-native SystemC model of the ARM926EJ-S processor which considerably slowed down simulation. In contrast, the simulation speed was not affected as much when the direct memory access (DMA) controller was added as the fourth master. This was because the DMA controller transferred data in multiple word bursts which can be handled very efficiently by the transaction-based T-BCA and CCATB models. The CCATB particularly handles burst mode simulation very effectively and consequently has the least degradation in performance out of the three models. Subsequent steps added the USB switch and another traffic generator which put considerable communication traffic and computation load on the

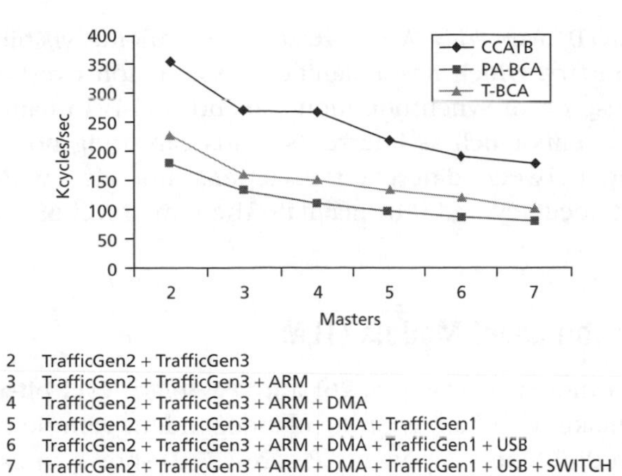

FIGURE 4.14

Simulation speed comparison [35]
© 2004 IEEE

Table 4.5 Comparison of speedup and modeling effort [35]

Model abstraction	Average CCATB speedup (x times)	Modeling effort
CCATB	1	~3 days
T-BCA	1.67	~4 days
PA-BCA	2.2	~1.5 weeks

© 2004 IEEE

system, resulting in a reduction in simulation speed. Overall, the CCATB abstraction level outperforms the other two models. Table 4.5 shows the average speedup of the CCATB over the PA-BCA and T-BCA models. On average, CCATB is faster than T-BCA by 67% and even faster than PA-BCA models by 120%. This speedup can be considerably higher if a native ARM926 processor (i.e., modeled in SystemC) is used.

Table 4.5 summarizes the speedup and also shows the time taken to model the communication architecture at the three different abstraction levels by a designer familiar with AMBA 2.0. While the time taken to capture the communication architecture and model the interfaces took just 3 days for the CCATB model, it took a day more for the transaction-based BCA, primarily due to the additional modeling effort to maintain accuracy at cycle boundaries for the bus system. It took almost 1.5 weeks to capture the PA-BCA model. Synchronizing and handling the numerous signals and design verification were the major contributors for the additional design effort in these models. CCATB models are thus faster to simulate and need less modeling effort compared to T-BCA and PA-BCA models.

As far as modeling accuracy is concerned, CCATB models are just as accurate as T-BCA and PA-BCA models, as demonstrated by the comparison with detailed waveforms in Fig. 4.10. It might appear that CCATB suffers from a loss of accuracy when compared to PA-BCA models, but that is not the case, since the speedup obtained by CCATB over PA-BCA is a result of sacrificing visibility of signals at every cycle boundary, which has a significant simulation overhead. Additionally, CCATB models use event synchronization semantics [14] to handle intra-transaction cycle level events (such as interrupts), thus capturing not only the system events occurring between different transactions, but also within transactions which can affect accuracy, and thus maintain the same level of accuracy as in PA-BCA models.

4.2.4 Transaction Level Models (TLM)

Transaction level models (TLM) [25, 26] are very high level bit-accurate models of systems that make use of a bus protocol-independent generic channel to transfer data between different components. Figure 4.7(d) shows an example of a TLM model. Note that the interface has no pins or signals, and is instead purely read/write transaction based. The timing for the computational behavior inside the master and slave components is also optional, and usually not specified. However, causality and functional correctness are still maintained, and synchronization events (e.g., wait statements in Fig. 4.7(d)) are still specified for this purpose. Whenever a master issues a read or a write transaction, it invokes a function call which instantly transfers the data item(s) to its destination. The read function call (the write function call is similar) is of the form:

read(address, data)

Note that only the destination address and data needs to be specified; control information need not be passed, since the communication channel is bus protocol independent (although information on data burst size and bit widths can be specified if the TLM model is more detailed). Such a scheme can be thought of as a P2P, zero-time interconnection between system components. Since there is no notion of timing or signals at the interface, or any bus protocol details (e.g., request, arbitration, and pipeline setup delays) inside the communication channel, TLM models are much faster to simulate than T-BCA models. Since communication protocol details are ignored in TLM models, it makes them unsuitable for communication architecture exploration. TLM models are typically used to create platforms for early embedded software development, as golden reference models to capture the application specification at an executable level to ensure functional correctness and for very high level, coarse grained system exploration [25, 26].

To enable communication architecture exploration at the TLM level, some approaches have been proposed that incorporate bus protocol structural and timing details at the TLM level. These approaches benefit from the fast simulation speed at the TLM level, but are not guaranteed to be completely accurate in estimating performance, due to the inherent inaccuracies in statically estimating and annotating delays corresponding to communication events in the high level TLM model.

An extension of the P2P TLM model described above is to add support for arbitration and shared buses, to capture contention during communication. This arbitrated-TLM (ATLM) has been used in a few exploration studies such as by Schirner and Domer [22] and Ariyamparambath et al. [39]. Because of the presence of the arbiter, a high level estimate of data traffic congestion can be obtained. This can be used to guide the allocation of SoC components onto appropriate buses. For instance, if an ATLM model is simulated and indicates excessive traffic conflicts on a shared bus with several components, then it might be useful to allocate some of the components to a separate bus, to improve performance. Conversely, if there is negligible traffic on a shared bus in a system, the bus can be removed and components on it can be migrated to another (existing) shared bus, to reduce wiring cost and improve bus utilization. While ATLM models can be used for such high level exploration, one needs to be aware of the fact that bus protocol-specific timing delays are not captured at this level. Such protocol-specific details can significantly alter traffic interactions on the buses, invalidating the results from the ATLM exploration.

To improve communication architecture performance estimation reliability, some approaches advocate annotating the ATLM models with bus-protocol-specific timing details. For instance, Ariyamparambath et al. [39] introduced the near cycle accurate (NCA) bus that has timing annotation to capture bus protocol-specific delays. Figure 4.15 shows how this timing annotation is applied to data communication in the NCA bus model. A master indicates a write transaction by calling the function *write(prio, addr,* T), where *prio* represents the masters priority for arbitration purposes, *addr* is the destination (slave) address to write the data to, and *T* is the data that needs to be transferred. The NCA abstract bus model automatically calculates the time delay t_D associated with the data transfer, and waits for time t_D before calling the slave interface and writing the data to it.

The NCA abstract bus model can be configured to capture delays, depending on the bus protocol being used. The delay t_D is calculated by annotating delay information for:

- *Internal cycles*: The bus protocol can require one or more cycles (e.g., bus request and grant cycles in AMBA 2.0) to complete the data transfer.
- *Pipeline cycles*: If data is transferred in a pipelined manner, then the number of cycles required to complete the transfer is different from a simple, non-pipelined case.
- *Burst mode*: Depending on the size of the data transfer and the burst length, it can take several cycles to complete the data transfer.

Such a timing annotation-based TLM model can be used for fast, bus-protocol specific, high level performance exploration, within reasonable bounds of accuracy

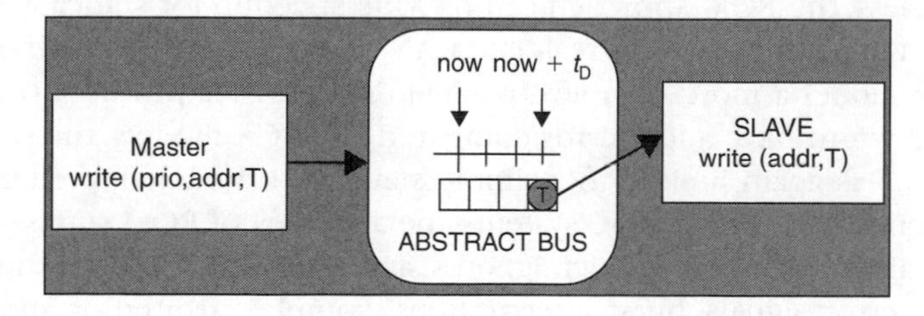

FIGURE 4.15

Bus transaction behavior for NCA bus [39]

for small SoC designs. Cycle accurate performance estimation becomes essential for more complex SoC designs, where small errors in timing can be propagated during simulation, and may create large discrepancies in the performance estimations over time.

Another timing annotated TLM model, was presented by Viaud et al. [40]. The proposed TLM/T (transaction level model with time) abstraction level uses ideas from parallel discrete event simulation (PDES) to model the parallel behavior of components in different manner compared to most other TLM approaches that use the SystemC library thread (*SC_THREAD*) constructs to model concurrent execution of components, and use *wait*() statements and the internal SystemC clock to model timing. In TLM/T, each component is also modeled as a thread, but has a local clock assigned to it. Communication occurs in the form of packets transferred on statically defined P2P channels, and the effect of arbitration is modeled by the global interconnect model, which includes all the P2P links interconnecting components. The local clocks of two threads are synchronized every time a packet is sent from one thread to the other. Viaud et al. [40] claim that the proposed scheme is more efficient than the traditional discrete event simulation kernel-based approach used in SystemC. Simulation speed is also improved because each (master) component has its own local clock, and there is no need for a global synchronization scheme at every system cycle. Experimental results on a test platform based on a generic OCP/VCI [41] communication architecture showed a speedup of 10× to 60× for the TLM/T model, when compared to a PA-BCA model, at a slight loss in accuracy of less than 1%. However, as discussed earlier, for more realistic SoC designs, the error in accuracy can be much larger, due to the limitation of the static timing annotation used in the communication model.

An approach that borrows the idea of trading-off intra-transaction visibility for simulation speedup from CCATB [34, 35] was proposed by Schirner et al. [36]. The modeling approach, called result oriented modeling (ROM) was shown to be highly accurate while obtaining a substantial simulation speedup for real-time serial communication architectures such as the CAN [37] bus architecture. The ROM approach initially predicts time taken to complete a transaction, and corrects this prediction if required at the end of the prediction period, to account for *disturbing influences* such as transactions from higher priority masters, which can lengthen transaction completion time. Due to the correction mechanism, the model complexity is higher than CCATB and other T-BCA based models. Nonetheless, the ROM approach can provide speedup for statically scheduled, predictable applications such as real-time CAN-based systems. While the approach was used to model a more complex, parallel data transfer protocol (AMBA AHB) [38], and shown to have a low error compared to T-BCA models, the scope of the experiments was again limited to a simple statically scheduled, predictable AHB-based testbench. For realistic SoC systems, the accuracy of ROM suffers due to the static prediction mechanism, which ignores slave delays, dynamic communication events (e.g., error signals, burst interruptions, complex arbitration mechanisms), interface delays (e.g., buffer overflows), as well as data dependencies inside computation blocks.

4.2.5 Multiple Abstraction Modeling Flows

All the modeling abstractions described previously have different strengths and weaknesses, stemming from the inherent trade-off between complexity of details captured, estimation accuracy, and simulation speed. While any one of these modeling abstractions can be used for communication architecture performance exploration in a traditional design flow, it is extremely useful to have a communication-centric exploration flow that spans and integrates several abstraction levels into an SoC design flow, to allow performance exploration with different levels of captured details, accuracy, and simulation speed. A few pieces of work have proposed such communication design space exploration flows that make use of multiple modeling abstractions. These approaches are described in more detail in the following section.

4.2.5.1 *Layered Abstraction Approaches*

Early work by Rowson and Sangiovanni-Vincentelli [42] illustrated the use of multiple abstraction levels for communication architecture exploration of a mixed HW/SW design communicating across a network using ATM packets. Figure 4.16 shows the three abstraction levels proposed for the purpose of exploring the communication space. At the highest level, even before HW/SW partitioning has been performed, is the abstract functional communication model, which is based on token passing. This model is used for functional validation of the system, but does not contain enough detail for communication architecture exploration. Once the first model is created, HW/SW partitioning is performed, and a transaction-based communication model is created, in which single and burst read and

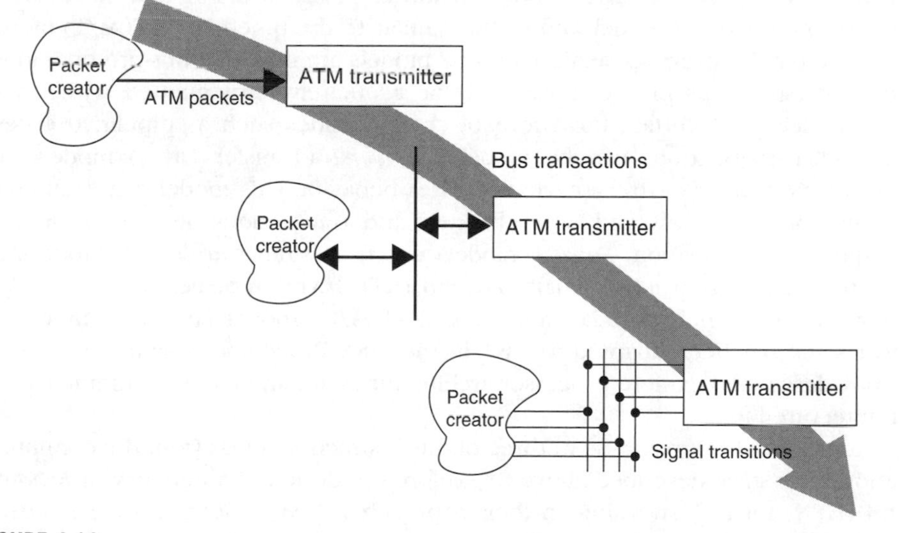

FIGURE 4.16

Communication abstraction levels proposed by Rowson and Sangiovanni-Vincentelli [42]
© 1997 IEEE

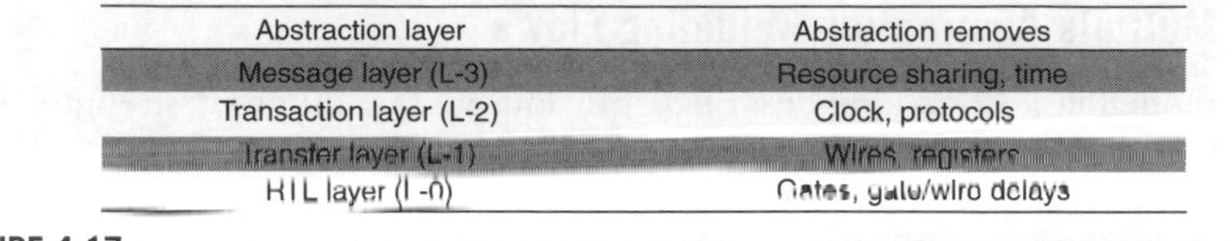

Abstraction layer	Abstraction removes
Message layer (L-3)	Resource sharing, time
Transaction layer (L-2)	Clock, protocols
Transfer layer (L-1)	Wires, registers
RTL layer (L-0)	Gates, gate/wire delays

FIGURE 4.17

Stack of communication abstraction layers

writes are modeled, along with arbitration. This corresponds to the ATLM model, as described in the previous section. Finally, this model is refined down to a PA-BCA model, after selecting an appropriate communication architecture, to perform detailed latency vs. throughput analysis for the SoC design.

The approach by Hines et al. [43] proposed using multiple levels of abstraction for communication architecture exploration, which the designer could switch between dynamically for greater exploration flexibility in terms of simulation speed and accuracy. The approach allows a designer to switch from a detailed PA-BCA model to less detailed TLM-like models (that directly copy data from master to slave, by passing the data values or simply a reference) to speed up exploration speed. A similar approach was proposed by Beltrame et al. [44], where dynamic switching between different communication abstraction levels (BCA, untimed TLM, timed TLM) was used to improve simulation speed for exploration.

Haverinen et al. [45] proposed a stack of communication abstraction layers, each having a different level of detail for modeling communication in a design flow. Figure 4.17 shows these abstraction layers, with the main features that each layer abstracts away. *Layer-3* (Message Layer) captures communication at an untimed, P2P abstract data transfer level, similar to the pure TLM level described in the previous section. *Layer-2* (Transaction Layer) adds timing and bit-accuracy details to the *Layer-3* model, and is thus similar to the timed TLM (TLM/T) model described in the previous section. *Layer-2* models are normally bus-protocol independent (since bus protocols can only be accurately implemented with cycle-true models), but certain bus protocol characteristics such as pipelined access and SPLIT transactions can be modeled. *Layer-1* (Transfer Layer) models are characterized by cycle-true (cycle accurate) behavior. This model can accurately capture bus protocols and uses channels and transactions, instead of signals, to speed up simulation. *Layer-1* models are thus similar to T-BCA models, as described in Section 4.2.3. Finally, *Layer-0* (RTL Layer) replaces the transaction interface with signals, to create a detailed PA-BCA/CA model. This communication layer stack has been adapted for use in the LISA Processor Design Platform, to jointly design and explore processor architecture with an on-chip communication architecture [46].

Kogel et al. [47] make use of three of the abstraction levels from the communication layer stack described above to explore the design of a network processing unit (NPU) for IP forwarding. In their approach, *Layer-3* (Message Layer) is used for function validation, *Layer-2* (Transaction Layer) is used for high level communication architecture selection and exploration, and *Layer-1* (Transfer Layer) is used for verifying the communication architecture selection decisions made with

Table 4.6 Communication modeling abstraction levels [48]

	Description	Features	Application area
TL0	Functional model	Functional description wrapped into a model. Delays for function and even communication are neglected. Fast to simulate, easy to deploy.	SW development
TL1	Parameterized transaction model	Parametric function delay (input-to-output delay). Strict communication protocol delay.	Top down design
TL2	Cycle accurate model	Internal structures are modeled strictly. Aiming for cycle and pin accuracy.	Architecture analysis

the *Layer-2* models. Another approach by Um et al. [48] also adapts three of the layers from the communication stack presented above, for the purposes of communication architecture exploration. Table 4.6 shows the abstraction levels used by the approach. It can be easily seen that *TL0* corresponds to *Layer-3* (Message Layer); *TL1* to *Layer-2* (Transaction Layer), and *TL2* to *Layer-0* (RTL Layer).

Pasricha et al. [34] presented another variant for integrating multiple communication-centric exploration abstractions into a design flow, as shown in Fig. 4.18. The flow starts at the specification model, which is a *C/C++* model that captures the algorithms and major functionality of the SoC application being designed. Once this model is completed, HW/SW partitioning is performed, to decide which portion of the functionality will be implemented in hardware, and which as software running on a processor. The bus protocol-independent TLM model consists of high level behavioral models of architectural components intended to be used in the SoC design. High level communication traffic estimation, functional validation, and embedded software development are some of the major activities that are performed at this level. The TLM model is refined down to the CCATB abstraction level [34, 35] which is used for fast and accurate communication architecture selection and exploration. The transaction interfaces at this level are refined to the pin-accurate level, to create the PA-BCA model, which is used for cycle and pin-accurate interface debugging and validation. Finally, the behavior inside the components is scheduled in a cycle accurate manner, and the CA model is created, which can be used for co-simulation with RTL or as an input to high level *C/C++* to RTL synthesis tools. This communication-centric exploration flow has been adapted for use in the work by Klingauf [49], which presents a design flow for embedded system design and exploration.

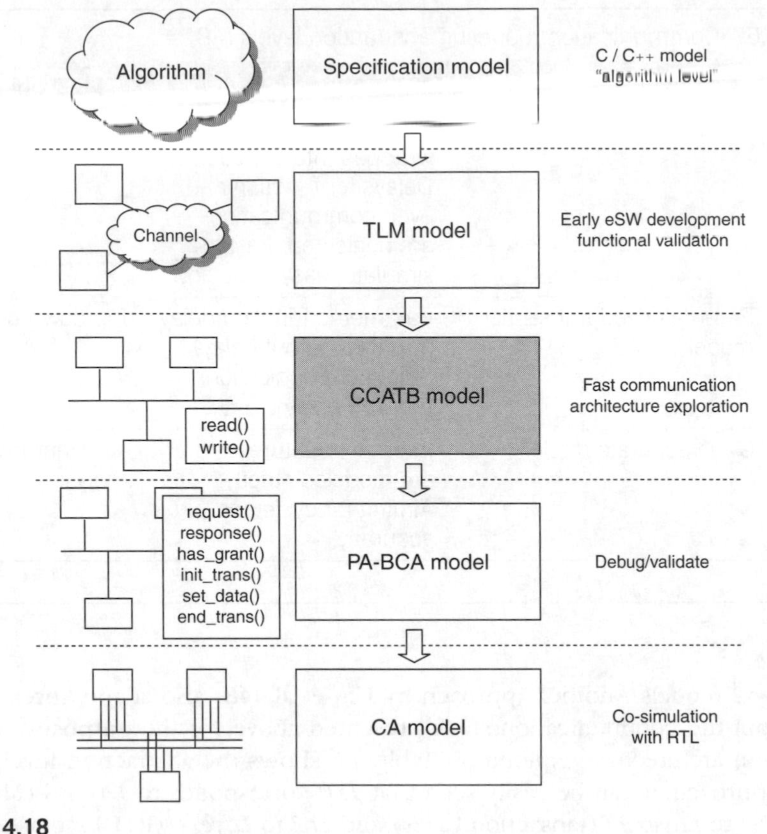

FIGURE 4.18

Communication-centric exploration flow [34]

© 2004 IEEE

4.3 HYBRID COMMUNICATION ARCHITECTURE PERFORMANCE ESTIMATION APPROACHES

As already mentioned earlier, static estimation techniques are unable to account for dynamic effects that can have a significant effect on performance. But the major advantage of using static techniques is that they enable fast performance estimation. Dynamic performance estimation approaches presented in the previous section provide accurate and reliable performance results, but even the fastest modeling abstractions can become time consuming because they require the application to be simulated. A third class of communication architecture performance estimation techniques attempts to combine the static estimation and dynamic simulation-based approaches, to speed up communication architecture performance estimation while generating accurate performance exploration results.

4.3.1 Trace-Based Approach

Lahiri et al. [50, 51] proposed a communication architecture exploration technique based on a hybrid trace-based performance analysis methodology that attempts to

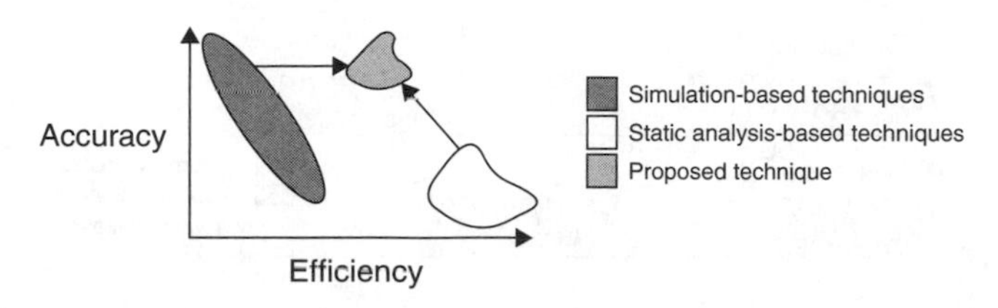

FIGURE 4.19

Accuracy and efficiency of hybrid technique proposed by Lahiri et al. [50, 51]
© 2000 IEEE

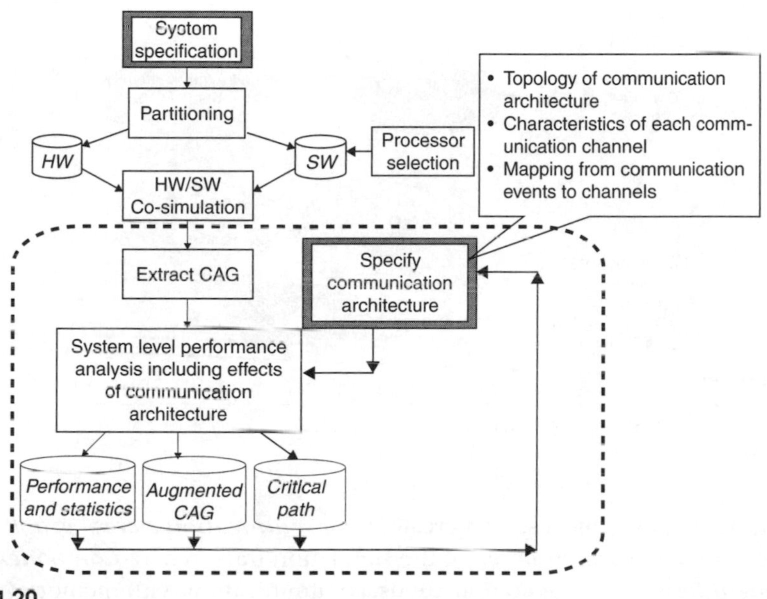

FIGURE 4.20

Hybrid performance estimation methodology [50, 51]
© 2000 IEEE

bridge the gap between slower dynamic simulation-based techniques, and faster, but less accurate static estimation techniques (Fig. 4.19).

A detailed flow of the proposed hybrid performance estimation methodology is shown in Fig. 4.20. The methodology is divided into two phases. The first phase (*dynamic phase*) is a pre-processing step, in which the entire system (consisting of both hardware and software components) is simulated, without considering the communication architecture. Communication in this phase is modeled as an abstract token passing mechanism, where the time taken to generate and consume a data token depends on the size of the token, and not on the number of concurrent communication flows. As such, the output of this phase is a timing inaccurate execution trace. The second phase (*static phase*), indicated by the dotted box in Fig. 4.20, performs performance analysis while considering the effect of the communication architecture. For this purpose, the computation, communication, and synchronization in the system are extracted from the simulation trace obtained

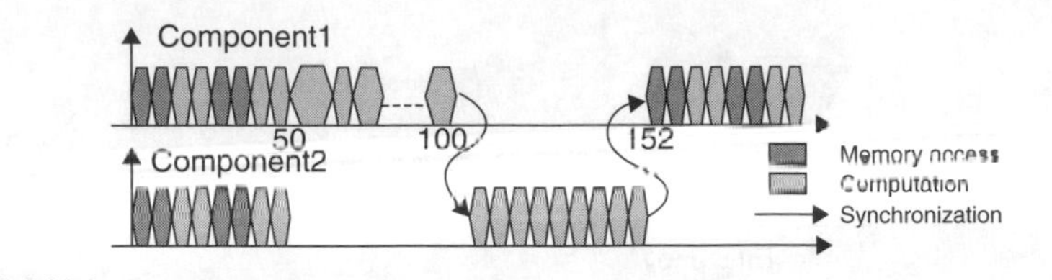

FIGURE 4.21(a)

Trace generated from simulation in first phase of methodology
© 2000 IEEE

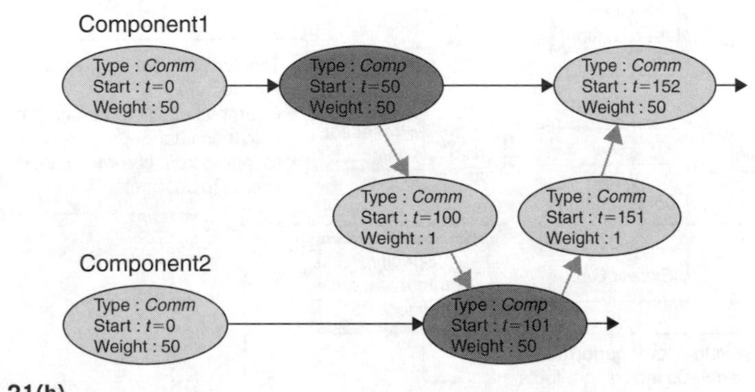

FIGURE 4.21(b)

CAG generated from simulation trace
© 2000 IEEE

from the first phase, and used to create a *communication access graph* (CAG). Figure 4.21(a) shows an example of the simulation trace where *Component1* and *Component2* each execute computations, communication with memory (via data transfers), and communication between each other (via synchronizing events). The exact values communicated or details of every computation are not captured—instead abstract vertices are created and collected into a CAG. Figure 4.21(b) shows an example of a CAG that consists of two types of vertices—computation and communication, and a set of directed edges representing timing dependencies. The CAG is acyclic because it is constructed from a simulation trace where all the dependencies have been unrolled in time. The graph is also compact, because each vertex represents a large number of simulation events.

The next step, after the creation of a CAG in the flow (Fig. 4.20) is to select a communication architecture and map the various communication events onto appropriate buses. The communication architecture consists of one or more types of buses, each of which is characterized by a set of parameters that specify details such as bus width, bus clock frequency, DMA burst size, and communication latency. The mapping of communication events to the buses of the chosen communication architecture is done manually. Once this is done, the communication architecture, the mapped communication events, and the CAG are sent as inputs to the performance analysis tool shown in Fig. 4.20. The tool augments

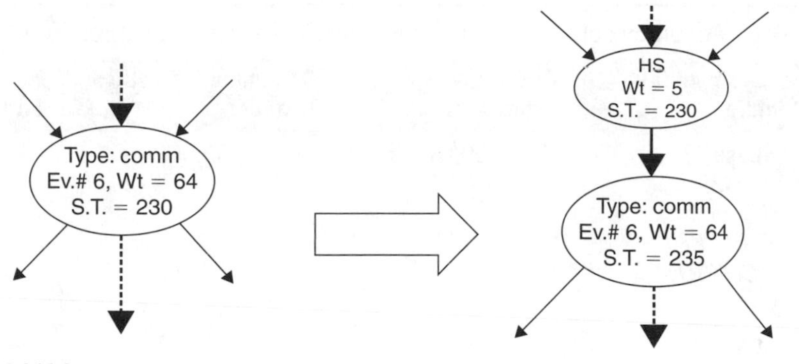

FIGURE 4.21(c)

Augmenting the abstract communication model in the CAG with bus protocol details [50, 51] *© 2000 IEEE*

the CAG (which uses an abstract, communication architecture independent communication model) by inserting new vertices that correspond to the latencies of the selected communication architecture protocol. Figure 4.21(c) shows a portion of a CAG where a single communication event (*event #6*) is augmented with communication architecture-specific detail. Since the communication *event #6* is mapped onto a shared bus, it becomes necessary to take into account the overheads due to the protocol, such as the overhead of handshaking. To do this, a new handshake (*HS*) vertex is introduced in the CAG. The starting time of this vertex (S.T. = 230) is the same as that of the original communication vertex. The weight of 5 cycles for the HS is derived from the specifications of the communication protocol, and the starting time of the original communication event is recalculated to 235, based on this weight. In a similar manner, other types of latencies due to contention and protocol overheads are introduced into the CAG by the tool. The outputs of the performance analysis tool are the following:

- An augmented version of the CAG that has all the latencies introduced as a result of moving from an abstract communication model to a well-defined communication architecture model.
- An estimate of the performance of the entire system (including statistics on bus usage, conflicts, etc.) that reflects the suitability of the communication architecture selection as well as the mapping of communication events to specific buses.
- Critical paths in the system, which can be a sequence of computation and communication.

Note that one underlying assumption of the approach by Lahiri et al. [50, 51] is that the communication architecture can affect the timing of the various computations and communications in the system, but cannot affect the system functionality. Several experiments were performed to illustrate the accuracy and simulation speedup of the proposed hybrid approach. Table 4.7 shows the estimation accuracy results from the exploration of three different communication

Table 4.7 Accuracy comparisons for alternative TCP-IP architectures [51]

TCP_IP Communication architecture	CAG analysis estimate (cycles)	Co-simulation estimate (cycles)	Error (%)
Config=**Case 1**, *Bus width*=32, *DMA_SIZE*=inf., *CREATE_PACKET>>IP_CHK>>CHKSUM*, *Latency*=1	69,013	69,508	−0.71
Config=**Case 2**, $bus_1=bus_2=bus_3$; *width*=32, *DMA_SIZE*=inf., *CREATE_PACKET>>IP_CHK>>CHKSUM*, *Latency*=1	35,079	36,075	−2.76
Config=**Case 2**, $bus_1=bus_2=bus_3$; *width*=16, *DMA_SIZE*=inf., *CREATE_PACKET>> IP_CHK>>CHKSUM*, *Latency*=1	67,645	66,636	1.51

architecture configurations, for a simple TCP-IP network interface card SoC subsystem. It can be seen that different communication architecture selection choices can have vastly different performance results, which demonstrates the need for such an exploration framework. The accuracy of the hybrid approach compares favorably to the results obtained from the detailed co-simulation of the entire system (including a model of the communication architecture), with a worst case error of under 3%. Table 4.8 compares the analysis time of the proposed hybrid approach to the time taken for analysis with a pure co-simulation-based approach. Three configurations of a simple four component SoC (*MEM4*) and one configuration of the TCP-IP network interface card SoC are considered. The first column reports the time taken for co-simulation in the first phase of the approach (Fig. 4.20) to generate the CAG from the simulation trace. The time is 0 for the second and third rows because the CAG has already been extracted for the *MEM4* system when exploring the first configuration, and does not need to be generated again for the configurations shown in rows 2 and 3. The second column shows the time taken to generate performance figures for a given CAG and communication architecture. The third column indicates the time taken for the co-simulation of the entire system, including the communication architecture (which is significantly more than the time taken for co-simulation with an abstract model of communication, as can be seen in the second column). The fourth column shows the speedup of the CAG-based estimation, compared to the co-simulation-based approach. A speedup of two orders of magnitude

Table 4.8 Simulation speed comparisons for alternative TCP-I architectures [51]

Case study	Co-simulation for CAG generation (seconds)	CAG-based analysis (seconds)	System co-simulation (seconds)	Speedup
MEM4 in Case 1 configuration	220	8.2	1325	162
MEM4 in Case 2 configuration	0	8.3	1895	228
MEM4 in Case 3 configuration	0	7.6	1863	245
TCP_IP in Case 1 configuration	84	3.0	688	229

is reported for each system explored. Note that since CAG generation is performed only once for a system, the exploration of several different communication architecture choices and configurations can be done much more rapidly using the proposed hybrid approach, than with a simulation-only approach. However, the accuracy of this approach can be affected if the complexity of the systems and communication architectures considered increases.

4.3.2 Queuing Theory-Based Approach

Another hybrid communication architecture exploration approach by Kim et al. [52, 53] proposed a static performance-estimation technique based on a queuing analysis as the first step in their design space exploration framework, to prune the design space before applying a simulation-based approach to the reduced design space. Figure 4.22 shows this hybrid exploration flow. It is assumed that the task schedule on each PE and memory access trace information for each PE are given. The memory traces are obtained using a cycle accurate ISS for each processor core and an RTL simulator for application-specific integrated circuit (ASIC) parts after the mapping is completed. The flow proceeds in an iterative manner, beginning with a single shared bus architecture, which becomes the only element in the set of architecture candidates. In subsequent iterations, additional architecture candidates are added to the set by selecting one of the PEs and moving it to another bus. Since communication between PEs is assumed to occur via shared memory blocks, there are different bus architecture configurations possible depending on which bus the shared memory blocks associated with the newly migrated PE go to. Queuing-based static analysis is performed on all these architecture candidates in the set. The PEs are the *customers* and a bus with its associated memory is the single *server* in the queuing analysis. The service request rate from each PE is extracted from the memory traces as a function of execution time. Bus contention is taken into account in the model, and a fast performance estimation made for the architecture candidates. Under the assumptions of the approach, the

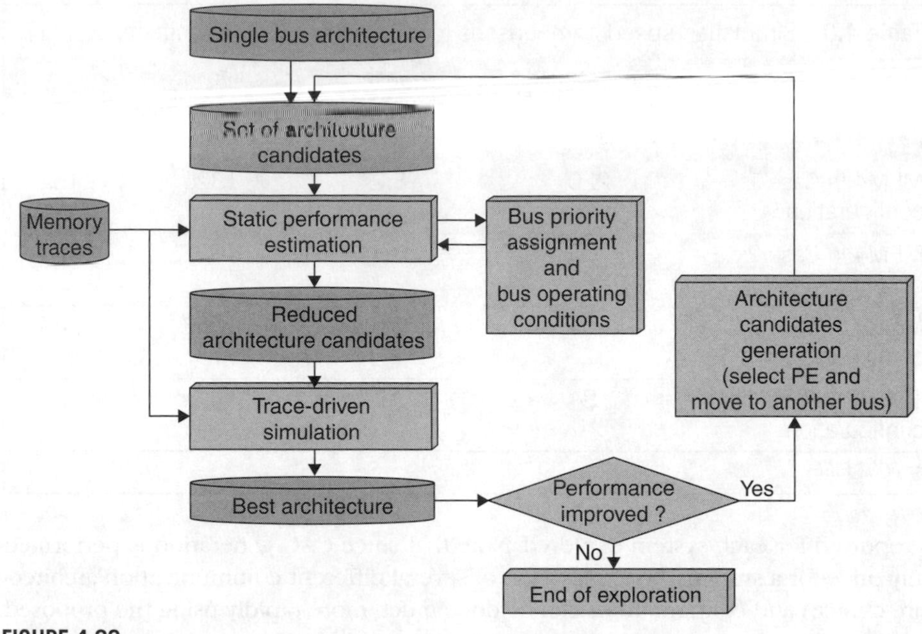

FIGURE 4.22

Communication architecture exploration flow [52, 53]
© 2003 IEEE

static estimation was shown to have an error of less than 10% compared to an accurate simulation-based approach [52]. As such, a selection of the top 10% of the candidates in terms of performance is made, and a reduced architecture candidate set is created (while the rest of the configurations are discarded). Trace-driven simulation is applied to this reduced architecture candidate set to generate accurate performance numbers for the candidate configurations. If the performance of the best configuration is not improved over the best configuration from a previous iteration, the exploration is terminated. Otherwise, more architectural candidates are generated by migrating another PE to a different bus. Despite the inability of the static queuing approach to handle complex bus protocol features (e.g., SPLIT transactions, OO transaction completion), the practical limitations of the assumptions made by the approach (e.g., schedule of PEs required to be known beforehand when there can be significant dynamic variation in realistic SoC designs), and concerns of scalability (a large number of candidates can considerably lengthen exploration time), this hybrid approach does manage to reduce the exploration time by using static estimation to prune the uninteresting communication architecture design space.

4.4 SUMMARY

In this chapter, we discussed the three major types of performance models used for communication architecture exploration. Static estimation models allow a fast estimation of communication architecture performance by assuming static delays for different communication events such as interface data packing/unpacking, HS synchronization, communication latency, pipelined access, burst data transfer, and arbitration. However, these static estimation approaches assume that computation and communication in an SoC design can be statically scheduled, which is not always true. Static approaches are also unable to predict dynamic component delays (e.g., memory access) as well as dynamic delays arising due to arbitration and traffic congestion, cache misses, burst interruptions, interface buffer overflows, and the effect of advanced bus features such as SPLIT transactions and OO transaction completion.

A more accurate (but slower) approach for performance estimation requires creating a model of the application that can be simulated. This allows a more accurate estimation of the dynamic data traffic behavior on the bus and the corresponding delays can be more reliably assessed. However, there are several ways in which simulation models can be created that differ based upon the level or kind of details that are captured (or omitted) by the model. We reviewed the different classes of dynamic performance estimation models that fall under four major categories in order of increasing simulation speed and decreasing accuracy: Cycle accurate (CA) models, PA-BCA models, T-BCA models, and TLM. We discussed different variants of these classes, and also looked at approaches that advocate using flows of multiple simulation models of varying accuracy to explore the communication architecture design space. Simulation-based performance exploration is absolutely essential to obtain reliable and accurate estimations. However, ever increasing levels of component integration in SoCs and the rising complexity of inter-component interactions means that simulation-based methods need to adapt to simulate only the necessary details required, to avoid a performance penalty. The development of faster simulation abstractions (e.g., transaction-based) is one solution to reducing exploration time that needs more research. Another interesting solution that we reviewed in this chapter is that of hybrid estimation techniques. These techniques combine the speed of static estimation with the accuracy of dynamic simulation to effectively traverse large exploration spaces, to generate reliable performance estimates quickly.

REFERENCES

[1] T.-Y. Yen and W, Wolf, "Communication synthesis for distributed embedded systems," ICCAD 1995, pp. 288–294.

[2] D.-T. Peng and K. G. Shin, "Static allocation of periodic tasks with precedence constraints," in *Proceedings, International Conference on Distributed Computing Systems*, 1989.

[3] S. Prakash and A. C. Parker, "SOS: Synthesis of application-specific heterogeneous multiprocessor systems," *Journal of Parallel and Distributed Computing*, 1992, vol 16, pp. 338–351.

[4] G. Gogniat, M. Auguin, L. Bianco and A. Pegatoquet, "Communication synthesis and HW/SW integration for embedded system design," CODES 1998, pp. 49–53.

[5] P. V. Knudsen and J. Madsen, "Communication estimation for hardware/software codesign," CODES 1998, pp. 55–59.

[6] P. V. Knudsen and J. Madsen, "Integrating communication protocol selection with partitioning in hardware/software codesign," ISSS 1998, pp. 111–116.

[7] P. V. Knudsen and J. Madsen, "Integrating communication protocol selection with hardware/software codesign," *IEEE Transactions on Computer-Aided Design*, Vol. 18, No. 8, August 1999, pp. 1077–1095.

[8] F.-M. Renner, J. Becker and M. Glesner, "Communication performance models for architecture-precise prototyping of real-time embedded systems," RSP 1999, pp. 108–113.

[9] F.-M. Renner, J. Becker and M. Glesner, "Automated communication synthesis for architecture-precise rapid prototyping of real-time embedded systems," RSP 2000, pp. 154–159.

[10] ARM AMBA Specification and Multilayer AHB Specification (rev2.0), http://www.arm.com, 2001.

[11] Y.-S. Cho, E.-J. Choi and K.-R. Cho, "Modeling and analysis of the system bus latency on the SoC platform," SLIP 2006, pp. 67–74.

[12] ARM AMBA AXI Specification, www.arm.com/armtech/AXI.

[13] SystemC initiative, http:www.systemc.org.

[14] T. Grötker, S. Liao, G. Martin and S. Swan, *System Design with SystemC*, Kluwer Academic Publishers, 2002.

[15] W. Müller, J. Ruf and W. Rosenstiel, *SystemC Methodologies and Applications*, Kluwer Academic Publishers, Norwell, MA, 2003.

[16] M. Loghi, F. Angiolini, D. Bertozzi, L. Benini and R. Zafalon, "Analyzing on-chip communication in a MPSoC environment," DATE 2004, pp. 752–757.

[17] "STBus Communication System: Concepts and Definitions," *Reference Guide*, STMicroelectronics, May 2003.

[18] Catapult C Synthesis, Mentor Graphics, http://www.mentor.com/products/c-based_design/catapult_c_synthesis/.

[19] Agility Compiler, Celoxica, http://www.celoxica.com/products/agility/default.asp.

[20] Cynthesizer, Forte Design Systems, http://www.fortedesignsystems.com/products/cynthesizer.asp.

[21] L. Séméria and A. Ghosh, "Methodology for hardware/software co-verification in C/C++," in *Proceedings of ASP-DAC*, 2000, pp. 405–408.

[22] G. Schirner and R. Domer, "Quantitative analysis of transaction level models for the AMBA Bus," DATE 2006, pp. 1–6.

[23] IBM CoreConnect Specification, http://www.ibm.com/chips/techlib/techlib.nsf/productfamilies/CoreConnect_Bus_Architecture.

[24] P. Kalla, X. S. Hu and J. Henkel, "A flexible framework for communication evaluation in SoC design," *Asia and South Pacific Design Automation Conference*, 2005, pp. 956–959.

[25] S. Pasricha, "Transaction level modeling of SoC with SystemC 2.0," Synopsys User Group Conference (SNUG), 2002.

[26] L. Cai and D. Gajski, "Transaction level modeling: An overview," *Proceedings of International Conference on Hardware/Software Codesign and System Synthesis (CODES+ISSS 2003)*, IEEE Press, 2003, pp. 19–24.

[27] M. Caldari, M. Conti, M. Coppola, S. Curaba, L. Pieralisi and C. Turchetti, "Transaction-level models for AMBA bus architecture using SystemC 2.0," DATE 2003, pp. 26–31.

[28] D. D. Gajski, J. Zhu, and R. Domer, "Essential issues in co-design," Irvine Technical Report, University of California, June 1997, http://www.ics.uci.edu.

[29] O. Ogawa, S. Bayon de Noyer, P. Chauvet, K. Shinohara, Y. Watanabe, H. Niizuma, T. Sasaki and Y. Takai, "A practical approach for bus architecture optimization at transaction level," DATE 2003, pp. 176–181.

[30] Y.-T. Kim, T. Kim, Y. Kim, C. Shin, E.-Y. Chung, K.-M. Choi, J.-T. Kong and S.-K. Eo, "Fast and accurate transaction level modeling of an extended AMBA2.0 bus architecture," DATE 2005, pp. 138–139.

[31] H.-O. Jang, M. Kang, M.-j. Lee, K. Chae, K. Lee and K. Shim, "High-level system modeling and architecture exploration with SystemC on a network SoC: S3C2510 Case Study," DATE 2004, pp. 538–543.

[32] Synopsys System Studio AMBA SystemC Library, http://www.synopsys.com/products/cocentric_studio/system_studio_sysclib.html.

[33] H. Boussctta, M. Abid, F. Layouni and C. Pistrito, "STBus transaction level models using SystemC 2.0," in *Proceedings of the 16th International Conference on Microelectronics, 2004*, ICM 2004, pp. 347–350.

[34] S. Pasricha, N. Dutt and M. Ben-Romdhane, "Extending the transaction level modeling approach for fast communication architecture exploration," *Design and Automation Conference (DAC 2004)*, San Diego, CA, June 2004, pp. 113–118.

[35] S. Pasricha, N. Dutt and M. Ben-Romdhane, "Fast exploration of bus-based on-chip communication architectures," *International Conference on Hardware/Software Codesign and System Synthesis (CODES+ISSS 2004)*, Stockholm, Sweden, September 2004, pp. 242–247.

[36] G. Schirner and R. Domer, "Accurate yet fast modeling of real-time communication," *Proceedings of CODES+ISSS*, 2006.

[37] Robert Bosch GmbH. CAN Specification, 2.0 edition, 1991, http://www.can.bosch.com.

[38] G. Schirner and R. Domer, "Fast and accurate transaction level models using result oriented modeling," *Proceedings of International Conference on Computer-Aided Design*, 2006.

[39] M. Ariyamparambath, D. Bussaglia, B. Reinkemeier, T. Kogel and T. Kempf, "A highly efficient modeling style for heterogeneous bus architectures," *Proceedings of International Symposium on System-on-Chip*, 2003, pp. 83–87.

[40] E. Viaud, F. Pecheux and A. Greiner, "An efficient TLM/T modeling and simulation environment based on conservative parallel discrete event principles," *Proceedings of Design, Automation and Test in Europe (DATE)*, 2006, pp. 1–6.

[41] Virtual Component Interface Standard. http://www.vsi.org.

[42] J. A. Rowson and A. Sangiovanni-Vincentelli, "Interface-based design," in *Proceedings of the Design Automation Conference (DAC)*, 1997, pp. 178–183.

[43] K. Hines and G. Borriello, "Dynamic communication models in embedded system co-simulation," *Proceedings of the Design Automation Conference (DAC)*, 1997, pp. 395–400.

[44] G. Beltrame, D. Sciuto, C. Silvano, D. Lyonnard and C. Pilkington, "Exploiting TLM and object introspection for system-level simulation," DATE 2006, pp. 1–6.

[45] A. Haverinen, M. Leclercq, N. Weyrich and D. Wingard. White Paper for SystemC Based SoC Communication Modeling for the OCP Protocol, http://www.ocpip.org/data/systemc.pdf, 2003.

[46] A. Wieferink, T. Kogel, R. Leupers, G. Ascheid, H. Meyr, G. Braun and A. Nohl "A system level processor/communication co-exploration methodology for multi-processor system-on-chip platforms," DATE 2004, pp. 1256–1261.

[47] T. Kogel, M. Doerper, A. Wieferink, R. Leupers, G. Ascheid, H. Meyr and S. Goossens, "A modular simulation framework for architectural exploration of on-chip interconnection networks," CODES+ISSS 2003, pp. 7–12.

[48] J. Um, W.-C. Kwon, S. Hong, Y.-T. Kim, K.-M. Choi, J.-T. Kong, S.-K. Eo and T. Kim, "A systematic IP and bus subsystem modeling for platform-based system design," DATE 2006, pp. 1–5.

[49] W. Klingauf, "Systematic transaction level modeling of embedded systems with SystemC," *Proceedings of Design, Automation and Test in Europe (DATE 2005)*, 2005, pp. 566–567

[50] K. Lahiri, A. Raghunathan and S. Dey, "Performance analysis of systems with multi-channel communication architectures," *VLSI Design*, 2000, pp. 530–537.

[51] K. Lahiri, A. Raghunathan and S. Dey, "System-level performance analysis for designing on-chip communication architectures," *IEEE Transactions on Computer-Aided Design of Integrated Circuits and Systems (TCAD)*, Vol. 20, No. 6, June 2001, pp. 768–783.

[52] S. Kim, C. Im, and S. Ha, "Schedule-aware performance estimation of communication architecture for efficient design space exploration," CODES+ISSS 2003, pp. 195–200.

[53] S. Kim, C. Im and S. Ha, "Schedule-aware performance estimation of communication architecture for efficient design space exploration," *IEEE Transactions on Very Large Scale Integration (VLSI) Systems*, Vol. 13, No. 5, May 2005, pp. 539–552..

CHAPTER

Models for Power and Thermal Estimation

5

In the previous chapter, we examined models for the early performance estimation of on-chip communication architectures. In addition to performance, power is another important architectural design constraint, that is beginning to dominate multiprocessor SoC (MPSoC) design, particularly for mobile applications such as cellular phones, MP3 players, and laptops. These portable devices run on batteries that have a limited energy budget between charges. MPSoC designs intended for use in such portable scenarios must have lower power consumption, to improve user experience. Reducing power consumption is also a priority for non-portable MPSoC applications, such as those used in server farms that tend to consume significant amounts of power (e.g., as much as 2 megawatts for a 25,000 square foot server farm with 8000 servers [1]). According to SIA [79] projections of future silicon technologies, the operating frequency and transistor density of MPSoCs will continue to increase, making power dissipation for these highly integrated and complex designs a major concern.

Excessive power dissipation has an undesirable thermal side effect of increasing device temperature that can adversely affect the *reliability* of MPSoCs. Reliability is the probability that the system will operate correctly at any given time of operation, and is measured by *mean time to failure* (MTTF). It is an important design criterion to extend MTTF beyond the expected useful life of a product, especially for critical systems such as those used in aircraft control, automotive control, medical equipment, and defense applications. Temperature cycles and spikes due to excessive power dissipation can induce mechanical stress and dielectric breakdown that can cause device failure. Commonly used techniques to reduce power dissipation such as voltage scaling, although beneficial, can also lead to issues with signal integrity (making it harder to guarantee error-free data transfers). As technology scales into the deep submicron (DSM) region, reduced voltage supply levels, increased leakage power, external electromagnetic interference (EMI), crosstalk, and soft errors will further reduce signal integrity. Voltage scaling will thus not be sufficient to mitigate the power dissipation problem. Expensive cooling and packaging equipment will be required to keep MPSoCs functioning at a reasonable temperature. Techniques to estimate and reduce power consumption will therefore become essential for lowering operating temperatures, so that the MTTF is increased and cooling and packaging costs are reduced.

Thus, with scaling trends in emerging technologies, and the increasing proliferation of the Internet and mobile computing, the power problem has assumed a critical status that cannot be ignored by MPSoC designers. On-chip communication architectures have a considerable impact on MPSoC power consumption. There are several reasons why the interconnect fabric is receiving so much attention with respect to power consumption [1, 2]. First, unlike transistors, interconnects have not scaled exponentially in DSM technologies, as a result of which interconnect capacitance forms a larger portion of total chip capacitance [3]. Second, the problem of modeling DSM effects could be largely ignored in pre-DSM technologies, where transistors were the main focus due to their large sizes. However, in DSM technologies, effects such as coupling capacitance between adjacent wires become increasingly dominant [4, 5]. Third, interconnects in today's designs are proportionally longer, which implies that interconnect delay has increased. Fourth, the use of a large number of repeaters and vias to reduce wire delay almost doubles power consumption in interconnects [6]. Finally, state of the art communication architectures consist not only of bus wires, but also significant amounts of hardware logic (e.g., bridges, arbiters, decoders, buffers, etc.) that is comparable to the amount of logic in embedded processors of moderate complexity [7]. It has been predicted that communication architectures will consume a larger portion of on-chip power in future technologies [8]. There is, therefore, a need to create models for estimating power consumption of on-chip communication architectures as early as possible in a design flow, to better design and optimize MPSoCs. Additionally, the thermal effects of power dissipation cannot be ignored, since they are beginning to have a significant impact on the power, performance, design, and reliability of on-chip buses [9].

In this chapter, we present models for on-chip communication architecture power and thermal estimation. To address the problem of excessive power dissipation, designers need such models to evaluate the impact of design decisions on chip power dissipation. Figure 5.1 shows the positioning of communication architecture power and thermal estimation models in a typical electronic system level (ESL) design flow. These models inevitably require extrapolating information from lower levels in the design flow up to the system level. Such estimation models allow designers to make optimizations to possibly reduce chip power dissipation early in the design flow, where design decisions have a greater impact. In Section 5.1, we present models for power estimation of bus wires, including DSM-aware bus wire power models that take into account crosstalk and coupling capacitance between adjacent wires. Section 5.2 elaborates on approaches that attempt to estimate the power consumption of the entire bus-based on-chip communication architecture, including that of bus logic components. Section 5.3 presents models for thermal evaluation of bus wires. Finally, Section 5.4 presents a discussion of PVT (process, voltage, temperature) variation-aware power estimation for on-chip communication architectures in ultra DSM (UDSM) technologies.

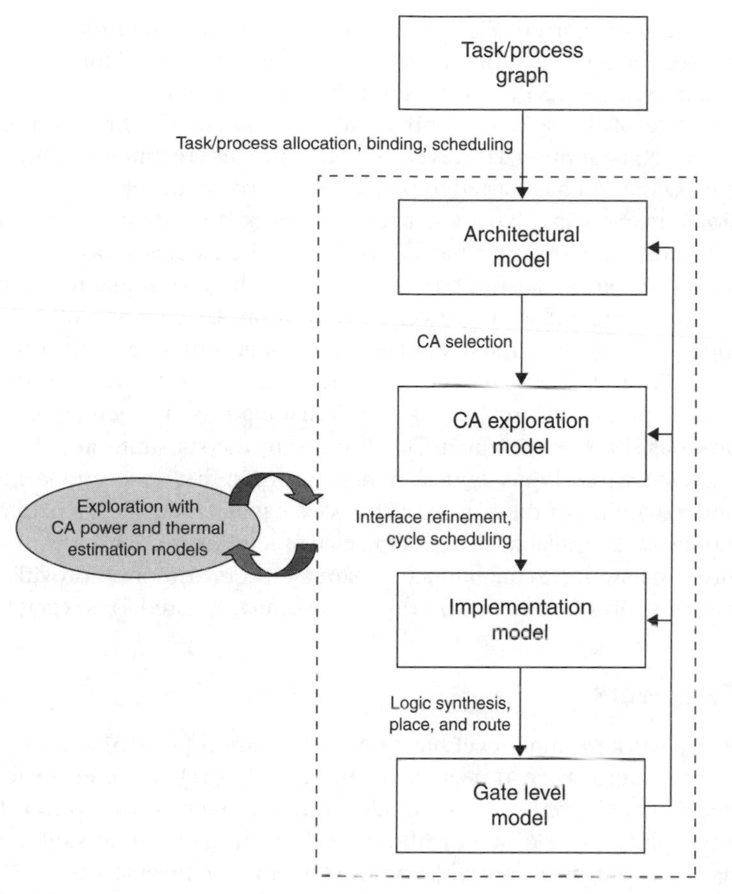

FIGURE 5.1

Communication architecture (CA) power and thermal estimation in a typical ESL design flow

5.1 BUS WIRE POWER MODELS

Complementary metal-oxide semiconductor (CMOS) is the dominant technology for designing system-on-chips (SoCs). Power consumption in CMOS logic circuits can be expressed by the following general equation [1]:

$$P = ACV^2 f + \tau AVI_{\text{short}} f + VI_{\text{leak}} \tag{5.1}$$

The equation has three components. The first term in the equation represents the *dynamic* power consumption due to the charging and discharging of the capacitive load on each logic gate's output. The dynamic power consumption is proportional to the frequency of the system's operation f, the activity of the gates of the system A, the square of the supply voltage V, and the total capacitance seen by the gate's output C. The second term in the equation represents power consumption due to *short circuit* current I_{short} that flows for a short instant of time T between the supply voltage and the ground when the output of a CMOS logic gate switches. The third term represents power consumption due to *leakage* current I_{leak} irrespective of the state of the system. In pre-DSM technologies, the first term dominated overall power consumption. However, as CMOS process technology shrinks toward DSM, the leakage term has started to play a more prominent role.

Like logic gates, wires have a capacitance associated with them, representing charge that must be added or removed to change the electrical potential on a wire. Whenever a data bit is transmitted on a bus wire, the charging and discharging of this wire capacitance results in power consumption. An accurate modeling of wire capacitance is, however, a non-trivial task and still an ongoing subject of advanced research [3]. Complications in estimation arise because the structure of a wire in contemporary integrated circuits (ICs) is 3-D, making the capacitance of such a wire a function of its shape, environment, its distance to the substrate, and its distance to the surrounding wires. Typically, designers use simple first order models to provide a basic understanding of the nature of the wire capacitance and its parameters. If a completed layout is available, designers use advanced extraction tools to obtain values of wire capacitance. Semiconductor manufacturers also often provide empirical data for various contributors to wire capacitance, as measured for several test dies.

5.1.1 Early Work

A lot of early work on high level bus power estimation [10–16] and bus encoding to reduce communication power consumption [17–21] (covered in more detail in Chapter 7) considered a simple model of bus power consumption that, as Eq. (5.2) shows, primarily focused on the contribution of dynamic switching power on a wire. The bus power consumption model can be expressed as:

$$P_{AVG} = \frac{1}{2} C_{bus} V_{dd}^2 n_{trans} f \tag{5.2}$$

where C_{bus} is the total bus capacitance (cumulative capacitance of all the wires in the bus), V_{dd} is the power supply voltage, n_{trans} is the average number of transitions on the bus, and f is the operating frequency of the system bus. Let B^t be the value of the bit string on the bus at time t, and let L be the total length of the data stream transmitted on the bus. Then the average number of transitions on the bus can be expressed as:

$$n_{trans} = \frac{\sum_{t=0}^{L-1} HD(B^t, B^{t+1})}{L-1} \tag{5.3}$$

where $HD(B^t, B^{t+1})$ is the Hamming distance between the words on the system bus times at times t and $t+1$. For example, for a 16 bit wide bus, if the word on the bus at time t is 1000100011001100, and the word on the bus at time $t+1$ is 1001100110001000, then the Hamming distance is the number of bit-flips between the two words, which is four.

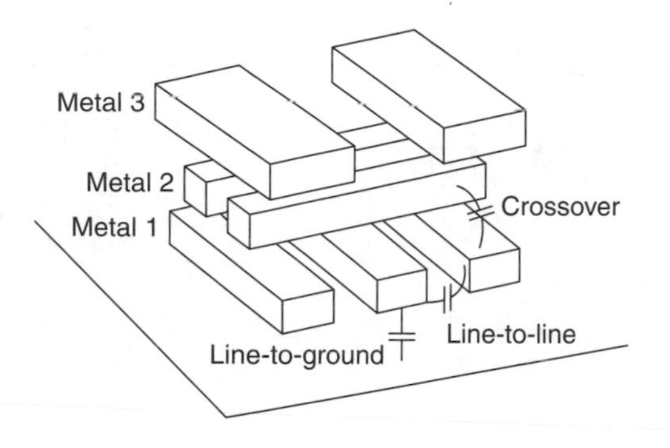

FIGURE 5.2

Three primary wire capacitance components: line-to-line, line-to-ground, and crossover capacitances [22]
© 1992 IEEE

An early model for wire capacitance was proposed by Chern et al. [22]. Figure 5.2 shows the three primary constituents of wire capacitance – line-to-line capacitance, line-to-ground capacitance, and crossover capacitance. The crossover capacitance is the capacitance between wires in different metal layers (since typical SoC designs are fabricated on multiple metal layers). The simplest of three constituents, the line-to-ground capacitance, was calculated as [22]:

$$\frac{C}{\epsilon} = \frac{W}{H} + 3.28\left(\frac{T}{T+2H}\right)^{0.023} + \left(\frac{S}{S+2H}\right)^{1.16} \tag{5.4}$$

where W is the metal width, S is the space between two lines (or wires), T is the thickness of the metal, H is the thickness of dielectric layer between metal layers, and ε is the dielectric constant. Similar expressions were used for the other two constituents of wire capacitance. The total capacitance of a wire is the sum of all these constituents. As an example, consider a wire in the metal 2 layer, in Fig. 5.2. The total capacitance of the metal 2 line C_2 is given by:

$$C_2 = C_{21} + C_{23} + C_{22} \tag{5.5}$$

where C_{21} is the capacitance from the metal 2 later to the metal 1 layer, C_{23} is the capacitance from the metal 2 later to the metal 3 layer, and C_{22} is the capacitance between wires in the metal 2 layer. C_{21} and C_{23} are crossover capacitances, and C_{22} is a line-to-line capacitance. Fairly complex, but comprehensive expressions that are functions of W, S, T, H, and ε (whose values can be obtained from technology library data sheets), for both of these constituent capacitances are presented in the appendix of [22]. Comparisons of the accuracy of these models with accurate numerical simulations and measurements by Sakurai et al. [23] indicated an accuracy of within 8% for capacitance values.

There are many analytical models that approximate the capacitance of a wire over a plane. More accurate models combine a bottom plate term with a fringing term to account for field lines originating from the edge and top of the wire [24]. Since wires today are becoming taller rather than wider, in order to reduce resistance

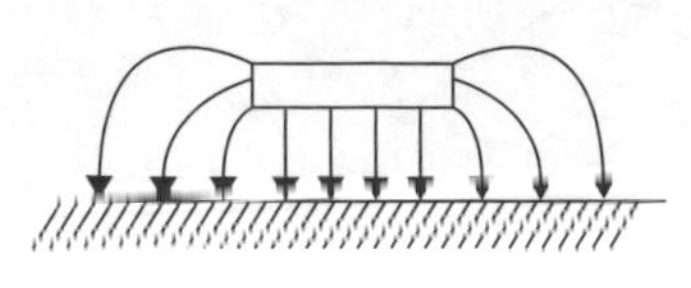

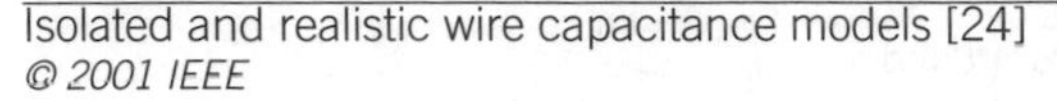
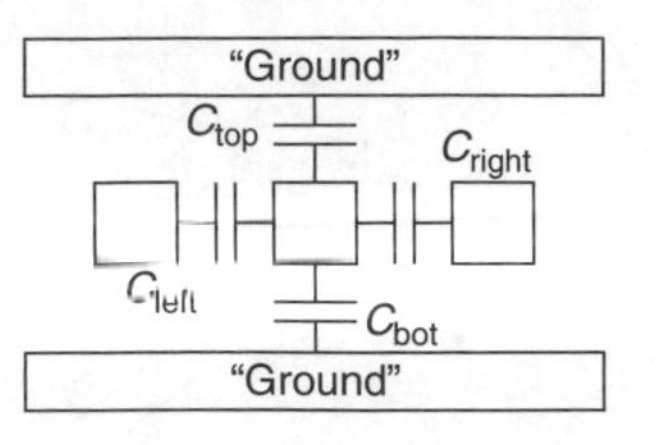

FIGURE 5.3

Isolated and realistic wire capacitance models [24]
© 2001 IEEE

with technology scaling, their side-to-side capacitances are growing and becoming more significant. As such, capacitance is better modeled by four parallel plate capacitors for the top, bottom, left, and right sides, plus a constant term for fringing capacitance, as shown in Fig. 5.3 [24, 25]. The vertical and horizontal capacitors can have different relative dielectrics for technologies that use *low K* materials, and the wire capacitance (C_{wire}) can be expressed as [24, 26]:

$$C_{wire} = \epsilon_0 \left(2K\epsilon_{horiz} \frac{\text{thick}}{\text{spacing}} + 2\epsilon_{vert} \frac{\text{width}}{ILD_{thick}} \right) + \text{fringe}(\epsilon_{horiz}, \epsilon_{vert}) \quad (5.6)$$

where ε_0 is the electric constant, ε_{horiz} and ε_{vert} are the horizontal and vertical relative dielectric constants, respectively, and ILD_{thick} is the thickness of the inter-layer dielectric (ILD). The plates at the top and bottom are typically modeled as grounded, since they represent orthogonally routed conductors, that averaged over the length of a wire, maintain a constant voltage (the capacitance would be multiplied by an appropriate factor if the orthogonal wires switched simultaneously and monotonically). Capacitances to the left and right on the other hand have data dependent effective capacitances that can vary. If the left and right neighbors of a wire switch in the opposite direction as the wire, then the effective side capacitances double, otherwise if they switch with the wire, the effective side capacitance approaches zero. This effect is termed as *Miller multiplication* and is modeled by varying the *K* parameter between 0 and 2 in Eq. (5.6). The *fringe* term depends weakly on geometry. Simplifications to the model can be made for top metal wires, for which only three parallel plates and fringing terms on the horizontal capacitors need to be considered.

5.1.2 Coupling-Aware Power Models

A significant amount of work has been done to model the capacitive (and inductive) parasitic interactions and coupling of bus wires in DSM technologies [2, 4, 5, 27–40]. The effects of coupling between wires not only tend to dominate line-to-ground coupling, but also introduce dependencies in the energy drawn from the power supply by their drivers. A compact energy model, based on a distributed circuit model for DSM buses was presented by Sotiriadis et al. [5]. Figure 5.4 shows a DSM bus that consists of several parallel lines driven by CMOS inverters and with repeaters inserted to reduce signal propagation delay. Figure 5.5 shows an

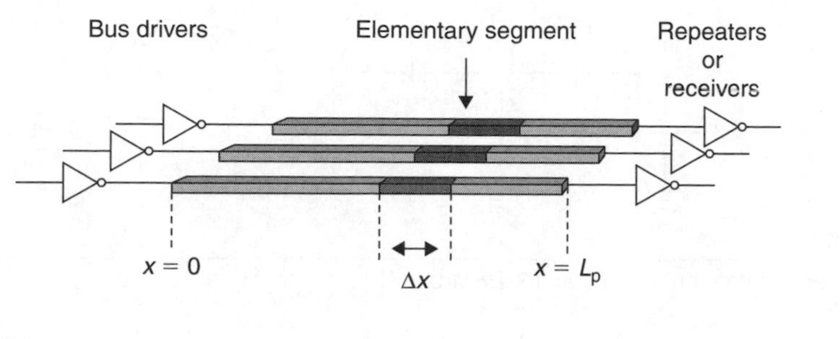

FIGURE 5.4

DSM bus [5]
© 2002 IEEE

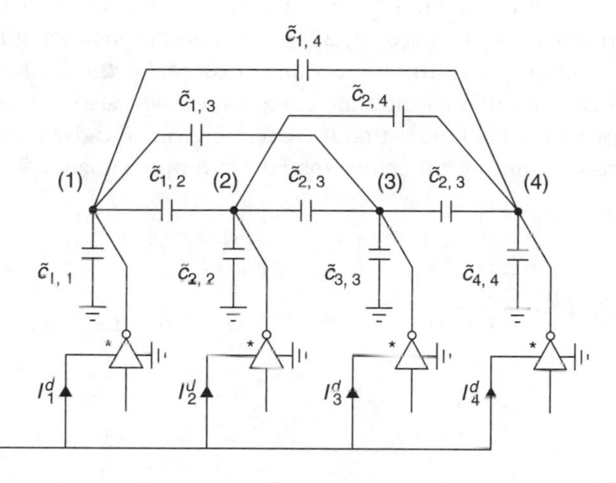

FIGURE 5.5

Lumped energy equivalent DSM bus model [5]
© 2002 IEEE

equivalent capacitive network model for the DSM bus used in [5], called the ***lumped energy equivalent DSM*** bus model, that represents all the capacitances for a set of wires (the figure shows a model for n=4 wires). The energy drawn during a transition, from the power supply by the bus drivers, for a bus with n lines is given by the expression:

$$E = \sum_{i=1}^{n} V_i^f e_i^T C^t (V^f - V^i) \qquad (5.7)$$

where $V^i = [V_1^i, V_2^i, \ldots, V_n^i]$ and $V^f = [V_1^f, V_2^f, \ldots, V_n^f]$ are vectors with n coordinates representing the initial and final voltages on the bus lines in the course of the transition, e_i is a vector with a 1 in the ith position, and 0 elsewhere, and C^t is the total capacitance conductance matrix (which represents the capacitances between the lines and ground). Since, traditionally, bus lines are laid parallel and co-planar, most of the electric field is trapped between the adjacent lines

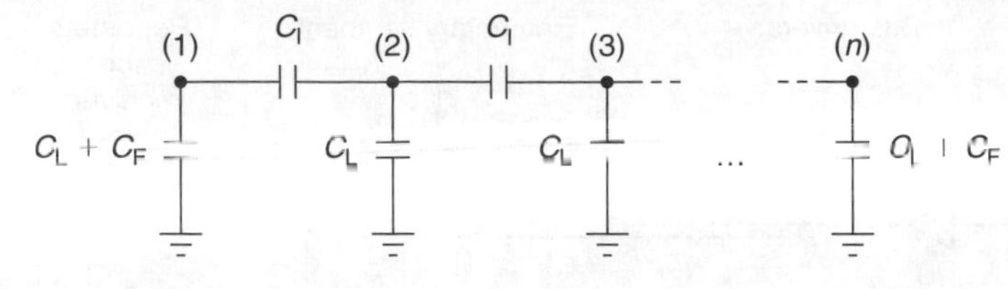

FIGURE 5.6

Simple approximate energy model for DSM bus [5]
© 2002 IEEE

and ground. The capacitance between non-adjacent lines is, thus, practically negligible compared to capacitance between adjacent lines, or the capacitance between the line and the ground. An approximate bus energy model can ignore parasitics between non-adjacent lines [3, 27, 30, 33]. Additionally, assuming that all the grounded capacitors have the same value (except for the boundary ones due to fringing effects) and that all interline capacitances are also the same, the approximate DSM bus model becomes that shown in Fig. 5.6 and the value of the approximate total capacitance conductance matrix C^{ta} (approximate C^{t}) is given as:

$$C^{ta} = \begin{bmatrix} 1+\lambda+\zeta & -\lambda & 0 & \cdots & 0 \\ -\lambda & 1+2\lambda & -\lambda & \cdots & 0 \\ 0 & -\lambda & 1+2\lambda & \cdots & 0 \\ \vdots & \vdots & \vdots & \ddots & \vdots \\ 0 & 0 & 0 & \cdots & 1+\lambda+\zeta \end{bmatrix} C_L \tag{5.8}$$

where

$$\lambda = \frac{C_I}{C_L}, \quad \zeta = \frac{C_F}{C_L} \tag{5.9}$$

C_L is the line capacitance (includes the capacitance of the driver and the receiver), C_F the fringing capacitance, and C_I the inter-line capacitance. The parameters λ and ζ depend on the particular CMOS technology used, as well as the specific geometry, metal layer, and shielding of the bus. Also, the C_F term can be ignored for wide buses since it does not contribute significantly to overall energy consumption.

A parameterizable analytical model for power dissipation on the bus was presented by Kretzschmar et al. [39]. This model was validated by power simulation with layout (including parasitics) of a particular bus implementation. The model of a bus line used is shown in Fig. 5.7. The line capacitance C_{line} of the wire encapsulates the ground and coupling capacitances of the wires (C_{wire}), in addition to the internal (C_{int}) and input capacitances (C_{inp}) of the active elements on the wire such as the drivers, repeaters, and receivers. As discussed earlier, due to DSM, the wire capacitance is determined not only by vertical capacitance $C_{vertical}$ but also increasingly by the coupling capacitances $C_{lateral}$ to adjacent wires. Due to the Miller effect, the power consumption of the wire becomes data dependent – if the neighboring wires switch in the same direction, then the $C_{lateral}$ is not charged; however for opposite switches of neighboring wires, twice $C_{lateral}$ is charged. Assuming an independent distribution of activity on all the bus lines, adjacent lines will switch as often in the same direction as in the opposite direction. On an average, the coupling capacitance between lines is charged once, and therefore it

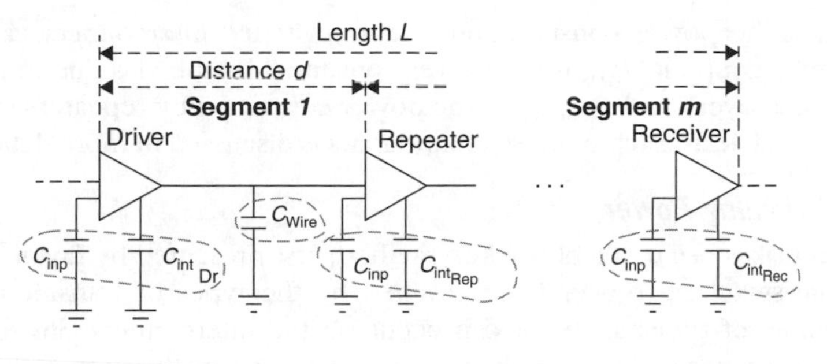

FIGURE 5.7

Model of bus line including driver, repeaters, and receiver [39]
© 2004 IEEE

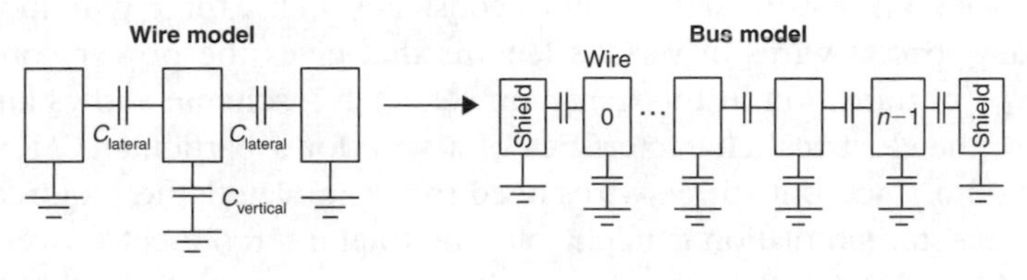

FIGURE 5.8

Bus wire model with two adjacent wires and extension to *n*-wire bus [39]
© 2004 IEEE

is assumed in the model that the two adjacent lines are connected to the ground, as shown in Fig. 5.8. Whenever there is a transition from 0 to 1 on the middle wire, the vertical capacitance and both the adjacent capacitances are charged. The wire capacitance per unit length is thus given by the relation:

$$C_{\text{wire}} = C_{\text{vertical}} + 2C_{\text{lateral}} \tag{5.10}$$

Assuming identical repeaters and drivers, the capacitance of an n-wire bus is given by:

$$C_{\text{bus}} = nL(C_{\text{vertical}} + 2C_{\text{lateral}}) + n(\frac{L}{d}C_{\text{int,Dr}} + \frac{L}{d}C_{\text{inp,Dr}} + C_{\text{int,Rec}} + C_{\text{inp,Rec}}) \tag{5.11}$$

where L is the length of the line and d is the inter-repeater distance. This equation can be used to calculate the average capacitance of a bus line per unit length, based on values from a CMOS technology library. The wire capacitance C_{wire} depends on the spacing to adjacent wires as well as the distance to the metal layers above and below. The bus power consumption can be obtained by plugging the capacitance value into Eq. (5.2).

5.1.3 High Level Power Models

A high level power model for the purposes of early design space exploration and incorporation in high level synthesis techniques [41] was proposed by Gupta et al. [2]. The total power consumption of a bus in this model is given by [39]:

$$P_{\text{total}} = P_{\text{sw}} + P_{\text{vias}} + P_{\text{repeaters}} \tag{5.12}$$

where P_{sw} is the power consumption due to switched interconnect capacitance and inter-wire coupling, P_{vias} is the power consumed by the vias due to the use of multiple metal layers, and $P_{repeaters}$ is the power consumed by repeaters inserted to minimize signal delay. Each of these components is discussed in more detail below.

5.1.3.1 *Switching Power*

The model makes use of a table-lookup method, first presented by Taylor et al. [42], where total switching power is determined by the types of transitions, instead of the number of transitions that can occur on the interconnect. Since coupling effects between wires decrease sharply the further apart they are, only an interconnect and its adjacent wires need to be considered. Table 5.1 shows the set of various transitions that are possible on three-wire interconnects. The general idea is to use low level transistor simulation to construct such a three-wire lookup table for minimally spaced wires of various lengths that gives the power consumption for each type of transition in the transition set. Such a scheme allows an accurate modeling of the electrical characteristics of a wire for a particular CMOS process technology. Also, since only three wires need to be simulated, the required time for low level transistor simulation is negligible. The total interconnect power can then be obtained by counting the types of transitions on the interconnect and performing a table lookup. The model does not, however, consider the effect of glitches and the authors point to techniques presented by Raghunathan et al. [43] as a means to suppress glitches. Also, instead of counting the transitions on the interconnect and performing a table lookup, the authors propose schemes [2] to estimate the types of transitions. This is done because the authors claim that high level design automation tools will consider multiple architectures to implement a design, many of which will be similar except for a few enhancements. As such, it will be useless and time consuming to run a full-fledged simulation on each architecture to characterize the switching activity on its interconnects. Instead, switching activity characterization is performed only once on an architecture that does not change drastically, and then first order estimates of switching activity are used on similar architectures.

5.1.3.2 *Power Due to Vias*

The purpose of vias is to (i) connect transistors residing on the substrate with the interconnect connecting these transistors, and (ii) connect interconnects running on multiple metal layers. The total power consumed by the vias is given by:

$$P_{vias} = V_N \cdot P_{via} \tag{5.13}$$

where V_N is the number of vias and P_{via} is the power consumption of a single via. The number of vias is estimated using interconnect layout, obtained with a

Table 5.1 Types of transitions on three-wire interconnects [2]

SSS	SXS	SSX	SXO	SXX
XXX	XSX	XXO	OXO	XSO

S = Stationary, X = Transition, O = Opposing transition
© 2003 IEEE

floorplanner and statistical methods, and then counting the number of times an interconnect changes direction. Although the via power P_{via} is dependent on the layer in which it resides, this is not taken into account in the model, which uses approximate values of P_{via} by taking its average for different configurations (or optionally using a weighting factor to each via configuration representing its proportional contribution to all the vias in the layout). The power consumed by vias used in repeaters is not estimated by Eq. (5.13), but is accounted for in the repeater power consumption term, described below.

5.1.3.3 *Power Due to Repeaters*

As mentioned earlier, with shrinking feature size, interconnect wires are getting proportionally longer and not scaling as well as transistors. A wire can be modeled as a simple *RC* network, for which the signal propagation delay is a quadratic function of length, because both resistance and capacitance are functions of length [3]. Inserting repeaters is a commonly used practice to reduce wire delay, since it reduces the quadratic dependence into a linear dependence on wire length [24, 44–46]. The starting and ending coordinates of an interconnect are obtained from a floorplanner, and statistical methods are used to obtain a layout for the interconnect. The number of repeaters (N_R) are obtained from formulations presented by Kapur et al. [6] and Bakoglu et al. [45], which give the optimal inter-repeater distance for a given CMOS process technology. The total number of repeater vias (V_R) are then calculated to be twice the number of repeaters, since paths are need to descend and ascend from the substrate where the repeaters reside. The total repeater power is then given as:

$$P_{rep} = C_{rep} \cdot V_{dd}^2 \cdot f \cdot \sum_{i \in I} \rho_i N_{Ri} + P_{via} \cdot \sum_{i \in I} V_{Ri} \qquad (5.14)$$

where the capacitance of a single repeater C_{rep} is given by the equations in [6], ρ_i is the switching activity, V_{dd} is the operating voltage, f is the clock frequency, and P_{via} was described earlier. The authors claim that Eq. (5.14) can be extended to incorporate leakage power in the repeaters.

5.2 COMPREHENSIVE BUS ARCHITECTURE POWER MODELS

In addition to wires, an important component of on-chip communication architectures is the bus logic, which consists of components such as bridges, arbiters, decoders, and buffer stages. A comprehensive, gate level power estimation methodology for the estimation of logic and interconnect power for the AMBA AHB/APB [47] hierarchical on-chip communication architecture was presented by Lahiri et al. [7]. The goal was to analyze the contributions of different components of the communication architecture to the overall communication architecture power consumption. This methodology and the results from the power estimation study are described below.

Figure 5.9 shows the methodology used for communication architecture power estimation. Synopsys CoreTools [48] was used to configure the parameters

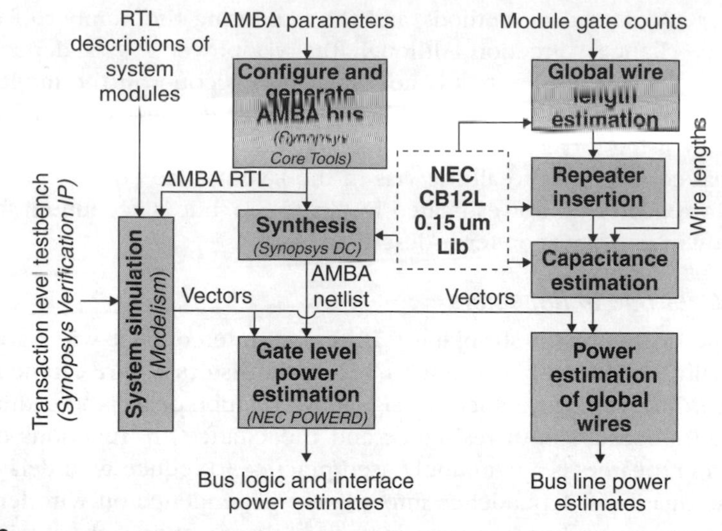

FIGURE 5.9

Methodology for communication architecture power estimation [7]
© 2004 IEEE

of the AMBA AHB bus such as address and data bus widths, number of masters, and number of slaves. CoreTools then generated synthesizable register transfer level (RTL) descriptions of the hardware. In the experiments, while the number of masters and slaves was varied, the address and data widths had fixed values (32 bits for addresses, 32 bits for AHB (advanced high performance bus) data bus, and 16 bits for APB (advanced peripheral bus) data bus). The arbiter was configured to use a static priority scheme. The bus hardware was synthesized to gate level using Synopsys Design Compiler [49] for the CB-12L low power 0.15μm CMOS technology library from NEC [50]. The Modelsim simulator [51] was used to obtain vectors for RTL and gate level simulations. The test system on which the AMBA (ARM microcontroller bus architecture) bus was integrated had a variable number of master and slave components, including peripherals such as interrupt controllers and memory controllers. Transaction level testbenches were constructed from programmable functional models of the masters, to generate bus traffic. The POWERD gate level power estimator [52] took the power simulation vectors and generated power measurements for different components. The global wire lengths were calculated using pre-layout wire length estimation techniques [8, 53] that take into account area and pin counts of components. Wire capacitances were obtained from the technology library, and repeater sizing and inter-repeater insertion distance was calculated using the delay optimal algorithm proposed by Bakoglu et al. [45]. The capacitances obtained were used to calibrate a transition count based power estimation model for bus wires [16]. A similar methodology was used to obtain power consumption for the AMBA AXI [54] bus architecture by Lee et al. [55].

This gate level power estimation methodology was used to determine the relative power consumption of different components of the AMBA on-chip communication architecture. For this purpose, a simple AMBA hierarchical shared

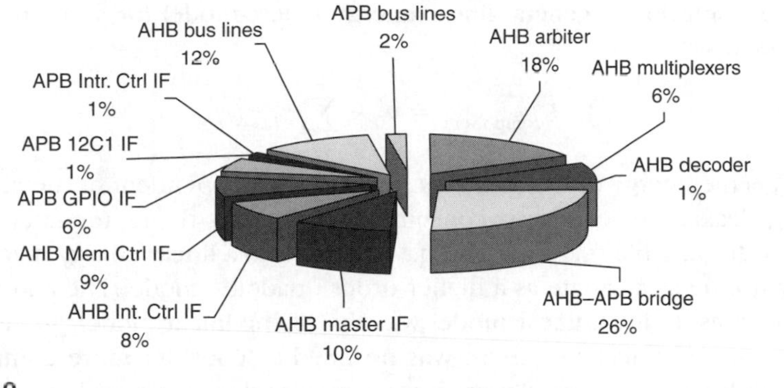

FIGURE 5.10

Breakdown of bus-based communication architecture power consumption for a simple AMBA AHB/APB hierarchical communication architecture [7]

bus-based system was considered, with two masters and two slaves (a memory controller and an interrupt controller) on the AHB, and three peripherals (general purpose I/O, interrupt controller, and I2C external bus interface) on the APB bus that were connected to the AHB bus via an AHB–APB bridge. The testbench had different transactions, including single transfers and bursts, with the addresses being generated randomly and uniformly. The global wire length was estimated to be between 1.5 and 3 mm, and the overall power consumption of the communication architecture was 12 mW. Figure 5.10 shows a breakdown of the power consumed by the various components of the AMBA bus. It can be seen that the power consumed by the bus lines is only 14% of the overall power, although it is often assumed that bus wires consume a large amount of power. This shows that it is important to consider bus logic when estimating power for an on-chip communication architecture.

5.2.1 Macro-Models for Bus Matrix Communication Architectures

A comprehensive macro-modeling-based power estimation approach for estimating the power consumption of the logic and wires in the AMBA AHB [47] bus matrix communication architecture was proposed by Pasricha et al. [56]. Unlike the approach proposed by Lahiri et al. [7] that estimates power at the gate level, this approach creates reusable power models of the on-chip communication architecture that can be used early in the design flow, at the system level, for fast and accurate power estimation.

The energy consumption of a bus logic (or for that matter any hardware) component can be obtained by identifying factors or events that cause a noticeable change in its energy profile. For this purpose, Pasricha et al. [56] proposed creating energy *macro-models* that can encapsulate factors having a strong correlation to energy consumption for a given component. A macro-model consists of variables that represent factors influencing energy consumption, and regression coefficients that capture the correlation of each of the variables with overall component energy consumption. A general linear energy macro-model for a component can be expressed as:

$$E_{component} = \alpha_0 + \sum_{i=1}^{n} \alpha_i \cdot \psi_i \quad (5.15)$$

where α_0 is the energy of the component which is independent of the model variables (e.g., leakage, clock energy consumption), and α_i is the regression coefficient for the model variable ψ_i. Note that Eq. (5.15) shows a linear energy macro-model that may not be as accurate as a higher order quadratic model. The authors' motivation for considering a linear model was that if the linear model provided good enough estimation accuracy, there was no need to consider more complex quadratic models (that are typically harder to create and evaluate, making power estimation more time consuming).

Three types of model variables - *control*, *data*, and *structural* - were considered in the energy macro-models. These variables represent different factors influencing energy consumption. The *control* factor represents control events, involving a control signal that triggers energy consumption either when a transition occurs from 1 to 0 or 0 to 1, or when it maintains a value of 0 or 1 for a cycle. Control variables can either have a value of 1 when a control event occurs, or 0 when no event occurs, in the energy macro-model relation Eq. (5.15). The *data* factor represents data events that trigger energy consumption on data value changes. Data variables take an integer value in Eq. (5.15) representing the Hamming distance (number of bit-flips) of successive data inputs. Finally, *structural* factors, such as data bus widths and number of components connected to the input also affect energy consumption of a component. They are represented by their integer values in Eq. (5.15).

A high level overview of the methodology used to create energy macro-models for bus logic components is shown in Fig. 5.11. Initially, a system testbench consisting of masters and slaves interconnected using the AMBA AHB bus matrix [47]

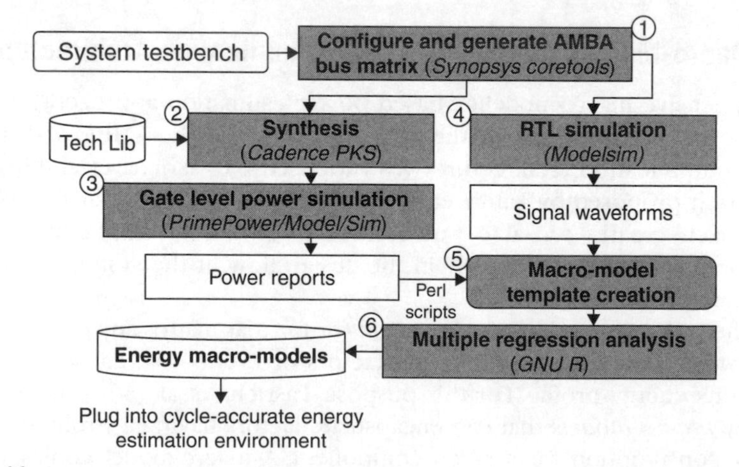

FIGURE 5.11

Energy macro-model generation methodology [56]
© 2006 IEEE

fabric is chosen. The testbench generates traffic patterns consisting of single and burst transactions of varying sizes, and different modes (e.g., SPLIT/RETRY, locked bus, etc.) that exercise the matrix under different operating conditions. Synopsys Coretools [48] is used to configure the bus matrix (i.e., specify data bus width, number of masters and slaves, etc.) and generate a synthesizable RTL description of the bus matrix communication architecture (Step 1). This description is synthesized to the gate level with the Cadence Physically Knowledgeable Synthesis (PKS) [57] tool, for the target standard cell library (Step 2). PKS pre-places cells and derives accurate wire length estimates during logic synthesis. In addition, it generates a clock tree including clock de-skewing buffers. The gate level netlist is then used with Synopsys PrimePower [58] to generate power numbers (Step 3).

In parallel with the synthesis flow, RTL simulation is performed to generate signal waveform traces for important data and control signals (Step 4). These signal waveforms are compared with cycle energy numbers, obtained after processing PrimePower generated power report files with Perl scripts, to determine which data and control signals in the matrix have a noticeable effect on its energy consumption. Only those signals are considered that change in value when an increase in bus matrix energy consumption of at least 0.01% is observed over the base case (i.e., no data traffic). Note that a finer grained selection criterion (e.g., 0.001%) will result in even more accuracy, but at the cost of more complex macro-models that take longer to create and evaluate. The selected data and control events become the variables in a macro-model template that consists of energy and variable values for each cycle of testbench execution (Step 5). Figure 5.12 shows an example of a macro-model template for one of the components of the bus matrix. The template consists of energy values (*cycle_energy*) and variable values (*S_load, S_desel, HD_addr, S_drive*) for each cycle of testbench execution. This template is used as an input to the GNU R tool [59] that performs multiple linear regression analysis to find coefficient values for the chosen variables (Step 6). Steps 1–6 are repeated for testbenches having different structural attributes such as data bus widths and number of masters and slaves, to identify structural factors (variables) that may influence cycle energy.

Statistical coefficients such as *Multiple-R*, *R-square*, and *standard deviation for residuals* [60] are used to determine the goodness of fit and the strength of the correlation between the cycle energy and the model variables. Once a good fit between cycle energy and macro-model variables is found, the energy macro-models are generated in the final step. These models can then be plugged into any system level cycle-accurate or cycle-approximate simulation environment, to get energy consumption values for the AMBA AHB bus matrix communication architecture.

To obtain the energy consumption for the entire AMBA AHB bus matrix communication architecture, the energy macro-model generation methodology was used to create macro-models for each of its components. The total energy consumption of a bus matrix can be expressed as:

$$E_{\mathrm{MATRIX}} = E_{\mathrm{INP}} + E_{\mathrm{DEC}} + E_{\mathrm{ARB}} + E_{\mathrm{OUT}} + E_{\mathrm{WIRE}} \tag{5.16}$$

where E_{INP} and E_{DEC} are the energy for the input and decoder components for all the masters connected to the matrix, E_{ARB} and E_{OUT} are the energy for arbiters and output stages connecting slaves to the matrix, and E_{WIRE} is the energy of all

cycle	cycle_energy	S_load	S_desel	HD_oddr	S_drive
10	0.54802	0	0	0	0
20	0.54802	0	0	0	0
30	0.54802	0	0	0	0
40	0.54802	0	0	0	0
50	0.54802	0	0	0	0
60	0.54802	0	0	0	0
70	0.54802	0	0	0	0
80	0.54802	0	0	0	0
90	0.54802	0	0	0	0
100	0.54802	0	0	0	0
110	0.54802	0	0	0	0
120	0.54802	0	0	0	0
130	1.632607	1	0	3	1
140	0.961418	1	0	0	1
150	0.56406	0	0	0	0
160	0.560536	0	0	0	0
170	0.601455	0	0	3	0
180	0.547972	0	0	0	0
190	1.721611	1	0	6	1
200	0.946274	1	0	0	1
210	0.56392	0	0	0	0
220	0.5604	0	0	0	0
230	0.611902	0	0	3	0

FIGURE 5.12

Energy macro-model template

the bus wires that connect the masters and slaves. Energy macro-models were created for the first four components, with E_{WIRE} being calculated separately.

The energy macro-models for the bus matrix components are essentially of the form shown in Eq. (5.15). Leakage and clock energy (which are the major sources of independent energy consumption) are considered as part of the static energy coefficient α_0 for each of the components. Based on experimental results, an approximately linear relationship between cycle energy and macro-model variables was observed for the components. The energy models for each of the components are presented below.

5.2.1.1 *Input Stage*

Every master connected to a bus matrix has its own input stage that buffers address and control bits for a transaction, if a slave is busy. The input stage model can be expressed as:

$$E_{\mathrm{INP}} = \alpha_{\mathrm{inp0}} + \alpha_{\mathrm{inp1}} \cdot \psi_{\mathrm{load}} + \alpha_{\mathrm{inp2}} \cdot \psi_{\mathrm{desel}} + \alpha_{\mathrm{inp3}} \cdot \psi_{\mathrm{HDin}} + \alpha_{\mathrm{inp4}} \cdot \psi_{\mathrm{drive}} \quad (5.17)$$

where ψ_{load} and ψ_{drive} are control signals asserted when the register is loaded, and when the values are driven to the slave, respectively; ψ_{desel} is the control signal from the master to deselect the input stage when no transactions are being issued; and ψ_{HDin} is the Hamming distance of the address and control inputs to the register.

5.2.1.2 *Decoder*

A decoder component is connected to every master, and consists of logic to generate the select signal for a slave after decoding the destination address of an issued transaction. It also handles multiplexing of read data and response signals from slaves. The decoder energy consumption model can be formulated as:

$$E_{\mathrm{DEC}} = \alpha_{\mathrm{dec0}} + \alpha_{\mathrm{dec1}} \cdot \psi_{\mathrm{slavesel}} + \alpha_{\mathrm{dec2}} \cdot \psi_{\mathrm{respsel}} + \alpha_{\mathrm{dec3}} \cdot \psi_{\mathrm{HDin}} + \alpha_{\mathrm{dec4}} \cdot \psi_{\mathrm{sel}} \tag{5.18}$$

where ψ_{slavesel} and ψ_{respsel} are control signals asserted in the cycle in which the slave select and the data/response MUX select signals are generated, respectively; ψ_{HDin} is the Hamming distance of the read data and response signals from the slave; and ψ_{sel} is a control signal where transition occurs when the decoder is selected or deselected.

5.2.1.3 *Output Stage*

Every slave is connected to the bus matrix through the output stage that handles multiplexing of address and control bits from the masters. It also calls the arbiter to determine when to switch between accessing masters. The energy consumption for the output stage is given by:

$$E_{\mathrm{OUT}} = \alpha_{\mathrm{out0}} + \alpha_{\mathrm{out1}} \cdot \psi_{\mathrm{addrsel}} + \alpha_{\mathrm{out2}} \cdot \psi_{\mathrm{datasel}} + \alpha_{\mathrm{out3}} \cdot \psi_{\mathrm{HDin}} + \alpha_{\mathrm{out4}} \cdot \psi_{\mathrm{noport}} \tag{5.19}$$

where ψ_{addrsel} and ψ_{datasel} are control signals asserted when address and data values are selected after a call to the arbiter results in a change in the master accessing the slave; ψ_{HDin} is the Hamming distance of address and data inputs; and ψ_{noport} is a control signal from the arbiter, which goes high when no masters access the slave in a cycle.

5.2.1.4 *Arbiter*

The arbiter is invoked by the output stage, and uses an arbitration scheme to grant access to one of the potentially several masters requesting for access to the slave. The cycle energy model for the arbiter is calculated as:

$$\begin{aligned} E_{\mathrm{ARB}} = {} & \alpha_{\mathrm{arb0}} + (\alpha_{\mathrm{arb1}} + n \cdot \alpha_{\mathrm{arb2}}) \cdot \psi_{\mathrm{arb}} + \alpha_{\mathrm{arb3}} \cdot \psi_{\mathrm{arb+1}} \\ & + (\alpha_{\mathrm{arb4}} + n \cdot \alpha_{\mathrm{arb5}}) \cdot \psi_{\mathrm{desel}} + \alpha_{\mathrm{arb6}} \cdot \psi_{\mathrm{desel+1}} \end{aligned} \tag{5.20}$$

where ψ_{arb} and $\psi_{\mathrm{arb+1}}$ are control signals representing the cycle when arbitration occurs, and the subsequent cycle when the master select signal is generated; ψ_{desel} and $\psi_{\mathrm{desel+1}}$ are control signals representing the cycle when the arbiter is not selected by any master, and the subsequent cycle when it generates the *noport* signal for the output stage; and n represents the number of masters connected to the arbiter.

5.2.1.5 *Bus Wires*

The bus wires that connect masters, slaves, and logic components in the bus matrix dissipate dynamic power due to switching, and leakage power due to the repeaters inserted in long wires to reduce signal delay. The expression for energy consumption of a bus wire from Kretzschmar et al. [39] is extended to include the effect of repeaters (to reduce wire delay), and is given as:

$$E_{\text{WIRE}} = 0.5\, V_{\text{dd}}^2 \left(\sum C_{\text{L}} + \frac{l}{d} \cdot C_{\text{REP}} + l \cdot (C_{\text{G}} + 2C_{\text{C}})\right) \cdot \alpha + \frac{l}{d} \cdot E_{\text{REP}} \quad (5.21)$$

where V_{dd} is the supply voltage, α is the switching factor representing bit transition activity on the wire, ΣC_{L} is the sum of load capacitances of all the components connected to the wire, including the driver and receiver, C_{REP} is the capacitance of a repeater, C_{G} is the wire-to-ground capacitance per unit length, C_{C} is the coupling capacitance per unit length of the wire to its adjacent wires, l is the length of the wire, d is the inter-repeater distance, and E_{REP} is the repeater internal energy. This single bus wire model is extended for an N bit bus. The Berkeley Predictive Technology Model (PTM) [61] is used to estimate values for ground (C_{G}) and coupling (C_{C}) capacitances. The static energy of the repeater (E_{REP}) and its capacitance (C_{REP}) are obtained from data sheets. The component load capacitance on the wire (C_{L}) is obtained after component synthesis. A high level simulated-annealing-based floorplanner (PARQUET) [62] is used to generate intellectual property (IP) block placement to obtain wire lengths and [44] is used to determine optimal-delay repeater spacing/sizing. Finally, the switching factor (α) is obtained from simulation.

In order to generate the energy macro-models for the AHB bus matrix communication architecture, we first selected a diverse set of bus matrix based system testbenches, having different number of masters and slaves (2–40 components), bus widths (32–128 bits), and data traffic profiles (single and variable sized burst transactions) that activate the logic components in the bus matrix in different states. These were used to generate the energy macro-models. The component regression coefficients were obtained using the methodology shown in Fig. 5.11, for the TSMC 180 nm standard CMOS cell library. The regression coefficients for the macro-models of each of the components of the bus matrix are shown in Table 5.2. Also shown are the R squared (R^2) values that measure the goodness of fit for each regression [60]. The value of R^2 is a fraction between 0.0 and 1.0. A value of 0.0 indicates that knowing the variable values will not allow us to predict the energy values, whereas a value of 1.0 indicates that knowing the variable values will allow us to predict the value of energy perfectly. From the table it can be seen that the R^2 values are close to 1.0, thus enabling reliable prediction of energy at the system level.

Next, the same system testbenches were targeted to the 130, 90, and 65 nm TSMC standard CMOS cell libraries and the regression coefficient generation process was repeated for the components in the bus matrix. The regression coefficients for the *Input Stage* bus matrix component, for each of the standard cell libraries, are shown in Table 5.3. It can be seen from the table that as the standard cell size is reduced, the coefficient values decrease, which is indicative of a decrease in overall power consumption. However, *note that the reduction in coefficient values is not linear* and the table indicates a change in relative importance of model variables with a change in standard cell library. For instance, the value of

Table 5.2 Coefficients for energy macro-models (180 nm)

Component	Variable	Energy coefficient (pJ)	Variable	Energy coefficient (pJ)	R^2 value
Input stage	α_{inp0}	7.78	α_{inp3}	0.96	0.97
	α_{inp1}	3.81	α_{inp4}	3.27	
	α_{inp2}	2.60			
Decoder	α_{dec0}	0.47	α_{dec3}	0.13	0.90
	α_{dec1}	3.04	α_{dec4}	0.38	
	α_{dec2}	2.17			
Output stage	α_{out0}	0.72	α_{out3}	0.14	0.94
	α_{out1}	2.61	α_{out4}	1.48	
	α_{out2}	1.53			
Arbiter	α_{arb0}	0.65	α_{arb4}	0.34	0.86
	α_{arb1}	0.76	α_{arb5}	0.48	
	α_{arb2}	0.30	α_{arb6}	0.52	
	α_{arb3}	0.60			

Table 5.3 Coefficients for input stage component (in pJ)

Standard cell library (in nm)	α_{inp0}	α_{inp1}	α_{inp2}	α_{inp3}	α_{inp4}
180	7.78	3.81	2.60	0.96	3.27
130	1.33	0.66	0.04	0.24	0.56
90	1.23	0.44	0.02	0.20	0.40
65	0.54	0.29	0.02	0.09	0.18

the coefficient for model variable α_{inp3} is less than one-third the value of the coefficient for α_{inp4} at 180 nm, but this ratio reduces to $\frac{1}{2}$ for the 65 nm library.

A major advantage of this energy macro-model approach proposed by Pasricha et al. [56] is the ease with which it can be retargeted to different standard cell libraries. Typically retargeting to a new standard cell library is a very time-intensive effort. However, the proposed approach simply requires the generation of gate level cycle energy numbers with the new library for the small system testbenches; these cycle energy numbers are used to update the cycle energy column in the macro-model template shown in Fig. 5.12, after which the regression analysis is repeated. Once the gate level cycle energy numbers are available for the standard cell library, it only takes a few minutes to obtain the new coefficients for each of the components in the bus matrix. This enables rapid retargeting of models for new cell libraries and results in an order-of-magnitude reduction in development time.

More experiments were performed to determine the macro-model accuracy for different technology libraries, by comparing the obtained macro-model energy estimates with detailed gate level energy estimates. Four system testbenches were selected for the experiments. These testbenches were separate and distinct from the ones used for characterization and generation of the energy macro-models, and had different bus matrix structures and traffic characteristics: (i) a 2 master, 3 slave bus matrix with 32 bit data bus width (2×3_32b), (ii) a 3 master, 4 slave bus matrix with 32 bit data bus width (3×4_32b), (iii) a 4 master, 5 slave bus matrix with 32 bit data bus width (4×5_32b), and (iv) a 2 master, 3 slave bus matrix with 64 bit data width (2×3_64b). These architectures were synthesized and cycle energy values were estimated using Synopsys PrimePower [58] at the gate level, for each of the 180, 130, 90, and 65 nm TSMC standard CMOS cell libraries. In parallel, the model variables and coefficient values in the energy macro-models were characterized for each of the components in the matrix, to obtain estimated energy values.

Figure 5.13 compares the average error in energy estimated at each cycle by the macro-models at the early system level, compared to gate level estimation, for the different standard cell libraries. It can be seen that the energy macro-models allow highly accurate cycle energy estimation that is unaffected by the scaling of technology libraries toward DSM. The maximum average cycle energy estimation error for the macro-models is only 4.19%.

Next, the energy macro-models were plugged into a fast transaction level bus cycle-accurate (BCA) simulation environment (cycle count accurate at transaction boundaries (CCATB) [63, 64]—described in more detail in Chapter 4) in SystemC [65, 66]. The CCATB simulation abstraction captures the bus matrix at a cycle-accurate granularity and uses function/transaction calls instead of signals to obtain simulation speed over BCA models that capture signal details, without sacrificing accuracy. The simulation model incorporates bus matrix energy macro-models

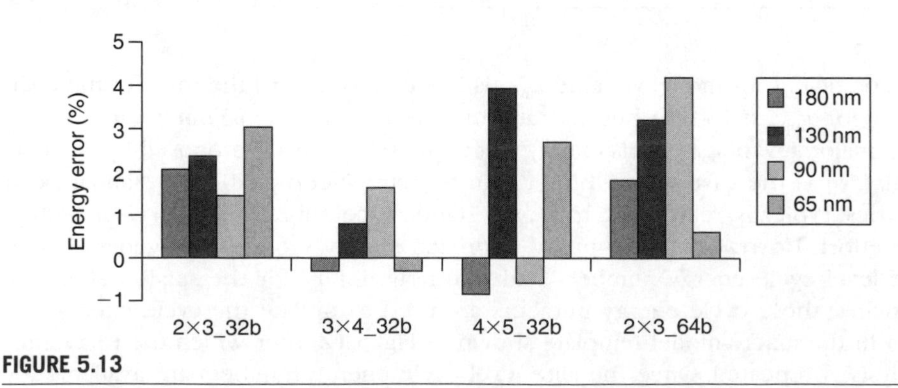

FIGURE 5.13

Average cycle energy estimation errors for various technology libraries

for cycle energy estimation. The macro-model equations for the bus matrix components described above are inserted in the code for each component, at points where a change in the value of an energy consuming event (i.e., dependent variable) can occur.

Figure 5.14 shows a snapshot of the power waveform generated by gate level simulation using Synopsys PrimePower, and the SystemC CCATB simulation using the energy macro-models for the 2×3_32b bus matrix system testbench. As can be seen, the power estimates using the system level CCATB simulation model are highly correlated to the actual power consumption. This high correlation highlights an additional benefit of the methodology in estimating peak energy (as opposed to average energy). Peak energy is important in the planning of power grids, which is becoming increasingly difficult to do in DSM technologies.

Finally, Table 5.4 compares the CPU time taken for the PrimePower simulation (for gate level cycle energy estimation) with the system level CCATB-based prediction, for the four system testbenches described earlier. The time taken for the gate level and system level power estimation does not depend on the CMOS technology library used. As can be seen from the table, the system level power estimation

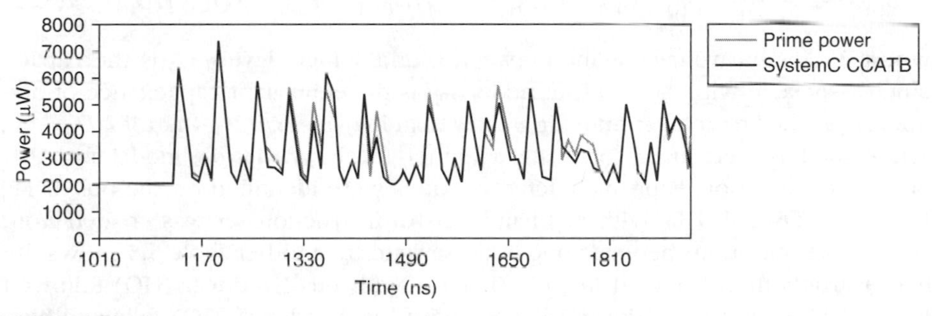

FIGURE 5.14

Predicted and measured power waveforms [56]

Table 5.4 Comparison of time taken for power estimation

Testbench	PrimePower gate-level CPU time (in hours)	SystemC T-BCA level CPU time (in seconds)	Speedup (X times)
2×3_32b	2.58	9.1	1021
3×4_32b	7.91	18.8	1515
4×5_32b	27.04	49	1987
2×3_64b	3.10	9.9	1127

gives a substantial speedup close to 2000× over gate level power estimation using PrimePower. It should be noted that the comparison does not include time taken for value change dump (VCD) file generation and parasitic RC extraction file generation needed for PrimePower-based estimation, which takes from several minutes to hours for each of the testbenches.

5.2.2 Other Macro-Model-Based Techniques

Similar macro-modeling techniques were utilized for estimating power consumption in an AMBA AHB shared bus by Caldari et al. [67] and the STBus communication architecture by Bona et al. [68]. Caldari et al. [67] considered a simple AMBA AHB bus, and decomposed it into the following components: an arbiter, a decoder, and multiplexing logic. Macro-models were created for these components using gate level analysis. For instance, assuming that the decoder has $n_O \geq 2$ outputs (i.e., the number of slaves connected to the bus) and defining HD_{IN} as the Hamming distance between two consecutive inputs, the macro-model for the decoder component was expressed as:

$$E_{DEC} = \frac{V_{DD}^2}{4}(n_I \cdot n_O \cdot C_{PD} \cdot HD_{IN} + 2 \cdot HD_{OUT} \cdot C_O) = f(n_O, HD_{IN}) \quad (5.22)$$

where V_{DD} is the voltage swing between 0 and 1 logic levels, C_O is the capacitance associated with each output node, C_{PD} is the equivalent capacitance of one node, n_1 is the first integer number greater than $\log_2(n_O\text{-}1)$, HD_{OUT} is 1 if $HD_{IN} \geq 0$. These models were used to create a higher level *instruction* model for AHB power consumption. Four main activity modes were identified on the bus: IDLE, READ, WRITE, and IDLE with bus handover. An instruction set was created from all possible transitions between one of these states to another. Table 5.5 shows the five instructions considered in [67]: (i) IDLE with bus handover (HO) followed by an IDLE with bus handover; (ii) IDLE with bus handover (HO) followed by a write; (iii) a READ followed by a WRITE; (iv) a READ followed by an IDLE with bus handover; and (v) a WRITE followed by a READ. The corresponding dynamic power consumption in the bus logic components for each of these instructions (obtained after running a testbench) is also shown in the table.

Table 5.5 Instruction energy [67]

Instruction	Average energy (pJ)
IDLE_HO_IDLE_HO	14.7
IDLE_HO_WRITE	16.7
READ_WRITE	19.8
READ_IDLE_HO	22.4
WRITE_READ	14.7

Unlike the macro-models proposed by Pasricha et al. [56], only dynamic energy is accounted for by the macro-models created by Caldari et al. [67]—leakage and clock energy consumption is ignored. Additionally, the energy consumption of the wires is also not considered during the power analysis of the AHB bus. Nonetheless, the results from [67] can give an approximate idea of power consumption for the bus logic components, in high level simulation environments. Similar approaches for high level approximate power estimation of the communication architecture were proposed by Dhanwada et al. [69] and Neffe et al. [70]. Dhanwada et al. [69] used gate level power estimation to estimate average energy for single and burst reads and writes on a bus. These average power numbers are then plugged into high level (transaction level) simulation models. Experiments on simple testbenches showed an error of around 12% for power estimation using transaction level models. Neffe et al. [70] used a gate level power estimation tool called *Diesel* [71] to estimate power consumption due to switching on bus wires. Diesel makes use of layout information to extract information about parasitic capacitances and resistances. The average switching values for each of the signals in the bus architecture were obtained using the tool, and then used in cycle-accurate and cycle-approximate transaction level simulation models in SystemC, for high level approximation of dynamic power consumption on bus wires. Experiments on a single shared bus-based system testbench indicated a power estimation error of around 8% for cycle accurate and 15% for cycle-approximate transaction level simulation models.

Another macro-model-based approach was proposed by Bona et al. [68] to estimate power consumption of the logic blocks in the crossbar (or shared bus) topology-based Type 2 and Type 3 STBus [72] on-chip communication architectures. In the proposed approach, a macro-model for the entire STBus communication architecture is created, from models of the node, type converter, and size converter subcomponents. The node represents the entire bus crossbar (or shared bus), while the type and size converter components are used to convert data to and from heterogeneous protocols (e.g., between Type 2 and Type 3) or data path widths. The authors describe the model for the STBus node and claimed that similar techniques can be used to design models for the type and size converters. An STBus node configuration n specifies a particular instance of the design space S, given as:

$$S = \{n \mid n = \langle i, t, rqr, rpr, p, C_L, dps, Type \rangle\} \tag{5.23}$$

where i is the number of initiators (or masters), t is the number of targets (or slaves), rqr is the number of request resources, rpr is the number of response resources, p is the type of arbitration policy, C_L is the total output pin capacitance, dps is the data path width (e.g., 32 or 64 bits), and $Type$ is the protocol mode (either Type 2 or Type 3). The energy model $E(n)$ of the STBus configuration n is then expressed as:

$$E(n) = P(n) \cdot C \cdot T_{clk} \tag{5.24}$$

where $P(n)$ is the average node power consumption during a simulation of C clock cycles, having a clock period T_{clk}. The average node power consumption

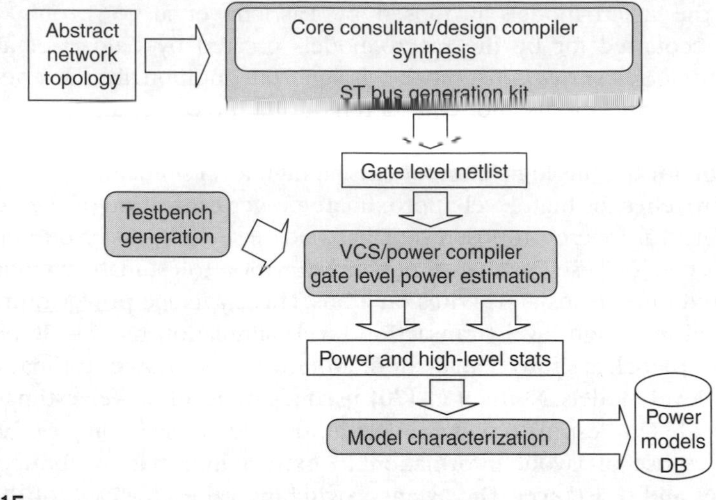

FIGURE 5.15

STBus power characterization flow [68]
© 2004 IEEE

$P(n)$ is calculated as a linear combination of three factors, according to the following equation:

$$P(n) = B(n) + P_{\text{sent}}(n) \cdot \frac{r_S}{C} + P_{\text{rec}}(n) \cdot \frac{r_r}{C} \tag{5.25}$$

where $B(n)$ is the average base cost that depends on the particular configuration n of the node, $P_{\text{sent}}(n)$ is the additive power due to data cells sent from the masters to the slaves, r_S is the total number of data cells sent to the slaves, $P_{\text{rec}}(n)$ is the power cost due to data received by the masters, r_r is the total number of data cells received by the masters, and C is the total number of clock cycles. The values of the coefficients $B(n)$, $P_{\text{sent}}(n)$, and $P_{\text{rec}}(n)$ are obtained using linear regression analysis [60], over power data obtained from gate level simulation.

Figure 5.15 shows the STBus power characterization flow used in the approach. The STBus Generation Kit [72] is used to synthesize a gate level netlist from a parameterized system level description of the STBus node. Gate level power simulations using VCS/PowerCompiler [73] are launched with traffic on the node generated by several testbenches, to determine the power consumption of the node. The testbench generator exposes the STBus node to a wide range of traffic conditions. In the final step, model characterization is done to compute the value of the coefficients—$B(n)$, $P_{\text{sent}}(n)$, and $P_{\text{rec}}(n)$ and the results are stored in a power model database.

The model for the STBus node was compared with power estimates at the gate level. A multiprocessor platform testbench consisting of four ARM7TDMI processor initiators and several targets was selected for the purpose. The platform was implemented in SystemC, with the processor instruction set simulator (ISS) models encapsulated in SystemC wrappers. The SystemC model was enhanced with the power models for the node from the power model database, obtained as

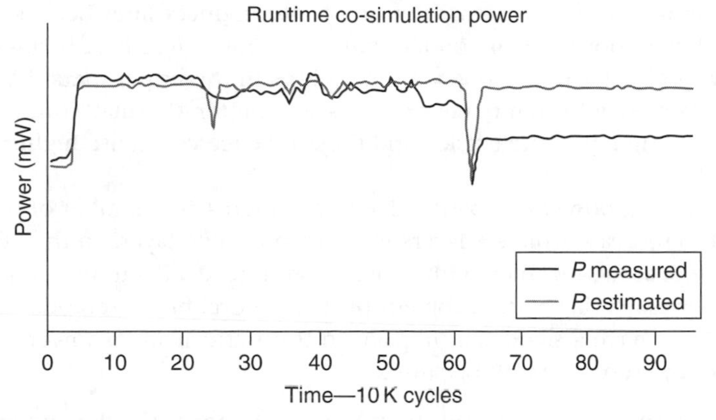

FIGURE 5.16

Estimated vs. measured average power in STBus [68]
© 2004 IEEE

shown in Fig. 5.15. A software benchmark executing on the processors generated the traffic on the STBus node. The SystemC simulation generated a trace file that contained comprehensive print-on-change information sampled on a clock cycle basis. This co-simulation file was used to drive the gate level Verilog simulation using VCS [73], and obtain a detailed power analysis of the mapped netlist using PowerCompiler [73]. Figure 5.16 compares the power estimated by simulating the SystemC framework that was enhanced by the power models ($P_{estimated}$), with the power measured by PowerCompiler at the gate level ($P_{measured}$). The system level estimation can be seen to have a good correlation with gate level power figures with an average total error of 9%. The overestimation in the last 200K cycles in Fig. 5.16 is due to a very high number of data cells being transmitted through the STBus node.

5.3 BUS WIRE THERMAL MODELS

Thermal effects are an inseparable aspect of signal propagation on interconnects. Current flowing through an interconnect causes power dissipation of I^2R, where I is the current flowing through the interconnect line, and R is the line resistance. Since the substrate, which is attached to the heat-sink, is typically far away from the interconnect lines (and especially the global-tier interconnects), the heat generated by the I^2R power dissipation cannot be efficiently removed, and leads to an increase in the interconnect temperature. This is the phenomenon of *Joule heating* or self-heating. Although the I^2R power dissipation is typically not a major portion of total on-chip power dissipation, this power dissipation occurs in the interconnects which happen to be separated from the substrate by a very low thermal conductivity dielectric. This can lead to a significant rise in interconnect temperature [74]. With the aggressive scaling down of CMOS process technology, low K inter-layer and inter-metal dielectrics (IMDs) have been introduced [75] to reduce RC delay, crosstalk, and dynamic power consumption [76, 93]. But this has led to an increase in the temperature of interconnect lines because (i) low *K* dielectrics have poor thermal conductivities leading to less heat being dissipated from the wires [77], and (ii) long via separations in the upper metal layers results in higher average wire temperatures (vias are better thermal conductors than their surrounding low *K* dielectrics, and their absence can cause higher wire temperatures) [78].

The increasing power dissipation due to the increasing number of metal layers (projected to increase from six layers in 180 nm to nine layers in the 50 nm node [79]) and repeaters, together with more pronounced effects of capacitive coupling, further exacerbates the problem of high interconnect temperatures. These thermal effects have a significant impact on the performance, power, design, and reliability of interconnects [9] because:

- interconnect lifetime (or reliability) has an exponential dependence on the inverse metal temperature;
- thermal effects put a limit on maximum allowed root mean square (RMS) current density through the interconnects to limit temperature increase, since the RMS value of current density is responsible for heat generation;
- interconnect metal resistivity is dependent on temperature (e.g., resistivity of Cu increases by 39% from 20°C to 120°C), and higher resistivity causes larger *RC* delay that results in performance degradation;
- leakage power has a significant dependence on temperature, and can be order of magnitudes greater at high temperatures [80]; and
- thermally induced open circuit metal failure under short duration high peak currents including electrostatic discharge (ESD) can introduce latent electromagnetic (EM) damage. ESD is in fact responsible for rapid thermal failures [81], and semiconductor surveys indicate that it is the single largest cause of failures in ICs [82].

Since all of the effects described above are expected to worsen [74, 83, 84] with increasing aspect ratios and decreasing wire spacing, and lead to even higher temperatures in future technologies, thermal models of bus wires are needed to better estimate and optimize the communication architecture to improve their thermal characteristics.

Thermal effects and their impact on performance, current density, and reliability have been studied by a few researchers [9, 74, 85]. A model for the energy dissipated as heat on bus lines was given by Sotiriadis and Chandrakasan [5], and later extended by Wang et al. [86]. The energy drawn from the power supply consists of two parts: (i) energy stored in the capacitances of the repeaters and bus, and (ii) energy dissipated as heat. The energy drawn by the *i*th driver on a bus, which is converted into heat during a transition, is given by the relation:

$$E_{\text{heat0}} = E - \Delta E_{\text{c}} \tag{5.26}$$

where E is the total energy drawn from the power supply, and is given as:

$$E = V_i^f e_i^T C^t (V^f - V^i) \tag{5.27}$$

and ΔE_c is the difference in energy stored in the capacitances after and before switching,

$$\Delta E_c = \frac{1}{2} V_i^f e_i^T C^t V^f - \frac{1}{2} V_i^t e_i^T C^t V^i \tag{5.28}$$

where V_f^i is the final voltage of wire i after switching, V_j^i is the initial voltage of wire i before switching, e_i represents a vector with a 1 at the ith position, and 0 at other positions, and V^f and V^i are voltage vectors of all the wires after switching and before switching. C^t is a matrix of the capacitances, having a form:

$$C^t = \begin{bmatrix} c_{1,1}+c_{1,2} & -c_{1,2} & 0 & 0 & 0 \\ -c_{1,2} & c_{2,2}+c_{1,2}+c_{2,3} & -c_{2,3} & 0 & 0 \\ 0 & -c_{2,3} & c_{3,4}+c_{3,3}+c_{2,3} & \ddots & 0 \\ \vdots & \vdots & \vdots & \ddots & -c_{31,32} \\ 0 & 0 & 0 & -c_{31,32} & c_{31,32}+c_{32,32} \end{bmatrix} \tag{5.29}$$

where $c_{i,j}$ is the coupling capacitance between two wires i and j; and $c_{i,i}$ is the total capacitance between wire i and ground, and includes the capacitance of the driver and receiver.

E_{heat0} represents the heat generated when a bus line is charged. Heat is also generated when a bus line is discharged [86]. The energy converted into heat during bus line discharging can be expressed as:

$$E_{\text{heat1}} = \Delta \tilde{E}_c \tag{5.30}$$

$$\Delta \tilde{E}_c = -\frac{1}{2} \tilde{V}_i^f e_i^T C^t \tilde{V}^f + \frac{1}{2} \tilde{V}_i^t e_i^T C^t \tilde{V}^i \tag{5.31}$$

where $\Delta \tilde{E}_c$ is the difference in energy stored in capacitance before and after the transition. The total energy dissipated as heat on the driver and bus lines is a summation of E_{heat0} and E_{heat1}.

A thermal model for the power lines with a worst case analysis using maximum current metrics was presented by Chiang and Saraswat [78]. Another interconnect thermal model based on average switching activities was presented by Huang et al. [87, 88]. These models are, however, not suitable for analyzing signal lines [76, 93], because (i) signal lines have much lower current densities than power lines, and (ii) activities of bus lines, and hence their thermal characteristics are tied to actual traffic patterns on the bus. To overcome these limitations, an energy and thermal model for buses was proposed by Sundaresan and Mahapatra [76, 93]. The energy model calculates energy dissipated due to switching (self and coupling) on bus lines. This model is then used to determine changes in wire temperature due to self-heating in the wire and heat conduction into the surrounding medium. The energy and thermal models proposed by Sundaresan and Mahapatra [76, 93] are presented below. Values for wire geometry (i.e., wire width, spacing, etc.) and technology and equivalence circuit parameters (i.e., capacitance and resistance of lines and repeaters) used in the models, for various technology libraries, are shown in Table 5.6. The top six values are from ITRS-2001 [79]. The values for c_{line} and c_{inter} were obtained from FastCap [89] simulations, and the value of r_{wire} was calculated using the formula:

$$r_{wire} = \rho \cdot l / (w_i \cdot t_i) \tag{5.32}$$

The energy dissipated in a wire i, due to the charging and discharging of its self-capacitance is termed as self energy, and can be expressed as:

$$E_i^s = 0.5 \times (c_{line} + C_{rep}) \cdot V_i^2 \tag{5.33}$$

where c_{line} is the self-capacitance of the wire, C_{rep} is the total capacitance of the repeaters on the line, and V_i is the difference between the final and initial voltages on the line. The total capacitance of repeaters can be expressed as:

$$C_{rep} = h \cdot k \cdot C_0 \tag{5.34}$$

where the repeaters are assumed to be h times the size of a minimum sized inverter, k is the number of repeaters needed for optimal delay on the wire, and

Table 5.6 Wire geometry and equivalent circuit parameters for top metal layer [76, 93]

Parameter	Technology node			
	130 nm	90 nm	65 nm	45 nm
Number of metal layers	8	9	10	10
Wire width, w_i, (nm)	335	230	145	103
Wire thickness, t_i (nm)	670	482	319	236
Height of inter-layer dielectric, t_{ild} (nm)	724	498	329	243
Relative permittivity of dielectric, ε_r	3.3	2.8	2.5	2.1
Thermal conductivity of dielectric, k_{ild} (W/mK)	0.6	0.19	0.12	0.07
Clock frequency, f_{clk} (GHz)	1.68	3.99	6.73	11.51
Supply voltage, V_{dd} (V)	1.1	1.0	0.7	0.6
Maximum current density in a wire, j_{max} (MA/cm^2)	0.96	1.5	2.1	2.7
Self-capacitance of wire, c_{line} (pF/m)	44.06	32.77	25.07	19.05
Coupling capacitance of wire, c_{inter} (pF/m)	91.72	76.84	68.42	58.12
Resistance of wire r_{wire} (kΩ/m)	98.02	198.45	475.62	905.05

C_0 is the capacitance of the minimum sized inverter. The optimal size and number of such repeaters for minimum wire delay [90] is given by the expressions:

$$h = \sqrt{\frac{R_0 \cdot C_{\text{int}}}{C_0 \cdot R_{\text{int}}}} \tag{5.35}$$

$$k = \sqrt{\frac{0.4(R_{\text{int}} \cdot C_{\text{int}})}{0.7(C_0 \cdot R_0)}} \tag{5.36}$$

where R_{int} is the resistance of the wire (r_{wire} in Table 5.6) and R_0 is the resistance of the minimum sized repeater (or inverter). C_{int} is the capacitance of the wire, which can be expressed as:

$$c_{\text{int}} = c_{\text{line}} + 2c_{\text{inter}} \tag{5.37}$$

In addition to energy dissipation due to self-switching, energy is also dissipated due to coupling between lines, due to the charging and discharging of the coupling capacitance between them. Consider two coupled wires i and j. Let V_i be the difference between final and initial voltages on line i, and let V_j be the corresponding difference for line j. If $V_i \neq 0$, and $V_i = -V_j$, then there is a toggle transition between the two lines, resulting in a doubling of the coupling capacitance, according to the Miller effect [3]. The coupling energy in line i is thus given as:

$$E_i^c = 0.5 \times c_{i,j}(V_i^2 - V_i \cdot V_j), \quad i \neq j \tag{5.38}$$

where $c_{i,j}$ is the same as c_{inter}, given in Table 5.6. As mentioned earlier, the values for coupling capacitance between any pair of wires were obtained using FastCap [89]. Using wire geometry information from various technology nodes (Table 5.6), a co-planar bus and a ground plane below the bus were modeled in FastCap. The equivalent electrical circuit for a five wire (5-bit) bus is shown in Fig. 5.17. The complete capacitance matrix for a 32-bit bus was computed (Eq. (5.29)). Simulation results showed that the traditional practice of ignoring coupling between non-adjacent lines can lead to inaccuracies, since the coupling capacitance between non-adjacent lines accounts for 8–10% of total capacitance for current and near future (up to 45 nm) CMOS technology nodes. Thus the coupling energy estimation included the effect of coupling with non-adjacent wires. Finally, the total energy dissipated in a bus is obtained by summing up the self and coupling (including non-adjacent coupling) energies of all the bus lines.

Once the energy on the bus lines dissipated as heat is determined, the temperature rise in each bus wire can be estimated by considering an equivalent thermal *RC* network, as shown in Fig. 5.18 for a 5-bit bus. To make an analogy between thermal and electrical quantities, the temperature difference between two nodes corresponds to a voltage difference, and the heat transfer rate corresponds to the current. Using an analogy with Kirchhoff's current law in electrical circuits, and equating the amount of heat flowing into a node with the amount of flowing out of it, in the thermal equivalent circuit, expressions for the instantaneous power dissipated in each wire of a bus can be developed as shown below.

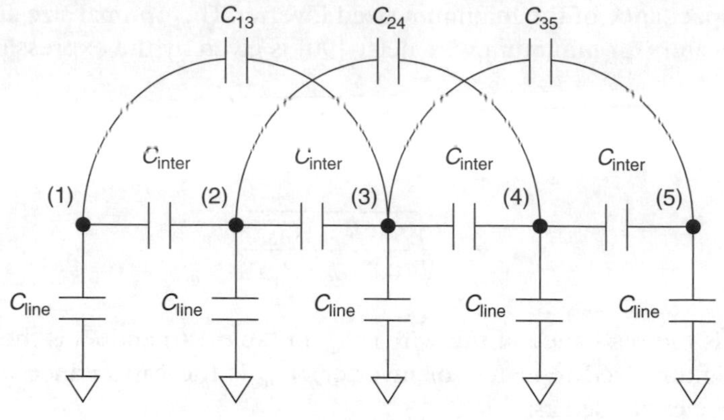

FIGURE 5.17

Equivalent electrical circuit of a 5-bit bus showing self and coupling capacitances [76]
© 2005 IEEE

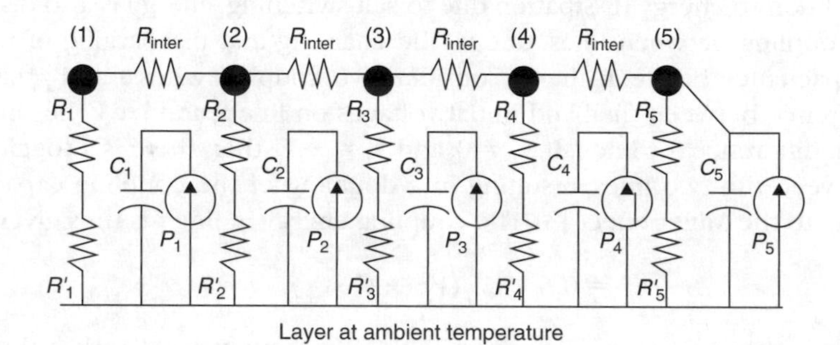

FIGURE 5.18

Equivalent thermal RC circuit of a 5-bit bus showing thermal resistances, capacitances, and heat sources modeled as current sources [76]
© 2005 IEEE

For wires at the edges,

$$P_i = C_i \cdot \frac{d\theta_i}{dt} + \frac{(\theta_i - \theta_0)}{R_i} + \frac{(\theta_i - \theta_{i+1})}{R_{\text{inter}}} \tag{5.39}$$

For wires in the middle,

$$P_i = C_i \cdot \frac{d\theta_i}{dt} + \frac{(\theta_i - \theta_0)}{R_i} + \frac{(2\theta_i - \theta_{i-1} - \theta_{i+1})}{R_{\text{inter}}} \tag{5.40}$$

where P_i is the instantaneous power dissipated in the ith wire, and θ_i is the ambient temperature (45°C or 318.15K). C_i is the thermal capacitance per unit length of the wire, given as:

$$C_i = C_s \cdot (t_i \cdot w_i) \tag{5.41}$$

where C_s is the specific heat per unit volume of the metal, w_i is the wire width, and t_i is the wire thickness, as shown in Fig. 5.19. R_i is the thermal resistance per

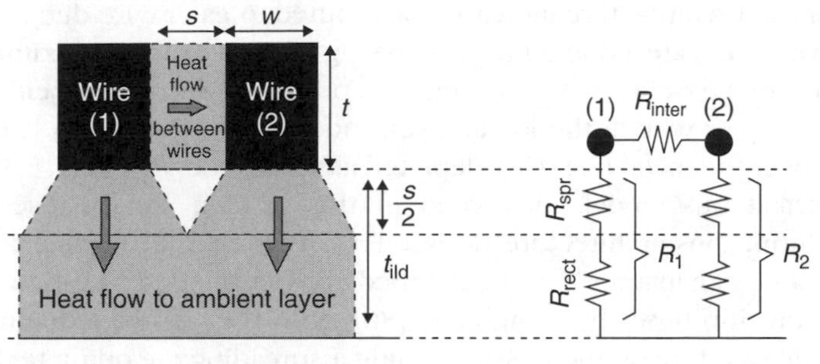

FIGURE 5.19

Geometry used to calculate equivalent thermal resistance for a wire in [76], based on [91]

unit length of the wire along the path of the heat transfer, as shown in Fig. 5.19. It can be calculated using knowledge of wire geometry and thermal conductivity of the *ILD*, as described by Chiang et al. [91]. R_i is essentially the sum of spreading resistance (due to the spreading of heat from the bottom face of the wire in a trapezoidal manner), and the thermal resistance due to rectangular heat flow, as depicted in Fig. 5.19. These relations can be expressed as,

$$R_i = R_{\text{spr}} + R_{\text{rect}} \tag{5.42}$$

$$R_i = \frac{\ln\left(\frac{w_i + s_i}{w_i}\right)}{2 \cdot k_{\text{ild}}} + \frac{t_{\text{ild}} - 0.5_{S_i}}{k_{\text{ild}}(w_i + S_i)} \tag{5.43}$$

where s_i is the wire spacing, w_i is the wire width, t_{ild} is the thickness of the ILD, and k_{ild} is the thermal conductivity of the ILD. Heat can also be transferred between adjacent wires due to a difference in activity rates of neighboring lines which creates a temperature gradient and hence a lateral heat flow. This effect is captured in the model using a lateral inter-wire thermal resistance per unit length (R_{inter}), whose value depends on wire geometry parameters such as wire spacing (s_i) and thickness (t_i), as shown in Fig. 5.18, as well as *IMD* thermal conductivity k_{imd}, and can be expressed as:

$$R_{\text{inter}} = s_i/(k_{\text{imd}} \cdot t_i) \tag{5.44}$$

With this information, Eqs. (5.39) and (5.40) can be solved to obtain the wire temperature θ_i. However, a correction needs to be applied to account for heat transfer from the underlying layers. In current C4/CBGA packages, a heat transfer path exists from the bottom-most to the top-most interconnect layer, from where the heat flows on through C4 bumps, the ceramic substrate, CBGA joints, and the printed circuit board to the ambient air [87]. A first order approximation for this heat transfer from lower layers can be expressed as:

$$\Delta\theta = \sum_{i=1}^{N} \frac{t_{\text{ild},i}}{k_{\text{ild},i} s_i \alpha_i} \cdot \left[\sum_{j=i}^{N-1} (j_{\max})^2 \rho_j \alpha_j t_j \right] \tag{5.45}$$

where a constant temperature increase is attributed to each wire due to this effect, and the lower layers are assumed to be carrying current at their maximum density j_{max}. α_i is a coverage factor that represents the probability of a wire being thermally coupled to another wire in the lower layer, and is given a value 0.5 [91], and ρ_j is the resistivity of the metal line (Cu), depicted as resistance R_i' in Fig. 5.18.

The thermal bus models presented in this section can be used to guide designers during bus architecture design. For instance, the thermal model from Sundaresan and Mahapatra [76, 93] described above was used to estimate the temperature of on-chip buses by Wang et al. [86], with the goal of reducing the transient peak temperature on the buses by using a spreading encoding technique.

5.4 DISCUSSION: PVT VARIATION-AWARE POWER ESTIMATION

In UDSM CMOS technologies below 90 nm, PVT variability has been shown to cause considerable fluctuation in power consumption for SoCs [94]. Consequently, power estimates derived early in the design flow are no longer valid, and considerable redesign effort is needed to account for these variability-induced fluctuations. Figure 5.20 illustrates the traditional SoC design approach, in which designers explore the power space of the design at the system level, and typically select the configuration with the least power consumption. In the figure, points *A* and *B* represent design configurations, and a designer would normally select configuration *B*, with the lower power consumption at the system level. Later in the design flow, at the gate level, designers encounter PVT variations that alter the power characteristics and behavior of the synthesized design. Each of the configuration points becomes a large region of uncertainty (representing possible power consumption under different PVT conditions), and it is no longer clear whether configuration *A* or *B* is the superior one in terms of lower power consumption. It is possible that an instance of the design configuration *A* (shown as *A*′ in the figure) is found to be superior to the best instance of design configuration *B* (shown as

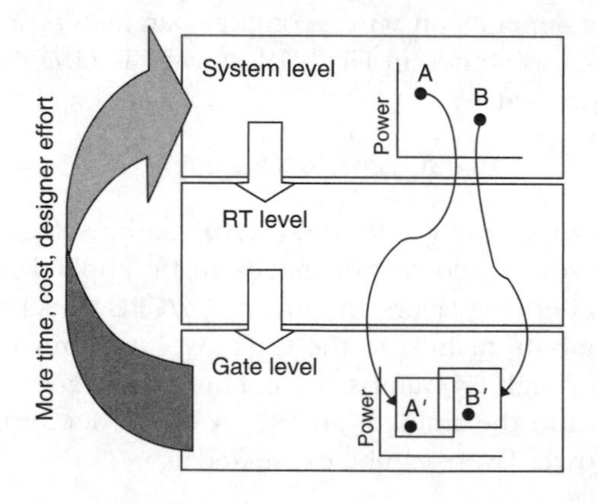

FIGURE 5.20

Traditional flow for on-chip communication architecture power estimation [95]

B' in the figure). As a result, designers end up spending considerable time and effort to explore design configurations at the gate level. It is also important to ensure that PVT variations do not cause a violation of design constraints for the selected design configuration. Design reiterations might be required if violations are detected (requiring changes in the design at the system level) that can severely influence design cost and time-to-market.

Traditionally, the most important means by which a foundry communicates PVT variations to designers is through library characterization at design corners, known as PVT corners, relating cell metrics (timing, power) to PVT variations. Up until the 130 nm CMOS technology node, design tools relied on three corners: *Typical*, *Worst*, and *Best* corners. The adjectives associated with these corners relate mainly to timing. The *Worst* corner combines high temperature, low V_{dd} (nominal − 10%), and a slow-slow (S-S) process that leads to worst case timing. The *Best* corner goes the opposite way, combining low temperature, high V_{dd} (nominal + 10%), and fast-fast (F-F) process to achieve maximum performance. The *Typical* performance corner lies between these two extreme corners. Up until the 130 nm node, synthesis tools used the *Worst* and *Best* corners during synthesis, guaranteeing that all the resulting functional chips would meet timing (by using *Worst* case corner to avoid setup time violations at register inputs, and using *Best* case corner to guarantee that no hold time violations occur). The *Typical* corner is usually used to characterize power consumption under nominal conditions. Since leakage was negligible up until the 130 nm node, the only factor affecting power consumption (mostly dynamic) was V_{dd}, and variations of about 20% were expected between the three corners.

For UDSM CMOS technologies under 90 nm, several factors—such as the drastic increase in the device leakage power, IR drop and use of power management strategies such as DVS/DFS and voltage islands—require IPs (and especially cell libraries, I/O, and memories) to be characterized at many more PVT corners, shown in Table 5.7. These corners become necessary for a variety of reasons: A *TypicalPerf* corner, for example, does not provide a realistic assessment of leakage power under typical conditions because when the application is running, a die would heat up to well above 25°C. With temperature being an exponential factor in leakage, a more realistic *TypicalLeakage* corner must be considered with typical-typical (T-T) process, nominal V_{dd}, and 125°C. A *WorstLeakage* corner is used to assess the absolute maximum leakage under F-F process (i.e., low V_t) and high V_{dd} (Nominal + 10%). Power management strategies such as voltage islands and discrete voltage scaling (DVS) cannot be validated at the chip level unless IPs are characterized under several low V_{dd} conditions. This requires another set of corners. IPs such as the Metro libraries from Artisan (ARM) [96] are characterized for V_{dd} increments of 100 mV for a range of possible V_{dd} values. *MaxPerfLowV*, *TypPerfLowV*, and *WorstPerfLowV* are needed for each V_{dd}. On the other hand, and under certain conditions, more performance may be needed. For that case, some cell libraries are characterized for higher than normal operating conditions. Thus, another set of corners are needed, including *MaxPerfHighV*, *TypPerfHighV*, and *WorstPerfHighV*. Since leakage can be significantly higher under those conditions, additional *TypicalLeakageHighV* and *WorstLeakageHighV* are sometimes

Table 5.7 PVT corners in UDSM technologies [95]

Nominal V_{dd} (in volts)	Corner	Process	Temperature	V_{dd}
1.0	MaxPerf	F-F	0	1.1
	TypPerf	T-T	25	1
	WorstPerf	S-S	125	0.9
	WorstLeakage	F-F	125	1.1
	TypLeakage	T-T	125	1
0.7	MaxPerfLowV	F-F	0	0.77
	TypPerfLowV	T-T	25	0.7
	WorstPerfLowV	S-S	125	0.7
1.2	MaxPerfHighV	F-F	0	1.32
	TypPerfHighV	T-T	25	1.2
	WorstPerfHighV	S-S	125	1.08
	WorstLeakageHighV	F-F	125	1.32
	TypLeakageHighV	T-T	125	1.2

available in order to assess typical and worst case leakage under high V_{dd} conditions. Note that the PVT corners shown in Table 5.7 do not constitute a maximal set. Many more corners can be added during library characterization, to support more elaborate design methodologies and possible operating environments and conditions. Alternatively, some technology libraries may not support *HighV* corners for reliability reasons.

Pasricha et al. [95] presented several experimental results for the power consumption of on-chip communication architectures across different PVT corners. Figures 5.21 and 5.22 show the normalized power (measured at the gate level) at the 90 and 65 nm nodes, respectively, for four different configurations of the AMBA AHB bus matrix [47] on-chip communication architecture. It can be observed from the figures that there is significant variability in estimated power for 90 and 65 nm libraries, especially between the *WorstLeakageHighV* and *TypPerfLowV* corners (more than a 10× difference). As mentioned earlier, just considering the traditionally used corners (e.g., *TypPerf*) is not realistic because there is a large variation in power consumption for sub-90 nm libraries due to DSM effects that can only be captured by additional corners. In order to meet power goals, designers thus need to consider multiple PVT corners to understand the power characteristics of a design.

From the experiments presented above, as well as experiments on larger SoC designs, Pasricha et al. [95] found that the power consumption for a PVT corner scales almost linearly with frequency. However, a much more interesting and important observation was that the power consumption numbers obtained for

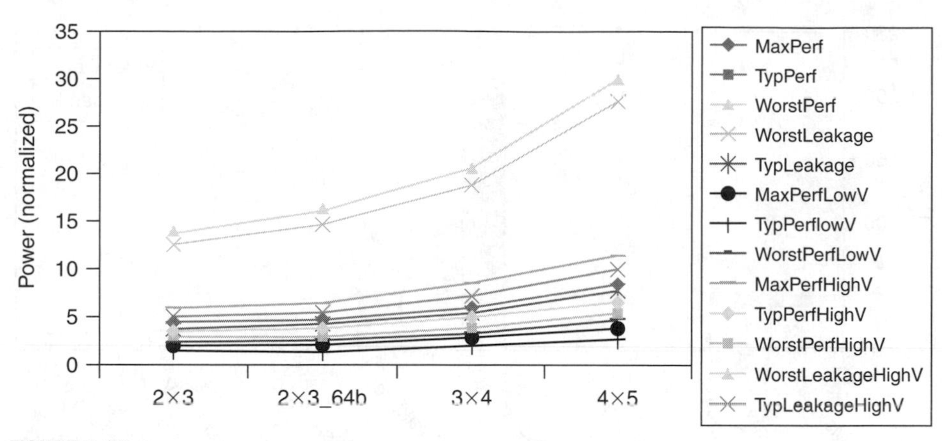

FIGURE 5.21

Normalized power for bus matrix configurations at 90 nm [95]
© 2008 IEEE

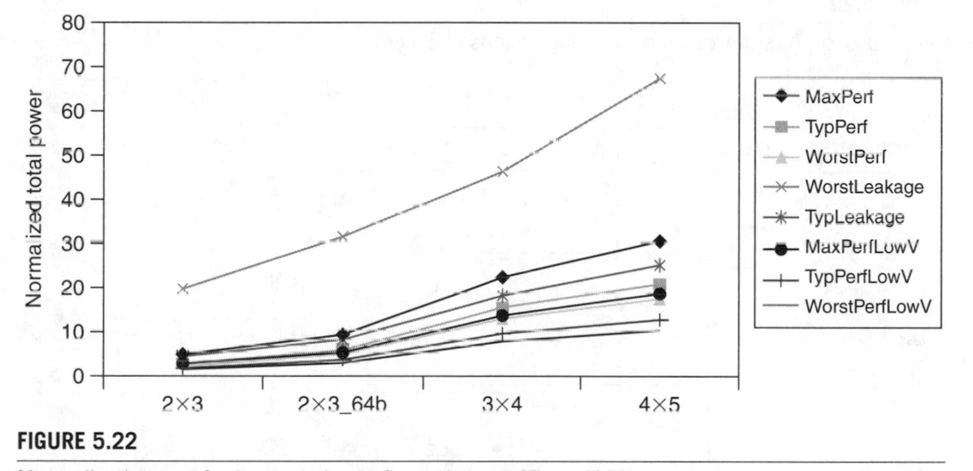

FIGURE 5.22

Normalized power for bus matrix configurations at 65 nm [95]
© 2008 IEEE

the different PVT corners show an almost constant ratio relative to each other. Thus if the power consumption of the bus matrix on-chip communication architecture for an implementation with PVT corner C1 is expressed as:

$$P_{C1} = P_{L1} + P_{D1} \times f \tag{5.46}$$

where P_{C1} gives the base level total power for a corner C1 that has base level leakage power P_{L1} and base level dynamic power P_{D1}, at frequency f; then the power consumption for an implementation under any other PVT corner C2 can be expressed as:

$$P_{C2} = \alpha_{1-2} \times P_{L1} + \beta_{1-2} \times P_{D1} \times f \tag{5.47}$$

where the total power P_{C2} for another corner C2 can be linearly scaled from the base level power relation in Eq. (5.46) by using scaling factors α_{1-2} and β_{1-2}

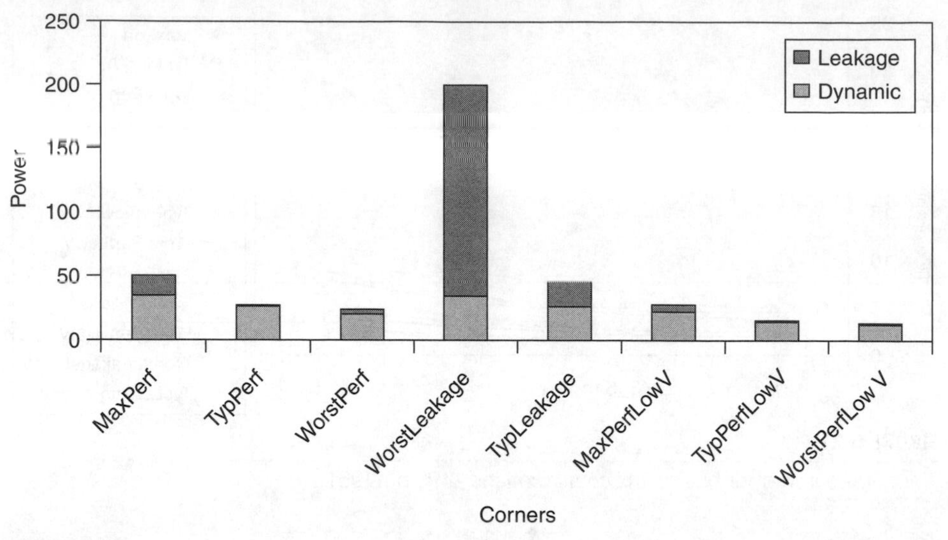

FIGURE 5.23

Decomposed power for bus matrix configurations at 65 nm [95]
© 2008 IEEE

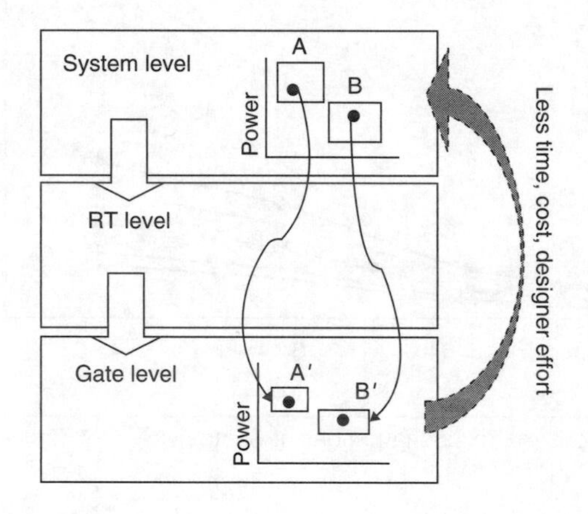

FIGURE 5.24

Advantages of PVT variation-aware power estimation at the system level [95]
© 2008 IEEE

for leakage and dynamic power, respectively. The scaling factors can be easily obtained by decomposing the total power into dynamic and leakage components (as shown in Fig. 5.23, for the 65 nm case shown in Fig. 5.22), and averaging the ratio values. Experiments to determine the accuracy of this scaling technique showed that it is possible to estimate power consumption for different PVT corners with extremely good accuracy for dynamic power (less than 5% in most cases), and fairly good accuracy for leakage power (less than 10% in most cases). This amenability to scaling for PVT corners is an extremely important result, and was obtained for cell libraries characterized with industrial strength numbers.

Thus, knowing the leakage and dynamic power for one PVT corner at the system level can allow a designer to obtain the power for other PVT corners using a simple linear model with good accuracy, which speeds up power exploration across corners considerably. Figure 5.24 shows how this technique can "scale up" and abstract the PVT variability at the system level, to provide a more realistic characterization of the design space, early in the design flow. The designer can then select a design configuration with greater confidence, after analyzing its behavior under PVT variations. This significantly reduces the exploration and redesign effort later in the design flow.

5.5 SUMMARY

Power is increasingly becoming a first class design objective, and the contribution of on-chip communication architectures to the overall chip power is rapidly increasing as technology is scaled down. It thus becomes essential for designers to create models for the early exploration of on-chip communication architecture power consumption. In this chapter, we presented a survey of several approaches that have been proposed to model the power consumption of bus wires, as well as bus logic components, at different levels of granularity. A major side effect of power dissipation on interconnect lines is an increase in interconnect temperature. These thermal effects have a marked impact on the power, performance, design, and reliability of interconnects. With shrinking process technology, these effects are only getting worse, motivating the need for thermal models of bus wires for designers to use during analysis and optimization, so that on-chip bus temperatures can be kept within reasonable limits. This chapter also presented thermal models of interconnects that have been proposed in the last few years. An interesting observation regarding the need to consider thermal effects in power-aware design methodologies was presented in [92], where the authors claim that thermal-aware estimations of power savings are 17–38% higher compared to conventional techniques that do not consider thermal effects. This implies that thermal effects must be considered while evaluating the effectiveness of power saving techniques, to prevent overly conservative design decisions. Finally, we presented a discussion on how it is becoming essential to consider PVT variations during power estimation, and how traditional power estimation techniques can be scaled to incorporate the effect of variations in UDSM CMOS technologies. All of the power and thermal estimation models (and PVT variation-aware power estimation techniques) described in this chapter are most useful when incorporated into larger frameworks used for communication architecture design. For instance, Pasricha et al. [56] make use of the power estimation models presented in Section 5.2 in their CAPPS on-chip communication architecture synthesis framework, to enable power-performance trade-offs during communication architecture synthesis. Pasricha et al. [95] also make use of the PVT variation-aware power estimation technique described in Section 5.4 to enable more reliable power estimation during on-chip communication architecture synthesis at the system level. CAPPS and other on-chip communication architecture synthesis frameworks are described in detail in the next chapter.

REFERENCES

[1] T. Mudge, "Power: A first-class architectural design constraint," *IEEE Computer*, Vol. 34, No. 4, April 2001, pp. 52–58.

[2] P. Gupta, L. Zhong and N. K. Jha, "A high level interconnect power model for design space exploration," *International Conference on Computer Aided Design (ICCAD)*, November 2003, pp. 551–558.

[3] J. Rabaey, A. Chandrakasan and B. Nikolic, *Digital Integrated Circuits*, Englewood Cliffs, Prentice Hall, NJ, 1996.

[4] M. Kuhlmann and S. S. Sapatnekar, "Exact and efficient crosstalk estimation," *IEEE Transactions on Computer-Aided Design (TCAD)*, Vol. 20, No. 7, July 2001, pp. 858–866.

[5] P. P.Sotiriadis and A. P. Chandrakasan, "A bus energy model for deep submicron technology," *IEEE Transactions on Very Large Scale Integration Systems (TVLSI)*, Vol. 10, No. 3, June 2002, pp. 341–350.

[6] P. Kapur, G. Chandra and K. C. Saraswat, "Power estimation in global interconnects and its reduction using a novel repeater optimization methodology," in *Proceedings Design Automation Conference (DAC)*, June 2002, pp. 461–466.

[7] K. Lahiri and A. Raghunathan, "Power analysis of system-level on-chip communication architectures," *International Conference on CODES+ISSS*, 2004, pp. 236–241.

[8] D. Sylvester and C. Hu, "Analytical modeling and characterization of deep-submicrometer interconnect," *Proceedings of IEEE*, Vol. 89, No. 5, May 2001, pp. 634–664.

[9] K. Banerjee, A. Mehrotra, A. Sangiovanni-Vincentelli and C. Hu, "On thermal effects in deep submicron VLSI interconnects," in *Proceedings of 36th ACM Design Automation Conference (DAC)*, 1999, pp. 885–891.

[10] W. Fornaciari, D. Sciuto and C. Silvano, "Power estimation for architectural exploration of HW/SW communication on system-level buses," *International Conference on HW/SW Codesign 1999*, Roma, Italy, April 1999, pp. 152–156.

[11] W. Fornaciari, D. Sciuto and C. Silvano, "Power estimation of system-level buses for microprocessor-based architectures: A case study," *International Conference on Computer Design (ICCD)* 1999, pp.131–136.

[12] D. Langen, A. Brinkmann and U. Ruckert, "High level estimation of the area and power consumption of on-chip interconnects," *ASIC/SOC Conference, 2000. Proceedings of 13th Annual IEEE International*, 2000, pp. 297–301.

[13] A. Raghunathan, N. K. Jha and S. Dey, *High-Level Power Analysis and Optimization*, Kluwer Academic Publishers, Norwell, MA, 1998.

[14] T. Givargis and F. Vahid, "Interface exploration for reduced power in core-based systems," *International Symposium on System Synthesis*, 1998, pp. 117–122.

[15] T. Givargis, F. Vahid and J. Henkel, "Fast cache and bus power estimation for parameterized SOC design," *Design, Automation and Test in Europe Conference*, 2000, pp. 333–339.

[16] T. D. Givargis, F. Vahid and J. Henkel, "Evaluating power consumption of parameterized cache and bus architectures in system-on-a-chip designs," *IEEE Transactions on Very Large Scale Integration Systems (TVLSI)*, Vol. 9, No. 4, August 2001, pp. 500–508.

[17] M. R. Stan and W. P. Burleson, "Bus-invert coding for low-power VO," *IEEE Transactions on Very Large Scale Integration Systems (TVLSI)*, Vol. 3, No. 1, March 1995, pp. 49–58.

[18] C. L. Su, C. Y. Tsui and A. M. Despain, "Saving power in the control path of embedded processors," *IEEE Design and Test of Computers*, Vol. 11, No. 4, Winter 1994, pp. 24–30.

[19] L. Benini, G. De Micheli, E. Macii, M. Poncino and S. Quer, "Power optimization of core-based systems by address bus encoding," *IEEE Transactions on Very Large Scale Integration Systems (TVLSI)*, Vol. 6, No. 4, December 1998, pp. 554–562.

[20] H. Mehta, R. M. Owens and M. J. Irwin, "Some issues in gray code addressing," *GLS-VLSI-96: IEEE 6th Great Lakes Symposium on VLSI*, Ames, IA, March 1996, pp. 178–180.

[21] E. Musoll, T. Lang and J. Cortadella, "Working-zone encoding for reducing the energy in microprocessor address buses," *IEEE Transactions on Very Large Scale Integration Systems (TVLSI)*, Vol. 6, No. 4, December 1998, pp. 568–572.

[22] J.-H. Chern, J. Huang, L. Arledge, P.-C. Li and P. Yang, "Multilevel metal capacitance models for CAD design synthesis systems," *IEEE Electron Device Letters*, Vol. 13, January 1992, pp. 32–34.

[23] T. Sakurai and K. Tamaru, "Simple formulas for two- and three-dimensional capacitances," *IEEE Transactions on Electron Devices*, Vol. ED-30, 1983, pp. 183–185.

[24] R. Ho, K. W. Mai and M. A. Horowitz, "The future of wires," *Proceedings of IEEE*, Vol. 89, April 2001, pp. 490–504.

[25] M. Bohr, "Interconnect scaling—The real limiter to high performance ULSI," in *Proceedings of IEDM*, 1995, pp. 241–244.

[26] Y. Nishi, "The trend of on-chip interconnects: An international perspective," presented at the *1998 Spring Seminar Series*, Stanford University.

[27] J. Davis and J. Meindl, "Compact distributed RLC interconnect models. Part II: Coupled line transient expressions and peak crosstalk in multilevel networks," *IEEE Transactions on Electron Devices*, Vol. 47, November 2000, pp. 2078–2087.

[28] K. Yamashita and S. Odanaka, "Interconnect scaling scenario using a chip level interconnect model," *IEEE Transactions on Electron Devices*, Vol. 47, January 2000, pp. 90–96.

[29] N. Menezes and L. Pillegi, "Analyzing on-chips interconnect effects," *Design of High Performance Microprocessor Circuits*, A. Chandrakasan, W J. Bowhill and F. Fox (Eds.), IEEE Press, Piscataway, NJ, 2001. Chapter. 16.

[30] S. Das, W. Smith and C. Paul, "Modeling of data bus structures using numerical methods," in *Proceedings of International Symposium on Electromagnetic Compatibility*, Dallas, TX, 1993, pp. 409–414.

[31] A. E. Ruehli (Ed.), *Circuit Analysis, Simulation, and Design*, Vols. 1 and 2, Elsevier, New York, 1986–1987.

[32] S. C. Wong, G.-Y. Lee and D.-J. Ma, "Modeling of interconnect capacitance, delay, and crosstalk in VLSI," *IEEE Transactions on Semiconductor Manufacturing*, Vol. 13, February 2000, pp. 108–111.

[33] C. K. Cheng, J. Lillis, S. Lin and N. Chang, *Interconnect Analysis and Synthesis*, Wiley, New York, 2000.

[34] L. Pileggi, "Coping with RC(L) interconnect design headaches," in *Proceedings of IEEE/ACM International Conference*, San Jose, CA, 1995, pp. 246–253.

[35] K.-W. Kim, S.-O. Jung, U. Narayanan, C. L. Liu and S.-M. Kang, "Noise-aware power optimization for on-chip interconnect," in *International Symposium on Low Power Electronics Design*, Italy, 2000, pp. 108–113.

[36] T. Uchino and J. Cong, "An interconnect energy model considering coupling effects," *IEEE Transactions on Computer-Aided Design of Integrated Circuits and Systems*, Vol. 21, No. 7, July 2002, pp. 763–776.

[37] P. Heydari and M. Pedram, "Interconnect energy dissipation modeling in high-speed ULSI circuits," in *Proceedings of Asia and South Pacific Design Automation Conference (ASPDAC)*, January 2002, pp. 132–140.

[38] N. Magen, A. Kolodny, U. Weiser and N. Shamir, "Interconnect-power dissipation in a microprocessor," in *Proceedings of the 2004 International Workshop on System Level Interconnect Prediction*, Paris, France, February 14–15, 2004, pp. 7–13.

[39] C. Kretzschmar, A K. Nieuwland and D. Muller, "Why transition coding for power minimization of on chip buses does not work," *Design, Automation and Test in Europe Conference and Exhibition (DATE)*, 2004. Proceedings, February 16–20, 2004, pp. 512–517.

[40] C. Chang, J. Cong, T. Uchinko and X. Yuan, "Power model for interconnect planning," *Proceedings of the Workshop on Synthesis and System Integration of Mixed Technologies*, October 2001, pp. 234–241.

[41] L. Zhong and N. K. Jha, "Interconnect-aware high-level synthesis for low power," in *Proceedings of International Conference on Computer-Aided Design*, November 2002, pp. 110–117.

[42] C. N. Taylor, S. Dey and Y. Zhao, "Modeling and minimization of interconnect energy dissipation in nanometer technologies," in *Proceedings of Design Automation Conference (DAC)*, June 2001, pp. 754–757.

[43] A. Raghunathan, S. Dey and N. K. Jha, "Register transfer level power optimization with emphasis on glitch analysis and reduction," *IEEE Transactions on Computer-Aided Design*, Vol. 18, No. 8, August 1999, pp. 1114–1131.

[44] J. Cong and Z. Pan, "Interconnect performance estimation models for design planning," *IEEE Transactions on Computer-Aided Design*, Vol. 20, No. 6, June 2001, pp. 739–752.

[45] H. B. Bakoglu, *Circuits, Interconnects, and Packaging for VLSI*, Addison-Wesley, Reading, MA, 1990.

[46] L. Zhong and N. K. Jha, "Interconnect-aware high-level synthesis for low power," in *Proceedings of International Conference on Computer-Aided Design (ICCAD)*, November 2002, pp. 110–117.

[47] ARM AMBA Specification and Multilayer AHB Specification (rev2.0), http://www.arm.com, 2001.

[48] Designware IP Reuse Tools, Synopsys Inc., http://www.synopsys.com/products/designware/ipreuse_tools.html.

[49] RTL Synthesis, Synopsys Inc., http://www.synopsys.com/products/logic/.

[50] CB-12 Family L/M/H Type for Cell Based ICs, http://www.necel.com/cbic/en/product/cb12.html.

[51] Modelsim 5.7e, http://www.model.com.

[52] NEC OpenCAD V 5.2 Users Manual. NEC Electronics, Inc., January 1999.

[53] W. E. Donath, "Placement and average interconnection lengths for computer logic," *IEEE Transactions on Circuits and Systems*, Vol. 26, April 1979, pp. 272–277.

[54] ARM AMBA AXI Specification, www.arm.com/armtech/AXI.

[55] I. Lee, H. Kim, P. Yang, S. Yoo, E.-Y. Chung, K.-M. Choi, J.-T. Kong and S.-K. Eo, "PowerViP: SoC power estimation framework at transaction level," *Asia and South Pacific Design Automation Conference (ASPDAC)*, January 24–27, 2006, pp. 551–558.

[56] S. Pasricha, Y. Park, F. J. Kurdahi and N. Dutt, "System-level power-performance trade-offs in bus matrix communication architecture synthesis," in *Proceedings of the 4th International Conference on Hardware/Software Codesign and System Synthesis (CODES+ISSS)*, Seoul, Korea, October 22–25, 2006, pp. 300–305.

[57] Cadence PKS, www.cadence.com/datasheets/pks_ds.pdf.

[58] Synopsys PrimePower, www.synopsys.com.

[59] GNU R, http://www.gnu.org/software/r/R.html.

[60] J. J. Faraway, *Linear Models with R*, CRC Press, 2004.

[61] Berkeley Predictive Technology Model, U.C. Berkeley, http://www-devices.eecs.berkeley.edu/~ptm/.

[62] S. N. Adya and I. L. Markov, "Fixed-outline floorplanning: Enabling hierarchical design," *IEEE Transactions on VLSI Systems*, Vol. 11, No. 6, December 2003, pp. 1120–1135.

[63] S. Pasricha, N. Dutt and M. Ben-Romdhane, "Extending the transaction level modeling approach for fast communication architecture exploration," in *Proceedings of Design and Automation Conference (DAC)*, San Diego, CA, June 2004, pp. 113–118.

[64] S. Pasricha, N. Dutt and M. Ben-Romdhane, "Fast exploration of bus-based on-chip communication architectures," *International Conference on Hardware/Software Codesign and System Synthesis (CODES+ISSS)*, Stockholm, Sweden, September 2004, pp. 242–247.

[65] SystemC initiative, www.systemc.org.

[66] T. Grötker, S. Liao, G. Martin and S. Swan, *System Design with SystemC*, Kluwer Academic Publishers, 2002.

[67] M. Caldari, M. Conti, M. Coppola, P. Crippa, S. Orcioni, L. Pieralisi and C. Turchetti, "System-level power analysis methodology applied to the AMBA AHB bus," *Design, Automation and Test in Europe Conference and Exhibition*, 2003, pp. 32–37.

[68] A. Bona, V. Zaccaria and R. Zafalon, "System level power modeling and simulation of high-end industrial network-on-chip," *Design, Automation and Test in Europe Conference and Exhibition*, February 16–20, 2004, pp. 318–323.

[69] N. Dhanwada, I. Lin and V. Narayanan, "A power estimation methodology for systemC transaction level models," in *Proceedings of the 3rd IEEE/ACM/IFIP International Conference on Hardware/Software Codesign and System Synthesis (CODES+ISSS)*, 2005, pp. 142–147.

[70] U. Neffe, K. Rothbart, C. Steger, R. Weiss, E. Rieger and A. Muhlberger, "Energy estimation based on hierarchical bus models for power-aware smart cards," *Design, Automation and Test in Europe Conference and Exhibition*, February 16–20, 2004, pp. 300–305.

[71] Philips Electronic Design & Tools, *Diesel 2.6 User Manual*, Eindhoven, The Netherlands, 2001.

[72] "STBus Communication System: Concepts and Definitions," *Reference Guide*, STMicroelectronics, May 2003.

[73] Synopsys Inc., "Core Consultant Reference Manual," "Power Compiler Reference Manual" and "VCS: Verilog Compiled Simulator Reference Manual," v2003.06, June 2003.

[74] K. Banerjee and A. Mehrotra, "Global interconnect warming," *IEEE Circuits and Devices*, September 2001, pp. 16–32.

[75] J. Ida et al., "Reduction of wiring capacitance with new low dielectric SiOF interlayer film for high speed/low power sub-half micron CMOS," *Symposium on VLSI Tech. Digest of Technical Papers*, 1994, pp. 59–60.

[76] K. Sundaresan and N. R. Mahapatra, "An accurate energy and thermal model for global signal buses," *18th International Conference on VLSI Design*, January 3–7, 2005, pp. 685–690.

[77] K. Banerjee, A. Amerasekera, G. Dixit and C. Hu, "The effect of interconnect scaling and low-k dielectric on the thermal characteristics of the IC metal," *Technical Digest. International Electron Devices Meeting*, 1996, pp. 65–68.

[78] T.-Y. Chiang and K. C. Saraswat, "Closed-form analytical thermal model for accurate temperature estimation of multilevel ULSI interconnects," in *2003 Symposium on VLSI Circuits Digest of Papers*, 2003, pp. 275–279.

[79] Semiconductor Industry Association, *International Technology Roadmap for Semiconductors (ITRS)*, 2001 Edition, 2001.

[80] K. Roy, S. Mukhopadhyay and H. Mahmoodi-Meimand, "Leakage current mechanisms and leakage reduction techniques in deep-submicrometer CMOS circuits," *Proceedings of the IEEE*, Vol. 91, No. 2, February 2003, pp. 305–327.

[81] C. Duvvury and A. Amerasekera, "ESD: A pervasive reliability concern for IC technologies," *Proceedings of IEEE*, Vol. 81, May 1998, pp. 690–702.

[82] T. Green, "A review of EOS/ESD field failures in military equipment," in *Proceedings of the 10th, EOS/ESD Symposium*, 1998, pp. 7–14.

[83] P. K. Chatterjee, W. R. Hunter, A. Amerasekera, S. Aur, C. Duvvury, P. E. Nicollian, L. M. Yang and P. Yang, "Trends for deep submicron VLSI and their implications for reliability," *Reliability Physics Symposium, 33rd Annual Proceedings., IEEE International*, 1995, pp. 1–11.

[84] S. Rzepka, K. Banerjee, E. Meusel and C. Hu, "Characterization of self-heating in advanced VLSI interconnect lines based on thermal finite element simulation," *IEEE Transactions on Components, Packaging and Manufacturing Technology—Part A*, Vol. 21, No. 3, 1998, pp. 1–6.

[85] K. Banerjee, "Trends for ULSI interconnections and their implications for thermal, reliability and performance issues (Invited Paper)," in *Proceedings of the 7th International Dielectrics and Conductors for ULSI Multilevel Interconnection Conference*, March 2001, pp. 38–50.

[86] F. Wang, Y. Xie, N. Vijaykrishnan and M. J. Irwin, "On-chip bus thermal analysis and optimization," *Design, Automation and Test in Europe (DATE)*, March 2006, pp. 1–6.

[87] W. Huang, M. R. Stan, K. Skadron, K. Sankaranarayanan, S. Ghosh and S. Velusamy, "Compact thermal modeling for temperature-aware design," in *Proceedings of the Annual ACM/IEEE Design Automation Conference*, June 2004, pp. 878–883.

[88] W. Huang, M. R. Stan, K. Skadron, K. Sankaranarayanan, S. Ghosh and S. Velusamy, "Compact thermal modeling for temperature-aware design," Technical Report CS-2004-13, Department of Computer Science, University of Virginia, April 2004.

[89] K. Nabors, S. Kim, J. White and S. Senturia, "Fast capacitance extraction of general three-dimensional structures," in *Proceedings of International Conference on Computer Design (ICCD)*, 1991, pp. 479–484.

[90] A. Naeemi, R. Venkatesan and J. D. Meindl, "Optimal global interconnects for GSI," *IEEE Transactions on Electron Devices*, Vol. 50, No. 4, April 2003, pp. 980–987.

[91] T.-Y. Chiang, K. Banerjee and K. Saraswat, "Compact modeling and SPICE-based simulation for electrothermal analysis of multilevel ULSI interconnects," in *Proceedings of the International Conference on Computer-Aided Design (ICCAD)*, 2001, pp. 165–172.

[92] J. C. Ku, M. Ghoneima and Y. Ismail, "The importance of including thermal effects in estimating the effectiveness of power reduction techniques," in *Proceedings of the IEEE Custom Integrated Circuits Conference*, September 18–21, 2005, pp. 301–304.

[93] K. Sundaresan and N. R. Mahapatra, "Accurate energy dissipation and thermal modeling for nanometer-scale buses," *International Symposium on High-Performance Computer Architecture (HPCA)*, February 12–16, 2005, pp. 51–60.

[94] S. Borkar et al., "Design and reliability challenges in nanometer technologies," *Design and Automation Conference*, 2004, p. 75.

[95] S. Pasricha, Y. Park, F. Kurdahi and N. Dutt, "PVT variation aware on-chip communication architecture synthesis," *International Conference on VLSI Design*, January 2008.

[96] http://www.arm.com/products/physicalip/metro.html.

CHAPTER

Synthesis of On-Chip Communication Architectures

6

The design of on-chip communication architectures is becoming more challenging as the number of components integrated into a single embedded multiprocessor system-on-chip (MPSoC) increases due to greater market demands for convergence devices. The increasing number of components in systems translates into more inter-component communication that must be handled by the on-chip communication infrastructure. The pull of multi-dimensional design constraints such as performance, power, reliability, cost, area, and time-to-market further complicates the communication architecture design process. Designs are expected to support high performance and reliability, along with lower power consumption, cost, area, and time-to-market. These design goals often conflict with each other—for instance, a communication architecture that supports high performance data transfers by adding more buses and increasing bus clock frequencies increases power consumption, cost, and area. The design of on-chip communication architectures thus requires techniques, not only to optimize for individual design goals, but also focus on making design decisions that can provide a good balance between the design goals.

The bus-based on-chip communication architecture design space is a union of two subspaces: (i) the *topology* space, which is concerned with the number of buses, how the buses are interconnected and the allocation of components on these buses, and (ii) the *protocol parameter* space, which is concerned with the values of bus protocol parameters such as arbitration schemes, bus widths, bus clock frequencies, buffer sizes, and DMA (direct memory access) burst sizes. *Bus architecture synthesis* can be defined as the process of designing a bus-based communication architecture topology and/or its protocol parameters to satisfy application constraints. The *inputs* to bus architecture synthesis are typically high level tasks mapped onto hardware and software components on the chip, that need to communicate data with each other (and possibly with external, off-chip components such as dynamic RAM (DRAM) memory). The bus architecture synthesis *constraints* can be related to performance, power dissipation, area, or reliability (or a combination of these). For instance, a streaming video application such as H.264

requires video frames to be displayed at a certain fixed frame rate (i.e., frames/second), which requires the communication architecture of an H.264 MPSoC design to guarantee a minimum data throughput to the display buffers. If this MPSoC is part of a mobile device, it will have further constraints limiting power dissipation (to maximize operation time on a limited battery budget) and chip area (to reduce size of the mobile device) that the communication architecture must not violate. Some of the constraint metrics can be practically represented in multiple forms. For instance, performance constraints can be represented as data latency or data throughput constraints, depending on the nature of the application. Similarly power constraints can be represented as average power, peak power, or overall energy constraints. Depending on the constraint type and its practical representation, as well as the nature of the application, different bus architecture *synthesis techniques* can be more effective for exploring the vast communication architecture design space. Since a completely manual traversal of this enormous communication architecture design space is inefficient and practically infeasible even for small systems (that may have millions of possible bus architecture configurations) these techniques must be semi- or completely automated. Synthesis techniques can broadly be classified into three categories: (i) *static* (heuristic-based), (ii) *dynamic* (simulation-based), and (iii) *hybrid* (combination of heuristics and simulation). The *output* of applying these techniques (i.e., bus architecture synthesis) is a bus-based communication architecture fabric, with a well-defined bus topology and protocol parameters, that connects all the hardware and software components on the MPSoC in a manner that satisfies all application constraints.

Figure 6.1 shows where bus-based on-chip communication architecture synthesis fits into a typical electronic system level (ESL) design flow. The synthesis process requires creating static, dynamic, or hybrid models of the system to estimate system behavior and explore different bus architecture configurations (these different modeling techniques are described further in Chapter 4) before arriving at a solution that satisfies application constraints. To explore the vast on-chip communication architecture design space during synthesis, efficient tools and techniques are needed. While there are commercial toolkits available for standard bus architectures, such as AMBA (AMBA Designer [1] and AMBA Design Kit [2]), STBus (STBus GenKit [3]), and Sonics Smart Interconnect (Sonics Studio [4]), that enable a graphical specification, customization, and even physical design (often using external plug-in CAD tools) of the bus architectures, these toolkits are not particularly useful in automating the exploration process and synthesizing communication architectures that satisfy diverse design constraints. In this chapter, we present research in the area of synthesis techniques that aid designers in efficiently exploring and designing bus-based on-chip communication architectures. Section 6.1 presents approaches for the synthesis of bus topology, while Section 6.2 is concerned with techniques for the synthesis of bus parameters. Section 6.3 presents techniques that synthesize both the topology and parameters of bus architectures. Section 6.4 describes novel approaches that synthesize bus architectures with some knowledge of the physical floorplan and wire layouts. Section 6.5 examines approaches that attempt to co-synthesize the bus architecture with the memory architecture. Finally, in

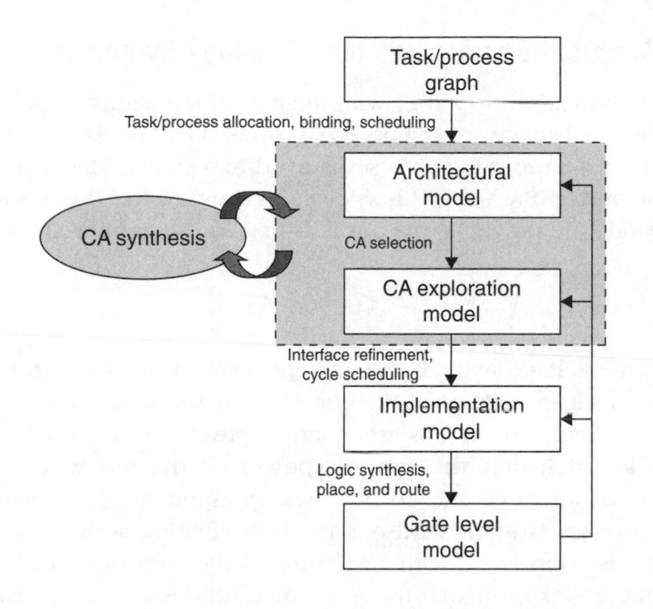

FIGURE 6.1

Bus-based on-chip communication architecture synthesis in a typical ESL design flow

Section 6.6, we wrap up with a brief discussion on some of the physical and circuit level design techniques for on-chip communication architectures.

6.1 BUS TOPOLOGY SYNTHESIS

The topology of a bus-based on-chip communication architecture determines the number of buses in the system, the manner in which they are interconnected to each other, and how the components in the system are allocated (connected) to the buses in the communication architecture. Synthesis approaches that attempt to generate a bus topology must therefore determine the number of buses, their manner of interconnection, and component allocation, while satisfying any of the constraints of the application being targeted.

Early work in bus topology synthesis [5, 6] focused on allocating inter-component communication to buses in heterogeneous distributed embedded systems. Assuming periodic tasks executing on processing elements (PEs), Yen and Wolf [5] proposed techniques to estimate communication delay on a bus using static analysis, and then either assigned a PE to an existing bus, or created a new bus, in order to meet task deadlines. Ortega and Borriello [6] explored the mapping of PEs in distributed real-time embedded applications to a set of off-chip bus architecture configurations. These configurations had different topologies (such as shared buses or point-to-point buses) and protocols (such as CAN or I^2C). The authors showed how the various configurations had different performance vs. cost tradeoffs that could be useful for designers.

6.1.1 Hierarchical Bus Architecture Topology Synthesis

A technique to design bus topology with the goal of reducing bus power consumption while meeting latency constraints was proposed by Liveris and Banerjee [7]. In their work, the arbitration on the buses is fixed to a simple FIFO (first-in-first-out) based scheme, and AMBA AHB [8] is selected as the target of the bus topology synthesis framework. The power on the bus is estimated using the well-known relation:

$$P = \frac{1}{2} V_{DD}^2 C f \alpha$$

where V_{DD} is the voltage swing between logic levels 0 and 1, C is the wire capacitance, f is the clock frequency on the bus, and α is the switching activity (a more comprehensive treatment of bus architecture power estimation is presented in Chapter 5). Note that only the dynamic power on the bus wires is considered here—power consumption due to bus logic components (e.g., arbiter, decoder, MUX) and repeaters is ignored. Also, only the switching activity on address and data bus wires is considered, with switching on the control wires being ignored. Wire capacitance C is estimated using the method described by Zhang et al. [9]. A floorplanning algorithm proposed by Murata et al. [10] is used to create an early floorplan and then an approximate wire length is obtained by using the bounding box of the centers of modules connected by the wire. The switching activity α is calculated separately for the address and data buses—for the data bus it is assumed that there is a probability of 0.5 that a wire on the bus being used will switch its value, while a lower probability is chosen for the address bus, due to lower address bit switching for synchronous burst transfers.

Each master is assumed to have latency constraints that determine the number of cycles available to complete a communication operation. To improve latency response of the communication architecture and potentially reduce the power consumption on the bus wires, Liveris and Banerjee [7] proposed using three different topology transformations. These transformations are illustrated in Fig. 6.2. The original architecture is shown in Fig. 6.2(a), and consists of three masters and four slaves on a single shared layer (or bus). The three possible topology transformations, to improve latency response and potentially reduce power consumption, are described below:

(i) *Private slave creation*: Making a slave i private to a master j is possible if j is the only master accessing slave i. In such a scenario, making the slave private allows the master to access the slave immediately, without having to request the arbiter for a bus grant. Additional decoder and MUX components are required for this transformation, as shown in Fig. 6.2(b). Private slave creation removes a slave from the shared bus, which reduces the fanout by one for all the signals driven by the AMBA logic.

(ii) *Slave isolation*: Moving a slave to another layer as shown in Fig. 6.2(c), especially a slave with a large latency, makes it easier to meet latency constraints because such a transformation reduces conflict on the shared bus and provides faster slave access to the masters. If a slave is heavily accessed by write operations, isolating it reduces the fanout of the wires accessed during the write operation.

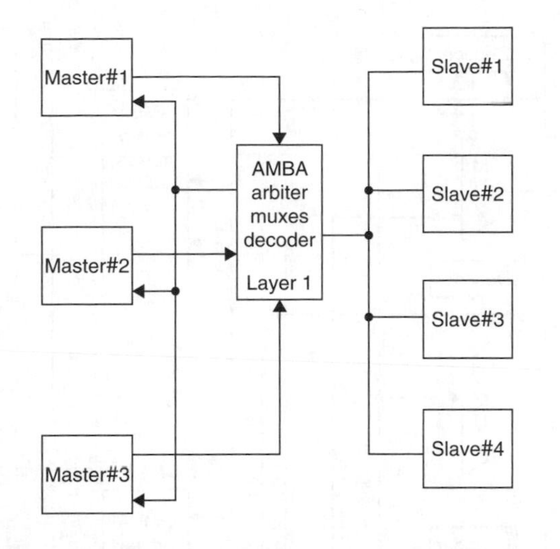

FIGURE 6.2(a)

Original single layer architecture with three masters and four slaves on a single layer [7]
© 2004 IEEE

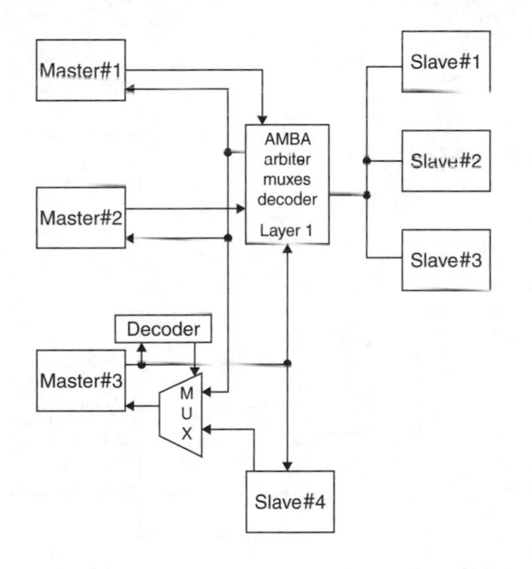

FIGURE 6.2(b)

Making slave 4 a private slave for master 3 [7]
© 2004 IEEE

(iii) *Grouping masters*: Moving masters to another layer, as shown in Fig. 6.2(d), can also reduce latency, due to a reduction in arbitration conflicts. This transformation reduces the fanout of the read data bus.

A reduction in the fanout due to the above transformations is assumed to reduce overall power consumption. However, this assumption holds only if the power consumption of the additional logic components due to the transformation is negligible relative to the overall power consumption.

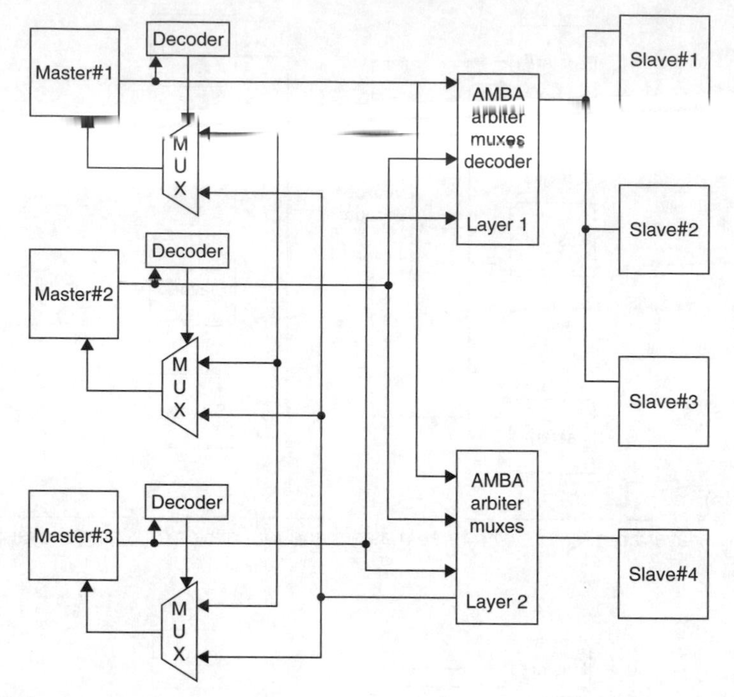

FIGURE 6.2(c)

Isolating slave 4 in another layer [7]
© 2004 IEEE

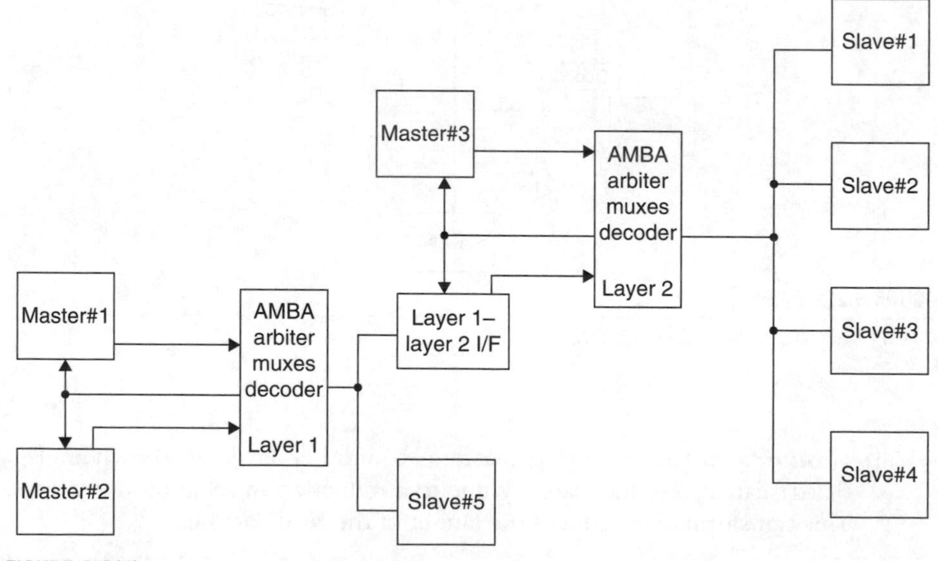

FIGURE 6.2(d)

Grouping masters on another layer [7]
© 2004 IEEE

These transformations form the core of the following synthesis heuristic. Initially, all masters and slaves are mapped to a single layer (advanced high performance bus) AHB bus. Next, the private slave creation transformation is applied for all eligible slaves, since it can potentially improve both performance (decreasing latency) and power. In case a latency violation exists for a master, the slave isolation transformation is applied to the slowest slave. If the violation persists, the grouping masters transformation is performed, by grouping and transferring the masters with less stringent latency requirements to a new layer. Once a solution that satisfies the latency constraints is obtained, the slave isolation and grouping masters transformations are performed to reduce power. At every iteration the power of the current solution is calculated, by invoking the floorplanner to obtain wire length and by using probability-based formulations to estimate switching activity on the wires. Any transformation that leads to an increase in power consumption is discarded. The potential to reduce switching activity guides the choice of transformation, and these transformations are carried out till no more improvement is obtainable.

The heuristic was implemented in *C++* and applied to two examples: a Sobel Transform SoC and a Vector Quantization Decoder SoC. Figure 6.3(a) shows the

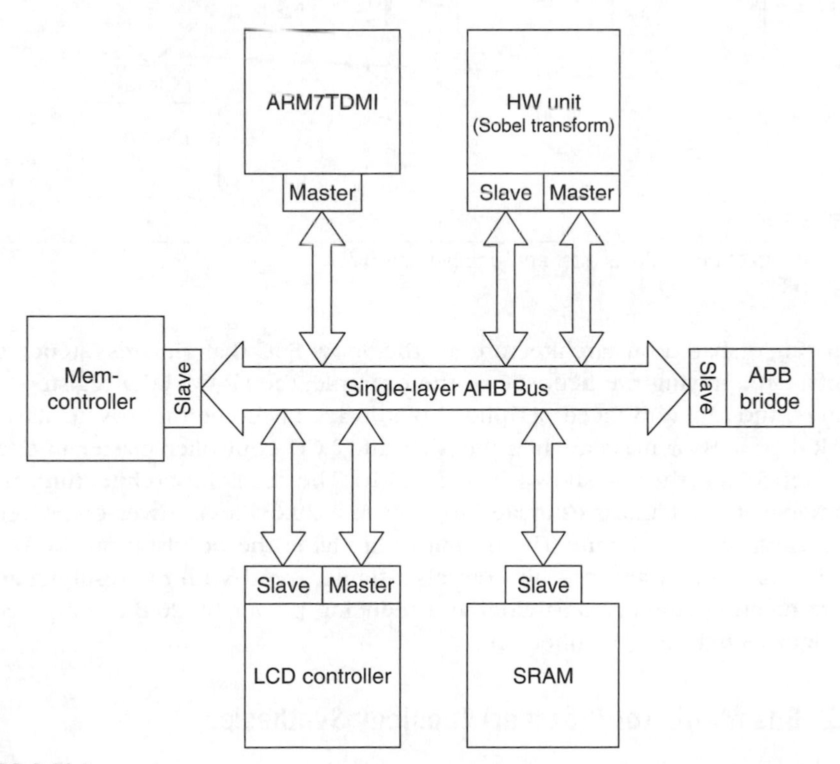

FIGURE 6.3(a)

Initial bus architecture for Sobel SoC [7]
© 2004 IEEE

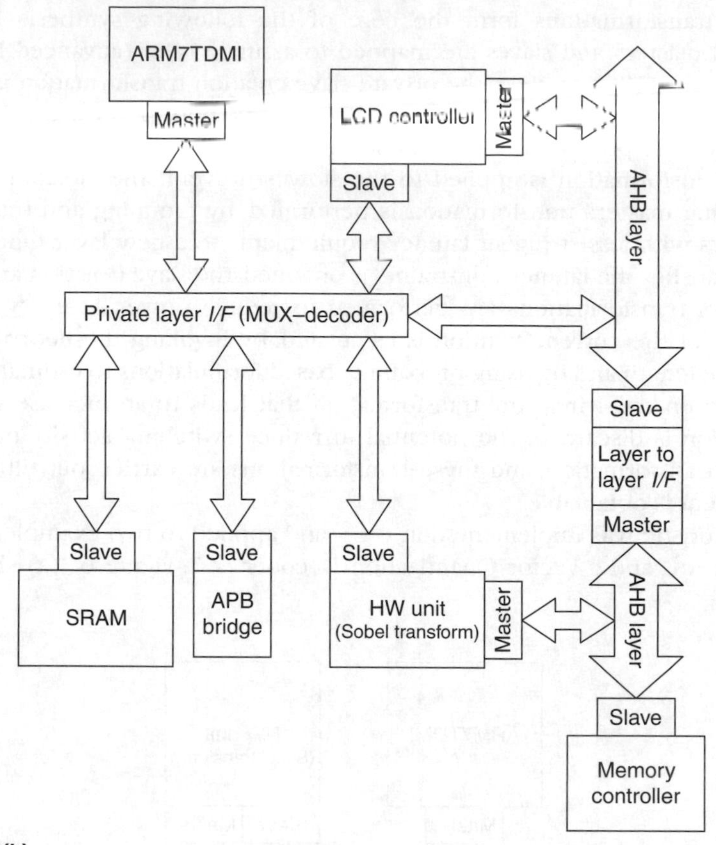

FIGURE 6.3(b)

Output architecture for Sobel after applying heuristic [7]
© 2004 IEEE

initial single shared bus architecture for the Sobel SoC that violates latency constraints. On applying the heuristic to the example, the SRAM, LCD registers, HW registers, and APB (advanced peripheral bus) bridge become the private slaves of the ARM processor master, while the ARM and LCD controller master interfaces are grouped together, as shown in Fig. 6.3(b). The resulting architecture meets latency constraints and is estimated to consume 29.6% less power compared to the single layer architecture. The output of applying the heuristic on the vector quantization SoC (system-on-chip) reveals a similar trend, with the resulting architecture meeting latency constraints, and reducing power by 26.8% compared to the original single layer architecture.

6.1.2 Bus Matrix (or Crossbar) Topology Synthesis

A methodology to design the topology for an STBus bus matrix (or crossbar) [11], with the goal of minimizing the number of buses while satisfying performance requirements of an application, was proposed by Murali and De Micheli [12]. The motivation for the approach is illustrated in Table 6.1, which shows the performance

Table 6.1 Comparison of crossbar performance and cost [12]

Type	Average latency (in cycles)	Maximum latency (in cycles)	Size ratio
Shared	35.1	51	1
Full	6	9	10.5
Partial	9.9	20	4

for an MPSoC application, running matrix multiplication benchmarks, and modeled in SystemC [13, 14], in terms of average and maximum packet latencies, and cost in terms of ratio of number of buses. Results are shown for three communication architecture topologies—shared bus, full crossbar, and partial crossbar. It can be seen that both average and maximum packet latencies are much higher for the shared bus, compared to the full and partial crossbars, which makes the shared bus unsuitable for the evaluated MPSoC system. The partial crossbar has a lower cost compared to the full crossbar, while having an almost comparable performance. A partial crossbar has the advantage of having fewer communication components (buses, arbiters, decoders, etc.), lower area, and reduced power consumption compared to a full crossbar. The goal of the methodology proposed by Murali and De Micheli [12] is therefore to design a minimal cost (i.e., having the least number of buses) partial crossbar bus architecture for a given MPSoC application, satisfying the performance constraints of the application. The performance constraints for the synthesized partial crossbar are represented as average and maximum packet latencies that must lie within acceptable bounds (as determined by the application designer) from the latencies obtained for a full crossbar.

Figure 6.4 shows the STBus crossbar methodology proposed in [12]. The application is initially designed using a full crossbar, and a SystemC simulation of the application is carried out at the cycle accurate granularity, using the MPARM [15] simulation environment (Phase 1). In order to capture local variations in traffic patterns and to perform traffic overlap calculations, a window-based traffic analysis scheme is used. The entire simulation period is divided into windows and traffic characteristics are recorded in each window. These traffic characteristics include amount of data received by target (slave), amount of pair-wise overlap between traffic streams to the targets, real time requirements of traffic streams, etc. The size of the window is parameterizable, and it is assumed that all the windows are of equal size. After this data collection phase, a preprocessing phase is performed, where overlapping critical traffic streams that need to be mapped to separate buses are identified (Phase 2). Targets with a large traffic overlap in a window and that need to be therefore mapped to separate buses are also identified in this phase. Additionally, the maximum number of targets that can be connected to a single bus (to bound maximum latency) is also identified. In the next phase an optimal partial crossbar configuration for the application is obtained, using the traffic information collected for each window (Phase 3). To obtain the optimal partial crossbar configuration, the performance constraints to be satisfied by the crossbar in each window are modeled as constraints of a mixed integer linear programming

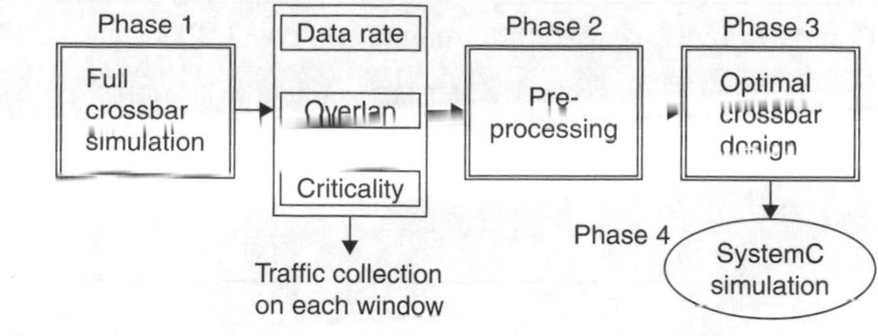

FIGURE 6.4

STBus crossbar design methodology [12]

Table 6.2 Component savings [12]

Application	Full crossbar bus count	Designed crossbar bus count	Ratio
Mat1	25	8	3.13
Mat2	21	6	3.5
FFT	29	15	1.93
QSort	15	6	2.5
DES	19	6	3.12

(MILP) problem. The formulations ensure that the bandwidth, overlap and criticality constraints are satisfied in each window. The CPLEX package [16] is used to solve the MILPs. Finally, the generated crossbars are instantiated in the MPARM simulation environment, and SystemC simulation is carried out to obtain performance metrics (Phase 4).

The design methodology is applied to MPSoCs running different benchmarks: Matrix suite-1 (Mat1-25 cores), Matrix suite-2 (Mat2-21 cores), FFT suite (FFT-29 cores), Quick Sort suite (QSort-15 cores), and DES encryption system (DES-19 cores). Table 6.2 shows the component savings for the proposed approach. Up to a 3.5× savings is obtained compared to a full bus matrix by using the partial crossbar generation methodology. The packet latencies of the generated partial crossbar are claimed to be within acceptable bounds of the minimum packet latencies in a full crossbar.

The MILP formulations make use of several parameters that can have an adverse affect on crossbar cost or performance. For instance, the choice of *window size* can affect the size of the crossbar generated by the methodology, as shown in Fig. 6.5. A small window size leads to finer control of application performance, resulting in crossbars with lower packet latencies. However, a very small window size can result in overdesign of the crossbar, resulting in a prohibitively high cost. A large window size on the other hand leads to less control over application performance, resulting in crossbars with higher latencies, but lower cost. Another parameter is the *traffic overlap threshold* that can also affect the cost

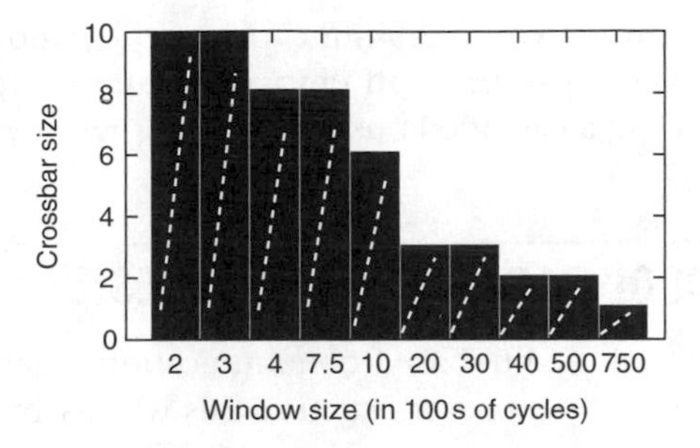

FIGURE 6.5

Crossbar size vs. window size [12]
© 2005 IEEE

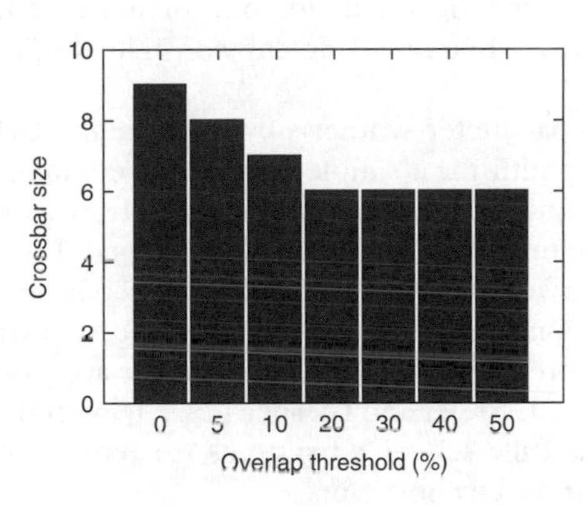

FIGURE 6.6

Crossbar size vs. overlap threshold [12]
© 2005 IEEE

of the generated crossbar, as shown in Fig. 6.6. The choice of parameters such as *window size*, *traffic overlap threshold*, and *maximum number of targets allowed on a bus*, is an important but non-trivial task during topology synthesis in the proposed methodology.

6.1.3 Summary of Other Topology Synthesis Research

Several other techniques have focused on bus topology synthesis. Thepayasuwan et al. [17–21] proposed a simulated annealing (SA)-based approach to synthesize a hierarchical shared bus architecture topology for an application. The topology space (consisting of number of buses and component allocation on the buses) is explored via an SA algorithm that selects a configuration based on the weights used in the cost function for criteria such as number of buses, communication conflict, and bus utilization. A larger weight for a particular criterion gives it more importance in the cost function, and the SA algorithm tries to optimize for criteria with larger weights. Drinic et al. [22] and Meguerdichian et al. [23] proposed a framework for determining optimal component mapping over a fixed number of buses, in order to maximize data throughput in the system. Topology synthesis for segmented buses was presented by Guo et al. [24, 25]. While the goal in

[24] was to obtain a solution with minimum wire energy, Guo et al. [25] proposed generating a set of solutions to trade-off chip area, energy, and delay. An SA-based approach for synthesizing a cascaded bus matrix was presented by Yoo et al. [26].

6.2 BUS PROTOCOL PARAMETER SYNTHESIS

In addition to their topology, bus-based communication architectures are characterized by several protocol parameters, such as bus widths, bus clock frequencies, transaction burst sizes, arbitration schemes, and buffer sizes. These parameter values can have a significant impact on the performance, power consumption, cost, and complexity of communication architectures. The process of bus parameter synthesis involves determining values for one or more of these parameter values for a fixed topology, while satisfying any constraints of the application being targeted.

Early work in bus parameter synthesis by Narayan and Gajski [27] focused on determining the bus width for a simple shared bus communication architecture on which multiple communication channels transferring possibly different sized data (between communicating processes) are mapped. The proposed approach showed trade-offs between bus widths and system performance. However, the design of the arbitration scheme was not considered, and traffic conflict on the shared buses was ignored. Another somewhat similar approach to determine bus width was proposed by Gasteier and Glesner [28, 29], with the goal of minimizing the bus width for statically scheduled systems (to reduce area and wiring cost) while satisfying performance constraints.

6.2.1 Component Mapping and Protocol Parameter Synthesis

An approach to determine bus protocol parameters as well as component mapping on buses to improve performance for a fixed bus topology was proposed by Lahiri et al. [30, 31]. Table 6.3 shows the effect of different component mappings for a hierarchical bus-based communication architecture with two buses and eight components (*C1–C8*). The three mapping configurations (*Arch1, Arch2, Arch3*) have very different performance, indicating the importance of selecting a suitable component mapping in improving system performance. Table 6.4 shows the effect of different arbitration priority schemes and DMA burst sizes on the performance

Table 6.3 Performance variation for different component mappings [30, 31]

Cases	*Bus1*	*Bus2*	Performances (clock cycles)
Arch1	*C1, C2, C3, C4*	*C5, C6, C7, C8*	11,723
Arch2	*C1, C3, C5, C7*	*C2, C4, C6, C8*	15,314
Arch3	*C1, C2, C5, C6*	*C3, C4, C7, C8*	9242

of the Arch3 configuration. It is clear from the table that bus protocol parameter values also have a significant influence on system performance. The goal of the synthesis methodology proposed by Lahiri et al. [30, 31] is to synthesize both the bus parameters and component mapping for a fixed bus topology.

Figure 6.7 shows the overall methodology for communication architecture synthesis proposed by Lahiri et al. [30, 31]. The inputs to this methodology are: (i) a system that has been partitioned and mapped onto hardware and software

Table 6.4 Performance variation for different bus protocol parameters [30, 31]

Case	*Bus1* protocol	*Bus2* protocol	Performance (cycles)
Arch3-subopt	*C1>C2>C5>C6* DMA = 5	*C3>C4>C7>C8* DMA=10	12,504
Arch3-opt	*C1>C6>C2>C5* DMA=10	*C3>C7>C4>C8* DMA=10	9242

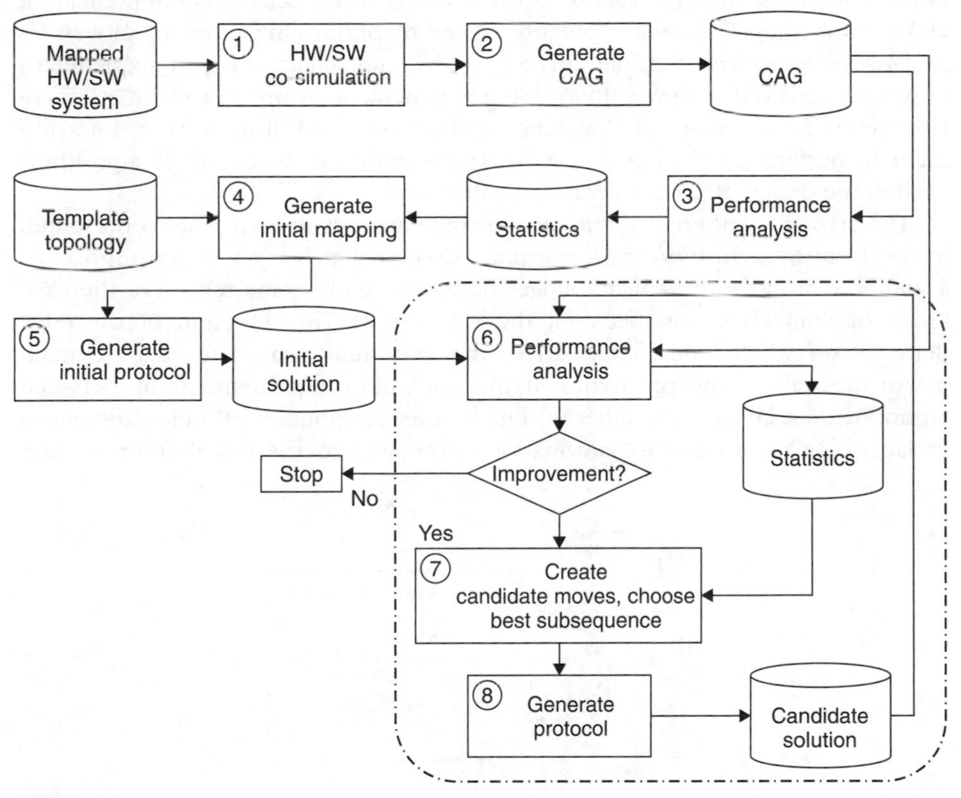

FIGURE 6.7

Communication architecture design methodology [30, 31]

components, and (ii) a communication architecture topology consisting of a fixed number of buses interconnected by bridges. The methodology automates the process of generating an optimal component mapping and protocol selection for the chosen communication architecture topology. In the first step, hardware/software (HW/SW) co-simulation of the entire system is performed, assuming completely parallel (conflict-free) communication between components. The execution traces from this co-simulation step are saved in the form of a communication analysis graph (CAG) that captures the behavior of the entire system, including computation, communication, and synchronization delays over the entire simulation trace [32]. Performance analysis is performed in the next step, using analysis algorithms for the CAG presented in [32] and [33], to generate various performance statistics on system performance and inter-component communication. Based on these statistics, and the topology template specified by the designer, an initial mapping of the components is performed using a heuristic. Optimal protocol parameters for each bus are obtained in the next step (Step 5), and the output of Step 4 and Step 5 results in an initial solution. Subsequently, the iterative part of the flow is invoked, and considers the effect of the communication architecture on the CAG. The performance analysis tool in Step 6 analyzes the CAG and re-evaluates the performance and communication statistics, based on the selected communication architecture. Step 7 evaluates potential gains of performing different moves, by moving a component from one bus to another, and selects the best sequence of moves to construct a new solution. Step 8 generates an optimal set of protocol parameters for the new solution. Steps 6–8 are repeated till no further improvement in performance is possible. A brief description of some of the algorithms used in the design methodology is presented next.

The statistics gathered by the performance analysis in Step 3 are represented in the form of a *communication graph* (CG), in which each vertex represents a component in the system, and edges between components represent the existence of communication between them (Fig. 6.8). The direction of the edge depends on which component drives the communication. The edges contain information about the properties of the communication transactions between components such as the number of transactions, distribution of their sizes (mean variance, etc.), critical path information (expressed as the distribution of their

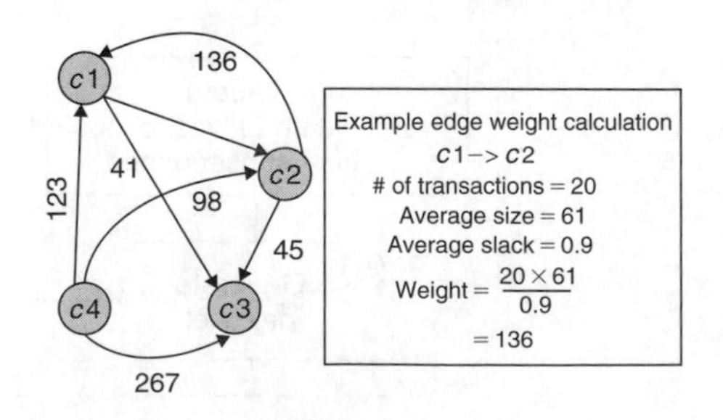

FIGURE 6.8

Example of communication graph (CG) [30, 31]
© 2004 IEEE

slacks), number of transactions with zero slack (critical transactions), etc. A single weight for each edge is derived by taking the product of the number of transfers and the average size of the transfer, and dividing it by the average slack, as shown in Fig. 6.8. This weight accounts for the frequency, volume, and criticality of the transfers between the components. This CG is analyzed by Step 4, to obtain a measure of the demand each component places on the communication architecture. For a component *i*, this demand is the sum of weights of the outgoing edges from v_i in the CG. The components are then arranged in a descending order of demand. Step 4 also ranks the buses in the communication architecture by analyzing the topology template. The topology is represented using a *template graph* (TG), consisting of a vertex for each bus, and bidirectional edges between vertices when the two buses are connected together by a bridge. The information on the edges represents properties of the bridge, such as the overhead of transmitting a single word and its frequency of operation. A self-looping edge on each vertex describes properties of the bus such as its width, operating frequency, and handshake overhead. The information on the vertices and edges in the TG is used to calculate delay for a set of communications involving a bus, which is then used to derive the connectivity metric to rank the bus. Thus higher rank is given to buses that have higher performance and are well connected to the rest of the buses. Once the components and the buses have been ranked, Step 4 selects the highest ranked component, and then attempts to select a bus for assignment. If a bus *P* has components *i* and *j* assigned to it, and component *k* is being considered for assignment, then its interaction level with bus *P* is calculated by summing the weights of the edges (v_k, v_j), (v_k, v_i), (v_i, v_k), and (v_j, v_k) in the CG. The bus with the maximum interaction level with component *k* is chosen as the target for its mapping. If there are multiple alternatives for mapping, a bus is chosen randomly. Once the mapping is complete, Step 5 generates bus protocol parameters. A static priority scheme is used for arbitration, with priorities being assigned based on the rank of components generated in the earlier step. The maximum block transfer size for a bus is calculated from the subgraph of the CG that consists of vertices representing components assigned to the bus. It is given by the weighted average of the size of transactions between components on the bus, with the weight incorporating criticality (derived from average value of slack).

In the iterative improvement portion of the flow, Step 7 calculates the potential gain of moving a component from its currently mapped bus to another bus, for every combination of component and potential destination buses. The potential gain of a move is estimated as follows. First, the lifetime of every communication transaction in the system is estimated. This lifetime consists of time spent waiting for handshake, or access to the bus, and the time for data transfer. The performance analysis tool generates the communication overlap (in cycles) for every pair of components in the system, as shown in Fig. 6.9(a), and creates a *communication conflict graph* (CCG), as shown in Fig. 6.9(b). From this graph, a congestion level for each bus *P* is obtained, by summing up weights of edges between components mapped to bus *P*. When moving a component from bus *P* to *Q*, the potential reduction in congestion on bus *P*, and the increase in congestion on bus *Q* is calculated. For instance, for the example in Fig. 6.9(b), if component *C3* is

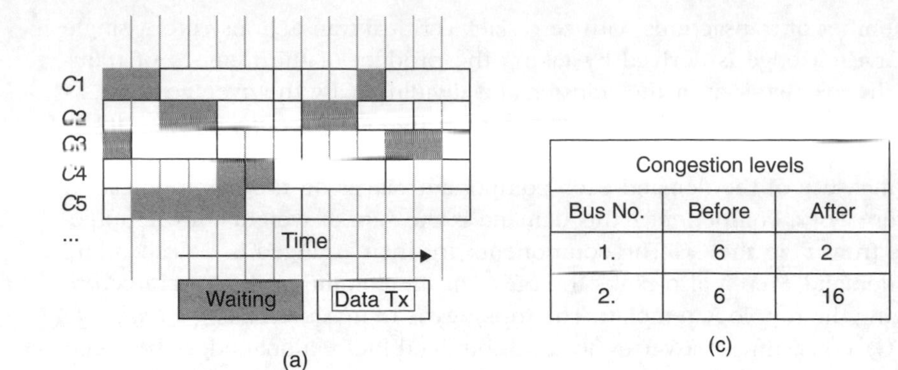

Congestion levels		
Bus No.	Before	After
1.	6	2
2.	6	16

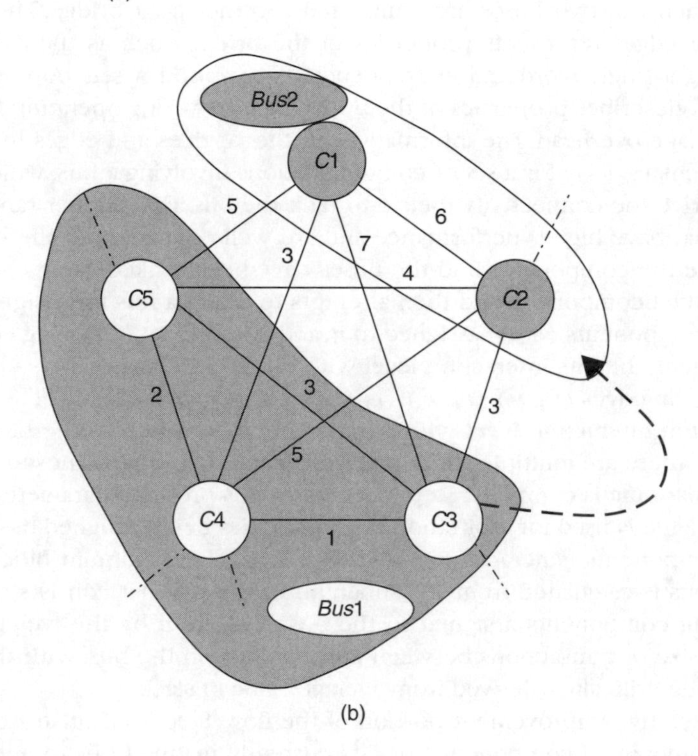

FIGURE 6.9

Estimating gain for potential move: (a) execution trace showing communication lifetimes (b) communication conflict graph (c) congestion level when *C3* is moved from *Bus1* to *Bus2* [30, 31]

moved from *Bus1* to *Bus2*, then the change in congestion levels on both buses is shown in Fig. 6.9(c). From Fig. 6.9(a), the number of overlapping time units on *Bus1* is 6 (1 between *C3* and *C4*, 2 between *C4* and *C5*, 3 between *C3* and *C5*). After the removal of *C3*, the number of overlap units on *Bus1* reduces to 2 (between *C4* and *C5*), while the number of overlap units on *Bus2* increases to 16. The changed congestion levels are used to recalculate time taken for transactions

Table 6.5 Experimental results [30, 31]

Case	ATM			SYS		
	Performance (cycles)	Speedup	CPU time (seconds)	Performance (cycles)	Speedup	CPU time (seconds)
Shared	24,654	1.00	10.3	32,328	1.00	6.8
Random	15,314	1.61	11.3	25,593	1.26	7.0
Initial	11,723	2.10	12.1	19,998	1.62	6.7
Opt	9242	2.67	23.5	18,139	1.78	11.8
Abstract communication	4992	4.94	138	9988	3.24	134

involving buses P and Q, and the difference in time (before and after the move) is the potential gain of a move. The move with the maximum gain is selected, and the procedure repeated iteratively, till no more gain is achievable.

The methodology proposed by Lahiri et al. [30, 31] was used to generate communication architecture component mapping and protocols for two case studies: (i) a cell forwarding unit of an output queued ATM switch, with a fixed topology having three buses connected by two bridges, and (ii) a simple communication system (SYS) with two buses connected by a single bridge. Table 6.5 shows the performance of the two case studies for different choices of communication architectures. *Row 1* represents the case when all components are mapped to a single shared bus. *Row 2* represents the case when component mapping and parameter values are chosen at random. *Row 3* represents the case when the initial solution (Fig. 6.7) is considered that ignores access conflicts. *Row 4* represents the output from the proposed methodology. Finally, *Row 5* represents the case when an abstract topology is considered, with infinite concurrency and bandwidth. For each system, in Columns 2 and 5, Table 6.5 reports the actual performance measurement (in terms of the number of clock cycles taken) to process 2000 input stimuli for the SYS case study, and to process 1000 cells for the ATM case study. In Columns 3 and 6, Table 6.5 reports the performance of each configuration as the *Speedup* relative to the case when all communication goes through a shared bus (*Row 1*). Columns 4 and 7 report CPU times, and these can be elaborated on as follows. In *Rows 1* and *2* (shared bus and random), the time spent in design space exploration is zero since the solution is predetermined. Hence the reported CPU times indicate the time spent on a single evaluation of the system performance. In *Row 3* (initial), CPU time includes the time spent in performance analysis as well as constructing the initial solution. In *Row 4* (optimized solution) it is the sum of the times spent in analysis, construction of the initial solution, and the iterative procedure. Finally, for the last row, CPU time indicates the time required to generate the initial system execution trace via detailed HW/SW co-simulation, performed only once for each system. It can be seen from the *Speedup* results in the table that the proposed methodology (*Row 4*) outperforms the other techniques. Additionally, the methodology is an order of magnitude faster (when comparing CPU times) than a technique that relies on complete system simulation (*Row 5*), for the case studies that were examined.

6.2.2 Arbitration Scheme Synthesis

A methodology to automatically determine the slot schedule for a time division multiple access (TDMA)-based arbitration scheme in a bus-based communication architecture was proposed by Shin et al. [34]. The approach makes use of a genetic algorithm (GA) [35] to automatically explore the TDMA-based arbitration parameter space. Figure 6.10 shows the automatic bus configurator (ABC) proposed by Shin et al. [34] that starts by creating a population of individuals for one generation of the GA. The GA then subsequently creates individuals by performing basic operations such as mutation, crossover, and reproduction. Once a population is formed, the parallel launcher concurrently executes the individuals, at the register transfer level (RTL). Each individual has an associated configuration file that contains values for the TDMA slot allocations for the masters on the buses. This file is used to generate RTL code, which is compiled, and then simulated. The results of simulation for each individual are used to compute its fitness (or objective) function, after analyzing the bandwidth allocated to the masters for a particular TDMA slot allocation schedule in the arbiter. Note that a prerequisite for using the GA is to carefully select and fine tune values of a few parameters such as

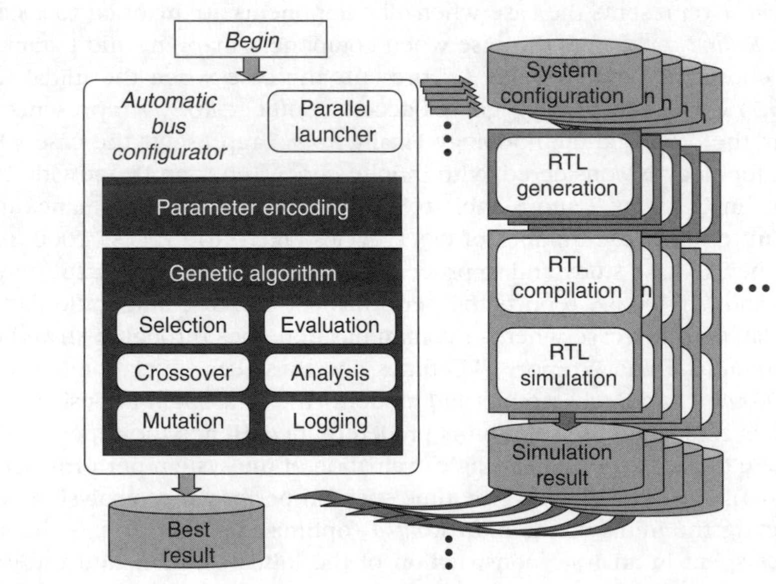

FIGURE 6.10

Automatic bus configurator [34]

population size [36], number of generations, crossover rate, and mutation rate to obtain correct operation.

The objective function used in the GA is aimed at meeting the throughput requirements for each master. Figure 6.11 shows this objective function that can be extended to encapsulate latency constraints as well, as shown in Fig. 6.12. The goal of the GA is to reduce the value of this objective function. The GA-based automated ABC approach was applied to a simple SoC in order to obtain the optimal TDMA slot schedule. The approach was compared against a manual exploration by an expert designer, with the same goal of obtaining an optimal TDMA slot schedule. Figure 6.13 shows the result of this experiment. The expert was able to

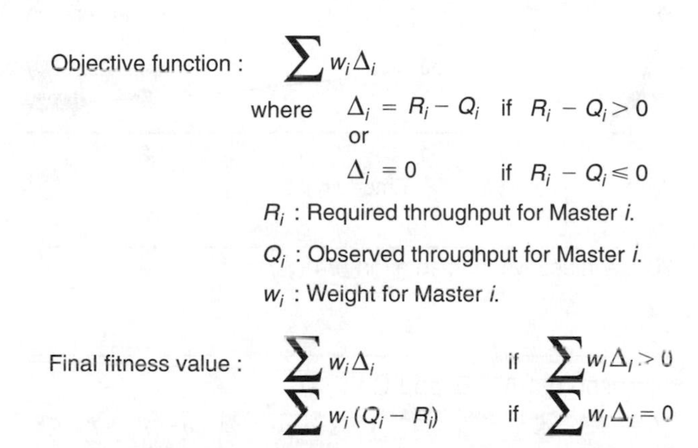

Objective function : $\sum w_i \Delta_i$
where $\Delta_i = R_i - Q_i$ if $R_i - Q_i > 0$
or
$\Delta_i = 0$ if $R_i - Q_i \leq 0$
R_i : Required throughput for Master i.
Q_i : Observed throughput for Master i.
w_i : Weight for Master i.
Final fitness value : $\sum w_i \Delta_i$ if $\sum w_i \Delta_i > 0$
$\sum w_i (Q_i - R_i)$ if $\sum w_i \Delta_i = 0$

FIGURE 6.11

Objective function aimed at meeting throughput requirements of masters [34]
© 2004 IEEE

Objective function : $w_B \Delta + w_L \sigma$

where $\Delta = B_{req} - B_{obv}$ if $B_{req} - B_{obv} > 0$
or
$\Delta = 0$ if $B_{req} - B_{obv} \leq 0$

where $\sigma = L_{obv} - L_{ent}$ if $L_{obv} - L_{ent} > 0$
or
$\sigma = 0$ if $L_{obv} - L_{ent} \leq 0$

B_{req} : Required bandwidth
B_{obv} : Observed bandwidth
L_{ent} : Required latency
L_{obv} : Observed latency
w_B : Weight for bandwidth
w : Weight for latency

FIGURE 6.12

Objective function aimed at meeting throughput and latency requirements of masters [34]
© 2004 IEEE

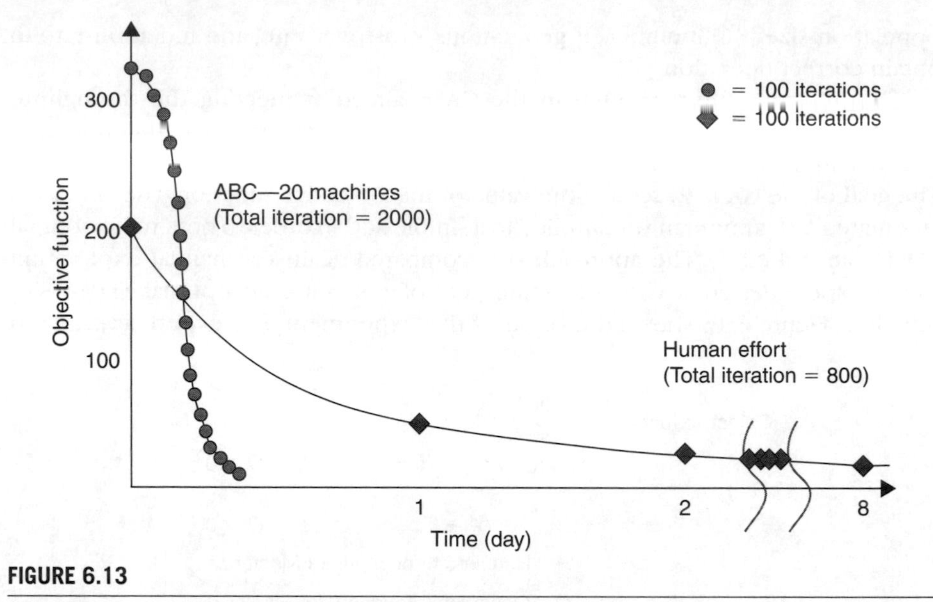

FIGURE 6.13

ABC (running on 20 machines) vs. human effort [34]
© 2004 IEEE

Table 6.6 Comparison of SA, SE, and GA [34]

	SA	SE	GA
Objective function (average)	8.81	8.5	8.21
Iterations (average)	4231.6	3663.7	1664.2

© 2004 IEEE

evaluate about 800 configurations in 8 days, in a serial manner, whereas the ABC approach was able to explore more than 2000 configurations in 8 hours. The authors claimed that the ABC approach was able to find a better solution than the expert designer, who took almost 80 hours for the manual exploration task.

The solution obtained using a GA was compared with solutions obtained using SA [37] and stochastic evolution (SE) [38]. The result of this comparison study is shown in Table 6.6. The results are obtained by averaging over 100 different simulation runs in which up to 5000 iterations are allowed. The number of iterations in the table refers to how fast the known optimum was found on an average. If the optimum value is not found within 5000 iterations during a simulation run, then the best suboptimal value (i.e., lowest objective function value found up to the point) is recorded. From the experiment, it was found that SA is not consistent with various initial configurations of parameters, and also has trouble handling too many parameters. SE tended to make decisions prematurely and resulted in a bad local optimum solution. The GA was the fastest and most consistent out of the three approaches, for the benchmarks that were considered. However, as mentioned previously, the GA is very sensitive to the choice of parameters such as crossover rate, mutation rate, and population size for correct operation. Experiments were conducted to obtain optimal values for these parameters in [34], and it was found that the best results were obtained for a crossover rate of 70%, a mutation rate of 25%, and a population size of 80%.

6.3 BUS TOPOLOGY AND PROTOCOL PARAMETER SYNTHESIS

In addition to approaches that have looked at addressing the problem of communication architecture topology and parameter synthesis separately, there have also been approaches proposed that consider both the topology and parameter value synthesis problems together. Initial work in the area of automated communication architecture topology and parameter synthesis [39–42] was primarily concerned with allowing a designer to manually specify topology and parameter values, and then automate the process of RTL code generation for the communication architecture. Lyonnard et al. [39] proposed a methodology where a designer manually specifies a shared bus or a point-to-point topology, as well as parameters such as synchronization protocol (handshake, FIFO based, etc.), and the RTL level description for the communication architecture is automatically generated. Ryu et al. [40–42] proposed a bus synthesis tool called *BusSynth* that could synthesize five different bus topologies, including a split bus, hierarchical shared bus, and a Bi-FIFO-based bus, together with parameter values such as address bus width, data bus width, and FIFO depth (and width). The tool requires a designer to manually select the bus topology and parameters, and generates custom Verilog RTL code for the chosen bus architecture configuration. In both these approaches, the onus is on the designer to explore the communication architecture design space, with the goal of determining a suitable choice for the bus architecture topology and parameter values for an application.

6.3.1 Hierarchical Shared Bus Topology and Protocol Parameter Synthesis

A technique to simultaneously synthesize the hierarchical shared bus topology and the width of the data buses was proposed by Pandey et al. [43], with the goal of finding a solution with the least number of buses satisfying application performance constraints. An integer linear programming (ILP) formulation was used to obtain the optimal bus width and minimum number of buses (ensuring maximum utilization) while satisfying the performance constraints in the system. A more comprehensive synthesis approach for hierarchical shared bus architectures was proposed by Pasricha et al. [44] that uses heuristics to automate the synthesis of not only the bus topology, but also several bus architecture parameters such as bus widths, bus clock speed, out-of-order (OO) buffer sizes [45], and DMA burst sizes. The goal in [44] is to generate a hierarchical shared bus architecture with the minimum number of buses satisfying the throughput constraints of the application.

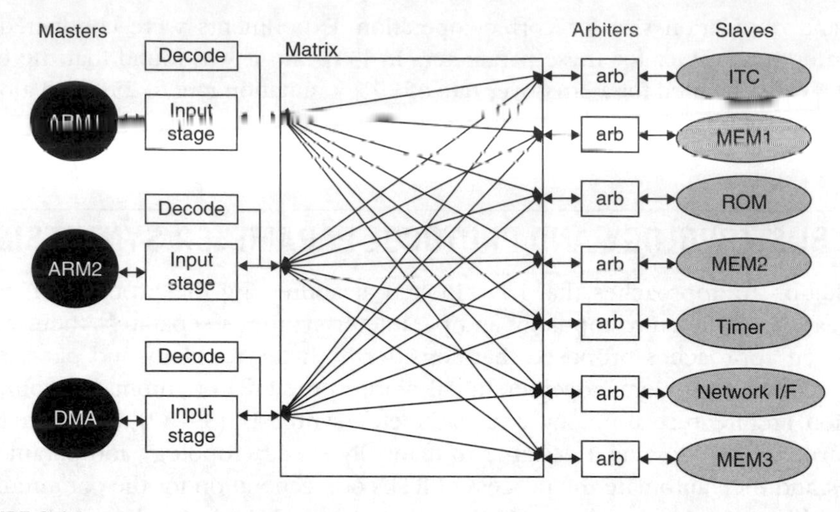

FIGURE 6.14

Full AMBA bus matrix architecture [46]
© 2006 IEEE

6.3.2 Bus Matrix Topology and Protocol Parameter Synthesis

A comprehensive bus matrix synthesis methodology was proposed by Pasricha et al. [46], to simultaneously synthesize the bus topology and protocol parameter values for bus widths, bus clock frequencies, OO buffer sizes, and arbitration schemes. The goal was to generate an optimal, minimum cost bus matrix (having the least number of buses) while satisfying application performance constraints. Figure 6.14 shows an example of a three master seven slave AMBA [8] bus matrix architecture for a dual ARM processor-based networking SoC subsystem application. The bus matrix consists of several buses in parallel that can support concurrent high bandwidth data streams. The *Input stage* is used to handle interrupted bursts, and to register and hold incoming transfers if receiving slaves cannot accept them immediately. The *Decode stage* generates select signal for appropriate slaves. Unlike in traditional shared bus architectures, arbitration (*arb*) in a bus matrix is not centralized, but rather distributed so that every slave has its own arbitration. One drawback of the full bus matrix structure shown in Fig. 6.14 is that it connects every master to every slave in the system, resulting in a prohibitively large number of buses in the matrix. The excessive wire congestion can make it practically impossible to route and achieve timing closure for the design [47]. To overcome this shortcoming, designers must tailor a full matrix structure down to a partial bus matrix, as shown in Fig. 6.15 that still meets the performance constraints of the application. Such a structure has fewer buses and consequently uses fewer components (arbiters, decoders, buffers), has a smaller area and also consumes less power.

The assumptions in [46] are as follows. It is assumed that HW/SW partitioning has taken place for the MPSoC design being considered, and that the appropriate functionality has been mapped onto hardware and software intellectual properties

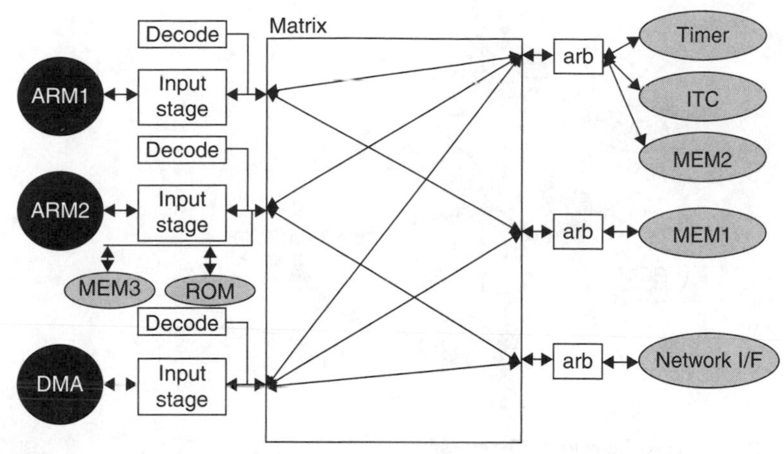

FIGURE 6.15

Partial AMBA bus matrix architecture [46]
© 2006 IEEE

(IPs). These IPs are standard "black box" library components that cannot be modified during the synthesis process, except for the memory components. The target standard bus matrix communication architecture (e.g., AMBA bus matrix [8]) that determines the pins at the IP interface and for which the matrix must be synthesized, is also specified. Typically, all buses within a bus matrix have the same data bus width, which usually depends on the number of data interface pins of the IPs in the design. It is assumed that this matrix data bus width is specified by the designer, based on the knowledge of the IPs selected for the design (which determines the pins at the IP interfaces, and consequently bus width). Performance constraints in the approach are represented via a *Communication Throughput Graph*, *CTG=G(V,A)* which is a directed graph, where each vertex $v \in V$ represents a component in the system, and an edge $a \in A$ connects components that need to communicate with each other. A *throughput constraint path* (TCP) is a subgraph of a CTG, consisting of a single master for which data throughput must be maintained and other masters, slaves, and memories that are in the critical path impacting the maintenance of the throughput. Figure 6.16 shows a CTG for a network SoC subsystem, with a TCP involving the *ARM2*, *MEM2*, *DMA*, and *Network I/F* components, where the rate of data packets streaming out of the *Network I/F* component must not fall below 1 Gbps (gigabits per second). The problem definition for the approach proposed by Pasricha et al. [46] is then expressed as follows. A bus B can be considered to be a partition of the set of components V in a CTG, where $B \subset V$. Then the problem is to determine an optimal component to bus assignment for a bus matrix communication architecture, such that V is partitioned onto a minimal number of buses **N** and satisfies all constraints in the design, represented by the TCPs in a CTG.

Since communication behavior in a system is characterized by unpredictability due to dynamic bus requests from cores, contention for shared resources, buffer overflows, etc., a simulation-based approach is necessary for accurate performance estimation. However, relying solely on simulation-based exploration can limit the

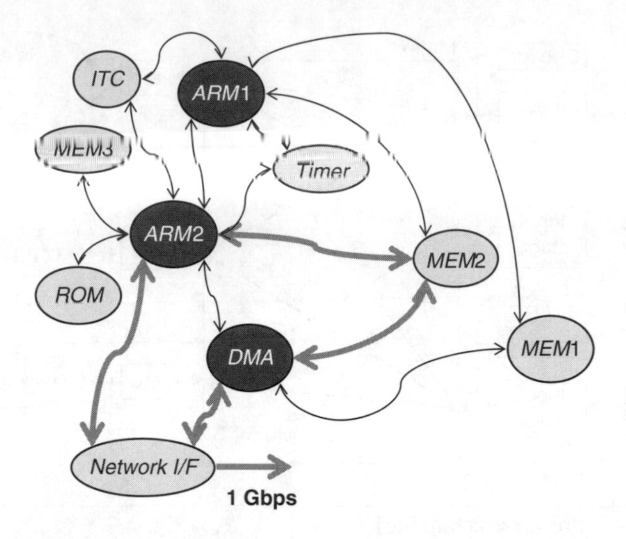

FIGURE 6.16

Communication throughput graph [46]
© 2006 IEEE

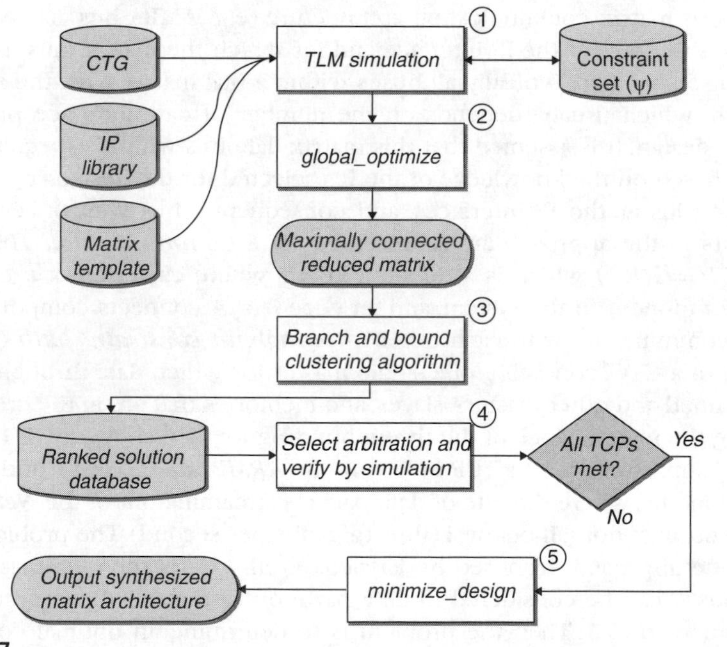

FIGURE 6.17

Automated bus matrix synthesis flow [46]
© 2006 IEEE

amount of space that can be explored in a reasonable amount of time. Figure 6.17 gives a high level overview of the flow proposed in [46], that makes use of a combination of static- and simulation-based dynamic analysis to speed up the synthesis process. The inputs to the flow include a CTG of the target application, a library of behavioral IP models, a target bus matrix template (e.g., AMBA bus matrix [8]), and a communication parameter constraint set (Ψ). The communication parameter constraint set is needed to ensure that the synthesis process generates realistic solutions, and allows designers to specify a set of discrete allowed values for bus parameters (e.g., specifying that the bus clock frequency for a bus can only be multiples of 33 MHz, up to a maximum of 330 MHz).

The general idea in the flow is to first perform a fast transaction level (transaction level model, TLM) [48] simulation of the application (modeled in SystemC [13]), to get application-specific data traffic statistics. This information is used in a global optimization phase to reduce the full bus matrix architecture, by removing unused buses and local slave components from the matrix. The resulting matrix is called a *maximally connected reduced matrix*. The next step is to perform a static branch- and bound-based clustering of slave components in the matrix that further reduces the number of buses in the matrix. The results of the static clustering analysis are ranked, from the best case solution (least number of buses) to the worst (most number of buses) and stored in a database. A fast bus cycle accurate simulation engine (CCATB [49]; described in Chapter 4) and heuristics are then used to validate and select the best solution that meets all the performance constraints, determine slave arbitration schemes, optimize the design to minimize bus clock frequencies and OO buffer sizes and, then, finally output the optimal synthesized bus matrix architecture.

A more detailed overview of the synthesis flow in Fig. 6.17 follows. In *Phase 1*, the IP library is mapped onto a full bus matrix and simulated at the TLM level, with no arbitration contention overhead since there are no shared channels and also because infinite ports are assumed at IP interfaces. The OO buffer sizes are set to the maximum allowed in Ψ. The *TLM simulation* phase produces application-specific data traffic statistics such as number of transactions on a bus, average transaction burst size on a bus, and memory usage profiles. Knowing the bandwidth to be maintained on a channel from the TCPs in the CTG, it is possible to estimate the minimum clock speed at which any bus in the matrix must operate, in order to meet its throughput constraint, as follows. The data throughput ($\Gamma_{TLM/B}$) from the TLM simulation, for any bus B in the matrix is given by:

$$\Gamma_{TLM/B} = (num\ T_B \times sizeT_B \times width_B \times \Omega_B)/\sigma$$

where *numT* is the number of data transactions on the bus, *sizeT* is the average size of these data transactions, *width* is the data bus width, Ω is the clock speed, and σ is the total number of cycles of TLM simulation for the application. The values of *numT*, *sizeT*, and σ are obtained from the TLM simulation in *Phase 1*. To meet the throughput constraint $\Gamma_{TCP/B}$ for the bus B,

$$\Gamma_{TLM/B} \geq \Gamma_{TCP/B}$$

Therefore,

$$\Omega_B \geq (\sigma \times \Gamma_{TCP/B})/(width_B \times numT_B \times sizeT_B)$$

The minimum bus clock speed thus found is used to create (or update) the local bus speed constraint set $\Psi_{L(speed)}$ for the bus B. As an example, if it was previously allowed to have bus speeds of any multiple of 33 MHz up to a maximum of 330 MHz, and Ω_B is found to be 90 MHz, the new constraint set would be any multiple of 33 MHz between 99 and 330 MHz.

In the next phase (*Phase 2* in Fig. 6.17), a global optimization (*global_optimize*) is performed on the matrix by using information gathered from the TLM simulation in *Phase 1*. In this phase, first all the buses that have no data traffic on them are removed from the full bus matrix. Next, the memory usage profile is analyzed from the simulation run and an attempt made to split those memory nodes for which different masters access non-overlapping regions. Finally, dedicated slave and memory components are clustered with their corresponding masters by migrating them from the matrix to the local buses of the masters in order to reduce congestion in the bus matrix. Note that memory splitting is performed before local node clustering because it enables generation of local memories that can then be clustered with their corresponding masters. After the *global_optimize* phase, the matrix structure obtained is termed as a maximally connected reduced bus matrix.

The next phase (*Phase 3* in Fig. 6.17) involves static analysis to determine the optimal reduced bus matrix for the given application. A branch- and bound-based clustering algorithm is used to cluster slave components, to reduce the number of buses in the matrix even further. Note that merging masters is not considered because it adds two levels of contention (one at the master end and another at the slave end) in a data path that can drastically degrade system performance. Before describing the algorithm, a few definitions are presented. A *slave cluster* $SC = \{s_1 \ldots s_n\}$ refers to an aggregation of slaves that share a common arbiter. Let M_{SC} refer to the set of masters connected to a slave cluster SC. Next, let $\Pi_{SC1/SC2}$ be a superset of sets of buses that are merged when slave clusters SC1 and SC2 are merged. Finally, for a *merged bus set* $\beta = \{b_1 \ldots b_n\}$, where $\beta \subset \Pi_{SC1/SC2}$, let K_β refer to the set of allowed bus speeds for the newly created bus when the buses in set β are merged, and is given by:

$$K_\beta = \Psi_{L(speed)}(b_1) \cap \Psi_{L(speed)}(b_2) \ldots \cap \Psi_{L(speed)}(b_n)$$

The branching algorithm starts out by clustering two slave clusters at a time, and evaluating the gain from this operation. Initially, each slave cluster has just one slave. The total number of clustering configurations possible for a bus matrix creates an extremely large exploration space, that cannot be traversed in a reasonable amount of time. In order to consider only valid clustering configurations and arrive at an optimal solution quickly, a bounding function is used, as shown in Fig. 6.18. This function is called after every clustering operation of any two slave clusters SC1 and SC2. In *Step 1*, a lookup table is used to see if the clustering operation has already been considered previously, and if so, the duplicate clustering is discarded. Otherwise the lookup table is updated with the entry for the new clustering. In *Step 2*, a check is performed to see if the clustering of SC1 and SC2 results in the merging of buses in the matrix, otherwise the clustering is not

Step 1: *if (exists lookupTable(SC1,SC2))*
discard duplicate clustering
else
updatelookupTable(SC1,SC2)

Step 2: *if* ($M_{sc1} \cap M_{sc2} == \phi$)
bound clustering
else
cum_weight + cum_weight + $| M_{sc1} \cap M_{sc2} |$

Step 3: *for each set* $\beta \in \Pi$ SC1/SC2 *do*

if $((K_\beta == \phi) \,||\, (\sum_{i=1}^{|\beta|} \Gamma_{TCP/i} < (width_B \times max\_speed_B)))$

FIGURE 6.18

Bound function [46]
© 2006 IEEE

beneficial and the solution can be bounded. If the clustering results in bus mergers, the number of merged buses is calculated for the clustering, and the cumulative weight of the clustering operation is stored in the branch solution node. In *Step 3*, a check is performed to see if the allowed set of bus speeds for every merged bus is compatible or not. If the allowed speeds for any of the buses being merged are incompatible (i.e., $K_\beta == \varphi$ for any β), the clustering is not possible and the solution is bounded. Additionally, a calculation is performed to estimate if the throughput requirement of each of the merged buses can be theoretically supported by the new merged channel. If this is not the case, the solution is bounded. The bounding function thus enables a conservative pruning process which quickly eliminates invalid solutions and allows a rapid convergence to the optimal solution.

The solutions obtained from the static branch and bound clustering algorithm are ranked from best to worst and stored in a solution database. The next phase (*Phase 4* in Fig. 6.17) validates the solutions by simulation. A fast transaction-based bus cycle accurate simulation engine [48] is used to verify that the reduced matrix still satisfies all the constraints in the design. The arbitration strategy selection is performed at this stage (from the allowed schemes in the constraint set Ψ). If a static-priority-based scheme for a shared slave (with priorities distributed among slave ports according to throughput requirements) results in TCP constraint violations, other arbitration schemes are used, in increasing order of implementation costs. A simpler arbitration scheme like round-robin (RR) is given preference, before resorting to the more elaborate schemes such as the two level TDMA/RR.

It is possible that even after trying these different arbitration schemes, there are TCP constraint violations. In such a case, the solution is removed from the solution database, and the next best solution selected, and this continues until a solution is obtained that successfully passes the simulation-based verification. This is the ***minimal cost solution***, having the least number of buses in the matrix, while still satisfying all TCP constraints in the design. Once such a solution is obtained, the ***minimize_design*** procedure is called (*Phase 5* in Fig. 6.17) where the bus clock speeds are minimized and OO buffer sizes pruned, to further reduce system cost. In this procedure, buses are iteratively selected in the matrix and an attempt

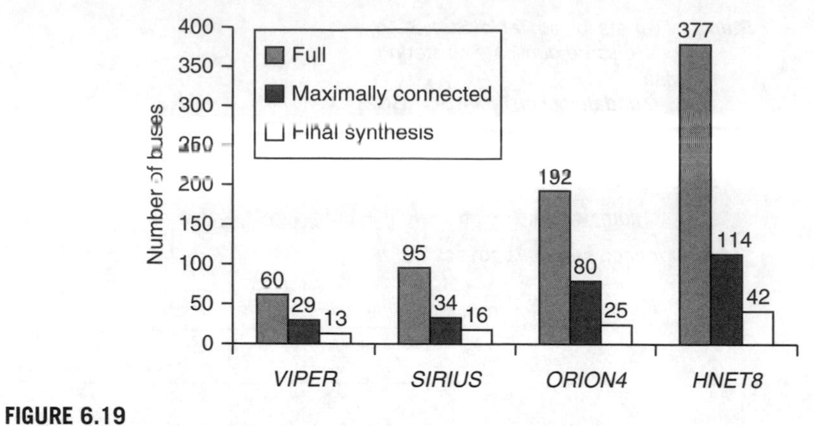

FIGURE 6.19

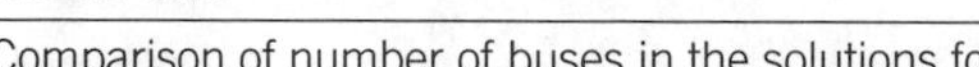

Comparison of number of buses in the solutions for four MPSoC applications [46]

made to arrive at the lowest value of bus clock frequencies (as allowed by Ψ) that does not violate any TCP constraint. Any changes made in bus frequencies are verified via simulation. After minimizing bus frequencies, the OO buffer sizes are pruned from the maximum values allowed to their peak traffic buffer count utilization values, obtained from simulation. Finally, the synthesized minimal cost bus matrix is generated, with a well-defined topology and parameter values.

The bus matrix synthesis methodology proposed by Pasricha et al. [46] was used to synthesize the bus matrix communication architecture for four MPSoC applications from the networking domain-*VIPER*, *SIRIUS*, *ORION4*, and *HNET8*. Figure 6.19 shows the number of buses in the full bus matrix, maximally connected reduced bus matrix (Fig. 6.17), and the final synthesized solution obtained from using the bus synthesis methodology, for the four MPSoC applications. It can be seen that the bus matrix synthesis methodology results in significant matrix component savings, ranging from 2.1× to 3.2× when compared to a maximally connected bus matrix, and from 4.6× to 9× when compared with a full bus matrix.

To compare the quality of the synthesis results from the approach proposed by Pasricha et al. [46], the closest existing piece of work dealing with automated matrix synthesis by Murali and De Micheli [12] was selected, with the aim of minimizing number of buses. Since [12] only generates matrix topology (while the approach in [46] generates both topology and parameter values), the comparison was restricted to the number of buses in the final synthesized design. The threshold-based approach proposed in [12] requires a designer to statically specify (i) the maximum number of slaves per cluster and (ii) the traffic overlap threshold, which if exceeded prevents two slaves from being assigned to the same bus cluster. The results of the comparison study are shown in Fig. 6.20. BMSYN is the bus matrix synthesis approach from [46] while the other comparison points are obtained from [12]. $S(x)$, for $x = 10, 20, 30, 40$, represents the threshold-based approach where no two slaves having a traffic overlap of greater than x% can be assigned to the same bus, and the x-axis in Fig. 6.20 varies the maximum number of slaves allowed in a bus cluster for these comparison points. The values of 10–40% for traffic overlap are chosen as per recommendations from [12]. It can be seen from Fig. 6.20

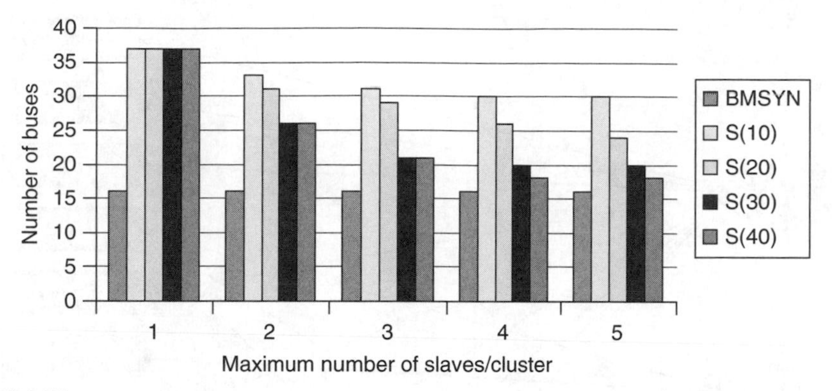

FIGURE 6.20

Comparison of [46] with threshold-based approach [12], for *SIRIUS* MPSoC [46]
© 2006 IEEE

Table 6.7 TCPs for MPSoC application [50]

IP cores in TCP	Throughput requirement (in Mbps)
ARM1, MEM1, DMA, MEM3, MEM5	320
ARM1, MEM3, MEM4, DMA, Network I/F2	240
ARM2, Network I/FI, MEM2	800
ARM2, MEM6, DMA, MEM8, Network I/F2	200
ARM3, ARM4, Network I/F3, MEM2, MEM7	480

© 2006 IEEE

that the synthesis approach by Pasricha et al. [46] produces a lower cost system (having lesser number of buses) than approaches that force the designer to statically approximate application characteristics.

The bus matrix synthesis methodology from [46] was subsequently extended by the authors in [50], with the incorporation of energy estimation models for bus wires and bus logic components (described in Section 5.2). Whereas the focus in [46] was on generating a single, optimal bus matrix solution with the minimal number of buses while satisfying performance constraints, the synthesis approach in [50] generates multiple candidate bus matrix solutions, on which to perform a power-performance trade-off analysis.

The power-performance trade-off methodology by Pasricha et al. [50] was applied to an MPSoC application from the networking domain to perform solution trade-off analysis. Figure 6.21 shows the CTG of this application, while Table 6.7 shows the TCPs for the application that must be satisfied by every solution in the final solution set. *ARM1* is a protocol processor (PP), while *ARM2* and *ARM3* are network processors (NPs). The *ARM1* PP is mainly responsible for setting up and closing network connections, converting data from one protocol type to another and generating data frames for signaling, operating, and maintenance. The *ARM2* and *ARM3* NPs directly interact with the network ports and are used for assembling incoming packets into frames for the network connections, network

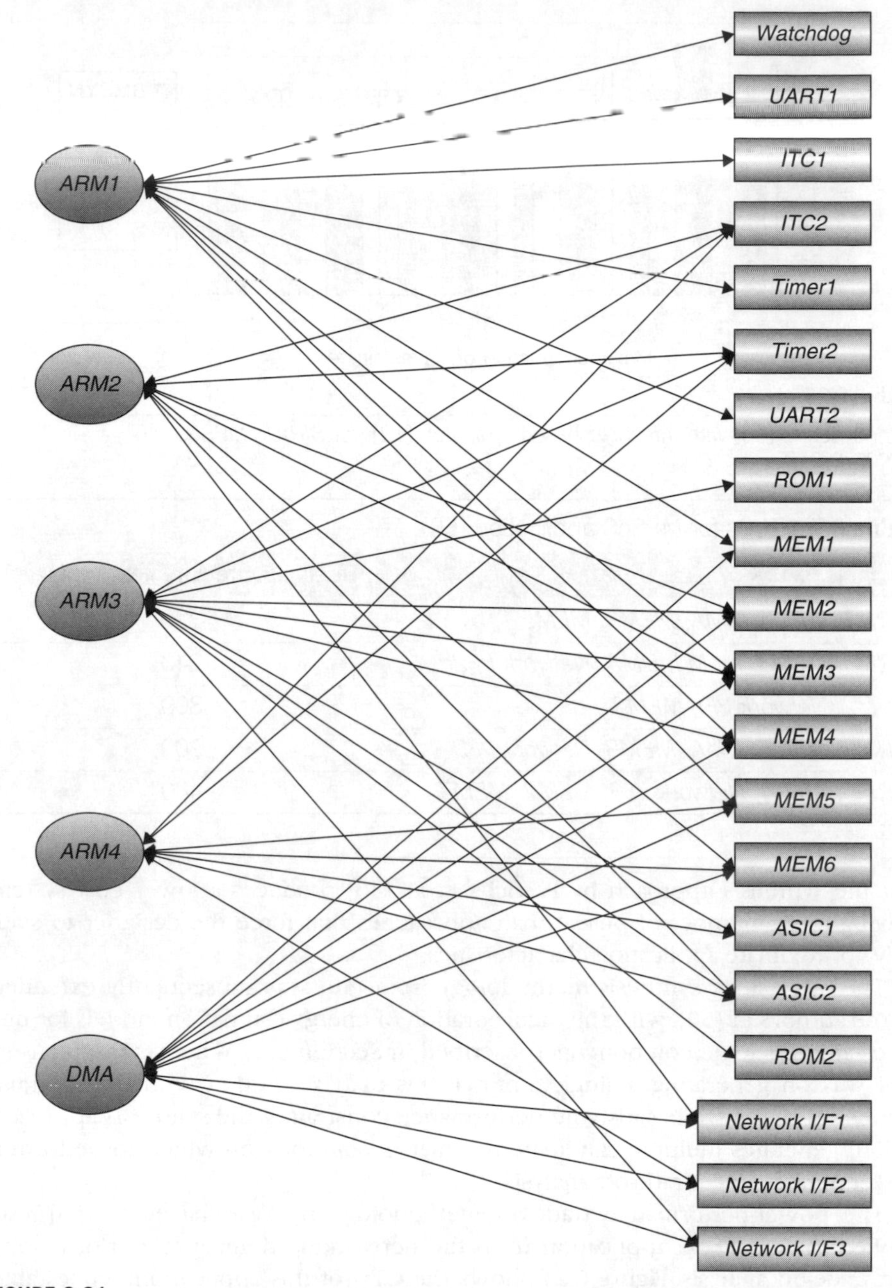

FIGURE 6.21

MPSoC application from networking domain [50]
© 2006 IEEE

port packet/cell flow control, assembling incoming packets/cells into frames, segmenting outgoing frames into packets/cells, keeping track of errors, and gathering statistics. *ARM4* is used for specialized data processing involving data encryption. The *DMA* is used to handle fast memory to memory and network interface data

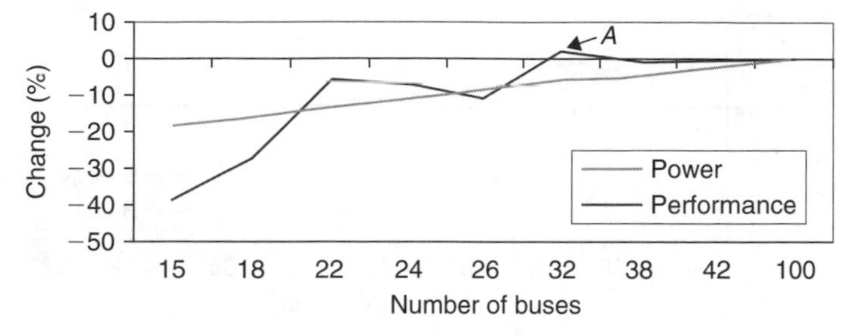

FIGURE 6.22

Power vs. performance trade-off graph for MPSoC [50]
© 2006 IEEE

transfers, freeing up processors for more useful work. Besides these master cores, the application also has a number of memory blocks, network interfaces, and peripherals such as interrupt controllers (*ITC1, ITC2*), timers (*Watchdog, Timer1, Timer2*), universal asynchronous receiver/transmitters, UARTs (*UART1, UART2*), and data packet accelerators (*ASIC1, ASIC2*).

Figure 6.22 shows the power-performance trade-off graph obtained from the output solution set using the methodology from [50], for the MPSoC application. The *x*-axis shows solutions in the output set having different number of buses, while the *y*-axis shows the percentage change in power and performance, using the original full bus matrix as the base case. It can be seen that reducing the number of buses reduces the average power dissipation, because of a smaller number of arbiters, output stages, and bus wires in the matrix that results in less static and dynamic power consumption. As far as the performance change is concerned (measured in terms of average percentage change in data throughput of constraint paths) a reduction in the number of buses increases traffic congestion and reduces performance due to an increase in arbitration wait time. However, it is interesting to note that there are certain points (e.g., point *A*) in Fig. 6.22 where reducing the number of buses actually improves performance. This happens because of a reduction in port switching at the master end as slave clusters grow, reducing re-arbitration delays for masters. In addition to the average percentage change in throughput, another useful metric to gauge application performance is its total runtime.

Figure 6.23 shows the corresponding total energy vs. application runtime trade-off graph for the MPSoC that is useful for cases where the application must execute within a given time or energy budget, such as for mobile battery-driven applications. Overall, the power-performance curves show a possible trade-off of up to approximately 20% for power and up to 40% for performance, enabling a designer to select the appropriate point in the solution space which meets the desired power and performance characteristics. The automated system level synthesis framework generated the solution set and statistics in a matter of a few hours, instead of days or even weeks it would have taken with a gate level estimation flow. Such a framework is invaluable for designers early in the design flow, for quick and reliable communication architecture design space exploration, to guide design decisions at the system level.

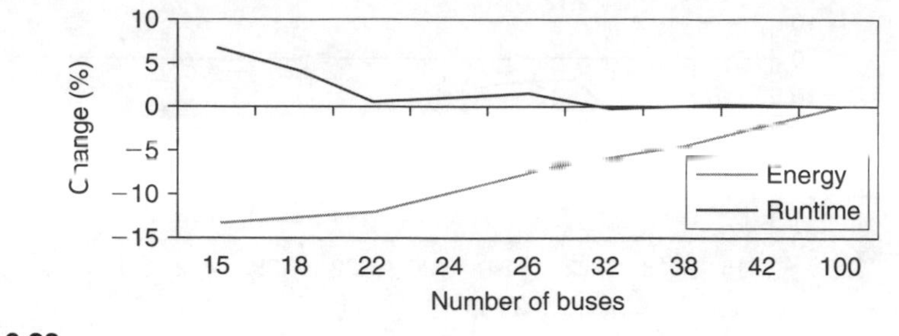

FIGURE 6.23

Energy vs. runtime trade-off graph for MPSoC [50]
© 2006 IEEE

Pasricha et al. [166] further extended the comprehensive synthesis approach presented above by incorporating a PVT (process, voltage, temperature) variation-aware power estimation technique (described in Section 5.4). Incorporating PVT variation-awareness in the system level bus matrix synthesis technique described above resulted in a set of curves for power and energy in the trade-off graph outputs (instead of a single curve for power in Fig. 6.22 and energy in Fig. 6.23), with each curve corresponding to a different PVT corner. This allowed for a more accurate power characterization in the face of PVT variations early in the design flow, at the system level, enabling designers to make more informed decisions when selecting a bus matrix configuration.

6.4 PHYSICAL IMPLEMENTATION AWARE SYNTHESIS

Until now, the synthesis approaches presented design the communication architecture without considering physical implementation issues that can influence performance, such as the layout of the components on the chip or the lengths and routing of the bus wires interconnecting the components. Such physical level information can be extremely important to factor into a synthesis effort, in order to guarantee that the synthesis results are reliable. However, such physical level information is typically available much later in the design flow, and therefore it can be challenging to abstract up this information too early in the design flow, during communication architecture design. A workaround is to make use of high level floorplanners to provide estimates of component placement on the chip, and then using routing estimates to obtain the layout and length of the interconnection wires between components. For IP-based designs, the area and form factor of the IP components is typically assumed to be known, for components that are being reused from earlier designs. This allows early block placement and floorplan generation of designs, to allow consideration of physical level issues during communication architecture synthesis.

Early work by Dick and Jha [51] used a high level floorplanner to create a block placement, and estimate global wiring delays to ensure that hard real-time communication deadlines between components were satisfied during bus topology synthesis. A GA was used to iterate over different bus topology configurations having low contention (and task assignments on components), with wire delay information from the floorplanner guiding the bus topology creation process. Drinic et al. [22] and Meguerdichian et al. [23] used a high level floorplanner to determine design feasibility during bus topology synthesis, by comparing estimates of wire length with an upper bound on wire length. However, an upper bound on wire length has the disadvantage of not accounting for varying capacitive loads of the components. A similar approach is used by Thepayasuwan et al. [17–21] to obtain wire lengths during bus topology synthesis. However, instead of comparing wire lengths against a statically determined upper bound as is done in [22, 23], the lengths are incorporated into an SA cost function that is used to synthesize the bus topology. SA subsequently minimizes the cost function, and selects a topology solution with low total wire length. Hu et al. [52] used a high level floorplanner to determine wire lengths during a point-to-point interconnect width parameter synthesis, with the aim of obtaining wire energy. The synthesis methodology proposed by Pasricha et al. [50] and described in the previous section used a floorplanning tool to obtain wire length for a bus matrix during simultaneous topology and parameter synthesis, with a similar goal of obtaining wire energy (including repeater energy). Guo et al. [24] used a floorplanner during segmented bus topology synthesis. The floorplanner aims to reduce the length of critical wires with high switching activity, in order to reduce wire energy consumption. This work is extended by the authors in [25], by changing the cost function of the SA-based floorplanner during segmented bus architecture topology synthesis, to obtain floorplans with varying wire lengths, and consequently varying energy, chip area, and wire delay. Trade-offs between energy, area, and delay are then output from the synthesis framework.

6.4.1 FABSYN: PHYSICALLY AWARE HIERARCHICAL SHARED BUS ARCHITECTURE SYNTHESIS

A framework to automate synthesis of the topology and parameters for hierarchical shared bus architectures, while detecting and eliminating bus clock cycle constraints early in the design flow, was proposed by Pasricha et al. [47, 53]. This framework, called FABSYN, attempts to synthesize a low cost hierarchical bus architecture (having the least number of buses), and makes use of an SA-based early floorplanner (PARQUET) [54] and wire delay estimation formulations [55] to identify bus clock cycle constraint violations, and eliminate them. An example of a bus clock cycle timing constraint violation is shown in Fig. 6.24. The figure shows a floorplan for an SoC where *IP1* and *IP2* are connected to the same bus as *ASIC1*, *MEM4*, *ARM*, *VIC*, and *DMA*, and the bus has a speed of 333 MHz, in order to meet performance constraints of the application. This implies that the bus cycle time is 3 ns. Now assume that *IP1* and *IP2* need to communicate with each other. For a 0.13 μm complementary metal-oxide semiconductor (CMOS) technology node, the floorplanner finds a wire length of 9.9 mm between pins connecting *IP1* and *IP2* to the bus. The wire delay for a signal traveling from *IP1* to *IP2*, after routing and layout, is found to be 3.5 ns, which clearly violates the bus clock cycle time

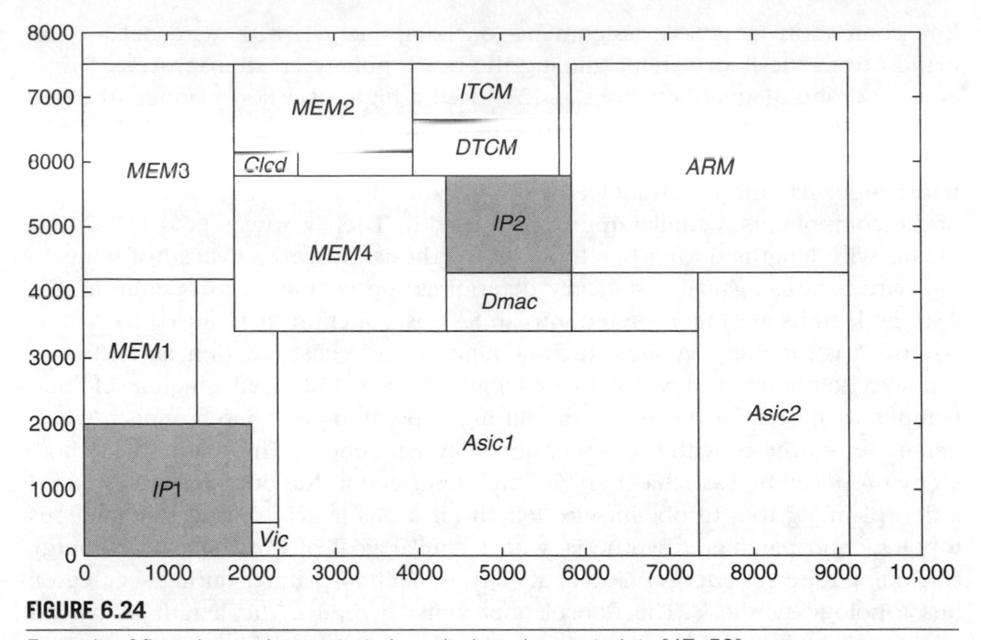

FIGURE 6.24

Example of floorplan to demonstrate bus clock cycle constraints [47, 53]
© 2005 IEEE

constraint of 3 ns. The excessive signal delay is an artifact of the excessive capacitive load on the bus that depends on the number and type of components on the bus. Typically, once such violations are detected at the physical implementation stage in the design flow, designers pipeline the buses by inserting latches, flip-flops, or register slices on the bus, in order to meet bus cycle time constraints. However, it was found that such pipelining of the bus can not only have an adverse effect on critical path performance, but also requires tedious manual reworking of RTL code and extensive re-verification of the design. This process can take from several weeks to months and adversely affect time-to-market. Since the synthesis process is responsible for determining the bus clock frequency (and hence the bus clock cycle period) and the capacitive load on the buses (since synthesis determines component allocation and number of buses), it is important for a synthesis framework to detect such violations during the synthesis process. FABSYN attempts to do precisely that—detecting potential clock cycle violations early in the design flow and then eliminating them.

Figure 6.25 shows the FABSYN synthesis flow, for hierarchical shared bus architecture synthesis. The goal of the synthesis framework is to synthesize a low cost hierarchical shared bus architecture (having the least number of buses), while satisfying application performance constraints. FABSYN synthesizes both the topology, and parameter values for bus widths, bus clock frequencies, arbitration schemes, and DMA burst sizes. The inputs to the flow include a CTG, a target hierarchical shared bus communication architecture standard (e.g., AMBA [8]), a set of Communication Parameter Constraints (Ψ), and a library of behavioral IP models. The CTG is a directed graph, where each vertex v represents a component in the system, and an edge a_{ij} connects components i and j that need to communicate

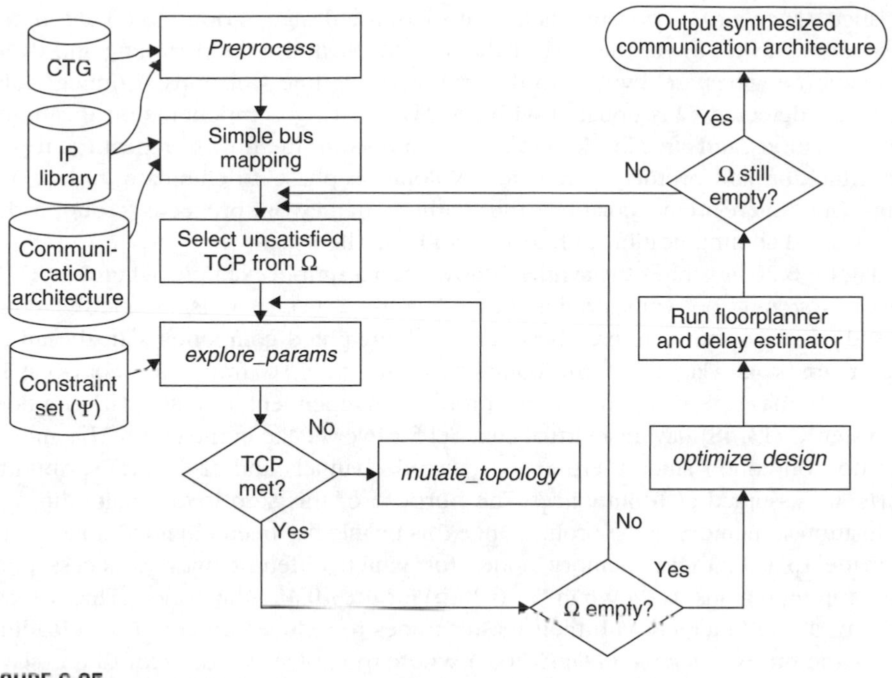

FIGURE 6.25

FABSYN synthesis flow for hierarchical shared bus architectures [47, 53]
© 2005 IEEE

with each other. Each vertex v contains information about the component it represents, such as its area, dimensions (fixed width/height or bounds on aspect ratio), capacitive loads on output pins and which bus type it can be connected to (e.g., a *main* high bandwidth bus like AMBA AHB [8]), a *peripheral* low bandwidth bus like AMBA APB [8], or both. Each edge a_{ij} is associated with a throughput constraint, in the form of a TCP, as dictated by the application. The superset of all TCPs in the application is referred to as set Ω. An example of a CTG was presented in the last section (Fig. 6.16). The set of Communication Parameter Constraints (Ψ) specifies a discrete set of valid values for communication parameters to be synthesized. This ensures that a realistic parameter value is chosen during synthesis, and also allows a designer to bias the synthesis process based on the knowledge of the design and technology being targeted. For example, a designer might decide that the synthesized design should only have data buses with 16, 32, 64, or 128 bit data bus widths, because the IPs in the design cannot support larger widths efficiently.

The general idea in the FABSYN synthesis flow (Fig. 6.25) is to first perform preprocessing transformations on the CTG to improve the performance of the entire system (*preprocess*) and subsequently map all the components from the CTG to a simple shared bus topology of the target bus-based communication architecture. Then, a TCP is iteratively selected from set Ω, starting from the TCP with the most stringent constraint, and a search performed on the communication protocol parameter space for a suitable parameter configuration (*explore_params*). If required, topology mutations are performed (*mutate_topology*) till the TCP constraint is satisfied. Once all TCP constraints are satisfied, the design is optimized (*optimize_design*) to further lower the cost of the system. Next, the floorplanning and delay estimation engines are invoked to detect bus cycle time violations. If timing violations are detected, Ω is updated with the TCPs having components on the buses with violations, and a feedback loop is used to re-enter the flow to repeat the topology mutation and protocol parameter exploration phase, to eliminate these violations. Once there are no violations, the synthesis framework proceeds to output the synthesized communication architecture and chip floorplan.

Figure 6.26 illustrates the synthesis flow with a simple example, where the CTG for the system to be synthesized is shown in Fig. 6.26(a). There is a single TCP comprised of edges shown in gray between the highlighted components. In the initial *preprocess* stage (Fig. 6.25), the components in the CTG from the behavioral IP library database are mapped to a bus protocol independent, TLM simulation model in SystemC [13, 48] having a virtual channel for every edge in the graph. This model has no contention since there are no shared channels and also because infinite ports are assumed at IP interfaces. The purpose of this step is to obtain, through simulation, a memory usage profile. Once this profile has been obtained, an attempt is made to split those memory nodes for which different masters access non-overlapping regions, as shown in Fig. 6.26(b) for the *MEM2* component. Finally, local slave nodes are merged with their master nodes to reduce contention and loading on shared buses, as shown in Fig. 6.26(c), where the leaf slave nodes for CPU1 (slave nodes *MEM2a* and S4) are merged with CPU1 into a *hypernode*. Memory splitting is performed before this step because it allows generation of local memories that can then be merged with their corresponding masters.

After the *preprocess* stage, all the components in the enhanced CTG and the selected bus architecture are mapped from the IP library database to the fast transaction-based bus cycle accurate simulation model [49] with a simple bus topology—a single shared *main* and a single shared *peripheral* bus. As mentioned earlier, every node in a CTG has information relating to the type of bus it can be connected to, which guides the mapping process. A bus *B* can be considered to

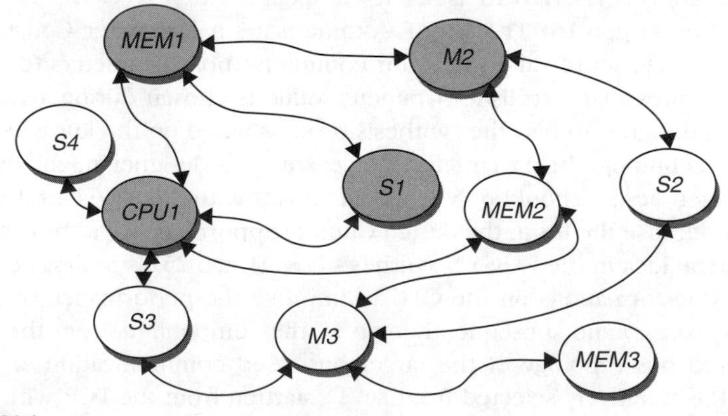

FIGURE 6.26(a)

FABSYN synthesis illustration: initial system CTG
© 2006 IEEE

be a partition of nodes V_{CTG} in a CTG, such that $B \subset V_{CTG}$. Figure 6.26(d) shows the mapped components on the main and peripheral bus partitions, for the preprocessed CTG in Fig. 6.26(c).

Once the simple topology has been created, the largest unsatisfied TCP constraint is selected from set Ω and a search is initiated for a suitable combination of communication parameter values to satisfy the constraint in the *explore_params* stage (Fig. 6.25). The *explore_params* searches for a suitable combination of protocol parameter values that satisfies the TCP constraint under consideration, for the current bus topology. The parameter values are bounded by the constraint set Ψ specified by the designer. However, the exploration space arising from the combinations of the bounded values can still be very large. In the interest of achieving practical running times, this space must be pruned further. This is done as follows. First, the bus widths and frequencies are decoupled from the arbitration schemes and DMA burst sizes. The bus widths and frequencies are set to the maximum

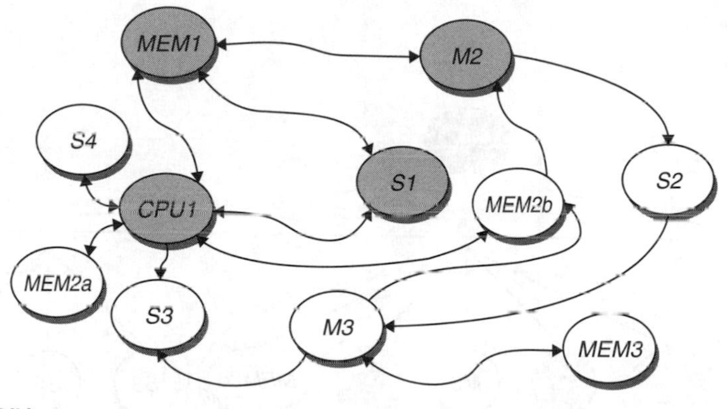

FIGURE 6.26(b)

After splitting *MEM2—preprocess*
© 2006 IEEE

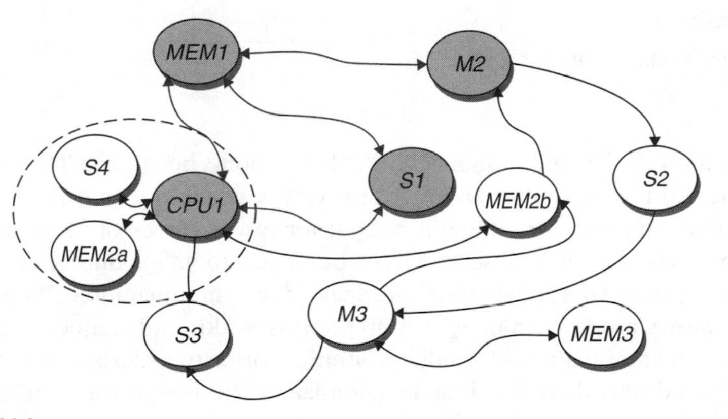

FIGURE 6.26(c)

Merging local slave/memory nodes to create hypernode—*preprocess*
© 2006 IEEE

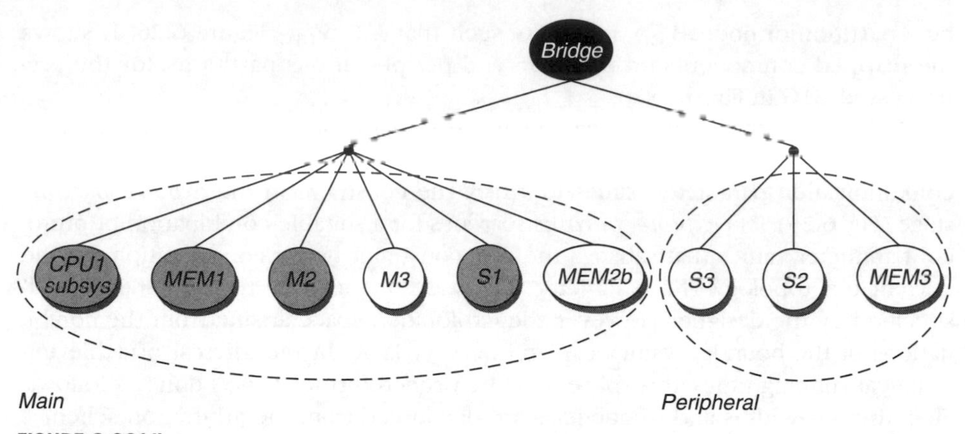

FIGURE 6.26(d)

After mapping nodes to main and peripheral bus
© 2006 IEEE

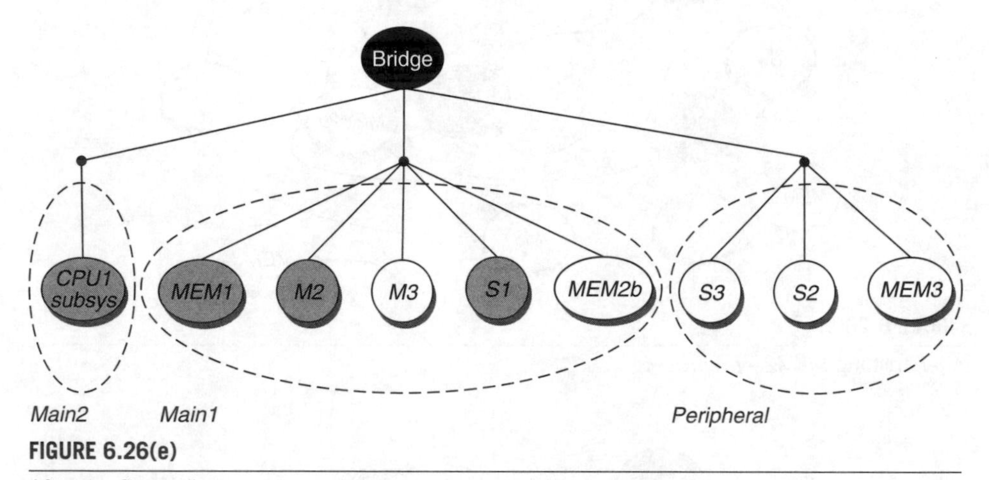

FIGURE 6.26(e)

After the first call to *mutate_topology*
© 2006 IEEE

allowed values set by the designer in Ψ. This is done because if TCP constraints are not met for the maximum values of bus widths and frequencies, it can be safely assumed that they will certainly not be met for lower values of these parameters. It is not possible, however, to set the DMA burst size to its maximum value and the arbitration priority to a fixed value, and make the same guarantee. Therefore, this first step allows a quick pruning of only the bus width and frequency parameter space. Next, a combination of a valid arbitration priority ordering and DMA burst size is selected, and then the design is simulated. The best result configuration is the combination of parameters for which the least number of TCP constraints are violated and the throughput for the TCP being considered is the highest. The set of valid arbitration priorities is governed by the following rules: (i) priorities of masters in TCPs with larger throughput constraints are always greater than priorities

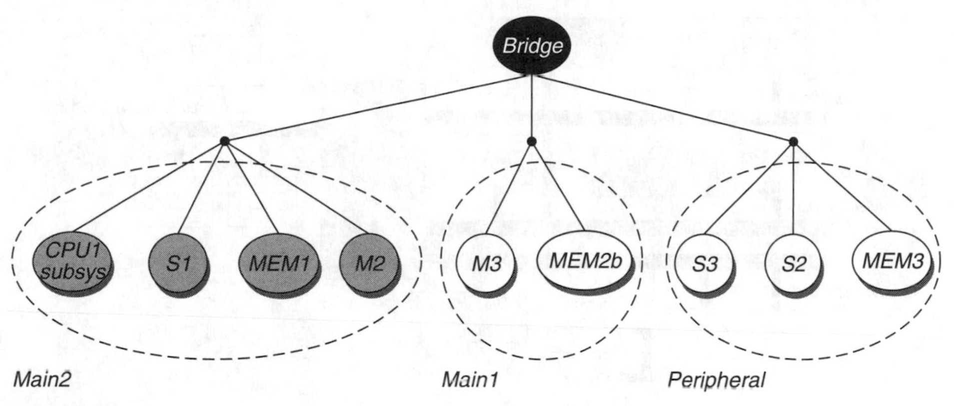

FIGURE 6.26(f)

After migrating all nodes in the TCP to *main2*
© 2006 IEEE

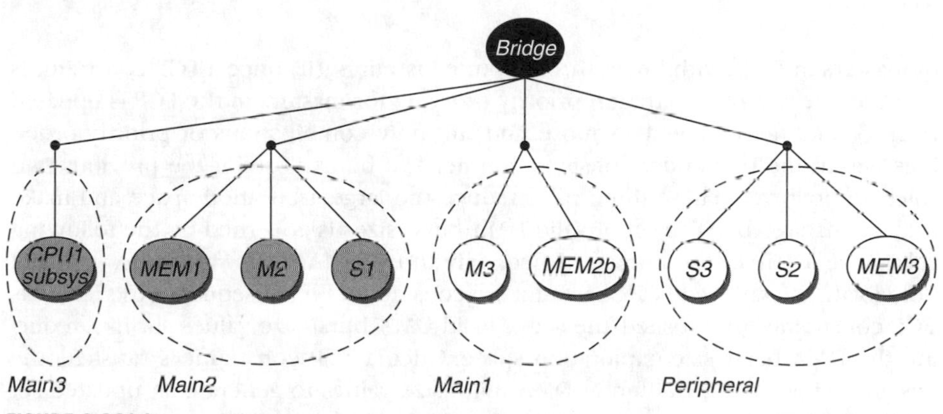

FIGURE 6.26(g)

Creating a new bus and migrating CPU1 to it
© 2006 IEEE

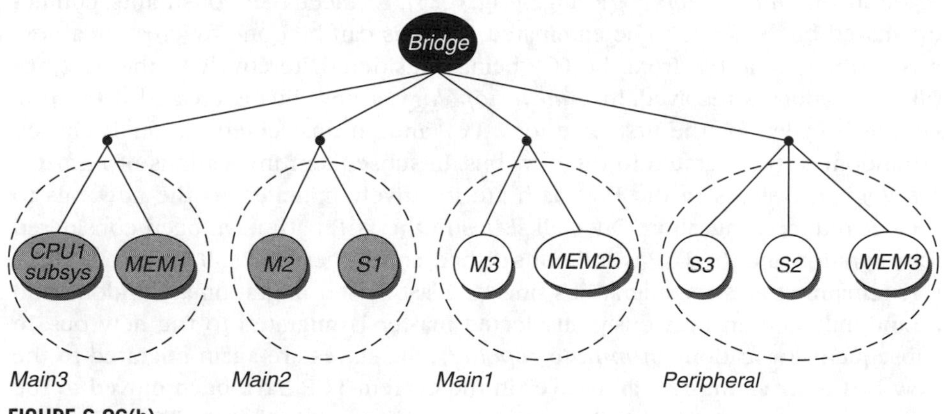

FIGURE 6.26(h)

Final topology with TCP satisfied
© 2006 IEEE

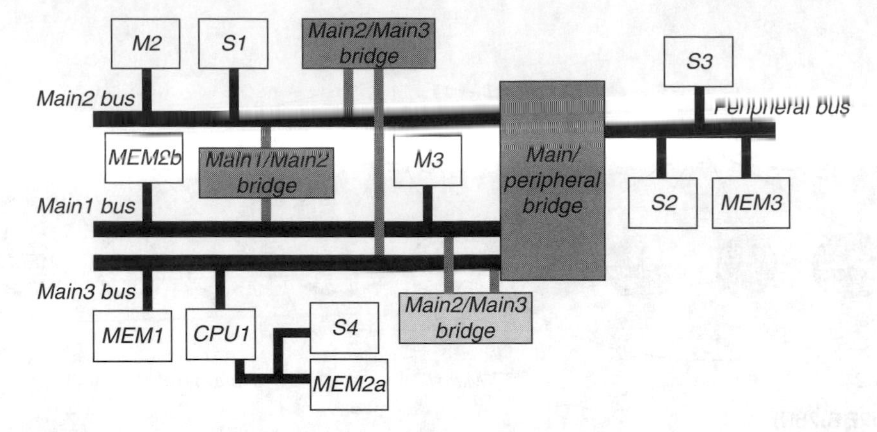

FIGURE 6.26(i)

Synthesized bus architecture [53]
© 2006 IEEE

of masters in TCPs with lower throughput constraints, (ii) once a TCP constraint is satisfied, the relative arbitration priority ordering for masters in the TCP is updated (Step 5) and not changed anymore, and (iii) only combinations of priority orderings within the TCP under consideration need to be explored if the previous two rules are followed. These three rules reduce the large arbitration space and make it more manageable. The set of valid DMA burst sizes is governed by the following rule: once a TCP constraint is satisfied, only those DMA burst size values that did not violate the satisfied TCP constraint are considered for subsequent TCPs. Thus, as TCP constraints are satisfied, the set of valid DMA burst size values shrinks, reducing the DMA burst size exploration space. Once a TCP constraint is satisfied, the design is simulated for different DMA burst size values to generate an updated set of allowed DMA burst sizes that will be used for subsequent TCP explorations.

If the TCP constraint is not satisfied for any combination of communication protocol parameter values, an attempt to change the communication topology is made in the *mutate_topology* stage (Fig. 6.25). To meet TCP constraints, conflict on shared buses needs to be eliminated, and this can be done by creating a new bus and migrating IPs, from the TCP being considered, iteratively to the new bus till the conflict is resolved. In *mutate_topology*, a new bus is created if the procedure is called for the first time for a TCP, and an unselected master is chosen at random to be migrated to the new bus. In subsequent invocations of *mutate_topology*, the slaves in the TCP path are iteratively migrated to the new bus to further reduce congestion. Once all slaves in the TCP path have been considered for migration and the TCP is still not satisfied, and if there are still unselected masters remaining, all slave migrations since the last master migration are undone, and a randomly chosen previously unselected master is migrated to the new bus. In subsequent invocations of *mutate_topology*, the slaves are again migrated to the new bus. After all masters and slaves in the current TCP have been moved to the new bus or at least considered for migration, it is possible that the TCP constraint is still not met. In that case, a new bus is created and the iterative component migration procedure is repeated. In this way, new buses are created till enough bandwidth is available to satisfy the TCP constraint. Note that if a topology mutation causes the best result configuration from *explore_params* to violate any previously satisfied TCP constraints, the mutation is rejected. Otherwise the mutation is considered, even if it deteriorates current TCP performance slightly. This allows overcoming falling into a local optima during the synthesis process. Figure 6.26(e)–(h) illustrate the topology mutation process, starting from the simple bus mapping in Fig. 6.26(d). The result of the first invocation of *mutate_topology* is shown in Fig. 6.26(e), which depicts a newly created bus onto which the *CPU1* master has been migrated. Subsequent calls to the procedure iteratively migrate the rest of the components in the TCP to the new bus. However, the TCP constraint is not satisfied for any of the intermediate topologies, due to data traffic conflicts on both the *main1* and *main2* buses, even when all the components in the TCP have been migrated to a separate bus, as shown in Fig. 6.26(f). Therefore another bus is created (*main3*) and a master (*CPU1*) is migrated to it, as shown in Fig. 6.26(g), followed by slaves in the TCP. For the configuration shown in Fig. 6.26(h), after *MEM1* has been migrated to the new bus, the throughput constraint is found to be satisfied, and no more topology mutation is required, unless there is a timing violation detected by the floorplanning and wire delay estimation engine later in the flow (Fig. 6.25).

Once all the TCP constraints are satisfied, the *optimize_design* procedure is called (Fig. 6.25). The purpose of this stage is to reduce the maximum values selected earlier for bus widths and bus clock frequencies. Each bus in the system is iteratively considered and an attempt made to lower the value for data bus width and bus clock frequency, without violating any TCP constraints. Reducing the bus width reduces the number of wires in a bus and lowers cost of the system. Reducing the bus frequency on the other hand reduces the probability of a bus cycle time violations, since it lengthens the bus clock cycle time period. The order in which the bus width or the bus frequency is reduced is flexible, and is left to the designer.

Next, the optimized system is passed through the floorplanning and wire delay estimator engine (Fig. 6.25). For the system shown in Fig. 6.26, the final modified CTG shown in Fig. 6.26(h) is passed to this engine. If a timing violation is detected, the set Ω is updated with TCPs that have components on the buses with violations, and a feedback loop is used to go back and attempt to eliminate these violations. Since the cumulative capacitive load of components directly contributes to increasing signal propagation delay, an attempt is made to reduce the number of components on the bus having a violation. Therefore, on going back into the flow using the feedback loop, first the TCP that has components on the violated bus with the largest load capacitance on its pins is selected from Ω, and the components are iteratively migrated to another existing bus (or a new bus if migration to existing buses causes TCP constraint violations). If there is still a violation, another TCP is selected from Ω and the components from that TCP are migrated away from the violated bus. Higher priority is given to reducing bus clock frequency over reducing data bus width in the *optimize_design* stage, since reducing bus clock frequency improves the probability of meeting the bus clock cycle period constraint. Note that the solution is guaranteed to converge when using a feedback path. This is because in the worst case the synthesis engine ends up creating a new bus (to migrate components away from the violated bus) which increases the cost of the system, but also improves system performance (even after considering bridge overhead delays) and the ability to meet bus cycle time constraints. Finally, after any violations have been resolved and all TCP constraints satisfied, the final synthesized bus topology, parameter values for bus speeds, data bus widths, DMA burst size, and arbitration priority ordering, along with the feasible floorplan is output. For the system shown in Fig. 6.26(a), the final synthesized architecture looks like the one shown in Fig. 6.26(i).

The *FABSYN* synthesis framework was applied to several industrial strength SoC designs [53] for low cost hierarchical bus architecture synthesis. Figure 6.27 shows the CTG for one of the case studies. There are two data manipulation related TCP constraints that must be satisfied in this system. The first TCP involves the *encryption engine* and includes the *ARM926*, *ASIC1*, *RAM3*, and *EXT_IF* blocks. The *EXT_IF* block fetches data and stores it in *RAM3*. The *ASIC1* and *ARM926* blocks fetch non-overlapping sections of the data, process them, and store them back in *RAM3*, from where the *EXT_IF* block fetches and streams them out at a minimum rate of 200Mbps. The second TCP involves the *USB subsystem*. Data packets received at the *USB* are routed to *RAM1*. The *ARM926* reads this data, processes it, and stores it back to *RAM1* from where the *DMA* engine

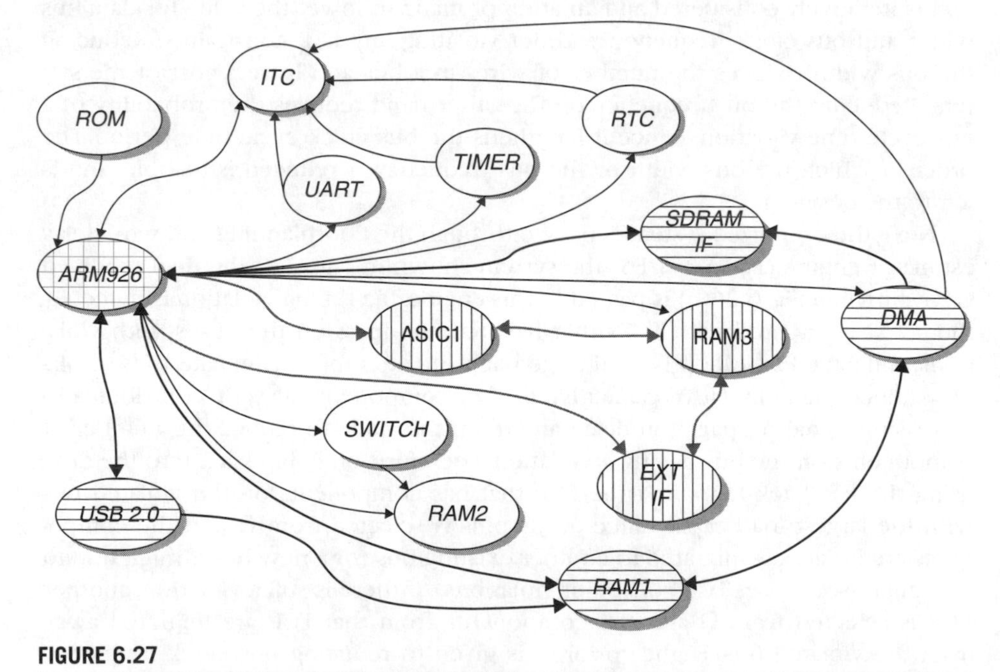

FIGURE 6.27

CTG for network communication subsystem SoC [53]

transfers it to *SDRAM_IF*, which streams it out at a minimum rate of 480 Mbps. There is also a third subsystem which involves the *SWITCH*, *RAM2*, and *ARM926* components. However, this is a very low priority data path which has no data rate constraint from the designer, and therefore not classified as another TCP to be satisfied. Table 6.8 shows the Communication Parameter Constraint set (Ψ) for this case study. The target communication architecture for the automated synthesis is the AMBA high performance AHB bus and a low bandwidth APB bus [8]. Figure 6.28 shows the final output of the synthesis flow—a synthesized architecture that meets all throughput and timing constraints. The values for the generated communication parameters are given in Table 6.9 and the final floorplan for this system is shown in Fig. 6.29. The automated synthesis engine initially created two AHB buses, with the *SWITCH* and *RAM2* components connected to *AHB1*, which was assigned a clock frequency of 200 MHz to meet the encryption path throughput constraint. However, the floorplanning engine detected a bus clock cycle time violation for the bus due to excessive capacitive loading. The *topology_mutate* stage then split the shared AHB bus and assigned the *ARM926*, *ASIC1*, and *EXT_IF* masters and their associated slaves to one bus, and the *SWITCH* and *RAM2* components to another AHB bus, to reduce capacitive loading. Finally, the *optimize_design* function reduced the bus frequencies for the AHB buses from 200 to 133 MHz and the APB bus to 66 MHz, to lower the cost of the system. Both the throughput constraints were still met at these lower bus speeds. The synthesis engine made a simple assumption and assumed a 133 MHz bus frequency for

Table 6.8 Communication Parameter Constraint set [53]

Parameter	Values
Bus width	8, 16, 32
Bus speed	33, 66, 100, 133, 166, 200
DMA burst size	1, 2, 4, 8, 16
Arbitration strategy	Static priority

Table 6.9 Communication parameter values [53]

Parameter	Values			
	AHB1	*AHB2*	*AHB3*	*APB1*
Bus width	32	32	32	32
Bus speed	133	133	133	66
DMA burst size	16			
Arbitration strategy	*ARM>USB>DMA>EXT_IF>ASICI>SWITCH*			

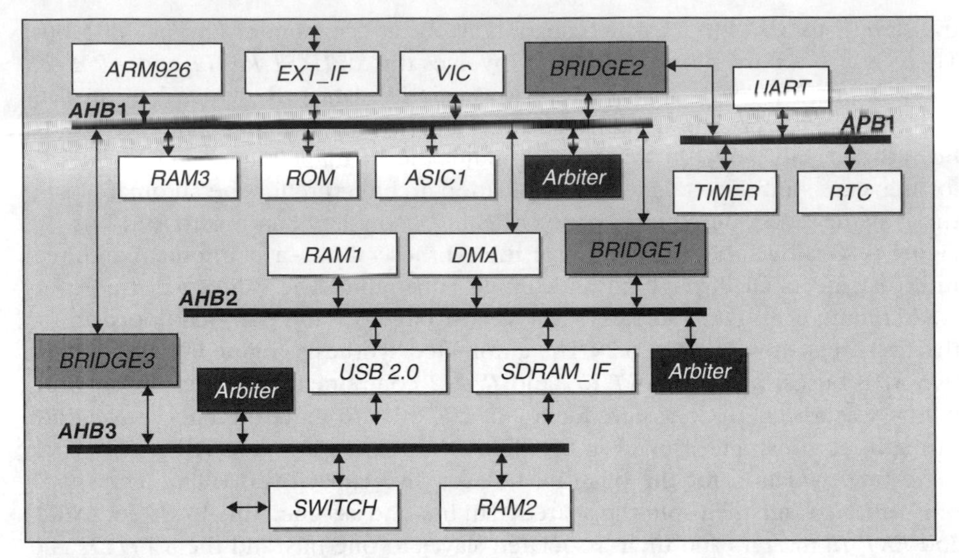

FIGURE 6.28

Synthesized SoC subsystem [53]
© 2006 IEEE

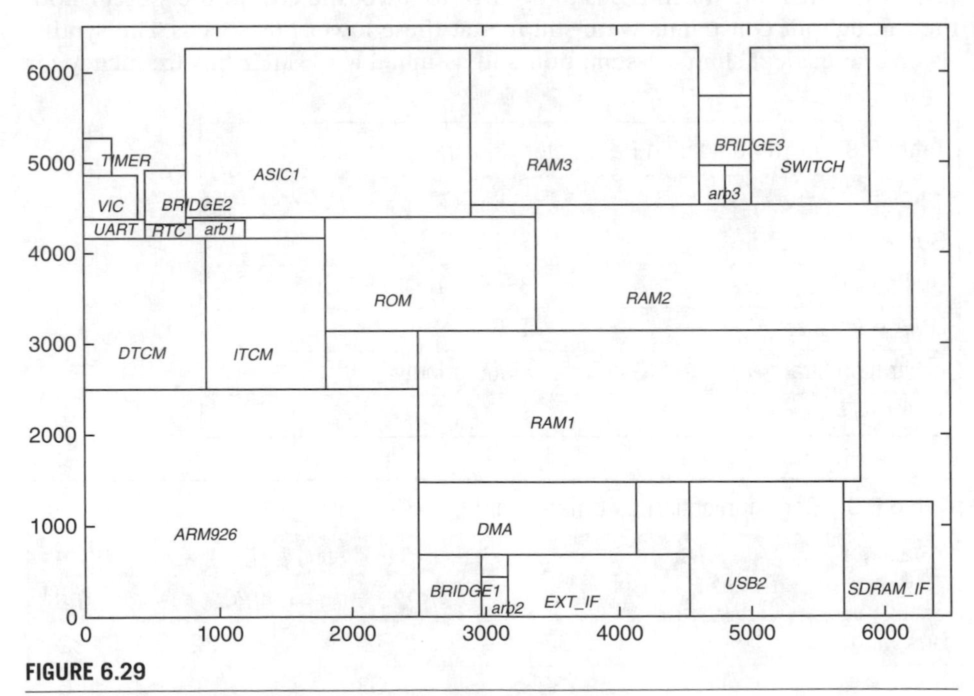

FIGURE 6.29

Final floorplan for SoC subsystem case study [53]
© 2006 IEEE

AHB3 to simplify the design of *BRIDGE3* to *AHB1*, but a designer can choose to further lower the *AHB3* bus speed if a more complex bridge is acceptable.

The quality of the *FABSYN* synthesis solution was compared with other synthesis approaches. Since none of the existing communication architecture

Table 6.10 Comparison study [53]

Case study 1 designs	Initial	ABS	Manual	FABSYN
Number of busses	2	3	5	4
TCP constraints satisfied	0/2	2/2, *not feasible*	2/2	2/2
Execution (millions)	49.76	24.51	18.80	20.32
Time to synthesize	~minutes	~hours	~days	~hours

synthesis approaches is aimed at detecting bus cycle time violations early in the design flow, there is no direct point of comparison. The quality of the synthesized designs from *FABSYN* was compared with an approach that maps all the components in the application to a single main/peripheral shared bus (*initial*), an automated bus architecture synthesis flow that does not use a high level floorplanner (*ABS*), and a manually intensive, high level synthesis effort by a designer that also makes use of a floorplanning, and wire delay estimation engine to detect timing violations (*manual*) just like the *FABSYN* floorplan-aware automated bus architecture synthesis approach. The *manual* synthesis approach involves a designer manually selecting a combination of bus topology and protocol parameter values, simulating the high level design models in SystemC, and then iteratively modifying the bus architecture and protocol parameter values based on the simulation results and designer intuition, till all constraints are found to be satisfied.

Table 6.10 presents the result from the comparison study. The *initial* approach is unable to satisfy any of the TCP constraints for the case study, because of excessive data traffic conflicts on its restricted number of buses. In contrast, the *ABS* approach does manage to satisfy TCP constraints for the case study, but synthesizes a bus architecture with bus clock cycle time violations that remain undetected, and thus the synthesized architecture is not feasible. The *manual* approach satisfies all TCP constraints and is also able to detect and eliminate bus clock cycle time violations in the design, just like the *FABSYN* approach. However, there are a few key differences between the *manual* approach and the *FABSYN* approach. Firstly, the *manual* approach generates bus architectures having a greater implementation cost (i.e., having a larger number of buses) when compared with architectures generated using *FABSYN*. This is because the automated *FABSYN* flow is able to traverse a much larger communication parameter exploration space in a reasonable amount of time than the *manual* approach, and avoids making conservative decisions to create a new bus like in the *manual* approach, unless all suitable combinations of communication parameters are unable to meet the TCP constraint for the existing bus topology. Secondly, the performance of the architecture generated by the *manual* approach is actually found to be better than the *FABSYN* approach. This is because of the larger number of buses used by the *manual* approach, which reduces data traffic conflict and improves concurrency, at the cost of increasing the implementation cost. But it is important to note that *FABSYN* is not concerned about the absolute performance of the system, but rather at satisfying all TCP constraints and minimizing the implementation cost of the synthesized architecture. Finally, and most importantly, the *manual* approach suffers from the major drawback of taking several days to come up with a bus architecture which is typically overdesigned and exceeds the requirements (resulting in a more expensive system), whereas the automated *FABSYN* approach generates a better quality architecture in a matter of a few hours.

6.5 MEMORY–COMMUNICATION ARCHITECTURE CO-SYNTHESIS

The memory architecture contributes to a significant amount of data traffic in a typical system that must be supported by the communication architecture. The process of memory architecture synthesis, which must determine the number, type, and size of the memories in the system, therefore directly impacts the design of the communication architecture, which must connect these memory modules together, to ensure that application constraints are satisfied. Memory also takes up a large chunk of on-chip area, as much as 70% in some cases [56]. Estimates indicate that this figure will go up to 90% in the coming years [57]. This indicates the need to consider memory architecture design, while synthesizing on-chip communication architectures. Traditionally, however, memory synthesis is performed separately from communication architecture synthesis (typically the memory architecture is decided before the communication architecture) [58–62]. This is done mainly due to tractability issues [63, 64], but it can be shown that designing the memory and communication architectures simultaneously can lead to better quality designs [65–67]. Consider the example of a networking MPSoC subsystem shown in Fig. 6.30(a). The figure shows the system after HW/SW partitioning, with all the IPs defined, including memory, which is synthesized based on data size and high level bandwidth constraint analysis. Figure 6.30(b) shows the traditional approach where communication architecture synthesis is performed after memory synthesis, while Fig. 6.30(c) shows the case where memory and communication architectures have been co-synthesized (using the COSMECA approach proposed by Pasricha and Dutt [65, 66]—described in more detail later in this section). Now consider the implications of using a co-synthesis framework, as shown in Fig. 6.30(b) and (c). Firstly, the co-synthesis approach is able to detect that the data arrays stored in *MEM1* and *MEM2* end up sharing the same bus, and automatically merges and then maps the arrays onto a larger single physical memory from the library, thus saving area. Secondly, the co-synthesis approach is able to merge data arrays stored in *MEM3* and *MEM5* onto a single memory from the library, saving not only area but also eliminating two buses, as shown in Fig. 6.30(c). However, *MEM5* cannot share the same bus as *MEM3* (or *MEM4*) in Fig. 6.30(b) because the access times of the pre-synthesized physical memories are such that they cause traffic conflicts which violate bandwidth constraints. Thirdly, due to the knowledge of support for OO transaction completion [45] by the communication architecture, the co-synthesis approach is able to add an OO buffer of depth 6 to *MEM4*, which enables it to reduce the number of ports from 2 to 1, thus saving

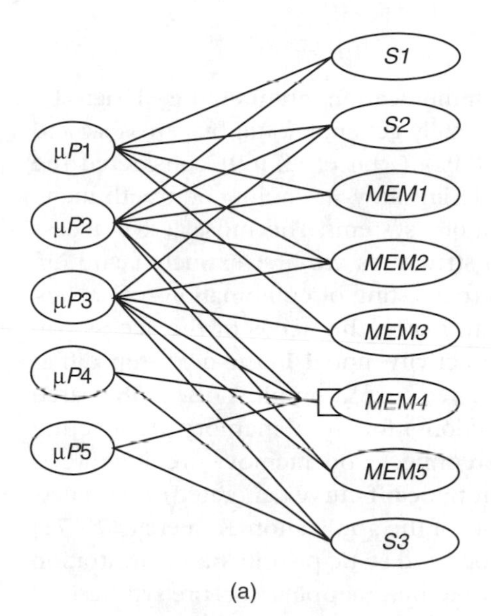

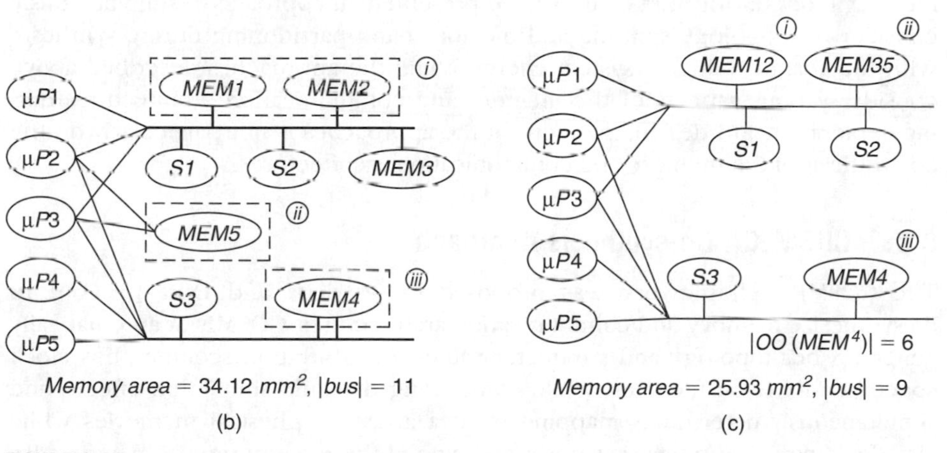

FIGURE 6.30

Comparison of traditional (separate synthesis) approach and co-synthesis approaches for MPSoC example: MPSoC system example (b) result of performing memory synthesis before communication architecture synthesis (c) result of performing co-synthesis of memory and communication architectures [65, 66]

area and cost (since dual port memories are more expensive), while still meeting performance constraints. It is thus apparent that a co-synthesis approach is able to make better synthesis decisions by exploiting the synergy and interdependence between the memory and communication architecture design spaces, to reduce the overall cost of the synthesized system.

6.5.1 Early Work

Early work considering memory and communication architecture design by Shalan et al. [68] proposed a tool to automatically generate a full bus crossbar and a dynamic memory management unit (DMMU). Grun et al. [69] considered the system connectivity topology early in the design flow, in conjunction with memory exploration, for simple processor-memory systems. The most active access patterns exhibited by the application data structures are first extracted, and different memory architecture configurations (consisting of combinations of caches, SRAM, and custom memory blocks) that can match the needs of the access patterns are obtained—assuming a simple connectivity model. In the next step, different communication architectures are considered (e.g., a simple AMBA AHB shared bus architecture, point-to-point interconnections) for these memory architecture configurations, and the most suitable interconnect and memory architecture is selected (from the pareto-optimal constraint trade-off curve generated), that meets the power, performance, and cost constraints of the application. Kim et al. [70, 71] explored the hierarchical shared bus topology and static-priority-based arbitration space, to determine the best memory port-to-bus mapping for pre-synthesized memory blocks. Srinivasan et al. [72, 73] presented an approach to simultaneously consider bus topology splitting and memory bank partitioning during synthesis, with the goal of reducing system energy. While the approaches described above consider various aspects of the memory and communication architectures during exploration and design, the focus of the approaches is not particularly on the co-synthesis of the memory and communication architectures.

6.5.2 COSMECA Co-synthesis Approach

The COSMECA framework was proposed by Pasricha and Dutt [65, 66] to co-synthesize memory and communication architectures. COSMECA automatically generates bus topology and parameter values for arbitration schemes, bus clock speeds, and OO buffer sizes, while considering dynamic simulation effects, and simultaneously determines mapping of data arrays to physical memories while also deciding the number, size, ports, and type of these memories, from a memory library. The goal of the framework was to obtain a least cost system, having minimal number of buses while satisfying performance and memory area constraints.

The communication architecture targeted for synthesis in COSMECA is the bus matrix architecture. The bus matrix synthesis approach from [46] (described in Section 6.3) is extended by adding support for simultaneous memory synthesis. The memory synthesis in COSMECA targets a memory library populated by on-chip SRAMs, on-chip DRAMs, EPROMs, and EEPROMs having different capacities, areas, ports, and access times. Application memory requirements are initially represented by abstract data blocks (DBs) in a *CTG* [46], which are collections of scalars or arrays accessed by the application, similar to basic groups in [74]. These DBs are initially grouped together into virtual memories (VMs) to reduce memory area cost. Only DBs with (i) similar edges (i.e., edges from the same masters) and (ii) non-overlapping access are merged, so as not to constrain mapping freedom later in the flow. A *data block dependency graph* (DBDG) is used to determine if

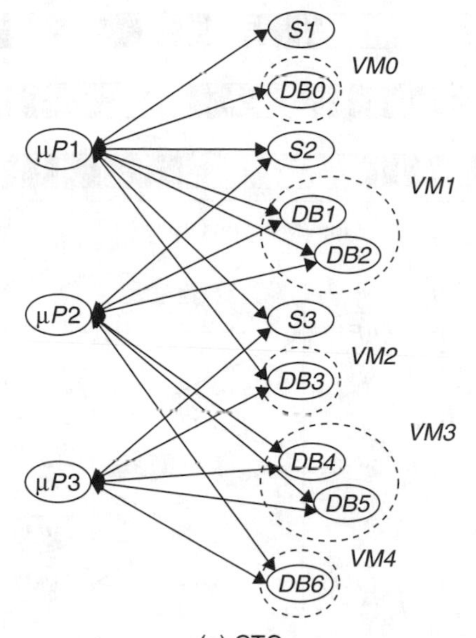

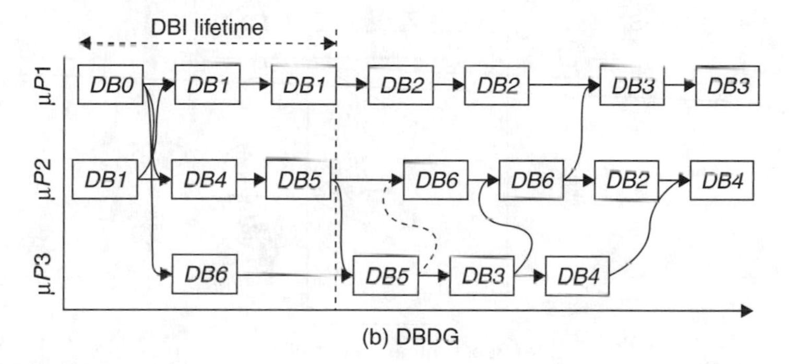

FIGURE 6.31(a, b)

COSMECA co-synthesis example [65, 66]
© 2006 IEEE

DBs have non-overlapping access. Figure 6.31(b) shows the DBDG for the example in Fig. 6.31(a). The DBDG shows the dependency of DB accesses on each other—a DB cannot be accessed till the source DBs of all its input edges have been accessed. If two DBs have similar edges and non-overlapping access, they are eligible for merger (e.g., *DB1*, *DB2* in Fig. 6.31(b)). The size of the VM created depends on the lifetime analysis of merged DBs—it is the sum of the sizes of the merged DBs, unless the lifetimes do not overlap, in which case it is the size of the larger DB being merged. Figure 6.31(b) shows the lifetime of *DB1*. It is possible for *DB2* to overwrite *DB1*, thus saving memory space.

Once the DBs have been merged into VMs, the standard bus matrix synthesis flow (derived from [46], and also presented in Section 6.3.2) is utilized. The CTG from Fig. 6.31(a) is mapped to the matrix template, and a branch- and bound-based clustering approach is applied to create a set of reduced bus matrix solutions,

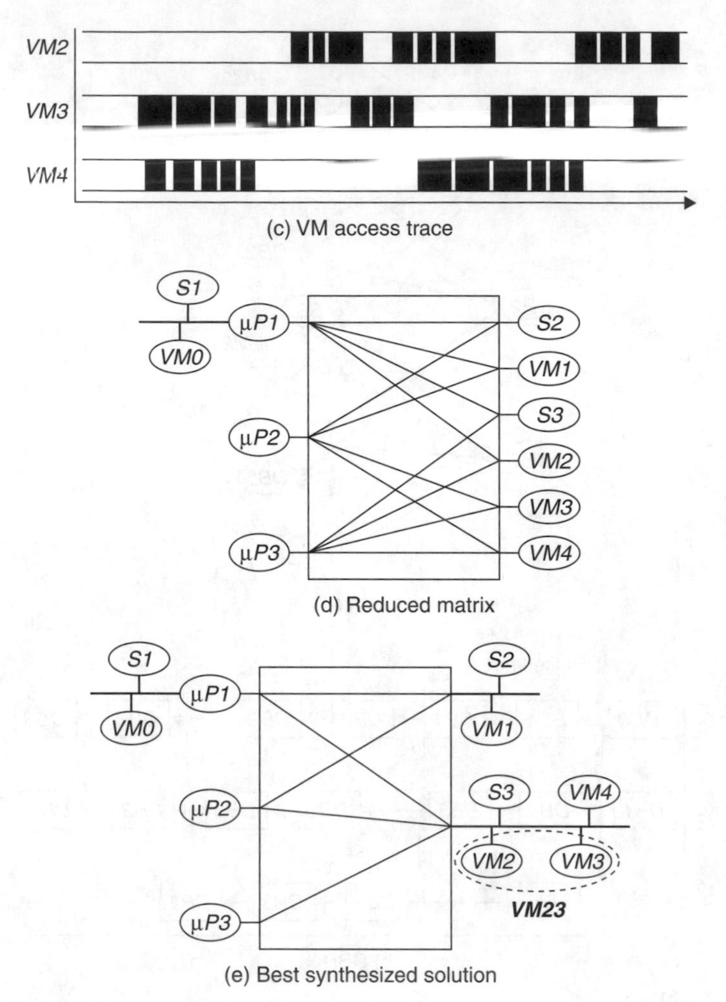

(c) VM access trace

(d) Reduced matrix

(e) Best synthesized solution

FIGURE 6.31(c, d, e)

(Continued)

one of which is shown in Fig. 6.31(d). Note that the memory space is still represented in an abstract manner, using VM blocks. Subsequently, a heuristic is used to guide the mapping of VMs to physical memories in the memory library. The goal is to find N solutions that satisfy memory area and performance constraints of the design. The general idea is to first simulate the best solution from the set of (ranked) matrix solutions, to generate memory access traces that are used to determine the extent of access overlap of VMs at each slave access point (SAP) in the bus matrix architecture (an SAP is defined as the point where arbitration occurs in a bus matrix structure, and which connects one or more slaves/memories). If the overlap is below a user defined *overlap threshold T*, the VMs are merged. Figure 6.31(e) shows how *VM2* and *VM3* are merged, since their memory access trace shown in Fig. 6.31(c) has an overlap less than the chosen value for T. The VMs are then mapped to physical memories from the memory library.

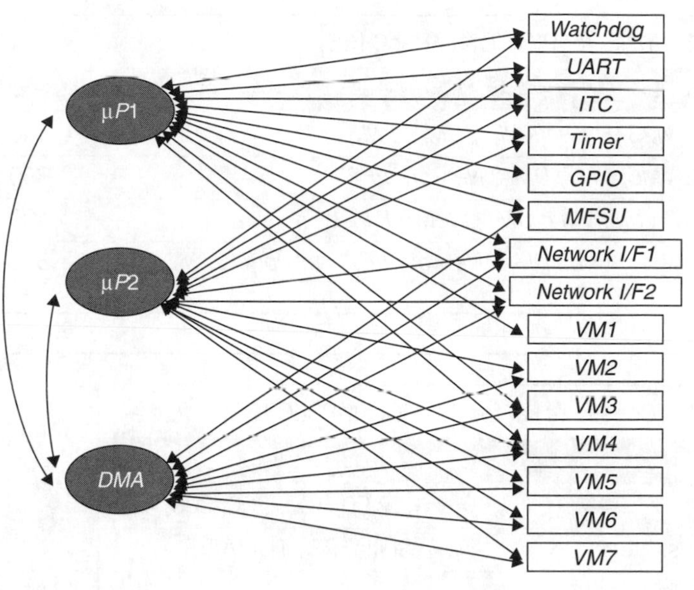

FIGURE 6.32

PYTHON CTG [66]

Initially, the best memory from the library which fits the size requirement and has the maximum port bandwidth is selected. If it is found that the throughput constraints are not met even for the memory with best performance, the matrix solution is discarded, and the next best matrix solution from the set of (ranked) matrix solutions is selected. Otherwise, if throughput constraints are met, and memory area constraints are also met, the solution is added to the final solution database. Next, an attempt is made to lower memory area by randomly selecting a VM at every SAP and replacing the mapped physical memory with one which meets the size requirements, but has lower area. If there is no performance violation, and if the area bounds are met, a solution has been found. This process is repeated till all VMs become ineligible for mapping optimization, or if the required N solutions have been found. If the former case is encountered, and the number of solutions found is less than N, the framework proceeds to select the next best solution from the set of (ranked) matrix solutions and repeats the process.

The COSMECA co-synthesis framework was applied to four industrial strength MPSoC applications, for which memory and communication architectures were required to be synthesized. Figure 6.32 shows the CTG for one of these applications (*PYTHON*), after the initial memory preprocessing phase in which DBs are merged into VMs. Not shown in the CTG, but included in the memory area analysis are the 32KB instruction and data caches for each of the two processors. For clarity, the TCPs are presented separately in Table 6.11. *μP1* is used for overall system control; generating data cells for signaling, operating, and maintenance; communicating and controlling external hardware; and to setup and close data stream connections. *μP2* interacts with data streams from external interfaces and performs data packet/frame encryption and compression. These processors interact with

Table 6.11 TCPs for PYTHON MPSoC [66]

IP cores in TCP	TCP constraint (in Mbps)
μP2, VM2, VM3, Network I/F1, DMA, VM6	400
μP2, VM2, VM6, VM7, DMA, Network I/F2	960
μP1, MFSU, VM3, VM4, DMA, Network I/F1	400
μP2, VM4, VM5, VM7, DMA, Network I/F1, Network I/F2	600

Table 6.12 Constraint set for PYTHON MPSoC [66]

Set	Values
Bus speed	25, 50, 100, 200, 300, 400
Arbitration strategy	Static, RR, TDMA/RR
OO buffer size	1–8
Memory mapping	VM1⇒ EEPROM

each other via shared memory and a set of shared registers (not shown here). The *DMA* engine is used to handle fast memory to memory and network interface data transfers, freeing up the processors for more useful work. *PYTHON* also has several peripherals such as a multi-functional serial port interface (*MFSU*), a universal asynchronous receiver/transmitter block (*UART*), a general purpose I/O block (*GPIO*), timers (*Timer, Watchdog*), an interrupt controller (*ITC*), and two proprietary external network interfaces.

Table 6.12 shows the constraint set for *PYTHON*. For the synthesis, the target bus matrix standard is the AMBA AXI [45]. A fixed bus width of 32 bits is selected, as per application requirements. The memory area constraint is set to 120 mm^2 and the estimated memory area numbers are for a 0.18 μm CMOS technology. The value for the overlap threshold T is set to 10% for the example. Figure 6.33 shows the best solution (least number of buses) with the least memory area for *PYTHON*. The figure also shows bus speeds, memory sizes, number of ports, and OO buffer sizes. Figure 6.34 shows the variation in memory area and number of buses in the matrix for the 10 best solutions ($N = 10$), for *PYTHON*. From the figure it can be seen that no solution having 7 buses in the bus matrix exists for *PYTHON*. The dotted line in Fig. 6.34 indicates the solution shown in Fig. 6.33. A significant variation in valid combinations of memory area and number of buses is observed, for the solution space. COSMECA enables a designer to trade-off memory area and bus count during the solution selection process.

During the course of the *COSMECA* co-synthesis flow, a threshold factor T was used to determine the extent to which VMs are merged at SAPs in the bus matrix. This parameter is specified by the designer. To understand the effect of this threshold factor T on the solution quality, the threshold value is varied and the

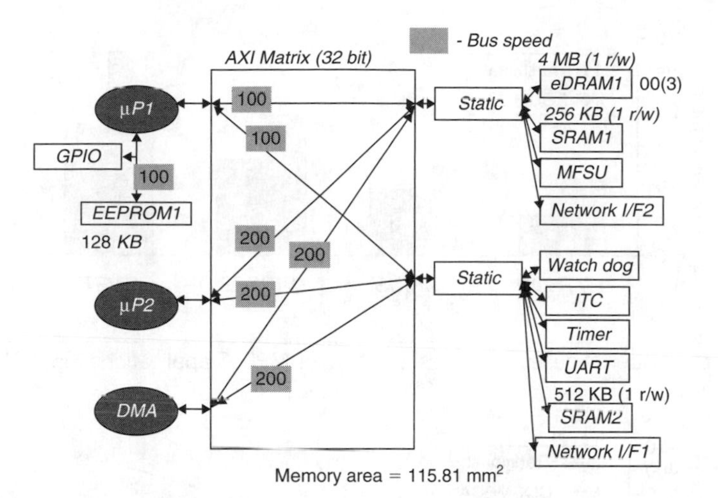

FIGURE 6.33

Synthesized output for PYTHON MPSoC [66]
© 2007 IEEE

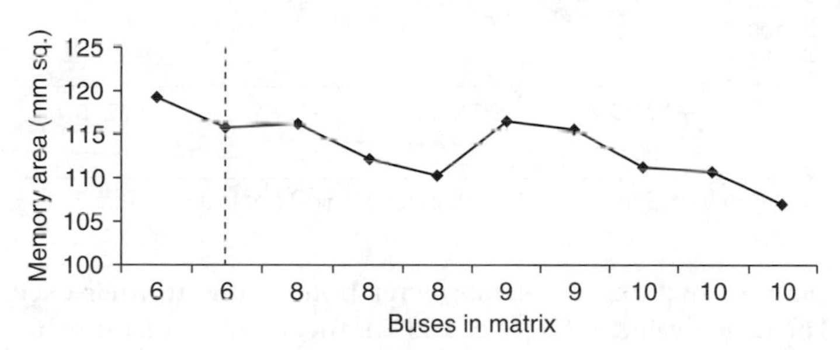

FIGURE 6.34

PYTHON output solution space (for $N = 10$) [66]
© 2007 IEEE

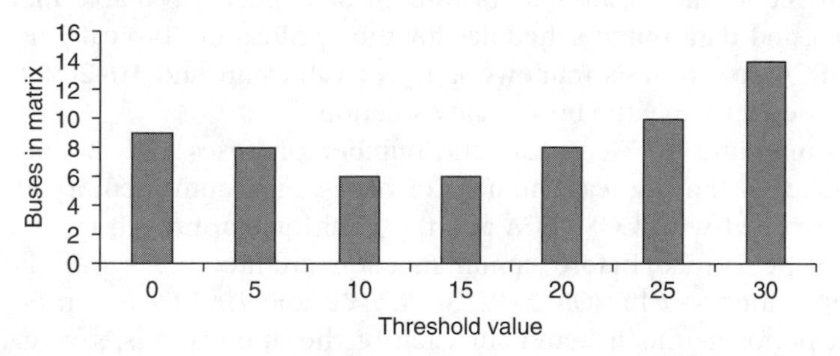

FIGURE 6.35

Effect of varying threshold value on solution quality for PYTHON [66]
© 2007 IEEE

co-synthesis flow repeated for the *PYTHON* MPSoC. The result of this experiment is shown in Fig. 6.35. It can be seen that for very low values of T (e.g., <10%), the number of buses in the matrix for the best solution is high. This is because low values of T discourage merger of VMs, which ends up creating a system with several

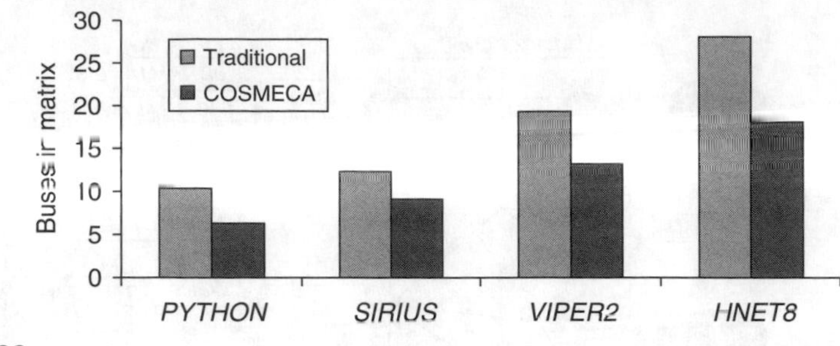

FIGURE 6.36

Comparison of number of buses in the best solution for MPSoC applications [65, 66]
© 2006 IEEE

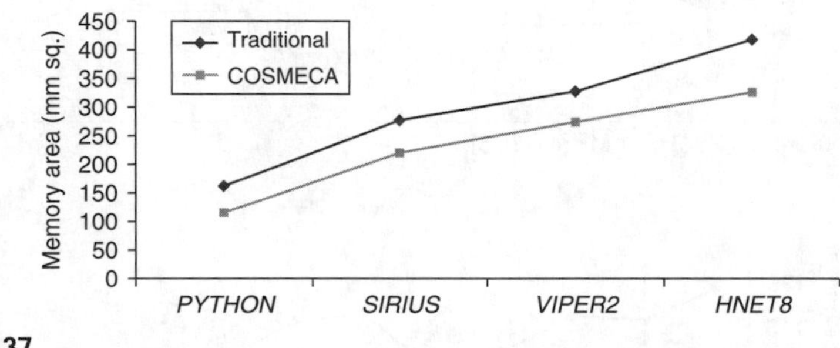

FIGURE 6.37

Comparison of memory area in the best solution for MPSoC applications [65, 66]
© 2006 IEEE

physical memories that exceed memory area bounds due to their excessive area overhead. For larger values of T (e.g., $\geq$20%), the number of buses for the best solution is also high, because it becomes harder to meet application throughput constraints with the large overlap. There might be slight variations to this trend, depending upon a complex amalgamation of factors such as stringency of throughput requirements, allowed maximum bus speeds, available memory port bandwidths, and data traffic schedules for the application. Typically however, for the COSMECA co-synthesis framework, lower values around 10–20% for overlap threshold T seem to give the best quality solutions.

Figures 6.36 and 6.37 compare the number of buses and memory area for the best solution (having least number of buses, minimum memory area for the solution) obtained with COSMECA and the traditional approach (where memory synthesis is performed before communication architecture synthesis) for four MPSoC applications—*PYTHON*, *SIRIUS*, *VIPER2*, and *HNET8*. It can be seen that COSMECA performs much better for each of the applications, saving from 25% to 40% in the number of buses in the matrix and from 17–29% in memory area, because it is able to make better decisions by taking the communication architecture into account while allocating and mapping DBs to physical memory components. Table 6.13 shows the total number of simulation runs and total simulation time in hours required by the COSMECA co-synthesis approach, for the four MPSoC applications. Note that the contribution of the static estimation phases such as the branch and bound clustering is almost negligible and simulation takes

Table 6.13 Co-synthesis time for MPSoC applications [66]

Applications	Simulation runs	Total co-synthesis time (in hours)
PYTHON	13	3.5
SIRIUS	19	8.6
VIPER2	26	17.8
HNETS	38	28.5

up most of the time during co-synthesis. It can be seen that the entire COSMECA flow took in the order of hours to generate the best solution for each of the four MPSoC applications considered. This is in contrast to the traditional semi-automated (or manual) communication architecture synthesis techniques which can take several days, and would take even longer with the added complexity of handling memory synthesis.

6.5.3 Other Co-synthesis Approach

Meyer and Thomas [161] attempt to extend COSMECA by adding layout-awareness during the co-synthesis phase. The memory/communication architecture co-synthesis is performed using an SA-based algorithm. However, the proposed co-synthesis approach oversimplifies the problem and is somewhat limited in its scope compared to COSMECA. Firstly, for the bus synthesis, only a simple shared bus is considered and is split into multiple buses by the SA algorithm to reduce excessive conflict if needed. Thus, only shared bus topology synthesis is performed—bus parameter synthesis is neglected. Secondly, the memory synthesis does not consider different memory types during data allocation like COSMECA does—only SRAM memories are supported in the proposed approach. The authors claim 20–27% cost reduction for a synthetic DSP (digital signal processor) software pipeline case study by using their approach compared to an approach that separately allocates memory and synthesizes buses. The runtime for the proposed approach (i.e., SA algorithm) is comparable to the time taken by COSMECA.

Sonics MemMAX [162] is a commercial memory scheduler solution that aims to reduce off-chip memory (DRAM) access latency and thus improve SoC performance by co-optimizing the on-chip bus and off-chip memory controller. The MemMAX memory scheduler module interfaces with the on-chip bus (Sonics SMX [162] or another open core protocol (OCP) [163] compliant bus) and a typical single threaded off-chip DRAM controller. The multi-threaded, pipelined MemMAX scheduler can be customized with information about the number, nature, and initiating sources of all memory accesses, quality of service (QoS) requirements of these sources, different scheduling algorithms, and bus interfaces. The MemMAX scheduler prefetches and reorders multiple data flows to off-chip memory while guaranteeing QoS requirements, and centralizes memory buffers, normally spread across component-bus interfaces, to reduce interconnect power and area.

6.6 DISCUSSION: PHYSICAL AND CIRCUIT LEVEL DESIGN OF ON-CHIP COMMUNICATION ARCHITECTURES

While this chapter till now has mainly focused on the design of on-chip communication architectures at the system level, the physical and circuit level issues associated with design of these communication architectures are worth mentioning. We present a brief overview of a few of the more important issues here. A more detailed treatment of these issues can be found in Chapter 11.

6.6.1 Performance Optimizations

One of the critical problems in interconnect design for synchronous designs is the minimization of signal propagation delay on the wires. Signal propagation is a function of the wire resistance (*R*) and capacitance (*C*), typically known as the *RC* delay. Since both the resistance and capacitance linearly increase with wire length, the signal propagation delay increases quadratically with the wire length. As the lengths of wires on which signals must propagate are increasing, and the wire widths decreasing with each successive CMOS technology generation, the wire resistance has been steadily increasing. Wire capacitance is also rapidly increasing due to coupling effects (crosstalk) in deep submicron (DSM) CMOS technologies. This has resulted in large signal delays on the interconnect wires that puts a limitation on the maximum performance achievable for a design [75–77]. Inserting repeaters (or buffers) on a long wire to partition it into shorter segments (each of which being driven by the repeater) reduces this quadratic dependency to a linear one, and is a commonly used technique to reduce signal propagation delay on interconnect wires. Wire shaping, which decreases wire resistance by changing the width (e.g., using *fat wires*) and thickness of the wire, is another way to reduce signal propagation delay. There is a large body of work dealing with the sizing and insertion of repeaters on interconnect wires [78–86, 159], as well as shaping the interconnect wires [87–92], to reduce signal propagation delay, or optimizing other criteria such as power consumption [80, 93–97]. Some approaches, such as [98–104] have proposed techniques for simultaneously solving the problems of repeater/buffer insertion (where to insert the buffers in a wire, and how many buffers to insert), buffer sizing, and wire sizing, to reduce signal propagation delay. Deodhar and Davis [105] proposed a technique for simultaneous optimal repeater insertion and voltage scaling, to reduce interconnect power consumption. The placement of these repeaters is becoming increasingly challenging, since today's designs require thousands of repeaters to meet timing and noise objectives [106]. Consequently, the area consumed by these repeaters is no longer insignificant, and must be considered during the floorplanning stage, especially since it may not be possible to place repeaters at any location inside a pre-designed functional block or at the top level of the hierarchy. The problem can be solved by making use of repeater block regions located around the chip during the floorplanning stage, which provide specified areas for the repeaters to be placed [107].

Even after repeater insertion, however, the signal propagation delay can be large enough to limit high frequency operation. This is because for high bus clock frequencies, clock cycle time is lower, and the signal must travel from source to destination in a shorter (clock cycle) duration of time. With shrinking CMOS process technology, wire delay has increased due to increase in parasitic capacitances, which further compounds the problem, and limits the bus to low frequencies of operation. A solution to this problem is to pipeline the interconnect wires, by inserting latches/flip-flops/register slices [45, 108–115] or FIFO buffers (FIFO queues, also known as elastic buffers) [116]. This allows the signal to propagate through a smaller distance (from one pipeline stage to another) in a single clock cycle, which eases timing constraints on the wire, and enables the bus to run at higher clock frequencies. Some approaches have proposed simultaneous flip-flop and repeater insertion for further interconnect optimization [117–120]. An approach to simultaneously optimize bus clock frequency, voltage, and FIFO sizes for low power operation was proposed by Chandra et al. [74]. Several techniques have been proposed for integrating interconnect pipelining with early floorplanning to better estimate and optimize interconnect performance [121–124]. Using retiming techniques [125–129] offers another solution, allowing already existing memory elements within logic blocks to be moved onto the wire, instead of inserting new latches to pipeline the wire. Such retiming techniques are, however, only applicable to soft IPs, and there is no guarantee that retiming allows wires to run at arbitrarily high frequencies.

Wave pipelining [130–133] is another approach that has been proposed to increase throughput on wires even further. This circuit level technique advocates the application of a new input signal even before the previous signal has reached its destination storage element. This results in multiple computation waveforms corresponding to successive signals co-existing on the same wire, which results in higher throughput. Sending multiple signals in the same clock period can result in the degradation of signal integrity, but there has been some work to determine the time after which the second signal can be sent in a clock cycle, without loss in signal integrity [134].

In DSM technologies, signal noise and integrity is increasingly becoming a problem, due to rising interference from inductive and coupling capacitance (crosstalk) effects [135]. Shielding is an effective technique that has been proposed to limit inductance effects and improve interconnect performance [136, 137]. The use of staggered repeaters [138] for buses was proposed to minimize the impact of coupling capacitance on delay and crosstalk noise. Khellah et al. [139] proposed skewed repeater insertion to reduce coupling capacitance and energy consumption on interconnects, by skewing consecutive repeaters in a bus line in the opposite direction. There have also been approaches that propose repeater insertion for simultaneous delay and crosstalk noise optimization [140, 141]. A technique to address simultaneous driver modeling, and repeater size and number optimization to reduce the effect of crosstalk on signal delay, was proposed by Sirichotiyakul et al. [142].

6.6.2 Power Optimizations

Numerous circuit and physical level techniques have been proposed to reduce interconnect power consumption in DSM technologies [143–156]. A survey of some circuit and physical level techniques to reduce interconnect power consumption was presented by Raghunathan et al. [145]. Several low voltage swing signaling techniques have been proposed [144, 150, 151] to reduce interconnect energy consumption. While lowering supply voltage results in a quadratic reduction in energy consumption and also reduces signal delay on wires [157], it leads to a reduction in the noise margin, and hence problems with signal integrity. Complex receiver circuitry can be used to improve reliability [146]. Alternatively, differential signaling [146] can be used, where the difference in voltage between two signal wires is used to convey data information. However, differential signaling entails additional interconnect and routing overhead. To reduce this overhead, pseudo-differential signaling techniques have been proposed, where the reference signal is shared among several bus lines. Guidelines on optimizing interconnect energy consumption through the use of differential signaling techniques are provided in the low voltage differential signaling (LVDS) standard [147]. Current mode signaling [148] is another promising approach to reduce energy consumption on long, global interconnects. A hybrid current/voltage mode signaling was proposed by Bashirullah et al. [158] which reduced interconnect power dissipation by up to 65% and 40% over current and voltage mode signaling techniques, respectively. There have also been some approaches that have proposed reordering of data allocation on wires and signal lines, in order to reduce coupling power in DSM [152–156, 160].

Increasingly, PVT variations are becoming dominant in Ultra DSM technologies, presenting an uncertainty that must be taken into account during interconnect design. A circuit level adaptive dynamic voltage scaling (DVS) scheme to reduce bus energy dissipation under PVT variations was proposed by Kaul et al. [165]. The scheme exploits the fact that on-chip buses are typically designed to meet performance constraints at worst-case conditions, including PVT variations, IR-drop, and neighboring net switching pattern. The probability of all worst-case conditions occurring simultaneously is usually small and hence, the bus is faster than it needs to be for more common case operating conditions. The proposed technique dynamically scales down the supply voltage for typical case conditions, resulting in energy reduction while still meeting delay constraints. The aggressive voltage scaling technique is based on dynamic detection and correction of delay errors. Error recovery is incorporated in the bus architecture by employing a modified flip-flop that samples its input at the normal clock as well as at a delayed clock, and is based on the *Razor* technique, proposed for use in logic pipeline designs by Ernst et al. [164]. If a difference between the two samples is detected, a control signal from the flop indicates that the data captured by the normal clock is incorrect and an error recovery mechanism ensures that the correct data value (that was sampled by the delayed clock) is propagated, while also ensuring that the incorrect data from the previous cycle is flushed out from the next stage. A major advantage with the proposed approach is that error recovery does not require retransmission on the bus, providing an opportunity for energy reduction on the bus with a smaller energy overhead for error recovery. The voltage is increased only when not doing so would result in an unacceptable number of delay error corrections, thereby improving the energy efficiency.

6.7 SUMMARY

Designing on-chip communication architectures is a non-trivial task, especially in the face of the multi-faceted design constraints of performance, power, cost, reliability, and area. Designers need techniques that can efficiently explore the increasingly intractable communication architecture design space, in order to satisfy and optimize constraints during communication architecture design. In this chapter, we presented research on techniques for efficient bus-based communication architecture synthesis. Some of these techniques focus on either bus topology design, or on the synthesis of bus protocol parameters such as arbitration schemes, bus widths, and clock frequencies for a fixed bus topology. However, the communication architecture design space is a combination of the topology and protocol parameter spaces, and there have been approaches that also comprehensively synthesize both topology and protocol parameter values, for different design constraints. Novel approaches have been proposed to couple bus-based communication architecture synthesis with some knowledge of the physical floorplan and wire layouts, as well as to couple the design of the memory and communication architectures, but there is scope for a lot more work in these open areas. Low level physical and circuit level approaches (discussed briefly in the discussion section above) can also significantly impact communication architecture design, to improve performance, reduce power, decrease area footprint, improve interconnect reliability, and reduce overall cost. However, a lot of open problems still remain to be solved, especially in the areas of wire metal layer assignment, wire sizing optimization, inductance estimation, timing-driven floorplanning, and shield wire insertion algorithms.

The next chapter will present the large body of work focusing on various encoding techniques to reduce the number of transitions on the bus wires, in order to minimize interconnect power consumption.

REFERENCES

[1] AMBA Designer, http://www.arm.com/products/solutions/AMBA_Designer.html.

[2] AMBA Design Kit (ADK), http://www.arm.com/products/solutions/AMBAADK.html.

[3] STBus GenKit, http://www.st.com/stonline/products/technologies/soc/stbus.htm.

[4] Sonics Studio, http://www.sonicsinc.com/.

[5] Y. Ti-Yen and W. Wolf, "Communication synthesis for distributed embedded systems," *IEEE/ACM International Conference on Computer-Aided Design (ICCAD)*, 1995, pp. 288–294.

[6] R. B. Ortega and G. Borriello, "Communication synthesis for distributed embedded systems," *IEEE/ACM International Conference on Computer-Aided Design (ICCAD)*, 1998, pp. 437–444.

[7] N. D. Liveris and P. Banerjee, "Power aware interface synthesis for bus-based SoC designs," *Design, Automation and Test in Europe Conference and Exhibition*, 2004, pp. 864–869.

[8] ARM AMBA Specification and Multilayer AHB Specification (rev2.0), http://www.arm.com, 2001.

[9] Y. Zhang, W. Ye and M. J. Irwin, "An alternative architecture for on-chip global interconnect: Segmented bus power modeling," *Conference Record of the Thirty-Second Asilomar Conference on Signals, Systems and Computers*, 1998, pp. 1062–1065.

[10] H. Murata, K. Fujiyoshi, S. Nakatake and Y. Kajitani, "Rectangle-packing-based module placement," *IEEE/ACM International Conference on Computer-Aided Design (ICCAD)*, 1995, pp. 472–479.

[11] "STBus Communication System: Concepts and Definitions," *Reference Guide*, STMicroelectronics, May 2003.

[12] S. Murali and G. De Micheli, "An application-specific design methodology for STbus crossbar generation," in *Proceedings of Design, Automation and Test in Europe (DATE)*, 2005, pp. 1176–1181.

[13] SystemC initiative, www.systemc.org.

[14] T. Grötker, S. Liao, G. Martin and S. Swan, *System Design with SystemC*, Kluwer Academic Publishers, 2002.

[15] M. Loghi, F. Angiolini, D. Bertozzi, L. Benini and R. Zafalon, "Analyzing On-Chip Communication in a MPSoC Environment," in *Proceedings of Design, Automation and Test in Europe (DATE)*, 2004, pp. 20752–20757.

[16] ILOG CPLEX, http://www.ilog.com/products/cplex/.

[17] N. Thepayasuwan, V. Damle and A. Doboli, "Bus architecture synthesis for hardware–software co-design of deep submicron systems on chip," *International Conference on Computer Design (ICCD)*, 2003, pp. 126–133.

[18] N. Thepayasuwan and A. Doboli, "Layout conscious bus architecture synthesis for deep submicron systems on chip," *Design, Automation and Test in Europe Conference and Exhibition (DATE)*, 2004, pp. 108–113.

[19] N. Thepayasuwan and A. Doboli, "Hardware–software co-design of resource constrained systems on a chip," *International Conference on Distributed Computing Systems Workshops*, 2004, pp. 818–823.

[20] N. Thepayasuwan and A. Doboli, "OSIRIS: Automated synthesis of flat and hierarchical bus architectures for deep submicron systems on chip," *IEEE Computer Society Annual Symposium on VLSI*, 2004, pp. 264–265.

[21] N. Thepayasuwan and A. Doboli, "Layout conscious approach and bus architecture synthesis for hardware/software codesign of systems on chip optimized for speed," *IEEE Transactions on Very Large Scale Integration (VLSI) Systems*, Vol. 13, No. 5, 2005, pp. 525–538.

[22] M. Drinic, D. Kirovski, S. Meguerdichian and M. Potkonjak, "Latency-guided on-chip bus network design," *IEEE/ACM International Conference on Computer Aided Design (ICCAD)*, 2000, pp. 420–423.

[23] S. Meguerdichian, M. Drinic and D. Kirovski, "Latency-driven design of multi-purpose systems-on-chip," *Design Automation Conference (DAC)*, 2001, pp. 27–30.

[24] J. Guo, A. Papanikolaou, P. Marchal and F. Catthoor, "Physical design implementation of segmented buses to reduce communication energy," *Asia and South Pacific Conference on Design Automation (ASPDAC)*, 2006, pp. 42–47.

[25] J. Guo, A. Papanikolaou, P. Marchal and F. Catthoor, "Energy/area/delay trade-offs in the physical design of on-chip segmented bus architecture," *International Workshop on System-Level Interconnect Prediction (SLIP)*, 2006, pp. 75–81.

[26] J. Yoo, D. Lee, S. Yoo and K. Choi, "Communication architecture synthesis of cascaded bus matrix," *Asia and South Pacific Design Automation Conference (ASPDAC)*, 2007, pp. 171–177.

[27] S. Narayan and D. D. Gajski, "Synthesis of system-level bus interfaces," *European Design and Test Conference*, 1994, pp. 395–399.

[28] M. Gasteier and M. Glesner, "Bus-based communication synthesis on system-level" *International Symposium on System Synthesis (ISSS)*, 1996, pp. 65–70.

[29] M. Gasteier and M. Glesner, "Bus-based communication synthesis on system level," *ACM Transactions on Design Automation of Electronic Systems*, Vol. 4, No. 1, 1999, pp. 1–11.

[30] K. Lahiri, A. Raghunathan and S. Dey, "Efficient exploration of the SoC communication architecture design space," *IEEE/ACM International Conference on Computer Aided Design (ICCAD)*, 2000, pp. 424–430.

[31] K. Lahiri, A. Raghunathan, G. Lakshminarayana and S. Dey, "Design of high-performance system on-chips using communication architecture tuners," *IEEE Transactions on Computer-Aided Design of Integrated Circuits and Systems*, Vol. 23, No. 5, 2004, pp. 620–636.

[32] K. Lahiri, A. Raghunathan and S. Dey, "Performance analysis of systems with multi-channel communication architectures," *International Conference on VLSI Design (VLSID)*, January 2000, pp. 530–537.

[33] K. Lahiri, A. Raghunathan and S. Dey, "Fast performance analysis of bus-based system-on-chip communication architectures," *International Conference on Computer-Aided Design (ICCAD)*, November 1999, pp. 566–572.

[34] C. Shin; Y.-T. Kim; E.-Y. Chung; K.-M. Choi; J.-T. Kong; S.-k. Eo, "Fast exploration of parameterized bus architecture for communication-centric SoC design," *Design, Automation and Test in Europe Conference and Exhibition (DATE)*, 2004, pp. 352–357.

[35] J. Koza, "Genetic programming: On the programming of computers by means of natural selection," Massachusetts Institute of Technology, Namco Ltd., 1998.

[36] J. T. Alander, "On optimal population size of genetic algorithms," CompEuro92, IEEE Computer Society Press, 1992, pp. 65–70.

[37] D. Wong, H. Leong and C. Liu, *Simulated Annealing for VLSI Design*, Kluwer Academic Publishers, Boston, 1988.

[38] Y. Saab and V. Rao, "Stochastic evolution: A fast effective heuristic for some generic layout problems," *ACM/IEEE Design Automation Conference (DAC)*, 1990, pp. 26–31.

[39] D. Lyonnard, S. Yoo, A. Baghdadi and A. A. Jerraya, "Automatic generation of application-specific architectures for heterogeneous multiprocessor system-on-chip," in *Proceedings of Design Automation Conference (DAC)*, 2001, pp. 518–523.

[40] K. K. Ryu, E. Shin and V. J. Mooney, "A comparison of five different multiprocessor SoC bus architectures," *EUROMICRO Symposium on Digital Systems Design*, September 2001, pp. 202–209.

[41] K. K. Ryu and V. J. Mooney, "Automated bus generation for multiprocessor SoC design," *Design, Automation and Test in Europe Conference and Exhibition (DATE)*, 2003, pp. 282–287.

[42] K. K. Ryu and V. J. Mooney, "Automated bus generation for multiprocessor SoC design," *IEEE Transactions on Computer-Aided Design of Integrated Circuits and Systems (TCAD)*, Vol. 23, No. 11, 2004, pp. 1531–1549.

[43] S. Pandey, M. Glesner and M. Muhlhauser, "On-chip communication topology synthesis for shared multi-bus based architecture," *International Conference on Field Programmable Logic and Applications*, 2005, pp. 374–379.

[44] S. Pasricha, N. Dutt and M. Ben-Romdhane, "Automated throughput-driven synthesis of bus-based communication architectures," *Asia and South Design Automation Conference (ASPDAC)*, 2005, pp. 495–498.

[45] ARM AMBA 3.0 AXI Specification, www.arm.com/armtech/AXI.

[46] S. Pasricha, N. Dutt and M. Ben-Romdhane, "Constraint-driven bus matrix synthesis for MPSoC," *Asia and South Pacific Design Automation Conference (ASPDAC)*, 2006, pp. 30–35.

[47] S. Pasricha, N. Dutt, E. Bozorgzadeh and M. Ben-Romdhane, "Floorplan-aware automated synthesis of bus-based communication architectures," *Design Automation Conference (DAC)*, 2005, pp. 565–570.

[48] S. Pasricha, "Transaction level modeling of SoC with SystemC 2.0," in *Proceedings of Synopsys User Group Conference (SNUG)*, 2002, pp. 55–59.

[49] S. Pasricha, N. Dutt and M. Ben-Romdhane, "Extending the transaction level modeling approach for fast communication architecture exploration," *Design Automation Conference (DAC)*, 2004, pp. 113–118.

[50] S. Pasricha, Y. Park, F. J. Kurdahi, and N. Dutt, "System-level power-performance trade-offs in bus matrix communication architecture synthesis," in *Proceedings of International Conference on Hardware/Software Codesign and System Synthesis (CODES+ISSS)*, 2006, pp. 300–305.

[51] R. P. Dick and N. K. Jha, "MOCSYN: multiobjective core-based single-chip system synthesis," *Design, Automation and Test in Europe Conference and Exhibition (DATE)*, 1999, pp. 263–270.

[52] J. Hu, Y. Deng and R. Marculescu, "System-level point-to-point communication synthesis using floorplanning information," *Asia and South Pacific Design Automation Conference (ASPDAC) and International Conference on VLSI Design*, 2002, pp. 573–579.

[53] S. Pasricha, N. D. Dutt, E. Bozorgzadeh and M. Ben-Romdhane, "FABSYN: Floorplan-aware bus architecture synthesis," *IEEE Transactions on Very Large Scale Integration (VLSI) Systems*, Vol. 14, No. 3, 2006, pp. 241–253.

[54] S. N. Adya and I. L. Markov, "Fixed-outline floorplanning: Enabling hierarchical design," *IEEE Transactions on Very Large Scale Integration (VLSI) Systems*, Vol. 11, No. 6, 2003, pp. 1120–1135.

[55] J. Cong and Z. (David) Pan, "Interconnect performance estimation models for design planning," *IEEE Transactions on Computer-Aided Design of Integrated Circuits and Systems*, Vol. 20, 2001, pp. 739–752.

[56] S. Meftali, F. Gharsalli, F. Rousseau and A. A. Jerraya, "An optimal memory allocation for application-specific multiprocessor system-on-chip," in *Proceedings of ISSS*, 2001, pp. 19–24.

[57] A. Allan et al., "2001 technology roadmap for semiconductors," *IEEE Computer*, Vol. 35, No. 1, 2002, pp. 42–53.

[58] S. Narayan and D. Gajski, "Protocol generation for communication channels," in *Proceedings of Design Automation Conference (DAC)*, 1994, pp. 547–551.

[59] I. Madsen and B. Hald, "An approach to interface synthesis," in *Proceedings of International Symposium on System Synthesis (ISSS)*, 1995, pp. 16–21.

[60] S. Wuytack, F. Catthoor, G. De Jong and H. J. De Man, "Minimizing the required memory bandwidth in VLSI system realizations," *IEEE TVLSI*, Vol. 7, No. 4, 1999, pp. 433–441.

[61] L. Cai, H. Yu and D. Gajski, "A novel memory size model for variable-mapping in system level design," *Proceedings of Asia and South Pacific Design Automation Conference (ASPDAC)*, 2004, pp. 813–818.

[62] T. van Meeuwen, A. Vandecappelle, A. van Zelst, F. Catthoor and D. Verkest, "System-level interconnect architecture exploration for custom memory organizations," *Proceedings of International Symposium on System Synthesis (ISSS)*, 2001, pp. 13–18.

[63] J. A. Rowson and A. Sangiovanni-Vincentelli, "Interface based design," *Proceedings of Design Automation Conference (DAC)*, 1997, pp. 178–183.

[64] K. Lahiri, A. Raghunathan and S. Dey, "System-level performance analysis for designing system-on-chip communication architecture," *IEEE TCAD*, Vol. 2, 2001, pp. 768–783.

[65] S. Pasricha and N. Dutt, "COSMECA: Application specific co-synthesis of memory and communication architectures for MPSoC," in *Proceedings of Design, Automation and Test in Europe (DATE)*, 2006, pp. 1–6.

[66] S. Pasricha and N. D. Dutt, "A framework for cosynthesis of memory and communication architectures for MPSoC," *IEEE Transactions on Computer-Aided Design of Integrated Circuits and Systems*, Vol. 26, No. 3, 2007, pp. 408–420.

[67] P. Knudsen and J. Madsen, "Integrating communication protocol selection with partitioning in hardware/software codesign," in *Proceedings of International Symposium on System Synthesis (ISSS)*, 1998, pp. 111–116.

[68] M. Shalan, E. Shin and V. Mooney, "DX-Gt: Memory management and crossbar switch generator for multiprocessor system-on-a-chip," in *Proceedings of SASIMI*, 2003, pp. 357–364.

[69] P. Grun, N. Dutt and A. Nicolau, "Memory system connectivity exploration," in *Proceedings of Design, Automation and Test in Europe (DATE)*, 2002, pp. 894–901.

[70] S. Kim, C. Im and S. Ha, "Efficient exploration of on-chip bus architectures and memory allocation," in *Proceedings of International Conference on Hardware/Software Codesign and System Synthesis (CODES+ISSS)*, 2004, pp. 248–253.

[71] S. Kim and S. Ha, "Efficient exploration of bus-based system-on-chip architectures," *IEEE Transactions on Very Large Scale Integration (VLSI) Systems*, Vol. 14, No. 7, 2006, pp. 681–692.

[72] S. Srinivasan, F. Angiolini, M. Ruggiero, L. Benini and N. Vijaykrishnan, "Simultaneous memory and bus partitioning for SoC architectures," in *Proceedings of SOCC*, 2005, pp. 125–128.

[73] S. Srinivasan, L. Li and N. Vijaykrishnan, "Simultaneous partitioning and frequency assignment for on-chip bus architectures," in *Proceedings of Design, Automation and Test in Europe (DATE)*, 2005, pp. 218–223.

[74] V. Chandra, H. Schmit, A. Xu and L. Pileggi, "A power aware system level interconnect design methodology for latency-insensitive systems," *IEEE/ACM International Conference on Computer Aided Design (ICCAD)*, 2004, pp. 275–282.

[75] M. Bohr, "Interconnect scaling—The real limiter to high-performance ULSI," *Technical Digest of the International Electron Devices Meeting*, December 1995, pp. 241–244.

[76] R. Ho, K. W. Mai and M. A. Horowitz, "The future of wires," in *Proceedings of IEEE*, Vol. 89, 2001, pp. 490–504.

[77] J. Cong, "An interconnect-centric design flow for nanometer technologies," in *Proceedings of IEEE*, 2001, pp. 505–527.

[78] H. B. Bakoglu, *Circuits, Interconnections, and Packaging for VLSI*, Addison-Wesley, 1990.

[79] C. Y. Wu and M. Shiau, "Accurate speed improvement techniques for RC line and tree interconnections in CMOS VLSI," *IEEE International Symposium on Circuits and Systems (ISCAS)*, 1990, pp. 2.1648–2.1651.

[80] V. Adler and E. B. Friedman, "Repeater design to reduce delay and power in resistive interconnect," *IEEE Transactions on Circuits and Systems-11, Analog and Digital Signal Processing*, Vol. 45, No. 5, 1998, pp. 607–616.

[81] Y. I. Ismail and E. G. Friedman, "Effects of inductance on the propagation delay and repeater insertion in VLSI circuits," *IEEE Transactions on VLSI Systems (TVLSI)*, Vol. 8, 2000, pp. 195–206.

[82] J. Rabaey, *Digital Integrated Circuits: A Design Perspective*, Prentice-Hall, 1996.

[83] A. Nalamalpu and W. Burleson, "Repeater insertion in deep sub-micron CMOS: Ramp-based analytical model and placement sensitivity analysis," in *Proceedings of IEEE International Symposium on Circuits and Systems (ISCAS)*, 2000, pp. 766–769.

[84] S. Srinivasaraghavan and W. Burleson, "Interconnect effort—A unification of repeater insertion and logical effort," *IEEE Computer Society Annual Symposium on VLSI*, 2003, pp. 55–61.

[85] L. P. P. P. van Ginneken, "Buffer placement in distributed RC-tree networks for minimal Elmore delay," *IEEE International Symposium on Circuits and Systems*, 1990, pp. 865–868.

[86] G. S. Garcea, N. P. van der Meijs and R. H. J. M. Otten, "Simultaneous analytic area and power optimization for repeater insertion," in *Proceedings of International Conference on Computer Aided Design (ICCAD)*, 2003, pp. 568–573.

[87] C.-P. Chen, Y.-P. Chen and D. F. Wong, "Optimal wire-sizing formula under the Elmore delay model," *Proceedings of ACM/IEEE Design Automation Conference (DAC)*, 1996, pp. 487–490.

[88] C.-P. Chen and D. F. Wong, "A fast algorithm for optimal wire-sizing under Elmore delay model," in *Proceedings of IEEE International Symposium on Circuits and Systems (ISCAS)*, Vol. 4, 1996, pp. 412–415.

[89] J. Cong and L. He, "Optimal wire sizing for inter-connects with multiple sources," *ACM Transactions on Design Automation of Electronic Systems*, Vol. 1, No. 4, 1996, pp. 478–511.

[90] J. Cong and D. Z. Pan, "Interconnect estimation and planning for deep submicron designs," in *Proceedings of ACM/IEEE Design Automation Conference (DAC)*, 1999, pp. 507–510.

[91] C. Svensson, "Optimum voltage swing on on-chip and off-chip interconnects," *IEEE Journal of Solid-State Circuits*, Vol. 26, No. 7, 2001, pp. 1108–1112.

[92] D. Sylvester and K. Keutzer, "Impact of small process geometries on microarchitectures in systems on a chip," in *Proceedings of the IEEE*, Vol. 89, No. 4, 2001, pp. 467–489.

[93] K. Banerjee and A. Mehrotra, "A power-optimal repeater insertion methodology for global interconnects in nanometer designs," *IEEE Transactions on Electron Devices*, Vol. 49, No. 11, 2002, pp. 2001–2007.

[94] D. C. Sekar, E. Demaray, H. Zhang, P. A. Kohl and J. D. Meindl, "A new global interconnect paradigm: MIM power-ground plane capacitors," *International Interconnect Technology Conference*, 2006, pp. 48–50.

[95] P. Kapur, G. Chandra and K. C. Saraswat, "Power estimation in global interconnects and its reduction using a novel repeater optimization methodology," in *Proceedings of Design Automation Conference (DAC)*, 2002, pp. 461–466.

[96] V. Wason and K. Banerjee, "A probabilistic framework for power-optimal repeater insertion in global interconnects under parameter variations," in *Proceedings of International Symposium on Low Power Electronics and Design (ISLPED)*, 2005, pp. 131–136.

[97] A. Youssef, M. Anis and M. Elmasry, "POMR: A power-aware interconnect optimization methodology," *IEEE Transactions on Very Large Scale Integration (VLSI) Systems*, Vol. 13, No. 3, 2005, pp. 297–307.

[98] C. C. N. Chu and D. F. Wong, "Closed form solution to simultaneous buffer insertion/sizing and wire sizing," in *Proceedings of International Symposium on Physical Design*, 1997, pp. 192–197.

[99] J. Cong and C.-K. Koh, "Simultaneous buffer and wire sizing for performance and power optimization," in *Proceedings of International Symposium on Low Power Electronics and Design*, 1996, pp. 271–276.

[100] J. Lillis, C.-K. Cheng and T.-T.Y. Lin, "Optimal wire sizing and buffer insertion for low power and a generalized delay model," *IEEE Journal of Solid State Circuits*, Vol. 31, No. 3, 1996, pp. 437–447.

[101] H. Shah, P. Shiu, B. Bell, M. Aldredge, N. Sopory and J. Davis, "Repeater insertion and wire sizing optimization for throughput-centric VLSI global interconnects," *IEEE/ACM International Conference on Computer Aided Design (ICCAD)*, 2002, pp. 280–284.

[102] R. Li, D. Zhou, J. Liu and X. Zeng, "Power-optimal simultaneous buffer insertion/sizing and wire sizing," in *Proceedings of International Conference on Computer Aided Design (ICCAD)*, 2003, pp. 581–586.

[103] R. Li, D. Zhou, J. Liu and X. Zeng, "Power-optimal simultaneous buffer insertion/sizing and uniform wire sizing for single long wires," in *Proceedings of IEEE International Symposium on Circuits and Systems (ISCAS)*, 2005, pp. 113–116.

[104] M. Tang and J.-F. Mao, "Optimization of global interconnects in high performance VLSI circuits," in *Proceedings of International Conference on VLSI Design*, 2006.

[105] V.V. Deodhar and J.A. Davis, "Optimization of throughput performance for low-power VLSI interconnects," *IEEE Transactions on Very Large Scale Integration (VLSI) Systems*, Vol. 13, No. 3, 2005, pp. 308–318.

[106] Y. Cao, C. Hu, X. Huang, A. B. Kahng, S. Muddu, D. Stroobandt and D. Sylvester, "Effects of global interconnect optimizations on performance estimation of deep submicron design," in *Proceedings of IEEE/ACM International Conference on Computer Aided Design (ICCAD)*, 2000, pp. 56–61.

[107] J. Cong, T. Kong and D. Z. Pan, "Buffer block planning for interconnect-driven floorplanning," in *Proceedings of IEEE/ACM International Conference on Computer Aided Design (ICCAD)*, 1999, pp. 358–363.

[108] L. Scheffer, "Methodologies and tools for pipelined on-chip interconnect," in *Proceedings of IEEE International Conference on Computer Design (ICCAD)*, 2002, pp. 152–157.

[109] J. Plosila, P. Liljeberg and J. Isoaho, "Pipelined on-chip bus architecture with distributed self-timed control," in *Proceedings of International Symposium on Signals, Circuits and Systems*, 2003, pp. 257–260.

[110] S. Hassoun, C. J. Alpert and M. Thiagarajan, "Optimal buffered routing path constructions for single and multiple clock domain systems," in *Proceedings of IEEE/ACM international Conference on Computer-Aided Design (ICCAD)*, 2002, pp. 247–253.

[111] M. R. Casu and L. Macchiarulo, "On-chip transparent wire pipelining," in *Proceedings of IEEE International Conference on Computer Design (ICCD)*, 2004, pp. 160–167.

[112] M. R. Casu and L. Macchiarulo, "Throughput-driven floorplanning with wire pipelining," *IEEE Transactions on Computer-Aided Design of Integrated Circuits and Systems*, Vol. 24, No. 5, 2005, pp. 663–675.

[113] D. K. Tong and E. F. Y. Young, "Performance-driven register insertion in placement," in *Proceedings of International Symposium on Physical Design (ISPD)*, 2004, pp. 53–60.

[114] V. Nookala and S. S. Sapatnekar, "Designing optimized pipelined global interconnects: algorithms and methodology impact," in *Proceedings of IEEE International Symposium on Circuits and Systems*, 2005, pp. 608–611.

[115] L. P. Carloni, K. L. McMillan and A. L. Sangiovanni-Vincentelli, "Theory of latency-insensitive design," *IEEE Transactions on Computer-Aided Design of Integrated Circuits and Systems*, Vol. 20, No. 9, 2001, pp. 1059–1076.

[116] V. Chandra, A. Xu, H. Schmit and L. Pileggi, "An interconnect channel design methodology for high performance integrated circuits," in *Proceedings of Design, Automation and Test in Europe Conference and Exhibition (DATE)*, 2004, pp. 1138–1143.

[117] P. Cocchini, "Concurrent flip-flop and repeater insertion for high-performance integrated circuits," in *Proceedings of International Conference on Computer-Aided Design (ICCAD)*, 2002, pp. 268–273.

[118] R. Lu, Z. G. C. Koh and J. Chao, "Flip-flop and repeater insertion for early interconnect planning," in *Proceedings of European Design and Test Conference*, March 2002, pp. 690–695.

[119] P. Cocchini, "A methodology for optimal repeater insertion in pipelined interconnects," *IEEE Transactions on Computer-Aided Design of Integrated Circuits and Systems*, Vol. 22, No. 12, 2003, pp. 1613–1624.

[120] W. Liao and L. He, "Full-chip interconnect power estimation and simulation considering concurrent repeater and flip-flop insertion," in *Proceedings of International Conference on Computer Aided Design (ICCAD)*, 2003, pp. 574–580.

[121] M. R. Casu and L. Macchiarulo, "Floorplanning for throughput," in *Proceedings of International Symposium on Physical Design*, 2004, pp. 62–69.

[122] M. Ekpanyapong, J. R. Minz, T. Watewai, H.-H. S. Lee and S. K. Lim, "Profile-guided microarchitecture floorplanning for deep submicron processor design," in *Proceedings of Design Automation Conference (DAC)*, June 2004, pp. 634–639.

[123] C. Long, L. J. Simonson, W. Liao and L. He, "Floorplanning optimization with trajectory piecewise-linear model for pipelined interconnects," in *Proceedings of Design Automation Conference (DAC)*, June 2004, pp. 640–645.

[124] A. Jagannathan, H. Honghua Yang, K. Konigsfeld, D. Milliron, M. Mohan, M. Romesis, G. Reinman and J. Cong, "Microarchitecture evaluation with floorplanning and interconnect pipelining," in *Proceedings of Asia and South Pacific Design Automation Conference (ASPDAC)*, 2005, pp. I8–I15.

[125] C. Leiserson, F. Rose and J. Saxe, "Optimizing synchronous circuitry by retiming," 3rd Caltech Conference on VLSI, 1983, pp. 87–116.

[126] A. Tabbara, R. K. Brayton and A. R. Newton, "Retiming for DSM with area–delay trade-offs and delay constraints," in *Proceedings of Design Automation Conference (DAC)*, 1999, pp. 725–730.

[127] J. Cong and X. Yuan, "Multilevel global placement with retiming," in *Proceedings of International Symposium on Physical Design*, 2003, pp. 208–213.

[128] C. Lin and H. Zhou, "Retiming for wire pipelining in system-on-chip," in *Proceedings of International Conference on Computer Aided Design (ICCAD)*, 2003, pp. 215–220.

[129] L. P. Carloni and A. L. Sangiovanni-Vincentelli, "Combining retiming and recycling to optimize the performance of synchronous circuits," in *Proceedings of Symposium on Integrated Circuits and Systems Design*, 2003, pp. 47–52.

[130] A. J. Joshi and J. A. Davis, "Wave-pipelined multiplexed (WPM) routing for gigascale integration (GSI)," *IEEE Transactions on Very Large Scale Integration (VLSI) Systems*, Vol. 13, No. 8, 2005, pp. 899–910.

[131] L. Zhang, Y. Hu and C.-P. Chen, "Wave-pipelined on-chip global interconnect," in *Proceedings of Asia and South Pacific Design Automation Conference (ASPDAC)*, 2005, pp. 127–132.

[132] W. P. Burleson, M. Ciesielski, F. Klass and W. Liu, "Wave-pipelining: A tutorial and research survey," *IEEE Transactions on Very Large Scale Integration (VLSI) Systems*, Vol. 6, No. 3, 1998, pp. 464–474.

[133] O. Hauck and S. A. Huss, "Asynchronous wave pipelines for high throughput datapaths," in *Proceedings of IEEE International Conference on Electronics, Circuits and Systems*, 1998, pp. 283–286.

[134] V. Deodhar and J. Davis, "Optimization for throughput performance for low power VLSI interconnects," *IEEE Transactions on Very Large Scale Integration (VLSI) Systems*, Vol. 13, No. 3, 2005, pp. 308–318.

[135] D. Sylvester and K. Keutzer, "A global wiring paradigm for deep submicron design," in *Proceedings of IEEE Transactions on Computer-Aided Design of Integrated Circuits and Systems*, Vol. 19, No. 2, 2000, pp. 242–252.

[136] Y. Massoud, S. Majors, T. Bustami and J. White, "Layout techniques for minimizing on-chip interconnect self-inductance," in *Proceedings of Design Automation Conference (DAC)*, 1998, pp. 566–571.

[137] S. P. Khatri, A. Mehrotra, R. K. Brayton, A. Sangiovanni-Vincentelli and R. H. J. M. Otten, "A novel VLSI layout fabric for deep submicron applications," in *Proceedings of Design Automation Conference (DAC)*, 1999, pp. 491–496.

[138] A. B. Kahng, S. Muddu and E. Sarto, "Tuning strategies for global interconnects in high-performance deep submicron IC's," in *Proceedings of International Conference on VLSI Design*, Vol. 10, No. 1, 1999, pp. 21–34.

[139] M. Khellah, M. Ghoneima, J. Tschanz, Y. Ye, N. Kurd, J. Barkatullah, S. Nimmagadda, Y. Ismail and V. De, "A skewed repeater bus architecture for on-chip energy reduction in microprocessors," in *Proceedings of IEEE International Conference on Computer Design (ICCD)*, 2005, pp. 253–257.

[140] C. J. Alpert, A. Devgan and S. T. Quay, "Buffer insertion for noise and delay optimization," in *Proceedings of IEEE Transactions on Computer-Aided Design*, Vol. 18, 1999, pp. 1633–1645.

[141] N. Menezes and C. P. Chen, "Spec-based repeater insertion and wire sizing for on-chip interconnect," *International Conference on Very Large Scale Integration (VLSI) Design*, 1999, pp. 476–483.

[142] S. Sirichotiyakul, D. Blaauw, C. Oh, R. Levy, V. Zolotov and J. Zuo, "Driver modeling and alignment for worst-case delay noise," in *Proceedings of Design Automation Conference (DAC)*, 2001, pp. 720–725.

[143] K.-W. Kim, S.-O. Jung, U. Narayanan, C. L. Liu and S. M. Kang, "Noise-aware power optimization for on-chip interconnect," in *Proceedings of International Symposium on Low Power Electronics and Design*, 2000, pp. 108–113.

[144] H. Zhang, V. George and J. Rabaey, "Low-swing on-chip signaling techniques: Effectiveness and robustness," *IEEE Transactions on VLSI Systems*, Vol. 8, No. 3, 2000, pp. 262–272.

[145] V. Raghunathan, M. B. Srivastava and R. K. Gupta, "A survey of techniques for energy efficient on-chip communication," in *Proceedings of Design Automation Conference (DAC)*, 2003, pp. 900–905.

[146] W. J. Dally and J. W. Poulton, *Digital Systems Engineering*, Cambridge University Press, NY, 1998.

[147] ANSI/TIA/EIA-644-1995 Standard, "Electrical characteristics of low voltage differential signaling (LVDS) interface circuits."

[148] A. Maheshwari and W. Burleson, "Current sensing techniques for global interconnects in very deep submicron (VDSM) CMOS," *IEEE Workshop on VLSI*, 2001, pp. 66–70.

[149] Y.-F. Tsai, V. Narayaynan, Y. Xie and M. J. Irwin, "Leakage-aware interconnect for on-chip network," in *Proceedings of Design, Automation and Test in Europe (DATE)*, 2005, pp. 230–231.

[150] A. Andrei, M. Schmitz, P. Eles, Z. Peng, B. M. Al Hashimi, "Simultaneous communication and processor voltage scaling for dynamic and leakage energy reduction in time-constrained systems," *IEEE/ACM International Conference on Computer Aided Design (ICCAD)*, 2004, pp. 362–369.

[151] H. Kaul, D. Sylvester, D. Blaauw, T. Mudge and T. Austin, "DVS for on-chip bus designs based on timing error correction," in *Proceedings of Design, Automation and Test in Europe (DATE)*, 2005, pp. 80–85.

[152] Y. Shin and T. Sakurai, "Coupling-driven bus design for low-power application-specific systems," in *Proceedings of Design Automation Conference (DAC)*, 2001, pp. 750–753.

[153] C.-G. Lyuh, T. Kim and K.-W. Kim, "Coupling-aware high-level interconnect synthesis for low power," *IEEE/ACM International Conference on Computer Aided Design (ICCAD)*, 2002, pp. 609–613.

[154] L. Macchiarulo, E. Macii and M. Poncino, "Wire placement for crosstalk energy minimization in address buses," in *Proceedings of the Design, Automation and Test in Europe Conference and Exhibition (DATE)*, 2002, pp. 158–162.

[155] E. Macci, M. Poncino and S. Salerno, "Combining wire swapping and spacing for low-power deep-submicron buses," in *Proceedings of Great Lakes Symposium on VLSI*, 2003, pp. 198–202.

[156] E. Macii, M. Poncino and S. Salerno, "Combining wire swapping and spacing for low-power deep-submicron buses," *ACM Great Lakes Symposium on VLSI (GLSVLSI)*, 2003, pp. 198–202.

[157] W. J. Dally and B. Towles, "Route packets, not wires: On chip interconnection networks," in *Proceedings of Design Automation Conference (DAC)*, 2001, pp. 684–689.

[158] R. Bashirullah, W. Liu and R. K. Cavin, "Low-power design methodology for an on-chip with adaptive bandwidth capability," in *Proceedings of Design Automation Conference*, 2003, pp. 628–633.

[159] D. Pamunuwa and H. Tenhunen, "Repeater insertion to minimise delay in coupled interconnects," in *Proceedings of International Conference on VLSI Design*, 2001, pp. 513–517.

[160] D. Li, G. Ma and G. Feng, "Optimized design of interconnected bus on chip for low power," *International Multi-Symposiums on Computer and Computational Sciences*, 2006, pp. 298–302.

[161] B. Meyer and D. Thomas, "Simultaneous synthesis of buses, data mapping and memory allocation for MPSoC," in *Proceedings of International Conference on Hardware/Software Codesign and System Synthesis (CODES+ISSS)*, 2007, pp. 3–8.

[162] Sonics MemMAX 2.0, SMX Interconnect, http://www.sonicsinc.com.

[163] Open Core Protocol International Partnership (OCP-IP). OCP Datasheet, Release Version 2.0/2.1, http://www.ocpip.org.

[164] D. Ernst, N. S. Kim, S. Das, S. Pant, R. Rao, T. Pham, C. Ziesler, D. Blaauw, T. Austin, K. Flautner and T. Mudge, "Razor: A low-power pipeline based on circuit-level timing speculation," *36th Annual IEEE/ACM International Symposium on Microarchitecture (MICRO-36)*, 2003.

[165] H. Kaul, D. Sylvester, D. Blaauw, T. Austin and T. Mudge, "DVS for on-chip bus designs based on timing error correction," *ACM/IEEE Design, Automation, and Test Europe (DATE)*, March, 2005.

[166] S. Pasricha, Y. Park, F. Kurdahi and N. Dutt, "Incorporating PVT Variations in System-Level Power Exploration of On-Chip Communication Architectures", in *Proceedings of International Conference on VLSI Design*, January 2008, pp. 363–370.

CHAPTER

7 Encoding Techniques for On-Chip Communication Architectures

The process of data word transmission on an on-chip bus leads to switching activity of data bits on the bus wires, which charges and discharges the capacitance associated with the wires and consequently leads to dynamic power dissipation, as discussed in Chapter 5. The effective capacitance associated with a wire is made up of not only the ground capacitance of the wire itself (self-capacitance), but also the coupling capacitance and coupling inductance due to switching activity on the adjacent wires. The latter phenomenon is referred to as crosstalk, and in addition to power consumption, it also leads to an increase in signal noise and signal propagation delay, which reduces data transmission reliability and performance. *Bus encoding* is a widely used technique to reduce dynamic switching power and the effects of crosstalk (signal noise, delay) during data transmission on buses. Low power encoding techniques aim to transform the data being transmitted on buses, so that the self-switching activity on buses is reduced. Crosstalk-aware encoding techniques also modify the switching patterns of a group of wires in such a manner that the crosstalk coupling effect is reduced. These techniques are invaluable for reducing power consumption, improving transmission reliability, and increasing system performance. Figure 7.1 shows where encoding techniques fit into a typical electronic system level (ESL) design flow. Since data transformation through encoding requires additional hardware logic at component interfaces (although some software-only encoding techniques have also been proposed), the implementation spans multiple levels in the design flow, starting from the communication architecture (CA) exploration level where the effectiveness of encoding techniques can begin to be evaluated.

For any encoding scheme, the encoder and decoder functions are the inverse of one another. Figure 7.2 shows a general encoder/decoder configuration between the sender and receiver on the bus. Bus encoding schemes can be classified according to several criteria, such as the type of code used (algebraic, permutation, or probability based [1]), the degree of encoding adaptability (static or dynamically adaptable encoding), the targeted capacitance for switching reduction (self, coupling, or both), the amount of extra information needed for coding (redundant or irredundant coding), and the method of encoding implementation (hardware, software, or a combination of the two). In this chapter, we classify encoding techniques

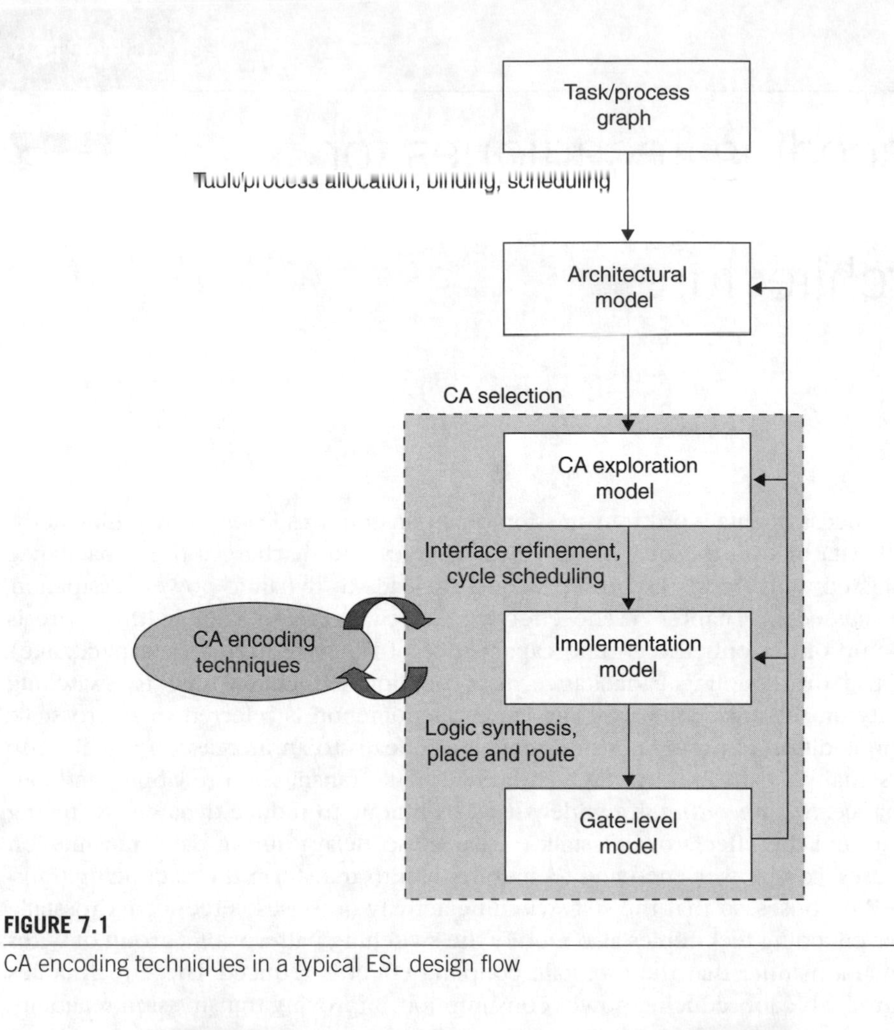

FIGURE 7.1

CA encoding techniques in a typical ESL design flow

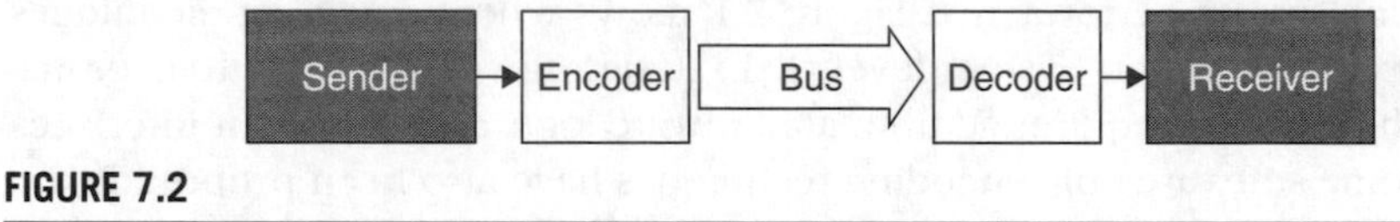

FIGURE 7.2

General encoder and decoder architecture

according to their intended benefits. We consider encoding techniques aimed at power reduction, signal transmission delay reduction and reliability improvement, or a combination of these. Certain optimizations such as crosstalk reduction can have multiple benefits associated with them, such as power reduction, signal delay reduction, and noise reduction. We attempt to classify the encoding techniques for crosstalk reduction as fairly as possible, based on their primary optimization goals. Section 7.1 presents encoding techniques aimed at reducing power consumption on bus wires. Section 7.2 describes encoding techniques that reduce signal propagation delay due to capacitive crosstalk, thus improving performance. Section 7.3

discusses encoding techniques that simultaneously reduce power consumption and capacitive crosstalk effects. Section 7.4 describes encoding techniques to reduce inductive crosstalk effects. Finally, in Section 7.5, we present encoding techniques for improving fault tolerance and reliability.

7.1 TECHNIQUES FOR POWER REDUCTION

One of the most common uses of encoding techniques has been to reduce the power consumption of bus wires. Power is dissipated on bus wires due to the charging and discharging of capacitances associated with a wire. There are two major sources of power consumption on a wire: self-switching and coupling (crosstalk). Power dissipated every time there is a bit flip (i.e., a transition from 0 to 1, or 1 to 0) on a wire due to its ground capacitance is referred to as self-switching power. In contrast, coupling power is due to the coupling capacitance of the wire created by switching in adjacent wires. It is possible that coupling power is dissipated even when there are no bit-flips on the wire itself! Low Power Coding (LPC) techniques attempt to modify the switching activity on the wires to reduce self switching and coupling power, by transforming the transmitted data word in some manner. The process of transforming transmitted data at the sender end is referred to as encoding. When the transmitted data reaches the receiver, a decoding step is performed, to retrieve the actual data. It is important that the savings obtained from reducing bus switching activity are not offset by the power dissipated by the encoding/decoding logic that needs to be added.

In the following subsections, we classify encoding techniques for low power according to whether they address self-switching or coupling power. The schemes in each of these classifications can be further divided into two categories: schemes for address bus encoding that are efficient for transmitted data which is sequential and has lots of repetition (typically the case for addresses generated by processors on their instruction address buses), and general schemes for address and data bus encoding that can handle random data (typically the case for addresses generated on data address buses, and data on data buses).

7.1.1 Schemes for Reducing Self-Switching Power

Self-switching is the phenomenon of bit flips from $0 \rightarrow 1$ or $1 \rightarrow 0$ on a wire over time that leads to charging and discharging of the capacitance associated with the wire with respect to its metal layer. This causes power dissipation on the wires that can be significant for on-chip buses, but even more critical for off-chip buses which typically have larger capacitances than on-chip buses. Several schemes have been proposed to reduce self-switching activity on buses. These are presented below.

7.1.1.1 *Address Buses*

Encoding schemes for address buses exploit the highly regular and sequential patterns associated with address streams. The effectiveness of these schemes depends on the transmitted data exhibiting spatial and temporal locality.

One of the earliest works in the area of encoding for address buses was the use of a *Gray code* [2], which is a permutation code (i.e., a one-to-one mapping from any set S to itself) that guarantees a single transition for every consecutive address streamed on the bus. Let the binary source word to be transmitted $B = \langle b_{n-1}, b_{n-2} \ldots b_1, b_0 \rangle$, and let the transformed gray sequence (called codeword) $G = \langle g_{n-1}, g_{n-2} \ldots g_1, g_0 \rangle$. Then the encoding function from Binary to Gray code is given as:

$$g_{n-1} = b_{n-1}$$
$$g_i = b_{i+1} \oplus b_i (i = n-2, \ldots, 0)$$

here denotes the exclusive-or (XOR) operation. The corresponding decoding function from Gray to Binary is given as:

$$b_{n-1} = g_{n-1}$$
$$b_i = b_{i+1} \oplus g_i (i = n-2, \ldots, 0)$$

For example, let B be a binary number $\langle 1,1,0,1 \rangle$. The values of b_3, b_2, b_1, and b_0 are 1, 1, 0, 1, respectively. The Gray code representation is then equal to $\langle b_3, b_3 \oplus b_2, b_2 \oplus b_1, b_1 \oplus b_0 \rangle$ which is equivalent to $\langle 1,0,1,1 \rangle$. For consecutive addresses transmitted over the instruction bus, the Gray code representation has less switching compared to the original binary representation. The asymptotic best performance of a single transition per transferred address is achieved for an infinite stream of consecutive addresses [3], and the code is optimum only in the class of irredundant codes (i.e., codes that do not add extra or redundant bits to the original data word).

If redundancy is allowed, more power savings can be achieved with the *T0 code* [4] that uses an extra line (INC) to signal when a pair of consecutive addresses is transmitted on the bus. When INC is high, the current value on the bus is frozen to avoid unnecessary switching activity, and the receiver becomes responsible for computing the new address. Thus, the decoder computes the current address as the previous address word plus a *stride* value ($s_i = s_{i-1} + S$; the stride value S being the difference between consecutive addresses in a sequential address mode). When two addresses are not consecutive, the INC line remains low, and the bus operates normally. When transmitting a sequence of consecutive addresses, this encoding requires no transition on the bus, compared to the single transition per transmitted word in the Gray code. On average the T0 code achieves a 60% reduction in switching activity on the address bus. One disadvantage of a redundant code, such as the T0 code, is that it changes standard bus widths and chip pin-outs, due to the need for an extra signal.

The need for an additional signal line is eliminated in the *T0-C encoding scheme*, proposed by Aghaghiri et al. [5], by observing that a new address transmission is sufficient to indicate to the receiver that the address increment mode is no longer in effect. This approach would however fail if a backward branch address is encountered whose value is the same as that currently on the bus. For instance, for consecutive address values 39, 40, 41 followed by a branch to 39, the frozen address bus would have 39, 39, 39, 39, which would lead the decoder to incorrectly decode the last address as 42. To address this scenario, the T0-C scheme transmits the value 42 on the fourth cycle. Since there would be no need to send 42 on the bus if there was no branch, the decoder recognizes this as a special case of a backward branch address whose value is the same as the previous value (39) on the bus. To generalize, for an address $b^{(t)}$ at the source to be transmitted at time t, the value on the bus $B^{(t)}$ is given as follows:

$$
\begin{aligned}
&\textit{if } (b^{(t)} == b^{(t-l)} + S) \\
&\quad B^{(t)} = B^{(t-l)}; \\
&\textit{else if } (B^{(t-l)} \,!= b^{(t)}) \\
&\quad B^{(t)} = b^{(t)}; \\
&\textit{else} \\
&\quad B^{(t)} = b^{(t-1)} + S;
\end{aligned}
$$

The T0-C code decreases switching activity on an address bus by an additional 14% over the T0 code.

The *T0-XOR*, *Offset*, *Offset-XOR*, and *T0-Offset codes*, introduced by Fornaciari et al. [6], also improve upon the T0 encoding scheme, by additionally reducing signal transitions for cases when there is a non-uniform address transition. The T0-XOR scheme combines the T0 code with an XOR operation, so that the T0 code is used for in-sequence values, while an XOR of the previous and current values is sent for out-of-sequence values. The T0-XOR code can be expressed as

$$B^{(t)} = \begin{cases} [b^{(t)} \oplus (b^{(t-1)} + S)] \oplus B^{(t-1)} & t > 0 \\ b^t & t = 0 \end{cases}$$

where $B^{(t)}$ is the encoded value transmitted on the bus lines at time t, $b^{(t)}$ is the original uncoded value at time t and S is the stride. The presence of the de-correlating XOR function obviates the redundant line INC, which is necessary for the T0 code. A fast encoding and decoding architecture for the T0-XOR code was presented by Parandeh-Afshar et al. [7]. In the Offset code, the difference of the current value $b^{(t)}$ with the previous value $b^{(t-1)}$ is transmitted on the bus. For highly correlated data, the switching on the bus lines is reduced, and for sequential addresses, the value on the bus is kept constant. The Offset-XOR code combines the Offset code with the XOR operation. The XOR of the offset with the previous value on the bus is sent over the bus. This also reduces the switching activity, since offsets are typically small numbers. The Offset-XOR code can be expressed as

$$B^{(t)} = \begin{cases} (b^{(t)} - b^{(t-1)}) \oplus B^{(t-1)} & t > 0 \\ b^t & t = 0 \end{cases}$$

In the T0-Offset code, the T0 code is combined with an Offset operation, so that the T0 code is used for in-sequence values, while the difference of the previous and current address values (i.e., Offset code) is sent for out-of-sequence values.

Note that unlike the other schemes proposed by Fornaciari et al. [6] (T0-XOR, Offset, Offset-XOR), the T0-Offset is a redundant code.

The *ALBORZ code* proposed by Aghaghiri et al. [8] introduces another encoding of the offset for the Offset-XOR code [6], to further reduce the switching activity on the bus. The encoder uses a codebook [9] to store recently used offsets. If the offset is present in the codebook, a limited-weight-code (LWC) [10, 11] associated with the offset is extracted from the codebook. An *N*-LWC is defined as a codeword with *N* 1s in it. Each entry of the codebook thus consists of two fields: the offset and its LWC code. The LWC is XOR'd with the previous value on the bus, and the resulting value is sent on the bus. Typically, LWCs are chosen such that they have the fewest number of 1s in them, to reduce switching on the bus (e.g., for a 32-bit bus, 32 1-LWC codes can be selected). This can significantly reduce transitions on the address bus. For instance, if a value of 0x1 h on the bus is followed by a BRANCH instruction with an offset 0xECh, the value is encoded with a fewer number of 1s, say 0x4 h, which results in reduced switching on the bus. The codebook can be implemented in a fixed (if the offset entries are fixed) or adaptive (if the offset entries are updated during program execution) manner. The overhead associated with the scheme is the additional area to store the codebook, as well as an additional bit added to the bus, to distinguish between the case when an encoded offset is being sent, and when the actual address is being sent (in the case of a codebook miss). Optimizations were proposed to remove the extra bit required in the encoding, to make the code irredundant. The ALBORZ scheme was shown to reduce switching activity by 89% on the instruction address bus, for the SPEC2000 benchmarks.

Two irredundant adaptive codebook-based address bus encoding schemes, similar in concept to the ALBORZ code, were proposed by Komatsu and Fujita [12]. The first scheme extends the T0-C code by adding an adaptive codebook to it, and is called *T0-CAC* (T0-C with adaptive codebook). In this scheme, the T0-C code is used for sequential addresses, but the adaptive codebook is used when a JUMP or a BRANCH operation results in a non-sequential address on the bus. The codebook contains recently accessed destination addresses of JUMP and BRANCH operations. This is particularly useful for programs with loops, where the same destination addresses of JUMP/BRANCH operations occur frequently. When a hit occurs on the codebook, an MSB one-hot encoding (i.e., there is only a single "1" and its position corresponds to the index in the codebook) is used to reduce switching activity on the bus. The second scheme extends the Offset-XOR code [6] with the adaptive codebook, and is called *OXAC* (Offset-XOR with adaptive codebook). Figure 7.3 shows the block diagram for the OXAC encoder. The OXAC code is practically identical to the ALBORZ code. An average switching activity reduction of 80–95% was shown for these schemes on the SPEC2000 benchmarks.

The *Beach code* proposed by Benini et al. [13] makes use of other types of temporal correlations than arithmetic sequentiality between transmitted patterns on the address bus. The code exploits the experimentally noted phenomenon of time-adjacent addresses showing particularly high block correlations. Given a typical execution trace of the address bus, some statistical information identifying possible block correlations is collected. Bus lines are then grouped into clusters

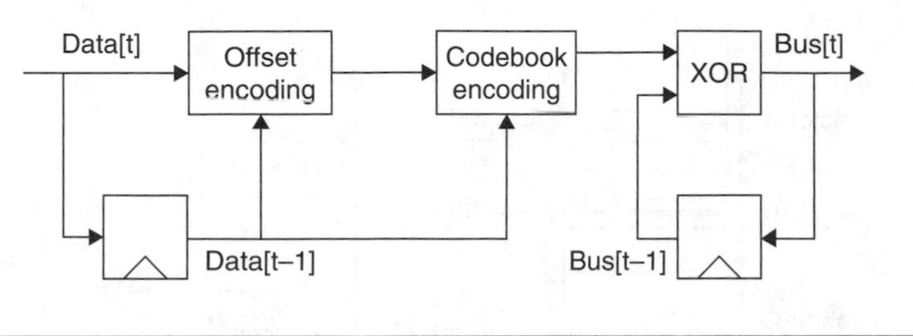

FIGURE 7.3

Block diagram of OXAC encoder [12]
© 2003 IEEE

according to their correlations—lines belonging to the same cluster are considered to be highly correlated. An encoding function is then automatically generated for each cluster, and each cluster of bits in the original cluster is transformed into a new bit configuration. The algorithm responsible for finding the encoding functions has the goal of minimizing switching activity. The output of this transformation is an encoded stream for which the average number of bus line transitions between two successive patterns is minimized. Thus, codes with small Hamming distance between them (i.e., codes with a small number of bit positions that are different) are assigned to data words that are likely to be sent on the bus in two consecutive clock cycles. Since the computation of the encoding functions is highly dependent on the selected execution trace, the code performs best for special purpose systems, where the same piece of code is repeatedly fetched and executed (e.g., most embedded systems). Ascia et al. [14] improved upon this technique by using a heuristic method based on genetic algorithms to search for an encoder to minimize switching activity. Similar to the assumptions made in the Beach scheme, the approach assumes that the trace of patterns transmitted on the bus is available in advance. The trace is used as an input for the genetic algorithm, which finds an encoder (represented as a truth table) that minimizes switching activity on the address bus. However, the runtime for such an approach can be large, even for reasonable sized systems.

The *Working Zone Encoding* (WZE) scheme proposed by Musoll et al. [15] exploits the locality of reference associated with most software programs. The proposed approach partitions the address space into working zones, and stores the starting addresses of each working zone in a set of registers. Figure 7.4 shows how an address space for an application with accesses to three vectors (or arrays) can be partitioned into three working zones, with every word being uniquely identified by its working zone address, and an offset within the zone. Applications typically favor a few working zones of their address space at any instant. A bit is used to denote a hit or a miss of the working zone. When there is a miss, the full address is transmitted on the bus. Otherwise, the bus is used to transmit the offset, which is one-hot coded. When the data access moves to a different working zone, the receiver is notified with a special codeword. The receiver subsequently changes the default reference address, after which the offset transmission can resume. For the case in which the number of zones is larger than the number of registers, a replacement policy is implemented. While the scheme is more flexible than the

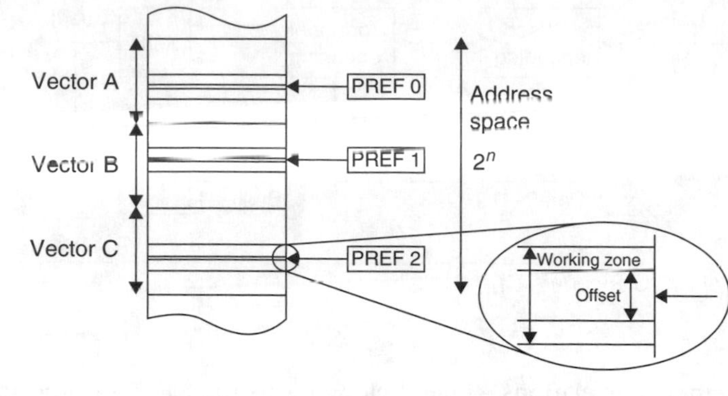

FIGURE 7.4

Address space with three vectors (arrays) which are each mapped to a working zone [15] *© 1997 IEEE*

Gray and T0 schemes, it relies heavily on certain assumptions of the patterns in the stream. If the data access policy is not array based, or if the number of working zones is too large, the scheme loses its effectiveness. Like the T0 code, the scheme also requires an additional wire for communicating a working zone change.

The *Pyramid code* proposed by Cheng and pedram [16, 17] reduces switching activity on Dynamic RAM (DRAM) address buses. DRAM is often laid out in a 2-D array. To identify a cell in the array, two addresses are needed—row and column. These addresses are usually multiplexed—the row address is sent first, followed by the column address. Since the switching activity on a DRAM address bus occurs in a time-multiplexed manner, the approaches above cannot be effectively applied, as they are meant for non-time-multiplexed buses. The Pyramid code minimizes switching activity on the DRAM address bus by formulating the problem as a Eulerian cycle problem [18] on a complete graph. Similar to the Gray coding scheme, the Pyramid coding technique is an irredundant coding scheme. The code remains quite effective in reducing the power dissipation of the multiplexed bus even when the sequentiality of the addresses is interrupted every four addresses. But the Pyramid technique can have a significant delay when the encoding is applied to larger address spaces.

An adaptive self-organizing list code was proposed by Mamidipaka et al. [19–21] to reduce switching activity on instruction and data address buses. The idea is to use a list to create a one-to-one mapping between addresses and codes. The list is reorganized every clock cycle to map the most frequently used addresses to codes with fewer 1s, to reduce switching activity on the bus. For multiplexed address buses, a combination of this approach with the INC-XOR approach (proposed by Ramprasad et al. [22]) is used. INC-XOR performs an XOR operation between the current address, and the summation of the previous address and the stride, and then sends the result over the bus (just like in the T0-C code). Obviously, when consecutive addresses grow by the stride, no transition occurs on the bus. However, the size of the list has an impact on performance—a long list is needed to achieve satisfactory results. The large hardware overhead of maintaining long lists, as well as the complex encoding/decoding hardware logic needed for the scheme makes it practically quite costly to implement.

A combination of *Unit Distance Redundant Codes* (UDRC) and an address reference caching mechanism was proposed by Givargis and Eppstein [23] to reduce switching activity on the address bus. UDRCs are codes that guarantee a Hamming distance of at most one between any pair of encoded symbols. A general construction for UDRCs is introduced, which provides multiple redundant encodings of each possible source symbol, in such a way that any arbitrary value can be encoded by a value of Hamming distance at most one from each previous codeword. The construction uses an optimal number of bits for a given set of symbols. Address reference caching exploits the fact that address references are likely to be made up of an interleaved set of short sequential address bursts. Reference caching isolates these streams by storing recently used address values and limits the communication to a UDRC encoded message that identifies a particular reference, at the cost of at most a single bit transition. Experiments on 14 applications from the PowerStone benchmark [24] showed an average of 60% reduction in switching activity, with the best and worst cases being 86% and 36%, respectively. The maximum performance penalty (i.e., critical-path delay) for the encoder and decoder was found to be 16 and 14 gates, respectively.

Compression, which is related to encoding, also provides significant energy savings. In a compression scheme for buses, data to be sent is compressed and transmitted in a single cycle on a narrow width bus if the compression is successful, or transmitted in multiple cycles if not. Early work [25, 26] proposed address compression schemes for processor–memory compression, using a small compression cache at the sender end, and a base register file at the receiver end. In the *Dynamic Base Register Caching* (DBRC) approach [25, 26], the original address is split into higher order and lower order components, and the former is stored in a compressor, which is a cache of base registers on the processor (sender) side. On a cache hit, the index and entry number to the BRC are transmitted on the bus with the uncompressed lower order part of the original address in the same cycle. A miss in the processor (sender) BRC results in sending a reserved bit pattern on the bus in the first cycle, followed by the missed address in the subsequent cycles. The memory (receiver) side consists of a register file that is loaded with the missed address. The bus extender (BE) scheme is similar to DBRC, but behaves differently on a cache miss, whereupon it starts sending the entire address immediately (instead of transmitting a reserved bit pattern) and a separate control signal line is used to indicate a hit or a miss in the sender cache. The benefits of using these buses for reducing switching activity for off-chip buses were studied by Basu et al. [27], while the energy benefits of using these compression schemes for on-chip address compression were studied by Liu et al. [28].

7.1.1.2 *Data Buses*

The applicability of the above encoding schemes is fairly limited for cases where the data being transmitted on the bus is highly uncorrelated. This is true for data buses, as well as some highly irregular access address buses, particularly the data address bus. The schemes presented below have been shown to be effective, even for data streams that have random data being transmitted.

Early work by Ramprasad et al. [29] established lower/upper bounds on the average power consumed in data buses using Shannon's channel coding theorem [30] and the concept of entropy. Additionally, an asymptotically optimal coding strategy based on the popular data-compression scheme, Lempel–Ziv coding [30] was used to reduce power consumption. However, the overhead of implementing a Lempel–Ziv encoder and decoder may easily offset the gains obtained from reduced switching on the bus. For applications where the transmission order of data is immaterial, large power savings were shown by Murgai et al. [31], by reordering data using a *minimum cost hamiltonian path* (MCH) formulation. The idea is to reorder the words to be transmitted in such a way so that the net switching activity in transmitting all the words is minimized. Unfortunately, hardware implementation of a MCH solver even using simplistic heuristics is a difficult proposition, and even the most basic implementations are likely to consume a lot of power, offsetting the benefits due to reordering of the transmitted words.

One of the most widely used encoding scheme to reduce bit-toggling on data buses is the *bus-invert code* (BI), which was proposed by Stan and Burleson [32]. The approach computes the Hamming distance between the current value and the previously transmitted value, and inverts the current value (i.e., performs a bit-wise negation) if the distance is greater than half of the bit-width of the bus. An additional bus wire is needed to signal the inversion to the receiver in this scheme. The BI code can be considered to be an example of what is called an LWC [10, 11], or a *starvation code* [33]. The code is a simple and effective technique for minimizing bus switching activity when the data stream exhibits randomness, as in the case of data buses.

The BI was extended by Shin et al. [34], who proposed a *partial bus-invert encoding* (PBI) scheme. The motivation for the scheme is that previous encoding approaches such as the BI code take the entire bus lines into account for encoding, which increases the overhead due to the encoding/decoding logic at the component interfaces. The PBI encoding is effective when certain bits of the source words in the data stream exhibit strong spatio-temporal correlations. The main idea is to identify such bits, group them together, and then apply the BI coding technique only to these bits. In particular, the PBI scheme partitions the bus lines into two parts, x and y, by analyzing the transition probabilities on the lines, and applies the BI method partially to one of the parts, to reduce switching and encoding implementation overhead. Yoo and Choi [35] extended this work by proposing an *interleaving PBI code* where the bit width and positions of the two parts x and y are dynamically changed. Field Programmable Gate Array (FPGA) devices are proposed to dynamically change the x and y bit groupings. A heuristic off-line algorithm divides a given input stream into several subsequences while considering the runtime and power cost overhead of reconfiguring the encoder circuit. Instead of using redundant bits, adding extra redundant bits can achieve the same effect of changing the x and y bit partitions at runtime even when using a fixed encoding function. The PBI approach was further extended by Hong et al. [36], who proposed the *decomposed bus-invert encoding* (DBI). The approach decomposed the bus lines into any arbitrary number of groups so that each partitioned group is considered independently for BI coding. A *look-ahead bus-invert code* was proposed by Chang et al. [37], which extends the general ideas described by Stan and Burleson [38]. Since the BI scheme only considers the current and previous value to make the bus inversion decision, it can lead to a local minimum encoding. The proposed look-ahead BI scheme uses a look-ahead table, to make better encoding decisions, especially for burst mode transfers.

Another interesting approach, called *transition signaling* was described by Stan and Burleson [10], in which a logical 1 is represented by a transition on the bus (from high to low, or low to high voltage levels), while there is no transition for the case of a logical 0. In contrast, in the normally used level signaling, a logical 1 is represented by a high voltage, and a logical 0 is represented by a low voltage. A transition signaling encoder produces the modulated symbol $b(t)$ to be transmitted from the source word $s(t)$ by using the following simple XOR operation: $b(t) = s(t) \oplus b(t-1)$. The decoder produces the source word by doing a similar XOR operation: $s'(t) = b(t) \oplus b(t-1)$. Transition signaling ensures that the number of transitions over a period of time is equal to the number of 1s present in the input to be transmitted, and is very useful for reducing switching activity on a bus. It is even more effective when used in combination with other schemes such as LWC [10]. Assuming a source word with K bits, an M-LWC is comprised of a set of codewords with the total number of 1s (or Hamming weight) $\leq M$. For minimum switching activity, the value of M must be as small as possible. However, there is a trade-off between the value of weight M and the size of the code N: the smaller the weight M (resulting in lower power dissipation), the larger is the size of the code N (resulting in more redundancy). This is shown in Table 7.1. As another example, Table 7.2 shows a *1*-LWC and *2*-LWC for $K = 4$. The proposed LWC approach thus uses transition signaling and redundant bus lines to achieve further reduction in the average Hamming distance between consecutive pairs of transmitted values.

Sundararajan and Parhi [39] proposed an improvement over the BI encoding scheme (which is a special form of limited weight coding) by incorporating *codeword slimming*. The proposed scheme transmits words in a bit-serial word parallel manner, rather than the traditional word serial bit-parallel transmission order. Each bit line is considered independently for transition activity minimization. Then a codeword is selected with a weight strictly less than a value d. If such a codeword is not found, then codeword slimming is performed, by taking the codeword with the shortest leading set of bits that have a weight d, and then throwing away the remaining bits in the codeword. The resulting pruned codeword gives a new

Table 7.1 M vs. N for a given value of K bits in the source word [11]

Information bits (K)	2	4	4	8	8	8	8	16
Maximum nr. of 1s (M)	1	1	2	1	2	3	4	8
Codeword length (N)	3	15	5	255	23	11	9	17

Table 7.2 2-LWC and 1-LWC for $K = 4$ [11]

Symbol	Minimally encoded	2-LWC	1-LWC
0	0000	00000	000000000000000
1	0001	00001	000000000000001
2	0010	00010	000000000000010
3	0011	00011	000000000000100
4	0100	00100	000000000001000
5	0101	00101	000000000010000
6	0110	00110	000000000100000
7	0111	11000	000000001000000
8	1000	01000	000000010000000
9	1001	01001	000000100000000
10	1010	01010	000001000000000
11	1011	10100	000010000000000
12	1100	01100	000100000000000
13	1101	10010	001000000000000
14	1110	10001	010000000000000
15	1111	10000	100000000000000

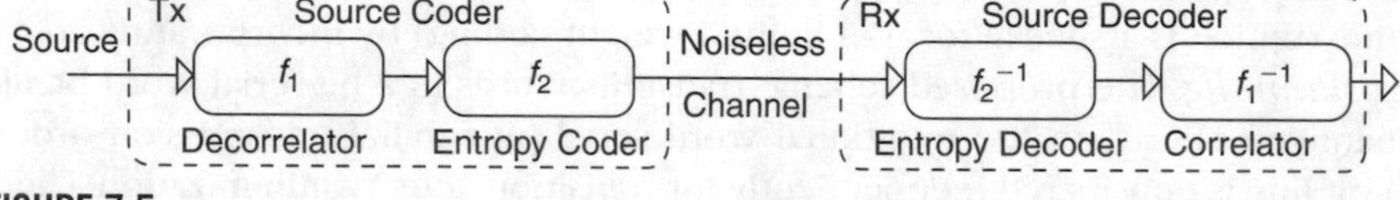

FIGURE 7.5

Generic communication system with source de-correlator/correlator and entropy coding/decoding steps [22]

codeword for transmission. The proposed scheme improves upon the BI code, but has a larger implementation overhead at the encoder and decoder (e.g., requiring shift registers for serial to parallel conversion, etc.).

A *hybrid encoding scheme* that combined multiple schemes was presented by Ramprasad et al. [22]. The authors proposed a two step framework, as shown in Fig. 7.5, assuming that the transmission medium is noiseless. In the first step, data is passed through a low complexity de-correlation source-coding function f_1. In the second step, a variant of an entropy coding function f_2 (that minimizes transitions instead of average number of bits at output) is employed to reduce switching activity. In the proposed approach, f_1 can be an *xor*, or a difference-based mapping (*dbm*—similar to the offset code) for the purpose of providing de-correlation. f_2 can be a *bus-invert* (*inv*), probability-based mapping (*pbm*—where codewords with fewer 1s are mapped to source words with higher probability of occurrence), or a value-based mapping code (*vbm*—where codewords with fewer 1s are assigned to lower values, based on the assumption that smaller values are assumed to have higher probability of occurrence, which is especially true if f_1 is *dbm*), for the purpose of activity reduction. Assuming that the probability distribution of the data is available a priori, the *xor-pbm* scheme can reduce switching activity on data buses by 30–40%. The *dbm-pbm* scheme reduces transition activity by about 5% more than the *xor-pbm* scheme because *dbm* skews the input probability distribution more than *xor*. These schemes were shown to outperform the BI and BI with compression schemes. A similar hybrid scheme was proposed by Benini et al. [40] that combines the BI and T0 codes to create the *T0_BI code* and the *dual_T0_BI code* (for time multiplexed buses), which improves upon the individual BI and T0 approaches in terms of switching activity reduction.

An *adaptive codebook-based encoding scheme* was proposed by Komatsu et al. [41], which extends the static codebook-based approach first proposed in [9]. The scheme uses a codeword with two parts: (i) the index of a pattern stored in a codebook and (ii) the *XOR* of the source word (i.e., data) with the pattern in the codebook. In the first step, a pattern in the codebook is selected so that the average Hamming distance between a source word and the best pattern in the codebook is minimized. The encoder compares a source word with all of the patterns in the codebook to find the pattern that has the minimum Hamming distance from the source word. In the second step, the index bits are transmitted using transition signaling (i.e., *XOR* of current value and previously transmitted value), along with the codeword (which is the *XOR* of the source word with the selected pattern). Figure 7.6(a) shows a block diagram of the encoder, while Fig. 7.6(b) shows the corresponding decoder. Both the encoder and the decoder know the codebook, which is updated dynamically with the most recent source word. This scheme has the overhead of using additional bus lines for sending the codeword, as well as the codebook tables at the transmitter and receiver, but was shown to improve upon the BI code.

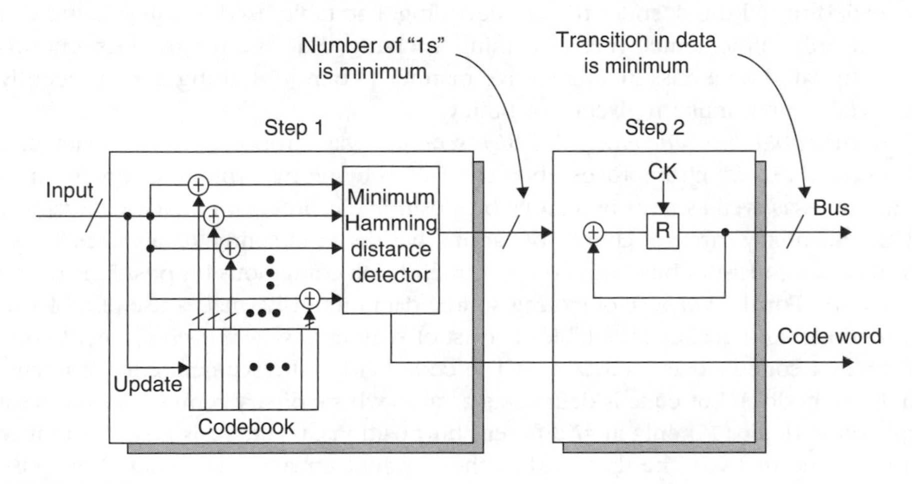

FIGURE 7.6(a)

Adaptive codebook scheme encoder implementation
© 1999 IEEE

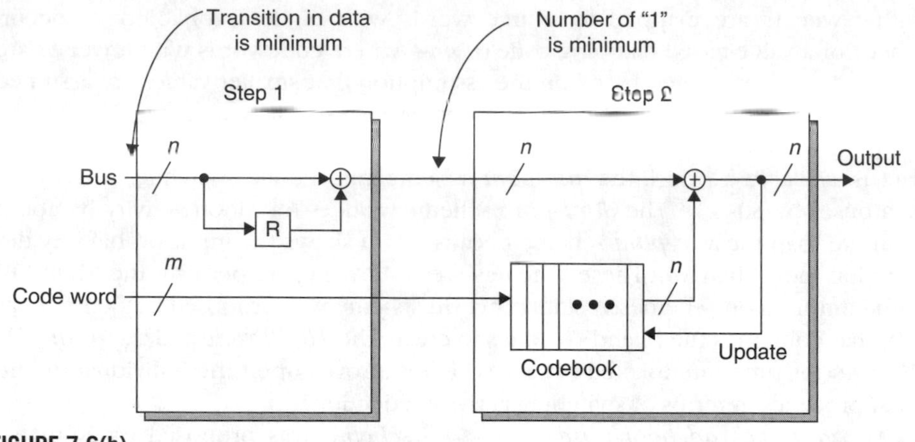

FIGURE 7.6(b)

Decoder implementation [41]

Frequent value encoding (FVE) schemes were proposed by Basu et al. [27] and Yang et al. [42, 43]. These schemes use value caches [27] or content addressable memories (CAM) [42, 43] at the sender and receiver ends for storing frequently transmitted data values. The contents of the tables at the sender and receiver are kept identical. The entire data value is cached for frequently occurring data values, and only the index of the table is transmitted on a hit, using a one-hot code. On a miss (for the case of an infrequent value), the value is transmitted in its un-encoded form. When the decoder receives the value and finds more than one hot wire, it concludes that the transmitted value is not encoded. It is possible that an infrequent value being transmitted in un-encoded form contains a single high bit and all of its remaining bits are zeros. To ensure that the decoder does not erroneously decode this value, an additional single bit control signal from the encoder is needed to tell the decoder to skip decoding. The table used in the scheme can be statically implemented (i.e., remaining unchanged during program execution) or be updated on a miss in an adaptive manner like in [44], using a least-recently used (LRU) or a similar replacement policy.

A *tuneable bus encoding (TUBE) scheme* was proposed by Sridhara and Shanbhag [45], which improves upon the FVE scheme by exploiting repetition in contiguous as well as non-contiguous bit positions in order to encode data values. TUBE essentially captures chunks of varying widths from source data values. These chunks can consist of bits from contiguous and non-contiguous bit positions of the data value. For the purpose of storing source data bits, TUBE makes use of tables at the encoder and decoder. The tables consist of segments, where each segment comprises of a code field and a data field. The code field in the segment entry contains an M-hot code. M-hot code is defined as a value whose binary representation has a high value (logic "1") only in M different bit positions, where M is a small number (usually one or two). The data field of the segment entry stores selected bit positions of the incoming source data value. For the first occurrence of a full or partial bit-pattern, the encoder transmits the un-encoded value over the bus and stores the bit-pattern in its segments. Upon receiving the un-encoded value, the decoder stores the bit-pattern from the data value in its segments. Thus at the end of every bus cycle, the encoder and decoder's table contents are exact replicas of each other. For subsequent occurrences of the repeating bit pattern, TUBE sends an *M*-hot code for the repeating portion. A correlator and de-correlator are added to the two ends of the buses, just like in the FVE scheme, to reduce correlation between successive values. A 3-bit timestamp is associated with the table entries, and replaces entries using the LRU replacement policy. The TUBE scheme requires one additional control signal to indicate the presence of encoded values on the bus. The scheme was shown to improve upon the FVE scheme by 21% on average for the MediaBench [46], NetBench [47], and SPEC2000 [48] benchmarks.

Benini et al. [49] proposed a ***probabilistic encoding scheme*** that is based on a detailed statistical characterization of the target data stream. More specifically, it is assumed that complete probability distribution of all pairs of possible consecutive values to be transmitted on the bus is available. A table is created and populated, with three columns—the current word, the previous transmitted word, and the code that aims to minimize transitions on the bus. But such a scheme is only practically feasible for buses of very small size. More realistic approximations of this approach for larger buses are created by partitioning a larger bus into smaller clusters and applying the above table-based approach. Of course there is a trade-off here between accuracy and complexity. The smaller the cluster size, the smaller is the saving in transition activity, because the spatial correlation between bits is partially lost. On the other hand, the larger the cluster size, the more complex is the design of the encoder and decoder, as well as the time to construct the code table. In contrast to the exact algorithm above (that uses an exponential table to store all possible pairs of source words and their joint occurrence probability to assign a minimum of transition activity codes to each pair of source words) an approximate algorithm is proposed by the authors for larger buses where only the *M* most probable pairs of consecutive words in the code are considered, for a small and manageable value of *M*. Comparisons of the exact clustered approach with the *xor-pbm* and *dbm-pbm* showed significant improvements, while the approximate algorithm performed at least as well as the approaches. This is an important result, because the exact algorithm, as well as *xor-pbm* and *dbm-pbm* are of limited practical applicability, due to the size and complexity of the encoders and decoders they necessitate. The authors also propose an adaptive version of these encoding approaches that does not require any prior knowledge of the input stream statistics. The proposed scheme works bit-wise, rather than word-wise for the purposes of a low-cost implementation. The basic idea in this case is to create the codes dynamically, based on the statistical information observed in a window of size *S*. Again, there is a trade-off between adaptation speed and delay for the choice of the window size. The adaptive approach was compared with the BI code (which is another code that does not require a priori information about the data stream) and shown to improve upon it.

A class of codes called *limited intra-word transition (LIWT) codes* was proposed by Salerno et al. [50] to reduce switching activity on LCD buses (i.e., buses that carry image data to the LCD screen). The scheme improved upon previous work

[51, 52] in this area in two ways. Firstly, the inter-pixel correlation existing in typical images is exploited and pixel differences are transmitted (instead of actual RGB values for the pixels) just like in [51]. However, unlike [51], only pixel differences with the largest occurrence probability are transmitted. Secondly, LIWT codes are characterized by codewords with increasingly larger value of intra-word transition count, which are mapped to values of pixel differences with increasingly smaller occurrence probability. This generalizes the type of codes used in [52], which only uses codewords with one intraword transition, thus encoding a larger set of values. This enables a reduction in energy in cases where smaller color depths (i.e., fewer bits per pixel) or less redundancy (i.e., fewer extra bits per symbol) is used, and can thus be used for different standard LCD protocols. Cheng et al. [53] proposed a software-only encoding approach for reducing power consumption on LCD buses. The approach implements the Gray code (for addresses), and the BI code (for more irregular data) in software. Modifying the software device driver to implement these codes allows saving in hardware implementation costs for these codes. Bahari et al. [54] proposed an inter-frame bus encoding technique where the pixel correlation between two consecutive frames is utilized to reduce switching on the bus.

A scheme similar to the WZE scheme, but more effective for data addresses, called *sector-based encoding* was proposed by Aghaghiri et al. [55]. The scheme partitions the source-word space into a number of sectors with unique identifiers called sector heads. These sectors can correspond to the address spaces for the code, heap, and stack segments of one or more application programs. Each source word is dynamically mapped to the appropriate sector and encoded with respect to the sector head. The sectors can be determined a priori, or be dynamically updated based on the source word that was last encountered in that sector.

7.1.1.3 *Serial Buses*

It has been shown [56–58] that serial buses consume less power compared to parallel buses. Although a serial bus needs a serializer and de-serializer, the number of buffers required by the serial bus is fewer than for a parallel bus [57]. Hatta et al. [56] proposed a technique for reducing the number of bus lines of the conventional parallel-line bus architecture by multiplexing each M-bits onto a single line. This bus architecture, which is shown in Fig. 7.7, transforms an N-bit conventional parallel-line bus into an N/M line (serial-link) bus. Parallel and serial link buses in a network-on-chip (NoC) are compared and it is shown that the parallel link bus has 5.5 times more power consumption and 17 times more area compared to the serial link bus. [57] HSPICE simulations performed by Ghoneima et al. [58] showed that, for the same throughput-per-unit-area as the conventional parallel line bus, the serial-link bus architecture reduces the energy dissipation by up to 31.42% for a 64-bit bus.

Lee et al. [59] proposed the SILENT technique to minimize transmission energy on a serial wire. The simple technique encodes words as the *XOR* between successive data words, that is, $B(t) = b(t) \oplus b(t-1)$. By serializing the encoded data words, the frequency of 0s on the wire increases, due to the correlation between successive data words. After de-serialization at the receiver, the original data word from a sender unit can be recovered by *XOR*ing the encoded word and a previously decoded word.

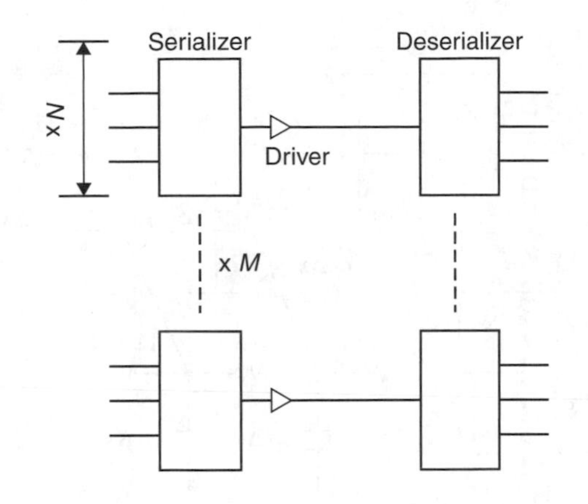

FIGURE 7.7

Circuit structure of serialized bus [56]

Another encoding technique, based on a lookup table (LUT), was proposed by Saneei et al. [60]. The scheme makes use of a lookup table for encoding the input data before the parallel to serial conversion and a lookup table after serial to parallel conversion in the decoder. Using the LUT, the input symbols are replaced with codewords that have some redundant bits compared to the original symbols. The idea is to map the data with a higher probability of occurring to codes with lower transitions. The scheme considers the last state of the bus and the probability distribution of the new data. To increase the transition reduction, redundant bits are added to codes and the n-bit data is mapped to a k-bit code ($k > n$), using a $2^n \times k$ LUT. Data with a higher probability of occurrence is mapped to codes with lower transitions, to save energy during the data transfer on the serial bus. In addition, for a k-bit code, there are 2^k codewords, half of which start with zero and the other half start with one. If the previous code on the bus ends with zero, the new data is mapped to a codeword that starts with zero. With this technique, transitions due to the first bits can be eliminated. Such a serial transmission of these codewords leads to lower switching activity compared to that of the original transmission of the input symbols.

Kuo et al. [61] proposed adapting BI coding for the serial bus to reduce transition activity. An additional bit is added to the tail of a packet to indicate if inversion was performed or not. In the same paper, another technique called *serial line encoding* (SLE) is proposed which requires two additional bits during code mapping. These bits are appended to the packet to reduce transitions.

7.1.2 Schemes for Reducing Coupling Power

All of the encoding schemes described in the previous subsections attempt to reduce the switching activity on bus lines, which translates into a reduction in self-switching power dissipation during communication. While reducing self-capacitance of bus lines is important, increasingly with the advent of deep submicron (DSM) technologies, inter-wire or crosstalk capacitance has been shown to lead to

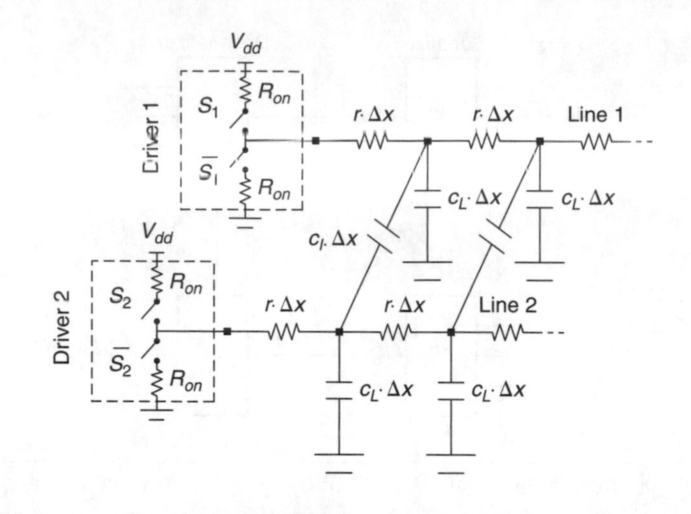

FIGURE 7.8

Coupled transmission line model in DSM [62]
© 2000 IEEE

significant power dissipation. Figure 7.8 shows a coupled transmission line model for bus lines in DSM [62]. In the figure, r is the series resistance per unit length of the lines, C_L is the capacitance to ground (or self-capacitance) per unit length, and C_I is the inter-wire capacitance (also called coupling or crosstalk capacitance) per unit length between adjacent lines. In addition to its dependence upon technology as well as structural factors such as wire spacing, wire width, wire length, wire material, coupling length, driver strength, and signal transition time, the coupling capacitance also depends upon the data dependent transitions and will increase or decrease depending upon the relative switching activity between adjacent bus wires. In fact, for a wire on the bus, simultaneous transition to opposite values (i.e., $0 \rightarrow 1$ and $1 \rightarrow 0$) of two adjacent bus lines dissipates about four times as much energy than without considering coupling effects, for technologies below 0.25 μm.

The coupling capacitance of a wire can be classified into four types *1C*, *2C*, *3C*, and *4C*, according to the coupling capacitance C_I of two wires [62]. As mentioned earlier, the crosstalk effect on a single wire (victim) depends on the signal transition of its neighboring wires (aggressors). Let (w_{i-1}, w_i, w_{i+1}) represent the signal pattern on three wires at any given time, where w_i is the victim, and w_{i-1}, w_{i+1} are the aggressors. Table 7.3 shows the different types of crosstalk and their relations with the signal transitions at time T_{t-1} and time T_t, where

$$(b, \bar{b}) \in \{0,1\} \qquad \bar{b} \text{ is the complement of } b$$

A *type-1* (or *1C*) crosstalk occurs if either w_{i-1} or w_{i+1} changes state, for example, a transition from (0,1,1) to (1,1,1) will cause a type-1 crosstalk. The coupling capacitance for type-1 crosstalk is C_I. A *type-2* (or *2C*) crosstalk occurs if w_i is in opposite state transition with one of its adjacent wires, whereas the other wire undergoes the same state transition as w_i. For example, a transition from (1,1,0) to (0,0,1) will cause a type-2 crosstalk. The coupling capacitance for type-2 crosstalk is $2C_I$. A type-2 crosstalk can also arise for the case where w_i changes state, for

Table 7.3 Different classes of crosstalk and their corresponding bit patterns [64]

Crosstalk type	Time	Bit pattern (w_{i-1}, w_i, w_{i+1})
1C	T_{t-1}	(b,b,b) (b,b,b) $(b,\bar{b},\bar{b})$ $(\bar{b},\bar{b},b)$
	T_t	$(b,\bar{b},\bar{b})$ $(\bar{b},\bar{b},b)$ (b,b,b) (b,b,b)
2C	T_{t-1}	(b,b,b) $(\bar{b},b,b)$ $(b,b,\bar{b})$ $(\bar{b},b,\bar{b})$$(b,b\bar{b})$ $(\bar{b},b,b)$
	T_t	$(b,\bar{b},b)$ $(\bar{b},\bar{b},b)$ $(b,\bar{b},\bar{b})$ $(\bar{b},\bar{b},\bar{b})$$(\bar{b},\bar{b},b)$ $(b,\bar{b},\bar{b})$
3C	T_{t-1}	$(b,\bar{b},b,)$ $(b,\bar{b},b)$ $(\bar{b},\bar{b},b)$ $(b,\bar{b},\bar{b})$
	T_t	$(b,b,\bar{b})$ $(\bar{b},b,b)$ $(\bar{b},b,b)$ $(\bar{b},b,\bar{b})$
4C	T_{t-1}	$(b,\bar{b},b,)$
	T_t	$(\bar{b},b,\bar{b})$

example, the transition from (0,0,0) to (0,1,0). A *type-3* (or *3C*) crosstalk occurs if w_i undergoes opposite state transition with one of the two adjacent wires, whereas the other is quiet. For example, a transition from (1,0,1) to (1,1,0) will cause a type-3 crosstalk and the coupling capacitance of w_i in this case will be $3C_I$. A *type-4* (or *4C*) crosstalk is caused when all three wires move to opposite states with respect to each other and their previous bus state. For example, a transition from (1,0,1) to (0,1,0) will cause a type-4 crosstalk. The coupling capacitance of w_i in this case rises to $4C_I$. Type-4, type-3, and type-2 are considered the worst crosstalk [63].

Consequently, it becomes essential for encoding schemes to take coupling effects into account for DSM technologies, to reduce power dissipation due to coupling (or crosstalk). Several approaches have been proposed that consider crosstalk capacitances between pairs of adjacent wires while encoding for low power consumption on buses. These approaches are classified into schemes aimed at minimizing power for address buses (i.e., regular access patterns) and for data buses (i.e., irregular access patterns), and are presented below.

7.1.2.1 *Address Buses*

A *permutation-based encoding* (PB) scheme that reduces crosstalk for buses without the need for any encoding/decoding circuitry was proposed by Macchiarulo et al. [65]. The proposed scheme performs permutations of the address bus lines at the layout level, during the physical design stage, to reduce coupling. Figure 7.9 shows how lines in a 4-bit bus are re-arranged, using vias that route the lines across multiple metal layers. The PB approach was shown to result in an average energy saving of 26% for address buses. Since the approach is orthogonal with respect to encoding techniques that target energy dissipated due to switching of self-capacitances, it can be combined with these techniques to yield even more average energy savings. The PB approach was combined with Gray coding to yield an average of 46% savings in energy. A very similar approach that proposed reordering

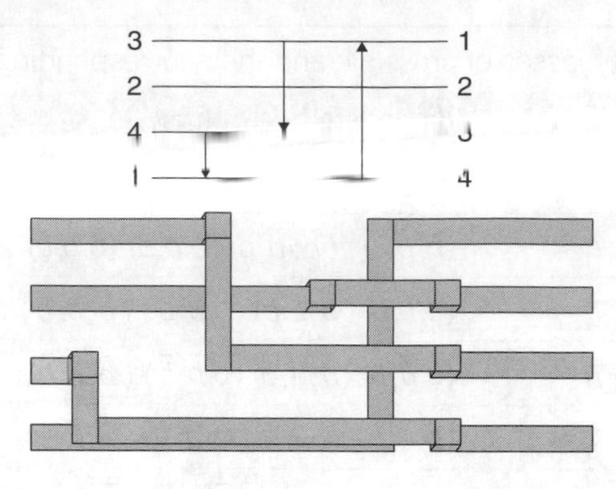

FIGURE 7.9

Via constructed permutation-based routing [65]
© 2001 IEEE

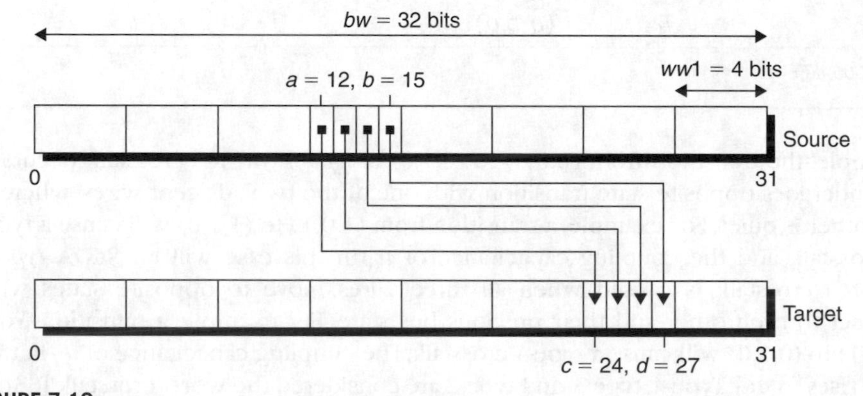

FIGURE 7.10

Example of window remapping [67]
© 2001 IEEE

bus lines to minimize opposite phase transitions on adjacent bus lines for crosstalk power reduction was proposed by Shin and Saburai [66].

An *adaptive address bus coding scheme* (A^2BC) proposed by Henkel and Lekatsas [67] used a similar concept to reduce self and crosstalk capacitance switching for address buses. The encoding scheme first identifies windows, which are essentially a collection of bus lines, and then remaps them so that the windows with high transition activities are adjacent to windows with little or no transition activity. Figure 7.10 shows an example of a 32-bit address bus for which a 4-bit window is re-mapped. While such a re-mapping step results in a shielding of high transition lines from each other, which reduces coupling between them, it does not reduce transition activities and coupling within a window. Thus, after the window remapping step, the self and coupling capacitances within a window are minimized by using a BI scheme [32] which inverts the bits in a window only if it reduces power due to both self-transitions and coupling.

Liu et al. [69, 70] presented address bus encoding schemes that extend compression techniques [25, 26, 44, 68], by using an LRU and XOR encoding of transmitted

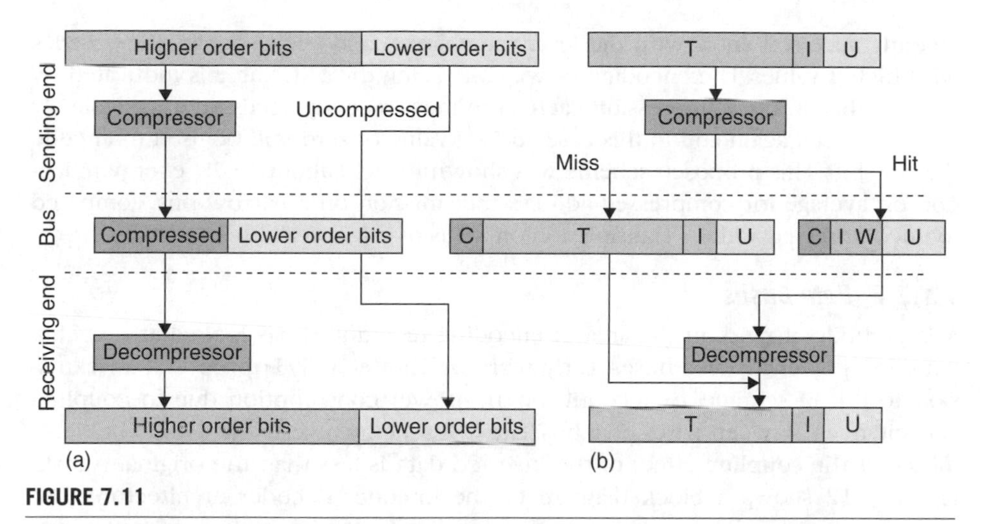

FIGURE 7.11

Dynamic address compression (a) schematic of dynamic address compression (b) address compression in bus expander (BE) [68] scheme [69]

words to reduce self-energy, and using bit rearrangement and idle bit insertion to reduce crosstalk coupling energy. Figure 7.11(a) shows a dynamic address compression scheme, while Fig. 7.11(b) shows the bus expander (BE) [68] compression scheme which is used in the approach. In the BE compression scheme, the sender has a compression cache that stores the higher order portion (tag or *T* field in Fig. 7.11(b)) of recently transmitted addresses. Due to the highly sequential nature of addresses, their lower order portions change much more frequently than their higher order portions. For this reason, only the higher order portion (*T* field) of the address is cached for compression and the index (*I* field) is used to search the compression cache in BE. On a hit in the compression cache, the sender transmits a control bit (*C* field) indicating a hit or a miss, index bits (*I* field) and way bits (*W* field) for the entry that was hit, and lower order bits (*U* field) of the original address. On a miss, the entire address is transmitted, along with the control bit. A narrow bus is sufficient to transmit a compressed address when a hit occurs, but in the case of a miss, the entire address is transmitted in multiple cycles. The receiver uses the control bit to determine if a compressed address has been received, in which case the *I* and *W* fields are used to select a tag *T* from its local copy of the compression cache. The degradation in performance due to compression, represented as percentage extra cycle penalty, is found to be less than 1%. This BE compression scheme is combined with bit rearrangement (similar to the A^2BC scheme) and idle bit insertion (extra bits added to ensure reduced crosstalk between adjacent wires) to reduce coupling power. Additionally, *XOR* operations (between current and previous values) are performed individually on the tag *T* for addresses in the case of a miss, and the *U* and *I* fields (for a miss or a hit), to reduce self and coupling power. The *W* field, which points to the way number of the tag that hit in a particular line of the compression cache, is also encoded using an LRU scheme. The LRU replacement policy, which is commonly used in caches, assigns the most recently accessed entry with the least value (zero) and successively older entries with higher values. The encoding of way bits using the LRU value is motivated by the fact that, in the compression cache, the most recently accessed entry is likely to be accessed again and in this case an LRU value of zero will be used to encode the way bits. This proposed scheme was shown to yield about 14.7% energy reduction on average for compressed address transmission on a narrow bus, compared to uncompressed address transmission on a bus of original width.

7.1.2.2 *Data Buses*

A large body of work in the area of encoding techniques has looked at reducing coupling power for data buses. Early work by Kim et al. [71] proposed an extension to the BI scheme, to account for the power consumption due to coupling capacitances between wires on a bus. The CBI scheme inverts the data to be transmitted, if the coupling effect of the inverted data is less than the original signals. Figure 7.12 shows a block diagram of the encoder–decoder architecture. The encoder consists of three components: a predictor, a CBI encoder, and a de-correlator. The predictor is a function of the past *K* input values, and is used by the CBI encoder to make a decision on whether to invert the bits or not. To keep hardware costs low, the value of *K* is chosen as 1. The CBI encoder considers different types of crosstalk when deciding to invert the bits, to save coupling power. The de-correlator employs a transition encoding scheme, to reduce transitions on the bus. The decoder is a mirror image of the encoder, and reverses these steps to recover the original word. To avoid using an additional signal to indicate inversion on the bus, the CBI scheme uses temporal redundancy, and uses an additional cycle to indicate if a transmitted word is inverted or not. The CBI scheme was shown to reduce power consumption about 30% with a one-cycle redundancy.

Another early work in the area of coupling-aware encoding by Sotiriadis and Chandrakasan [72] proposed the ***transition pattern coding scheme*** (TPC), which also tries to minimize inter-wire coupling transitions. The scheme adds additional bus lines to the word meant for transmission, and creates a transition matrix to select codeword patterns in which the neighboring bus lines change values in the

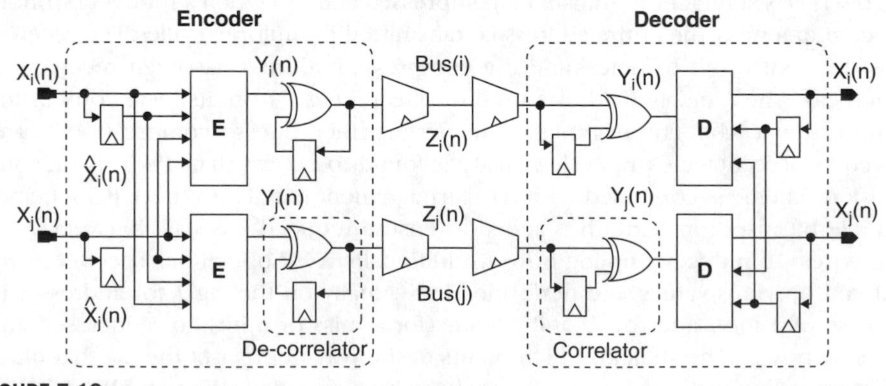

FIGURE 7.12

Low power CBI encoder–decoder framework [71]

same direction, to reduce effective coupling capacitance and inter-wire energy consumption. Since the proposed technique is practical only for smaller buses, the authors propose splitting up a larger bus into smaller groups and applying the scheme to each of the groups (partitioned TPC), for larger buses. Xie et al. [73] studied the bus partitioning schemes used in TPC, and used a genetic algorithm-based partitioning approach to obtain more energy savings than the random partitioning approach used in [72].

Another approach that considers coupling while extending the BI scheme is the *odd/even bus-invert code* (OE-BI), proposed by Zhang et al. [74]. The coding technique is based on the observation that coupling capacitances are charged and discharged by activity on the neighboring lines, where one line has an odd number and the other an even number (if bus lines are numbered in order). The coupling activity can be reduced by independently controlling the odd and even bus lines, with two additional lines—the odd invert and the even invert. The two invert lines support four possible cases—no bus lines are inverted (00), only odd lines are inverted (10), only even lines are inverted (01), or all lines are inverted (11). The scheme selects the case with the smallest coupling activity to transmit on the bus. Even after this encoding however, there can be toggling sequences 01→10 and 10→01 that result in 4 times more energy dissipation than other coupling events. A *targeted two-phase transfer method* (TPTM) is proposed to eliminate such toggling events, at the expense of extra delay. If two adjacent lines x and y have opposite transitions, TPTM delays x by half a cycle so that the transitions are not simultaneous. Experimental results showed that the OE-BI can reduce the coupling transitions by 36% compared to only 17% for the original BI. By applying TPTM to a simple OE-BI scheme that does not explicitly take coupling into account during inversion (i.e., similar to partial BI [34]), a 32% reduction in coupling transitions can be achieved with very little hardware overhead.

An *adaptive dictionary-based encoding scheme* (ADES) was proposed by Lv et al. [44], that considers inter-wire coupling. The dictionary technique identifies recurring patterns in the source data and stores them in a dictionary, similar to the FVE [27] scheme. These patterns can then be represented by a fewer number of bits. The approach assumes that there is not sufficient information about the data to be transmitted, in which case an adaptive approach becomes suitable. The original data word is divided into three parts: non-compressed part (typically the lower bits of the word, which were found to vary a lot), index part (used to lookup the dictionary), and the upper part (which is stored in the dictionary and identified by the index part). Whenever there is a hit, the index and the non-compressed parts are transmitted. The decoder uses the index to recreate the entire word from its dictionary. The dictionary is adaptively updated on a miss, while it remains unchanged on a hit. The ADES scheme was shown to improve upon the BI, working-zone and TPC techniques.

An encoding scheme that extends the work on bus wire reordering [65] (described earlier in Section 7.1.2.1) for reducing coupling power was proposed by Wong and Tsui [75]. While [65] was concerned with reordering bit lines to reduce coupling power for address buses, the proposed scheme is applicable to data buses with fewer correlations than address buses. The proposed approach is motivated by the observation that inverting bit lines before rearranging their order enhances the efficiency of the bit line reordering. A two-phase algorithm is proposed to generate the optimal flipping and reordering pattern. Phase 1 finds the optimum set of bit lines to be inverted, while Phase 2 rearranges the order of bit lines (similar to [65]) so that cross-coupling switching is minimized. For the static variant of the encoding scheme, the whole two-phase algorithm and encoding is performed at compilation time. For the dynamic variant of the encoding scheme, only the reordering phase is performed at compile time, to reduce the overhead of storing the flipping pattern. Unlike the static scheme for which only a single permutation is generated for a program, multiple permutations are generated for each program in the dynamic scheme. The decoding information for both the schemes is stored in the program header, and fetched/loaded into an LUT before execution. It was shown that the proposed dynamic scheme obtains a 15% improvement over [65]. A similar approach was proposed by Deng and Wong [76], which combines BI coding with bus wire reordering [66] to optimize the energy consumption due to self and inter-wire capacitance. Based on statistical information about the data stream, the bus lines are divided into groups and BI coding is applied selectively on the groups to reduce switching activity. Then the coupling energy is optimized by reordering the bus lines [66]. A fast simulated annealing algorithm that uses table-lookup was used to solve the net reordering problem. The hybrid scheme was shown to be more effective in reducing power than either BI or net-reordering.

An *application-specific encoding scheme* to minimize power consumption (self and coupling) on buses communicating program code between programs and memories was proposed by Petrov and Orailoglu [77]. The focus of the encoding scheme is on optimizing application hot-spots—heavily executed program segments that are tractable due to their limited size and quite independent of each other. Figure 7.13 shows the proposed encoding methodology. After the compilation phase, the application binary code is analyzed with particular emphasis on the major application loops (i.e., hot-spots). A cost-efficient, energy-aware encoding is identified and applied on the bit streams formed by each bit position of the particular instruction sequence, with the objective of minimizing the net effect of both the single and the coupled bit transition events. The transformed binary code is stored into the instruction memory, while the particular application-specific functional transformations identified by the proposed post-compile coding algorithm are used to create the decode circuitry on the processor end to restore the original bit sequence. Thus all of the required algorithmic support for identifying the optimal transformations and obtaining the power optimized instruction code is performed off-line when compiling and linking the program code. The only required hardware support is the decoding circuitry on the processor front-end. The absence of any encoding hardware on the critical memory side and of any bus modifications whatsoever gives the approach an advantage in terms of performance impact and hardware area overhead compared to the other low-power bus encoding techniques. The hardware support is also programmable and accessible via software, enabling optimal per-application program encoding for power reductions achieved in a completely reprogrammable way with no need for costly design iterations.

A *variable length coding compression-based scheme* was proposed by Muroyama et al. [78]. The scheme extends the variable length coding scheme

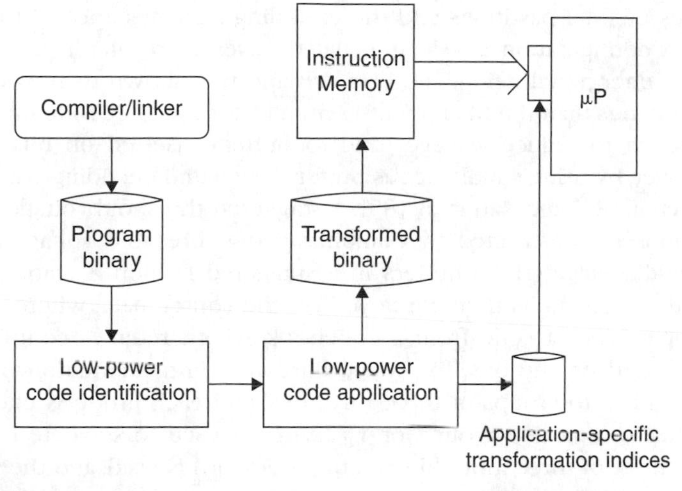

FIGURE 7.13

Application-specific encoding methodology [77]

approach (proposed by the same authors [79]) for reducing self-capacitance switching power by additionally considering coupling power. The variable length encoding compresses the N-bit data with variable length codes ranging from a minimum length of 1 to a maximum length of M ($M > N$), and transmits it over an M-bit bus. Probabilistic information about the source data is exploited during the assignment of codes, to assign smaller length codes to more frequently occurring data values. Typically though, the variable length encoding process can significantly increase the bit width, which may not be very practical.

Brahmbhatt et al. [80, 81] proposed an adaptive low power encoding algorithm based on *weighted code mapping* (WCM). The WCM transforms the n-bit original data into an m-bit low energy code, where $m = n + a$ (with a being the redundant bits). The code mapping is determined by the probability distribution of the source data stream. A window-based adaptive encoding algorithm is proposed to improve the energy saving by adaptively changing the code mapping for different data probability characteristics. The WCM algorithm is combined with the delayed bus algorithm [82] to achieve further energy reduction. The delayed bus algorithm delays all the 0-to-1 transitions on a bus by a certain amount of time to avoid opposite transitions on adjacent bus wires. The hybrid WCM, and the adaptive hybrid WCM schemes were shown to improve upon the BI, OE-BI, delayed bus algorithm and the forbidden pattern encoding schemes.

Jayaprakash and Mahapatra [83] proposed a *partitioned hybrid encoding* (PHE) technique in which the bus is partitioned optimally and the most energy-efficient encoding scheme is independently applied to each partition. The hybrid technique considers the BI, OEBI, and not encoding as the possible options. The partitioning and choice of applying a particular scheme to a partition is solved using dynamic programming, and requires data traffic statistics for tuning. Traffic from a set of SPEC2000 benchmarks is used to determine energy savings for all the combinations of valid partitions and the encoding schemes applied to the partitions. Then, a configuration is selected that on average results in maximum self and coupling energy reduction. The PHE scheme was shown to provide 1 to 13 times more savings than the BI and OEBI schemes for the SPEC2000 benchmarks.

An approach to reduce leakage (due to buffers inserted on bus lines) and crosstalk power, by using simultaneous buffer design and encoding was proposed by Deogun et al. [84] and Rao et al. [85]. A staggered threshold voltage (STV) buffer was proposed, constructed by combining low threshold voltage (LTV) and high threshold voltage (HTV) buffers in a staggered fashion. A static codebook-based encoding scheme is then employed on the source data, where the source data is mapped onto a code from a codebook which reduces coupling power when transmitted on the bus. To reduce hardware implementation overhead, a large bus is split into groups of three bus lines, and each group is encoded separately. A shield (dedicated ground or V_{dd} wire) is used to separate each group. Thus, for each set of three input bits, a 4-bit codeword is used, and the size of the codebook is $2^3 = 8$. A heuristic is used to analyze the memory trace of the application program and create state and transition probability tables (codebooks) for the groups of 3-bit data bits. The heuristic generates a mapping from the 3-bit input to the 4-bit codeword that minimizes power consumption. On average, the proposed method provided a saving of 26% in total power and an average leakage saving of 42% with a small increase ($<5\%$) in dynamic power, due to the additional bus lines. Another approach from Sridhara and Shanbhag [45] proposed joint buffer (repeater) insertion and crosstalk avoidance coding as a low-power alternative to simple buffer insertion for buses. A methodology is developed to calculate the repeater size and separation that minimizes the total power dissipation for joint repeater insertion and coding for a specific delay target. This methodology is combined with the forbidden transition overlapping code (FTOC) [86] and the one lambda code (OLC) [87], and shown to provide power savings for 130-nm, 90-nm, 65-nm, and 45-nm technology nodes.

7.2 TECHNIQUES FOR REDUCING CAPACITIVE CROSSTALK DELAY

With shrinking device sizes, increasing chip complexity and faster clock speeds, wire delay is becoming increasingly significant [88, 89]. The propagation delay through long cross-chip buses is proving to be a limiting factor in the speed of some designs, and this trend is only expected to get worse. It has been shown that the delay through a long bus is strongly dependent on the coupling capacitance between the wires. In particular, the crosstalk effect when adjacent wires simultaneously transition in opposite directions is particularly detrimental to the delay. When the cross-coupling capacitance is comparable to or exceeds the loading capacitance on the wires, the delay of such a transition may be twice or more than that of a wire transitioning next to a steady signal. This delay penalty is commonly referred to as the *capacitive crosstalk delay*. The capacitive crosstalk delay strongly depends on the transition activities of the adjacent signals, and hence the crosstalk type. Type-4 and type-3 crosstalk have the worst delay characteristics, followed by type-2 and then type-1. A few techniques involving selective skewing of bus data signals [90], transistor sizing [91], and repeater sizing [92] to reduce capacitive crosstalk induced delay have been proposed. Encoding is one of the more effective ways to reduce capacitive crosstalk delays. The techniques described in the previous section for reducing the inter-wire coupling capacitance do not particularly eliminate any type of worst case crosstalk, and will result in low power but will not necessarily reduce the delay penalty that limits the performance of high speed on-chip buses. Here we present encoding techniques that focus on reducing crosstalk delay.

Several pieces of work [93–98] have proposed models for delay on bus wires, for the case of a single wire and a pair of wires. But these approaches did not study the interaction patterns of multiple, capacitively coupled and independently driven lines. A crosstalk-aware delay model was proposed by Sotiriadis et al. [99], which extends the Elmore delay model [100] to account for a distributed model with distributed coupling components and an arbitrary number of lines driven by independent sources. It was shown that the impact of crosstalk on the delay is data-dependent and different combinations of transition directions of a specific wire and its adjacent wires induce different delays. As mentioned earlier, bus data patterns can be classified as type-4, type -3, type-2, and type-1 patterns, based on the maximum amount of crosstalk that they can exhibit. The effect of data patterns is taken into account, enabling an estimation of the delay on a sample-by-sample basis instead of making a worst case assumption. Using this delay model, the authors suggest using coding techniques to speed up communication by encoding the data being transmitted through the bus, with the goal of eliminating the type-4 and type-3 types of transitions that result in a large signal propagation delay.

Victor and Keutzer [101] introduced a *self-shielding code* (SSC) for wire delay reduction by eliminating type-4, type-3, and part of type-2 crosstalk. The scheme takes the original code whose data width is n, and encodes it using a code whose minimal bit width is m, such that $m < 2n$ and there is no worst case crosstalk switching in the encoded sequence. The approach extends the concept of inserting shield wires [102, 103] to minimize crosstalk, which can be considered to be a kind of encoding. In the simplest case, an additional shield wire would be needed for every bus wire to eliminate crosstalk. For instance, a data bit of "0" is encoded as a "00" signal on the wires, and a "1" is encoded as "10." The purpose of this encoding is to prevent adjacent wires from transitioning in opposite directions, and this particular encoding achieves that goal by forcing every other wire to a steady value. The proposed approach attempts to determine other encodings that can achieve the same goal with fewer wires. A fundamental theoretical limit on the performance of codes with and without memory is presented. Specifically, it is found that a 32-bit bus can be encoded with 40 wires using a code with memory or 46 wires with a memory-less code, in comparison to the 63 wires required with a simple shielding scheme.

Duan and Tirumala [63] proposed the *forbidden pattern coding* that reduces crosstalk delay by eliminating certain high energy patterns in the transmitted data (which account for type-4 and type-3 crosstalk) and by increasing the bus width. A forbidden pattern on an n-bit bus $b_1, b_2, b_3 \ldots b_{n-1}, b_n$ was defined as follows:

$$\begin{aligned} b_i &= v \\ b_{i+1} &= \overline{v} \\ b_{i+2} &= v \end{aligned}$$
$$\text{where } 1 \leq i \leq n-2, \text{ and } v \in \{0,1\}$$

It was shown that if these forbidden patterns are eliminated from the bus, type-4 and type-3 crosstalk delays can be eliminated. This can be achieved by adding m redundant bits to the n-bit bus to create codes of length $n+m$ that do not have any forbidden patterns. Mathematical bounds on the number of m additional bus lines needed for this purpose were also calculated. For ease of implementation, large buses were partitioned into groups, and this coding scheme was applied individually to each group. However, in such a case, it is possible for forbidden patterns to occur across groups. For instance, for a 16-bit bus partitioned into groups of four signals, and using 5 bits to encode each 4-bit word in a group, two adjacent groups can have the sequence {111**10**}{**1**1100}. This creates a forbidden pattern ("101") across group boundaries. To prevent this, the bits of the second code are inverted if the boundary bits are different with respect to the first code. So the same sequence would then become {11110}{00011}, which eliminates the forbidden pattern. An additional group complement bit needs to be transmitted for every group to enable the decoder to correctly decode the transmitted word. The entire set of group complement bits are transmitted in a separate group. Thus for a 16-bit bus, 26 bits are required to implement the scheme. A less aggressive crosstalk delay reduction variant of the scheme was also proposed, which only eliminates type-4 crosstalk, but requires fewer additional bits. This work was further extended by the authors in [104] where similar crosstalk free codes were proposed to eliminate type-2 and type-1 crosstalk and speedup signal propagation on buses. It was shown that using the type-2 crosstalk canceling techniques, the bus can be sped up by a factor of 6, with an area overhead of 200%. However, applying the same technique for type-1 crosstalk elimination was found to not be practically feasible.

Li et al. [105] proposed a *crosstalk aware interconnect* (called DYN) that uses a faster clock and dynamically controls the number of cycles required for transmission based on the estimated delay of the data pattern to be transmitted. For this purpose, a crosstalk analyzer circuit is incorporated into the sender side of the bus, which supports a variable cycle transmission mechanism, and is shown in Fig. 7.14 for a 32-bit bus. The crosstalk analyzer (called X-analyzer) first compares the previous data (stored in a register) and current data to be transmitted, with the help of pattern recognition (PR) components to determine whether a 3-bit transition pattern belongs to the type-4, type-3, or type-2 crosstalk class. OR trees are used to combine the signals indicating the patterns recognized for each of the bus wires into signals $G4$, $G5$, and $G6$. These signals indicate whether patterns in class type-2, type-3, or type-4 (respectively) exist in the whole 32-bit data. The signals $G4$, $G5$, and $G6$ are then used for variable cycle transmission, where the basic idea is to use multiple short clock cycles instead of the original long clock

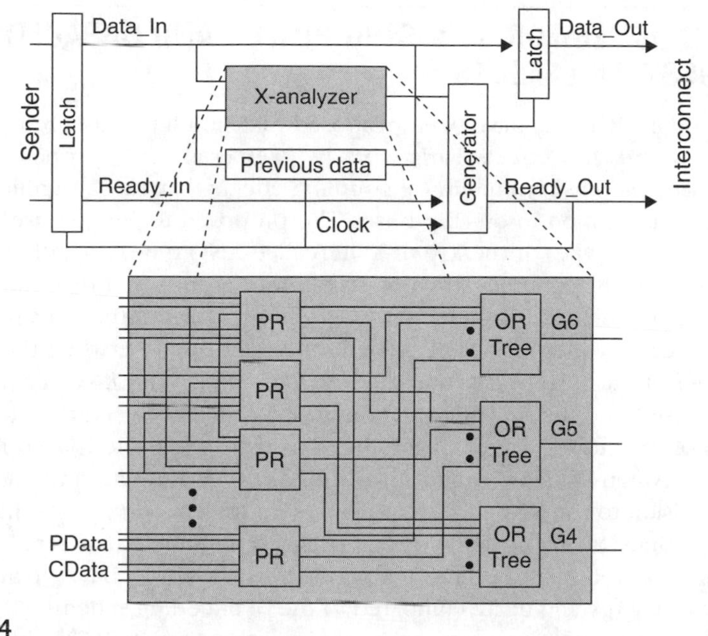

FIGURE 7.14

DYN encoder implementation [105]
© 2004 IEEE

cycle (designed for worst case type-4 crosstalk), and where the number of clock cycles is determined by the X-analyzer based on the crosstalk class. By controlling a ***Ready_Out*** signal, the cycles used to transmit data can be changed dynamically. For data patterns with crosstalk class type-2, type-3, and type-4, the ***Ready_Out*** signal is placed one, two, and three cycles after the data is placed on the bus, respectively. Otherwise, ***Ready_Out*** is placed on the bus in the same cycle as the data. Once the ***Ready_Out*** is placed on the bus, the next data to be transmitted is fed into the X-analyzer and the process repeats. Since the data is transmitted as it is without any coding in DYN, this technique does not provide any energy benefits. The DYN technique was shown to have much less overhead, compared to the SSC [101], a double spacing scheme (DBS) [106] that increases wire spacing to reduce coupling and a shielding method (SHD) [106] that inserts a V_{dd} or ***Ground*** wire (neither of which has any transitions) between every two wires to eliminate type-4 and type-3 coupling. In terms of performance, DYN provides a performance improvement of 31.5% on average, compared to 15.9%, 40.5%, and 46.3% for the SSC, DBS, and SHD schemes, respectively. The advantage of DYN over these techniques is its low area overhead of only 13% for a 5-mm long bus, compared to 70%, 49%, and 98% for the SSC, DBS and SHD schemes, respectively. To further improve the performance of the DYN approach, BI coding is used to transform many of the data transitions into those that have a lesser impact due to crosstalk, thereby enabling them to be transmitted in fewer cycles of the multi-cycle transmission. A 10.5% performance improvement was obtained by using the BI coding with the DYN scheme, compared to the original DYN scheme.

7.3 TECHNIQUES FOR REDUCING POWER AND CAPACITIVE CROSSTALK EFFECTS

Several encoding schemes have been proposed that simultaneously reduce power and capacitive crosstalk effects. Early work by Baek et al. [107] proposed the *low energy set scheme* (LESS) to minimize coupling effects that cause significant delay and power consumption in on-chip buses. The proposed technique uses an XOR-XNOR (called XON type) or the XNOR-XOR (called XNO type) operations to transmit data. In the XON technique, the bus is split into groups of 4 bits, and the XOR operation is performed between the most significant 2 bits of the current data and the previous data, while the XNOR operation is performed between the least significant 2 bits of the current data and the previous data. In the XNO technique, the order is reversed, and the XNOR operation is performed between the most significant 2 bits of the current data and the previous data, while the XOR operation is performed between the least significant 2 bits of the current data and the previous data. The decision to use XNO or XON depends on the encoding rules that analyze the characteristics of the bit sequences, in order to minimize the energy-delay and self-switching on the bus. The LESS technique was shown to have greater reduction in power, energy and delay, compared to the BI encoding scheme.

Lyuh and Kim [108] proposed an encoding scheme called *EN_shield-lp* that extends the SSC [101], to enable simultaneous power minimization and crosstalk delay reduction. The encoding algorithm takes a transition probability graph as an input, which describes the probability of transitions between all possible n-bit source words. Next, assuming a code length of m ($m > n$), a codeword graph is created, with edges between codes that do not result in type-4 crosstalk when there is a transition between them. Then the low power self-shield encoding problem becomes a problem of finding a clique of size 2^n from the codeword graph. Since this is an NP-complete problem, a greedy heuristic is used to assign a code that minimizes transition cost (which factors in self/coupling capacitance and probability of transition for the mapped source words) in the codeword graph. The proposed scheme was shown to consume less power than the original data as well as some other schemes such as BI and a naive shielding scheme [106] while eliminating worst case crosstalk delay.

Subrahmanya et al. [109] proposed the memory-less *no adjacent transition coding scheme* (NAT)—another bus encoding technique that simultaneously reduces power consumption and eliminates worst case crosstalk. The NAT coding scheme uses a combination of transition signaling [10] and LWC schemes [10, 11]. The proposed scheme is a variant of the m-limited LWC technique that limits the number of ones in every transmitted codeword to a constant m and also avoids all binary codewords with adjacent ones. This approach, in conjunction with transition signaling (for which the number of transitions over a period of time is equal to the number of 1s present in the input to be transmitted) ensures that the number of transitions per transmission is bounded by a constant and at the same time eliminates crosstalk by avoiding transitions on adjacent lines. A block diagram of the proposed (n, b, t)-NAT code is shown in Fig. 7.15, where b refers to the number of bits in the input to be encoded, n is the number of bits in the encoded

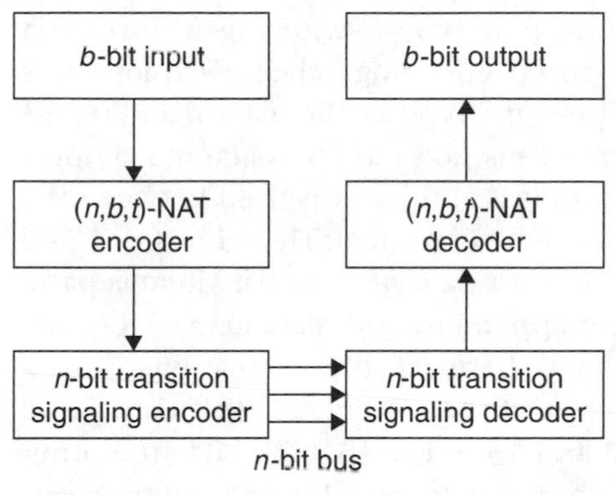

FIGURE 7.15

(*n, b, t*)-NAT encoder/decoder scheme [109]
© 2004 IEEE

Input	NAT code	Input	NAT code
0000	000000	0100	001000
0001	000001	0101	010000
0010	000010	0110	100000
0011	000100	0111	000101
1000	001001	1100	010100
1001	001010	1101	100001
1010	010001	1110	100010
1011	010010	1111	100100

FIGURE 7.16

(6, 4, 2)-NAT code [109]
© 2004 IEEE

output and t decides the maximum number of 1s allowed in the output. Here the encoded n-bit output can have less than or equal to t number of ones, with no two adjacent 1s in it. An example of a (6, 4, 2)-NAT code is shown in Fig. 7.16. While the NAT scheme does not reduce power by as much as the LWC scheme, it does remove worst case crosstalk completely, unlike the LWC scheme.

Overlapping coding was proposed by Sridhara et al. [86], which is a type of partial coding technique that divides the bus (channel) into subchannels. In the proposed overlapping code, two adjacent subchannels are overlapped at their boundary. If n and k are the number of code bits and data-bits in the subchannel, respectively, then k data bits are mapped to the central n-2 bits of the codeword, and the boundary bits of the data word from the boundary bits of the codeword. This coding technique avoids crosstalk delay by using forbidden pattern overlapping codes (derived from [63]) to prevent overlapping from causing crosstalk delay at the boundary bits, and by using FTOCs (derived from [101]) to ensure that a mapping with unchanged boundary bits exists from data words to codewords. The overlapping encoding scheme reduces delay and energy on the bus, and also takes up less area since it does not use any shield wires. Specifically, it was shown that for a 1-cm 32-bit bus in 0.13-μm complementary metal-oxide semiconductor (CMOS) technology, a 48-wire solution is obtained that has 1.98 × speedup, 10% energy savings, and requires 20% less area than shielding.

Khan et al. [110] proposed an encoding scheme that transforms the incoming data in such a way as to eliminate two of the worst crosstalk types (type-4 and type-2). Unlike existing schemes for simultaneous power and crosstalk minimization, the proposed scheme does not require probabilistic information of the data stream, and can be applied to data whose statistical properties are not known beforehand, making it suitable for generic SoC systems. The encoding scheme is based on an intrinsic property exhibited by 4-bit binary sequence which is explained as follows: Consider a 4-bit bus that represents a maximum of sixteen 4-bit binary sequences (4-tuples). If any one of the 4-tuples is taken, modulo-2 summed with two basis functions Z_1 (0101) and Z_2 (1010) (alternate bit complement) and compared with the remaining 4-tuples, it is observed that one of the two XOR'd data (either data

XOR'd with Z_1 or data XOR'd with Z_2) will have no type-4 switching with respect to the remaining fifteen 4-tuples. The proposed encoding scheme exploits this property for low power, worst crosstalk (type-4 and type-3) elimination and type 3 crosstalk minimization. The encoding scheme thus takes a 4-bit data and outputs a 4-bit code, along with an extra bit for decoding. Another bit is used between the decode bit and the code to reduce crosstalk between them. Thus $4 + 1 + 1 = 6$ bits are required to encode a 4-bit data. For an 8-bit data size, the data item is partitioned into two 4-bit data and the procedure applied to both data items; 3 decode bits are used (instead of 2) to reduce crosstalk between the decode bits. Thus a total of $4 + 1 + 4 + 1 + 3 = 13$ bits are output for encoding an 8-bit data. For a 16-bit bus, the same procedure is repeated and $13 + 1 + 13 = 27$ bits are output by the scheme. The encoding scheme was shown to reduce bus energy, with savings ranging from 24% for highly de-correlated uniformly distributed data, to 35% for highly correlated data application-specific data (such as that produced in image processing applications), for the 0.18-μm CMOS technology.

Mutyam et al. [111] improved upon the variable cycle transmission technique from [105] that was aimed at minimizing crosstalk delay by applying temporal redundancy (VCTR) to eliminate type-4 and type-3 crosstalk, and reduce delay even further, as well as achieve energy savings. The proposed scheme reduces the cost of the hardware crosstalk analyzer at the encoder, and when a type-4 or type-3 crosstalk is encountered, then instead of delaying the transmission for the required number of cycles (as proposed in [105]), the data is encoded into two words. Each of the words is transmitted with a smaller delay, so that the overall delay is less than in [105]. An additional *temporal_redundancy* signal is needed to signal this special case to the decoder. It was shown for the SPEC2000 INT benchmarks that the technique achieved delay improvement of about 30% and delay improvement of around 9–13% for address and data buses. Although the VCTR scheme does not reduce delay as much as the SSC [101] and the shielding scheme [106], it has lower area and energy overhead. VCTR also achieves lower delay and energy consumption than the original variable cycle transmission technique from [105].

7.4 TECHNIQUES FOR REDUCING INDUCTIVE CROSSTALK EFFECTS

In addition to capacitive coupling, inductive coupling is becoming increasingly significant for on-chip interconnects, especially for buses in DSM operating at high frequencies [112]. Inductive crosstalk effects can have a major impact on communication performance by increasing wire delay [113, 114], similar to capacitive crosstalk. Long-range inductive crosstalk can also cause serious signal integrity related problems [114, 115]. Signal overshoots and undershoots due to wire inductance may damage devices. Finally, inductance in power and ground grids can increase the noise in the supply and ground voltages when large currents flow. This is also known as the *ground-bounce problem*. Inductance effects therefore cannot be neglected in today's high performance circuit designs, especially for global interconnects such as clock wires and signal buses.

Consider the case of two adjacent bus lines as shown in Fig. 7.17, with current flowing down one of the lines ((a) on aggressor line). This current will couple into the adjacent line (victim line) and create two different noise signals. One of the noise signals will flow in the victim line in the same (forward) direction as the aggressor current and is referred to as *forward crosstalk*. The other noise signal will flow in the victim line in the opposite (backward) direction and is known as *backward crosstalk*. Inductive coupling causes backward crosstalk ((b) in Fig. 7.17) and capacitive coupling causes both types of crosstalk ((c) in Fig. 7.17). In circuits operating at higher frequencies (near GHz ranges), inductive coupling is expected to dominate over capacitive coupling.

There are some other important differences between capacitive and inductive effects that need to be considered [117]. The worst case switching pattern that incurs the largest delay for an *RC* circuit model (commonly used by encoding techniques for reducing capacitive crosstalk) is when adjacent wires simultaneously switch in opposite transition directions. However, considering the *RLC* circuit model for the bus structure, shown in Fig. 7.18, it is found that the worst case switching pattern with the largest on-chip bus delay is when all wires simultaneously switch in the same direction. Thus this worst case pattern is the best case pattern of a coupling *RC* model. Further, the best case pattern with the *RLC* model is that the central wire of the bus switches in a different direction from all other wires that all switch in the same direction. This best case pattern in the *RLC* model is thus actually the worst case pattern with the *RC* model! There has been some work on encoding techniques for reducing inductive effects for off-chip buses [118, 119]. Here we present some encoding schemes that attempt to reduce inductive crosstalk effects for on-chip buses.

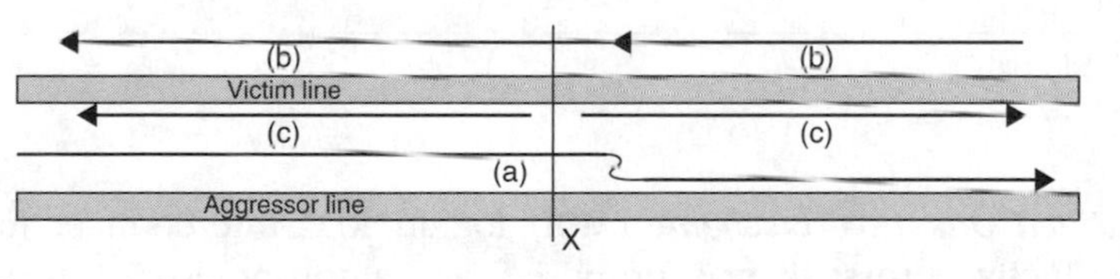

FIGURE 7.17

Forward and backward crosstalk [116]
© 2007 ACM Press

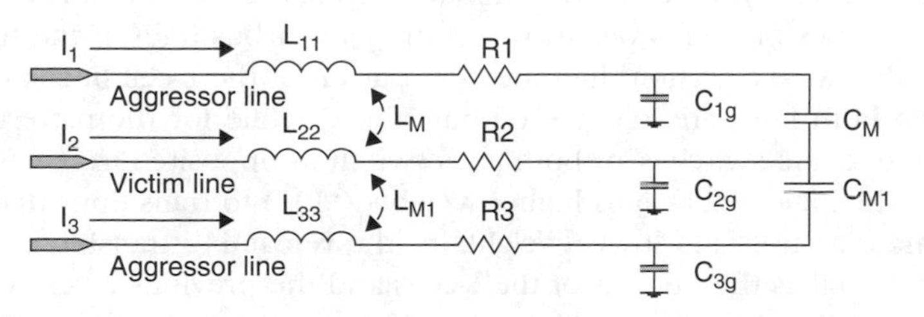

FIGURE 7.18

RLC circuit model for three-wire bus structure [116]
© 2007 ACM Press

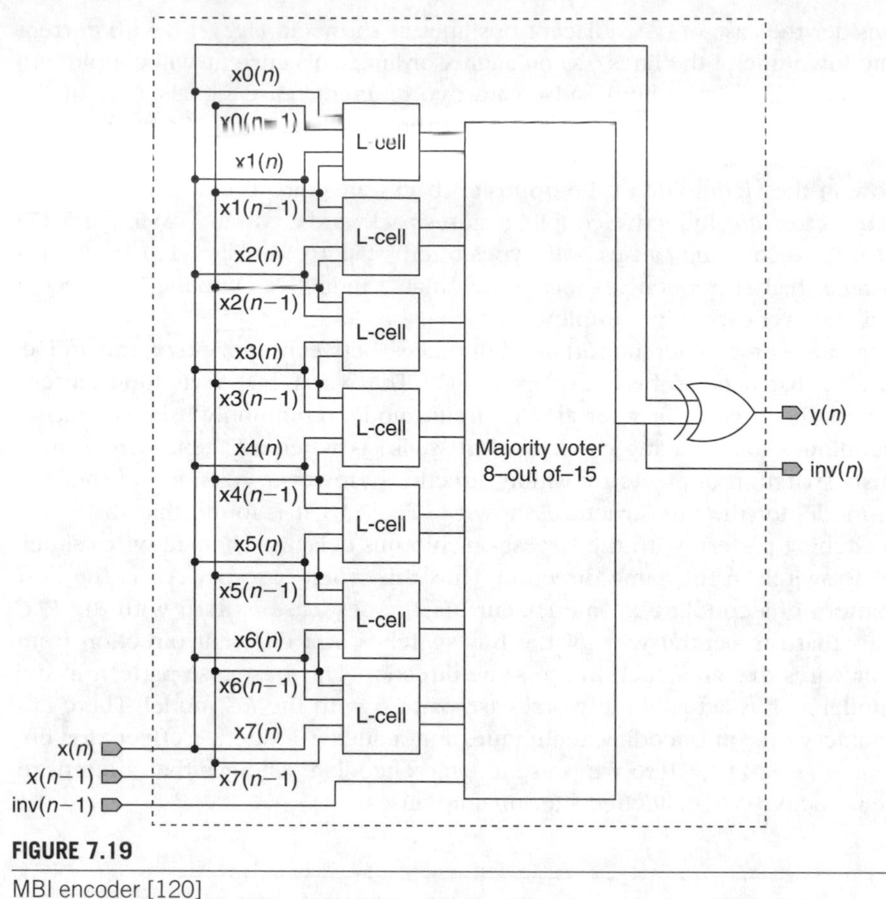

FIGURE 7.19

MBI encoder [120]
© 2004 IEEE

A ***modified bus invert scheme*** (MBI) for an RLC interconnect model that reduces inductive crosstalk was proposed by Lampropoulos et al. [120]. The scheme selectively inverts data patterns such that the number of transitions in the same direction is minimized. In the encoder, shown in Fig. 7.19, the bus lines are partitioned into pairs and each pair of adjacent lines as well as their values from the previous clock cycle drives the inputs of a logic cell (L-cell). The logic cell encodes the types of events occurring on the pair of bus lines. If the transitions happen in the same direction then the 2-bit output of the L-cell becomes "11." In cases when both lines are idle, the output is "00," while for the patterns where either only one line switches, or both lines switch in opposite directions the output is "01." The logic cell assigns higher weights ("11") to transitions that happen in the same direction and lower weights to the remaining transition types. The majority voter takes the outputs of the L-cells and the previous invert signal, and sets the invert signal to 0 when the count of 1s on the majority voter inputs is less than n and to 1 otherwise. It was shown for an 8-bit bus that the MBI scheme provides the most reduction in crosstalk noise, when compared to the BI and CBI

[71] schemes. Tu et al. [117] proposed a similar scheme utilizing the BI method to reduce the on-chip bus delay due to coupling effects while inductance effects dominate. The proposed BI-based method inverts the input data when the number of bits switching in the same direction is more than half of the number of signal bits. Simulation results indicated that the proposed encoding method can significantly reduce the worst coupling delay of a bus. The encoding scheme is recommended for cases when buses or parallel signal wires are about thousands of micrometers long and work above GHz frequencies.

Raghunandan et al. [116] proposed an encoding scheme that uses selective bit inversion and shield bit insertion to reduce inductive crosstalk delay and simultaneous switching noise (SSN). In the encoding scheme, input data is split into chunks of 4 bits, and encoded as two 3-data tuples that are transmitted in two successive cycles. For every 4-bit chunk, an SSN check is performed by comparing the data to be transmitted with the data already on the bus. The SSN check is used to determine if two or more simultaneous transitions are occurring for a set of 3-bit transitions ($\uparrow\uparrow-$, $\downarrow\downarrow-$, $-\uparrow\uparrow$, $-\downarrow\downarrow$, $\downarrow\downarrow\uparrow$, $\uparrow\uparrow\downarrow$, $\downarrow\uparrow\uparrow$, $\uparrow\downarrow\downarrow$, $\uparrow\uparrow\uparrow$, $\downarrow\downarrow\downarrow$). Here the symbols $\uparrow$, $\downarrow$, $-$ are used to indicate $0\rightarrow1$, $1\rightarrow0$, and $1\rightarrow1$ (or $0\rightarrow0$) transitions, respectively. If the SSN checker does not detect two or more simultaneous transitions, then the first three bits of the 4-bit chunk are sent in the first cycle, and in the second cycle the fourth bit, a shield bit ($-$, no transition) and a coding bit with a value of 0 are transmitted. If however the SSN checker does detect two or more simultaneous transitions, it inverts the first three bits and sends them in the first cycle, and a high impedance state (Z) will be sent as the coding bit value. The high impedance state on the bit (realized using a tri-state buffer) indicates inverted bits to the decoder. Using a high impedance state instead of a normal 1 for the coding bit will essentially not allow any current to flow, bringing down both power and inductive coupling noise (contribution of that single bit) on the bus lines. Simulation experiments were carried out to find out the delay and SSN reduction for interconnect lines of different lengths (2, 5, and 10 mm) at various technology nodes (180, 130, 90, and 65 nm). Results obtained showed that the proposed bus encoding scheme provided a delay reduction of about 54–73% with respect to the worst case delay. For SSN, the encoding scheme was tested with various SPEC'95 benchmarks and it was found that SSN is reduced by about 33% on an average compared with the un-encoded data. Finally, energy minimization of about 13% on an average is achieved by the application of proposed scheme for the SPEC'95 benchmarks.

7.5 TECHNIQUES FOR FAULT TOLERANCE AND RELIABILITY

With the trend of shrinking technology nodes, supply voltage scaling, increasing interconnect density, and greater integration levels, designs are exposed to a large number of noise mechanisms such as capacitive and inductive crosstalk, power supply noise, leakage noise, process variations, charge sharing, soft errors, etc. [121–124]. In this context, the on-chip interconnect fabric plays an important role in ensuring reliable system operation for future ICs [88, 125]. Fault tolerance of computing systems is usually ensured by incorporating redundancy (i.e., additional resources) so that computation can be correctly carried out even in presence of faults [126, 127]. For fault tolerant data communication *error detecting* and *error correcting codes* are used to append check bits to the data bits in order to enable detection and correction of erroneous data bits that have been corrupted during transmission. Error detecting codes are widely deployed for the implementation of self-checking circuits (SCCs), because of lower cost implementation, and also because they provide flexibility for error recovery to be carried out either in hardware or in software [128]. Error correcting codes comparatively have a larger overhead associated with them, in terms of performance penalty and the additional circuitry required for implementation. Two frequently used codes in SCC are the parity check and two rail codes. The parity check code adds a single parity bit to the transmitted data bits and can detect all error patterns of an odd number of bits, but is not able to detect double errors that are relevant in a crosstalk-dominated scenario. The two rail code represents a signal as a pair of two complementary codes, thus doubling the number of bus lines. However, this large overhead may not be acceptable in spite of the high error detection efficiency of the code [129]. Other codes such as m-out-of-n and Berger codes [130] have been found to be very useful in detecting unidirectional errors (i.e., $0 \rightarrow 1$ or $1 \rightarrow 0$ errors) that frequently occur in very large scale integrated (VLSI) circuits. The Berger code is in fact an optimally separable all-unidirectional error detecting code—no other separable code is able to detect all unidirectional errors with fewer number of check bits.

Unidirectional errors are, however, unable to describe the effects of noise sources in DSM technologies, where multiple bidirectional errors (for instance, due to crosstalk causing two coupled lines to switch in the opposite direction and both transitions being delayed, inducing sampling errors) need to be detected. Many solutions have been proposed to overcome the limitation of traditional unidirectional error control schemes. These include schemes for changing layout (i.e., wire spacing, shielding, line crossing, keeping two complementary bits as far apart in a two rail code, etc.), using layout information to create weight-based codes (e.g., extensions of Berger and m-out-of-n codes [131, 132]) and unbalancing bus line drivers to increase probability of single errors with respect to bidirectional ones. However, these techniques require low level layout or electrical information, which is a drawback. A more general approach such as redundant encoding is much more desirable for this purpose, since it can be applied earlier in the design flow, independent of the technology and the final layout knowledge.

Block coding is a class of redundant encoding that first segments a binary sequence into message blocks consisting of k bits, and then transforms each block into a codeword of length n bits ($n > k$) according to some rules. A binary block code is linear if and only if the modulo-2 sum of two codewords is also a codeword [133]. An example of a widely deployed class of linear codes is the Hamming code, which is a popular choice for error control in digital communications and data storage. The Hamming code, which is exactly a distance-3 code (i.e., the smallest distance between two distinct codewords is at least three bits) is essentially an error correcting (SEC) code. A distance-3 Hamming code can be modified to increase its minimum distance to 4, adding one more check bit, chosen so that the parity of all of the bits, including the new one, is even [134]. This version of the Hamming code, that features seven check bits instead of the six in SEC, is traditionally used for single error correction and double error detection (SECDED). It is also possible to use Hamming code for error detection (ED) purposes only, which can detect not only all single and double errors, but also a large amount of multiple errors. In general, if the minimum Hamming distance between any two codewords is d, then it is possible to either correct up to $\lfloor (d-1)/2 \rfloor$ errors or detect up to $d-1$ errors [135]. Hamming codes are promising for on-chip bus encoding because of their implementation flexibility, low complexity and multiple bidirectional error detecting capability.

Cyclic codes are another class of linear codes with the property that any codeword shifted cyclically (an end-around carry) will also result in a codeword. Cyclic redundancy check (CRC) codes are the most widely used cyclic codes in computer networks, and are suitable candidates for on-chip encoding as well. While a Hamming code can detect multiple errors arbitrarily located all over the codeword, a CRC code is particularly suitable for dealing with errors affecting lines that are close to each other (within the detectable burst length).

The problem of reliable data transmission on buses is closely tied to the issue of low power design—the practice of reducing voltage supply to reduce transmission power, low swing signaling [136], or the inevitable voltage scaling due to technology node scaling also leads to greater susceptibility to noise sources because of decreased noise margins. Consequently designers must trade off power consumption with reliability (or fault tolerance). Several encoding techniques for on-chip transfers have been proposed that combine low power operation with error detection/correction codes (EDC/ECC).

Hegde and Shanbhag [137] proposed using linear codes such as Hamming codes [133] and Reed-Muller (RM) codes [138] for noise tolerant coding on buses in the presence of DSM noise. The goal was to meet predefined requirements on communication reliability, and to minimize energy consumption. It was found that RM codes lead to about a 4× reduction in energy dissipation at the target performance of bit error rate (BER) = 10^{-14}. The better performance of RM codes over Hamming codes was due to the better error detection capability of RM codes. The authors claimed that more sophisticated encoding schemes could lead to even more energy savings.

Bertozzi et al. [125] investigated the efficiency of several Hamming code implementations—such as single error correction Hamming code (SEC), SEC and double error detection Hamming code (SECDED), and error detection-only Hamming code (ED), and compared them to a single parity bit code (PAR). For a 32-bit bus, SEC and ED require 6 additional bits, SECDED requires 7 bits, while PAR only requires 1 additional bit. The error correcting schemes (SEC, SECDED) were found to have the largest delays at the decoder due to the complexity of error detection, whereas PAR had the lowest delay due to its simple decoder. For long interconnects with large capacitive loads, the energy cost of bus transitions is dominant over encoder/decoder energy, and it was found that retransmission-based schemes (ED, SECDED, PAR) consume lower energy per useful bit transferred than error correction schemes (SEC). For shorter wires with much smaller load capacitance, the bus transitions are less important and the impact of encoder/decoder energy becomes more dominant. In this case, ED and PAR consume the lowest energy per useful bit transferred, while SEC and SECDED consume much larger energy because of their error correction circuitry. Thus it was concluded that retransmission of erroneous words (error detection with retransmission) was the most efficient error recovery technique from an energy viewpoint, whereas error correction was found to suffer from much larger hardware overheads. This work was extended by the authors [139], by considering low swing buses where the bus drivers are powered at a lower power supply while the encoder and decoder are powered at standard voltage levels, and by considering an additional class of codes: CRC, which exhibit a burst error detection capability. It was again shown that retransmission-based error detection strategies (ED, PAR, SECDEC, CRC) consume less power than using correction. In particular, CRC burst error detecting codes were found to be more energy efficient than random detection oriented schemes such as Hamming codes for the case of short wires, while the opposite holds for longer wires.

Lin et al. [140] proposed an *adaptive error protection scheme* for energy efficiency, where the type of the coding scheme is modulated dynamically. Three error detection schemes are considered—parity (PAR), double error detection (DED) Hamming code, and triple error detection (TED) Hamming code. The idea behind this strategy is to monitor the dynamic variations in noise behavior and use the least powerful (and hence the most energy efficient) error protection scheme required to maintain the error rates below a pre-set threshold. The simplest error detection coding method (PAR), while being most energy-efficient, will lead to high number of undetectable errors. At the other extreme, adopting the worst case approach and employing the most powerful scheme (TED) catches most of the errors but incurs unnecessary energy consumption most of the time due to its conservative nature. The proposed scheme monitors the change in noise behavior and switches to the least powerful error detection scheme that can maintain the undetected error rates below specified levels. This allows a minimization of the energy consumption while maintaining the required protection levels. The main area overhead of the adaptive method is the size of the encoder and decoder, due to the need for implementing the PAR, DED, and TED functionality.

Patel and Markov [141] introduced *boundary-shift codes* for error correction and crosstalk avoidance through self-shielding [101]. Data words are mapped to codewords which are self-shielding [101] and consequently have no invalid transitions (i.e., transitions from one codeword to the other that cause adjacent bits to switch in opposite directions). This allows crosstalk to be minimized. Additionally, the codewords are selected so that they have a large enough Hamming distance between them in order to differentiate codewords reliably even in the presence of errors (bit flips). The authors define a *dependent boundary* in a word as a place where two adjacent bits differ and denote it by the position of the leftmost bit of the boundary. Two words sharing no dependent boundaries cannot form an invalid transition. For instance, consider the two words

$$c_1 \rightarrow 01100111$$
$$c_2 \rightarrow 11001110$$

Here, c_1 and c_2 have dependant boundaries {1, 3, 5} and {2, 4, 7}, respectively. Since there is no overlap, this transition is valid. If a codebook has codewords with only even dependent boundaries, then performing a 1-bit circular right shift yields a new codebook with no even dependent boundaries. Since the two codebooks share no dependent boundaries, one can alternate between the two to obtain an SSC. This is called a ***boundary-shift code***. Compared to SSCs, these codes are not only able to prevent crosstalk but also tolerate errors. The additional protection comes at the cost of rate (performance) reduction. Compared to shielding, the proposed technique provides error-correction in addition to self-shielding without appreciably reducing the rate, though some additional encoding and decoding logic is required.

Sridhara and Shanbhag [87] proposed several joint codes by combining different error correcting codes such as the Hamming code (HC), the duplicate-add-parity code (DAP) and the DAP with shielding (DSAP), with different crosstalk avoidance codes such as the forbidden transition code (FTC) [101], one lambda code (OLC, a code in which two adjacent bit boundaries in the codebook cannot both be of 01-type or 10-type.) and the forbidden overlap code (FOC, a code in which the codebook cannot have both 010 and 101 appearing centered around any bit position). The benefits of joint crosstalk avoidance and error correction are quantified in terms of reduction in delay and energy dissipation. The achieved improvements vary with bus length, ratio of coupling capacitance to bulk capacitance, bus width, and the process technology. For a bus length of 10 mm and a 32-bit bus in standard 130-nm CMOS technology, the FOC+HC joint code was found to be a *zero-latency* error correction code, in the sense that the reduction in bus delay completely masks coding latency for global buses. Compared to other codes, DAP and DSAP reduce maximum coupling to achieve significantly more speedup and also provide high energy savings as they achieve reduced swing with low codec power overhead. Sridhara and Shanbhag [142] extended this work to propose a more general two stage encoding framework that first passes the data to be transmitted through a non-linear source coder that reduces self and coupling transition activity and imposes a constraint on the peak coupling transitions on the bus, and then uses a linear error control coder to add redundancy, for error detection and correction. Additional joint codes were considered, such as a combination of BI and Hamming code (BIH), and the combination of BI with DAP (DAPBI). DAP-based codes were found to provide the best trade-off between bus delay, codec latency, power, area, and reliability.

Komatsu and Fujita [143] proposed a combined ECC/EDC and low power code. The proposed scheme uses parity code (PAR), single error correction (SEC) code, and the single error correction and double correction code (SECDEC) code for reliable data transfers. If an error is detected at the receiver end and cannot be corrected, retransmission of data occurs. A modified BI code, called BI* is proposed for low power transmission. In case the number of 1s in a word is greater than half the bits in the word, the word is inverted and XOR'd with the previous transmitted value on the bus, otherwise if the number of 1s is less than half the bits in the word, the word is not inverted, and simply XOR'd with the previous value on the bus. For wider buses, a partitioned BI* (PBI*) scheme is proposed which acts on different segments of the bus, to improve signal transition reduction capability of the approach. Different combinations of an error correction/detection scheme, and a low power encoding scheme were evaluated. The combination the PBI* and parity (PAR) codes was found to reduce both self and coupling transitions better than other combinations.

7.6 SUMMARY

In this chapter, we described encoding techniques for on-chip communication architectures that are used to transform the data to be transmitted. Encoding is a commonly used and beneficial technique which can lead to reduction in switching power, propagation delay, crosstalk noise (both inductive and capacitive), and transmission errors. While several encoding techniques have been proposed that individually aim for low power, low latency and high reliability data transfers, it is becoming increasingly important for schemes to jointly optimize for several design goals, and provide trade-offs to the designer between power, performance, and reliability, to ensure practical usefulness. With the advent of ultra-DSM (UDSM) CMOS technology, better encoding schemes will be needed, that can trade off signal reliability (i.e., provide resilience against crosstalk noise, and other sources of error) with low power (average, as well as peak power) and improved performance, without incurring a significant overhead in terms of power, area, and latency of the encoder/decoder circuitry.

REFERENCES

[1] W.-C. Cheng and M. Pedram, "Memory bus encoding for low power: A tutorial," in *Proceedings of International Symposium on Quality Electronic Design (ISQED)*, 2001, pp. 199–204.

[2] C.-L. Su, C.-Y. Tsui and A. M. Despain, "Saving power in the control path of embedded processors," *IEEE Design and Test of Computers*, Vol. 11, No. 4, 1994, pp. 24–30.

[3] H. Mehta, R. M. Owens and M. J. Irwin, "Some issues in gray code addressing," in *Proceedings of IEEE/ACM Great Lakes Symposium on VLSI (GLS-VLSI)*, Ames, IA, March 1996, pp. 178–180.

[4] L. Benini, G. De Micheli, E. Macii, D. Sciuto and C. Silvano, "Asymptotic zero-transition activity encoding for address busses in low-power microprocessor-based systems," in *Proceedings of IEEE/ACM Great Lakes Symposium on VLSI (GLS-VLSI)*, Urbana, IL, March 1997, pp. 77–82.

[5] Y. Aghaghiri, F. Fallah and M. Pedram, "Irredundant address bus encoding for low-power," in *Proceedings of IEEE International Symposium Low-Power Electronics and Design (ISLPED)*, August 2001, pp. 182–187.

[6] W. Fornaciari, M. Polentarutti, D. Sciuto and C. Silvano, "Power optimization of system-level address buses based on software profiling," in *Proceedings of International Conference on Hardware/Software Codesign (CODES)* 2000, pp. 29–33.

[7] H. Parandeh-Afshar, A. Afzali-Kusha and A. Khakifirooz, "A very high performance address BUS encoder," in *Proceedings of ISCAS*, 2006.

[8] Y. Aghaghiri, F. Fallah and M. Pedram, "ALBORZ: Address level bus power optimization," in *Proceedings of International Symposium of Quality Electronic Design (ISQED)*, 2002, pp. 470–475.

[9] M. Ikeda and K. Asada, "Bus Data Coding with Zero Suppression for Low Power Chip Interfaces," in *Proceedings of International Workshop on Logic and Architecture Synthesis*, December 1996, pp. 267–274.

[10] M. R. Stan and W. P. Burleson, "Limited-weight codes for low-power I/O," in *Proceedings of International Workshop on Low Power Design*, Napa, CA, April. 1994, pp. 209–214.

[11] M. R. Stan and W. P. Burleson, "Low-power encodings for global communication in CMOS VLSI," *Transactions on Very Large Scale Integration (TVLSI)*, Vol. 5, No. 4, December 1997, pp. 444–455.

[12] S. Komatsu and M. Fujita, "Irredundant address bus encoding techniques based on adaptive codebooks for low power," in *Proceedings of the Conference on Asia South Pacific Design Automation (ASPDAC)*, 2003, pp. 9–14.

[13] L. Benini, G. De Micheli, E. Macii, M. Poncino and S. Quer, "System-level power optimization of special purpose applications: The beach solution," in *Proceedings of ACM/IEEE International Symposium Low Power Electronics and Design (ISLPED)*, Monterey, CA, August 1997, pp. 24–29.

[14] G. Ascia, V. Catania, M. Palesi and A. Parlato, "Switching activity reduction in embedded systems: a genetic bus encoding approach," *IEE Proceedings on Computers and Digital Techniques*, Vol. 152, No. 6, 4 November 2005, pp. 756–764.

[15] E. Musoll, T. Lang and J. Cortadella, "Exploiting the locality of memory references to reduce the address bus energy," in *Proceedings of ACM/IEEE International Symposium Low Power Electronics and Design (ISLPED)*, Monterey, CA, August 1997, pp. 202–207.

[16] W.-C. Cheng and M. Pedram, "Power-optimal encoding for DRAM address bus," in *Proceedings of ACM/IEEE International Symposium on Low Power Electronics Design (ISLPED)*, 2000, pp. 250–252.

[17] W.-C. Cheng and M. Pedram, "Low power techniques for address encoding and memory allocation," in *Proceedings of the Conference on Asia South Pacific Design Automation (ASPDAC)*, 2001, pp. 245–250.

[18] D. E. Knuth, Fundamental Algorithms, "The Art of Computer Programming" Vol. 1, Addison-Wesley, 1973.

[19] M. Mamidipaka, D. Hirschberg and N. Dutt, "Low power address encoding using self-organizing lists," in *Proceedings of International Symposium Low-Power Electronics and Design (ISLPED)*, 2001, pp. 188–193.

[20] M. Mamidipaka, N. Dutt and D. Hirschberg, "Efficient power reduction techniques for time multiplexed address buses," in *Proceedings of the 15th International Symposium on System Synthesis (ISSS)*, 2002, pp. 207–212.

[21] M. Mamidipaka, N. Dutt and D. Hirschberg, "Adaptive low-power address encoding techniques using self-organizing lists," *IEEE Transactions on Very Large Scale Integration Systems (TVLSI)*, Vol. 11, No. 5, October 2003, pp. 827–834.

[22] S. Ramprasad, N. R. Shanbhag and I. N. Hajj, "A coding framework for low-power address and data busses," *IEEE Transactions on Very Large Scale Integration Systems (TVLSI)*, Vol. 7, No. 2, June 1999, pp. 212–221.

[23] T. Givargis and D. Eppstein, "Reference caching using unit distance redundant codes for activity reduction on address buses," in *Proceedings of International Workshop on Embedded System Hardware/Software Codesign (ESCODES)*, San Jose, September 2002, pp. 43–48.

[24] A. Malik, B. Moyer and D. Cermak, A lower power unified cache architecture providing power and performance flexibility, in *Proceedings of International Symposium on Low Power Electronics and Design (ISLPED)*, June 2000, pp. 241–243.

[25] A. Park and M. Farrens, "Address compression through base register caching," in *Proceedings of the Annual ACM/IEEE International Symposium on Microarchitecture*, November 1990, pp. 193–199.

[26] M. Farrens and A. Park, "Dynamic base register caching: A technique for reducing address bus width," in *Proceedings of the Annual International Symposium on Computer Architecture*, June 1991, pp. 128–137.

[27] K. Basu, A. Choudhary, J. Pisharath and M. Kandemir, "Power protocol: Reducing power dissipation on off-chip data buses," in *Proceedings of the Annual ACM/IEEE International Symposium on Microarchitecture*, 2002, pp. 345–355.

[28] J. Liu, K. Sundaresan and N. R. Mahapatra, "Dynamic address compression schemes: A performance, energy, and cost study," in *Proceedings International Conference on Computer Design (ICCD)*, 2004, pp. 458–463.

[29] S. Ramprasad, N. R. Shanbhag and I. N. Hajj, "Achievable bounds on signal transition activity," in *Proceedings of International Conference on Computer Aided Design (ICCAD)*, San Jose, CA, USA, October 1997, pp. 126–129.

[30] T. M. Cover and J. A. Thomas, *Elements of Information Theory*, John Wiley & Sons, 1991.

[31] R. Murgai, M. Fujita and A. Oliveira, "Using complementation and resequencing to minimize transitions," in *Proceedings of the 35th ACM/IEEE Design Automation Conference (DAC)*, June 1998, pp. 694–697.

[32] M. R. Stan and W. P. Burleson, "Bus-invert coding for low power I/O," *IEEE Transactions on Very Large Scale Integration Systems (TVLSI)*, March 1995, pp. 49–58.

[33] J. Tabor, "Noise reduction using low weight and constant weight coding techniques," Master's thesis, EECS Dept., MIT, May 1990.

[34] Y. Shin, S. Chae and K. Choi, "Partial bus-invert coding for power optimization of application-specific systems," *IEEE Transactions on Very Large Scale Integration Systems (TVLSI)*, Vol. 9, No. 2, 2001, pp. 377–383.

[35] S. Yoo and K. Choi, "Interleaving partial bus-invert coding for low power reconfiguration of FPGAs," in *Proceedings of the Sixth International Conference on VLSI and CAD*, 1999, pp. 549–552.

[36] S. Hong, K. Chung, U. Narayanan and T. Kim, "Decomposition of bus-invert coding for low-power I/O," *J. Circuits Systems and Computers*, Vol. 10, No. 1, 2, 2000, pp. 101–111.

[37] N. Chang, K. Kim and J. Cho, "Bus encoding for low-power high-performance memory systems," in *Proceedings of the 37th Conference on Design Automation Conference (DAC)*, 2000, pp. 800–805.

[38] M. R. Stan and W. P. Burleson, "Two-dimensional codes for low power," in *Proceedings of International Symposium on Low Power Electronics and Design (ISLPED)*, 1996, pp. 335–340.

[39] V. Sundararajan and K. K. Parhi, "Reducing bus transition activity by limited weight coding with codeword slimming," in *Proceedings of the 10th Great Lakes Symposium on VLSI (GLSVLSI)*, 2000, pp. 13–16.

[40] L. Benini, G. DeMicheli, E. Macii, D. Sciuto and C. Silvano, "Address bus encoding techniques for system level power optimization," in *Proceedings of Design, Automation and Test in Europe (DATE)*, Paris, France, February 1998, pp. 861–866.

[41] S. Komatsu, M. Ikeda and K. Asada, "Low power chip interface based on bus data encoding with adaptive code-book method," *Ninth Great Lakes Symposium on VLSI (GLS-VLSI)*, 1999, pp. 368–371.

[42] J. Yang and R. Gupta, "FV encoding for low-power data I/O," *ACM/IEEE International Symposium on Low Power Electronic Design (ISLPED)*, 2001, pp. 84–87.

[43] J. Yang, R. Gupta and C. Zhang, "Frequent value encoding for low power data buses," *ACM Transactions on Design Automation of Electronic Systems*, Vol. 9, No. 3, July 2004, pp. 354–384.

[44] T. Lv, W. Wolf, J. Henkel and H. Lekatsas, "An adaptive dictionary encoding scheme for SOC data buses," in *Proceedings of the Conference on Design, Automation and Test in Europe (DATE)*, 2002.

[45] S. R. Sridhara and N. R. Shanbhag, "A low-power bus design using joint repeater insertion and coding," in *Proceedings of the International Symposium on Low Power Electronics and Design (ISLPED)*, 2005, pp. 99-102.

[46] C. Lee, M. Potkonjak and W. Mangione-Smith, "MediaBench: A tool for evaluating and synthesizing multimedia and communications systems," *International Symposium on Microarchitecture*, 1997, pp. 330-335.

[47] G. Memik, W. H. Mangione Smith and W. Hu, "NetBench: A benchmarking suite for network processors," *International Conference on Computer Aided Design (ICCAD)*, San Jose, California, November 2001, pp. 39-42.

[48] SPECINT2000, http://www.specbenh.org/cpu2000.

[49] L. Benini, A. Macii, M. Poncino and R. Scarsi, "Architectures and synthesis algorithms for power-efficient bus interfaces," *IEEE Transactions on Computer-Aided Design of Integrated Circuits and Systems*, Vol. 19, No. 9, September 2000, pp. 969-980.

[50] S. Salerno, A. Bocca, E. Macii and M. Poncino, "Limited intra-word transition codes: An energy-efficient bus encoding for LCD display interfaces," in *Proceedings of the International Symposium on Low Power Electronics and Design (ISLPED)*, 2004, pp. 206-211.

[51] W.-C. Cheng and M. Pedram, "Chromatic encoding: A low power encoding technique for digital visual interface," in *Proceedings of Design, Automation and Test in Europe (DATE)*, March 2003, pp. 694-699.

[52] S. Salerno, A. Bocca, E. Macii and M. Poncino, "Energy-efficient bus encoding for LCD displays," in *Proceedings of ACM/IEEE Great Lakes Symposium on VLSI (GLS-VLSI)*, March 2004, pp. 240-243.

[53] W. Cheng, J. Liang and M. Pedram, "Software-only bus encoding techniques for an embedded system," in *Proceedings of the 2002 Conference on Asia South Pacific Design Automation/VLSI Design*, 2002.

[54] A. Bahari, T. Arslan and A. T. Erdogan, "Interframe bus encoding technique for low power video compression," in *Proceedings of VLSI Design*, 2007, pp. 691-698.

[55] Y. Aghaghiri, F. Fallah and M. Pedram, "Transition reduction in memory buses using sector-based encoding techniques," *IEEE Transactions on Computer-Aided Design of Integrated Circuits and Systems*, Vol. 23, No. 8, August 2004, pp. 1164-1174.

[56] N. Hatta, N. D. Barli, C. Iwama, L. D. Hung, D. Tashiro, S. Sakai and H. Tanaka, "Bus serialization for reducing power consumption," IPSJ SIG Technical Reports 2004-ARC-159, 2004.

[57] A. Morgenshtein, I. Cidon, A. Kolodny and R. Ginosar, "Comparative analysis of serial vs parallel links in NoC," in *Proceedings of International Symposium on System-on-Chip*, November 2004, pp. 185-188.

[58] M. Ghoneima, Y. Ismail, M. Khellah, J. Tschanz and D. Vivek, "Serial-link bus: A low-power on-chip bus architecture," *IEEE/ACM International Conference on Computer-Aided Design (ICCAD)*, November 6-10, 2005, pp. 541-546.

[59] K. Lee, S. Lee and H. Yoo, "SILENT: serialized low energy transmission coding for on-chip interconnection networks," in *Proceedings of the IEEE/ACM International Conference on Computer-Aided Design (ICCAD)*, 2004, pp. 448-451.

[60] M. Saneei, A. Afzali-Kusha, Z. Navabi, "Serial bus encoding for low power application," *International Symposium on System-on-Chip*, 2006.

[61] C.-H. Kuo, W.-B. Wu, Y.-J. Wu, J.-H. Lin, "Serial low power bus coding for VLSI," *International Conference on Communications, Circuits and Systems*, 2006, pp. 2449-2453.

[62] P. P. Sotiriadis and A. Chandrakasan, "Bus energy minimization by transition pattern coding (TPC) in deep sub-micron technologies," in *Proceedings of the 2000 IEEE/ACM*

International Conference on Computer-Aided Design (ICCAD), San Jose, California, November 05–09, 2000, pp. 322–328.

[63] C. Duan and A. Tirumala, "Analysis and avoidance of cross-talk in on-chip buses," in *Proceedings of the Ninth Symposium on High Performance Interconnects (HOTI)*, 2001.

[64] W.-W. Hsieh, P.-Y. Chen, T. T. Hwang, "A bus architecture for crosstalk elimination in high performance processor design," in *Proceedings of International Conference on Hardware/Software Codesign and System Synthesis (CODES+ISSS)*, 2006, pp. 247–252.

[65] L. Macchiarulo, E. Macii and M. Poncino, " Low-energy for deep-submicron address buses," in *Proceedings of the 2001 International Symposium on Low Power Electronics and Design (ISLPED)*, 2001, pp. 176–181.

[66] Y. Shin and T. Sakurai, "Coupling-driven bus design for low-power application-specific systems," in *Proceedings of Annual ACM/IEEE Design Automation Conference (DAC)*, 2001, pp. 750–753.

[67] J. Henkel and H. Lekatsas, "A2BC: Adaptive address bus coding for low-power deep submicron designs," in *Proceedings of Annual ACM/IEEE Design Automation Conference (DAC)*, June 2001, pp. 744–749.

[68] D. Citron and L. Rudolph, "Creating a wider bus using caching techniques," in *Proceedings of International Symposium on High Performance Computer Architecture*, January 1995, pp. 90–99.

[69] J. Liu, K. Sundaresan and N. R. Mahapatra, "Energy-efficient compressed address transmission," in *Proceedings of the 18th International Conference on VLSI Design (VLSID)*, 2005, pp. 592–597.

[70] J. Liu, K. Sundaresan and N. R. Mahapatra, "Efficient encoding for address buses with temporal redundancy for simultaneous area and energy reduction," in *Proceedings of the 16th ACM Great Lakes Symposium on VLSI (GLS-VLSI)*, 2006, pp. 111–114.

[71] K.-W. Kim, K.-H. Baek, N. Shanbhag, C. L. Liu and S.-M. Kang, "Coupling-driven signal encoding scheme for low-power interface design," in *Proceedings of International Conference on Computer Aided Design (ICCAD)*, 2000, pp. 318–321.

[72] P. P. Sotiriadis and A. Chandrakasan, "Low power bus coding techniques considering inter-wire capacitances," *Custom Integrated Circuits Conference*, 2000, pp. 507–510.

[73] L. Xie, P. Qiu and Q. Qiu, "Partitioned bus coding for energy reduction," in *Proceedings of Asia and South Pacific Design Automation Conference (ASP-DAC)*, 2005, pp. 1280–1283.

[74] Y. Zhang, J. Lach, K. Skadron and M. R. Stan, "Odd/even bus invert with two-phase transfer for buses with coupling," in *Proceedings of the International Symposium on Low Power Electronics and Design (ISLPED)*, 2002, pp. 80–83.

[75] S.-K. Wong and C.-Y. Tsui, "Re-configurable bus encoding scheme for reducing power consumption of the cross coupling capacitance for deep sub-micron instruction bus," *Design, Automation and Test in Europe Conference and Exhibition (DATE)*, 2004, pp. 130–135.

[76] L. Deng, and M. D. Wong, "Energy optimization in memory address bus structure for application-specific systems," in *Proceedings of the 15th ACM Great Lakes Symposium on VLSI (GLS-VSLI)*, 2005, pp. 232–237.

[77] P. Petrov and A. Orailoglu, "Low-power instruction bus encoding for embedded processors," *IEEE Transactions on Very Large Scale Integration Systems*, Vol. 12, No. 8, August 2004, pp. 812–826.

[78] M. Muroyama, K. Tarumi, K. Makiyama and H. Yasuura, "A variation-aware low-power coding methodology for tightly coupled buses," in *Proceedings of Asia and South Pacific Design Automation Conference (ASP-DAC)*, 2005, pp. 557–560.

[79] M. Muroyama, A. Hyodo, T. Okuma and H. Yasuura, "A power reduction scheme for data buses by dynamic detection of active bits," *IEICE Transactions on Electronics*, Vol. E87-C, No. 4, April 2004, pp. 598–605.

[80] A. R. Brahmbhatt, J. Zhang, Q. Wu and Q. Qiu, "Adaptive low-power bus encoding based on weighted code mapping," in *Proceedings IEEE International Symposium on Circuits and Systems (ISCAS)*, 2006.

[81] A. R. Brahmbhatt, J. Zhang, Q. Wu and Q. Qiu, "Low-power bus encoding using an adaptive hybrid algorithm," in *Proceedings of the 43rd Annual Conference on Design Automation (DAC)*, 2006, pp. 987–990.

[82] M. Ghoneima and Y. Ismail, "Delayed line bus scheme: A low-power bus scheme for coupled on-chip buses," in *Proceedings of International Symposium on Low Power Electronics and Design (ISLPED)*, 2004, pp. 66–69.

[83] S. Jayaprakash and N. R. Mahapatra, "Partitioned hybrid encoding to minimize on-chip energy dissipation of wide microprocessor buses," in *Proceedings of International Conference on VLSI Design (VLSID)*, 2007, pp. 127–134.

[84] H. Deogun, R. Rao, D. Sylvester and D. Blaauw, "Leakage- and crosstalk-aware bus encoding for total power reduction," in *Proceedings of Design Automation Conference (DAC)*, 2004, pp. 779–782.

[85] R. R. Rao, H. S. Deogun, D. Blaauw and D. Sylvester, "Bus encoding for total power reduction using a leakage-aware buffer configuration," *IEEE Transactions on Very Large Scale Integration (VLSI) Systems*, Vol. 13, No. 12, December 2005, pp. 1376–1383.

[86] S. R. Sridhara, A. Ahmed and N. R. Shanbhag, "Area and energy-efficient crosstalk avoidance codes for on-chip buses," in *Proceedings of the IEEE International Conference on Computer Design (ICCD)*, 2004, pp. 12–17.

[87] S. R. Sridhara and N. R. Shanbhag, "Coding for reliable on-chip buses: fundamental limits and practical codes," in *Proceedings of the 18th International Conference on VLSI Design (VLSID)*, 2005, pp. 417–422.

[88] J. A. Davis, et al., "Interconnect limits on gigascale integration (GSI) in the 21st century," in *Proceedings of the IEEE*, 89, December 2001, pp. 305–324.

[89] M. T. Bohr, "Interconnect scaling—The real limiter to high performance ULSI," in *Proceedings of IEEE Electron Devices Meeting*, 1995, pp. 241–244.

[90] K. Hirose and H. Yasuura, "A bus delay reduction technique considering crosstalk," in *Proceedings of Design, Automation and Test in Europe (DATE)*, 2000, pp. 441–445.

[91] T. Xiao and M. Sadowska, "Crosstalk reduction by transistor sizing," in *Proceedings of Asia and South Pacific Design Automation Conference (ASPDAC)*, 1999, pp. 137–140.

[92] D. Li, A. Pua, P. Srivastava and U. Ko, "A repeater optimization methodology for deep submicron, high-performance processors," in *IEEE International Conference on Computer Design: VLSI in Computers and Processors (ICCD)*, Austin, TX, October 1997, pp. 726–731.

[93] T. Sakurai, "Closed-form expressions for interconnect delay, coupling and crosstalk in VLSI's," *IEEE Transactions on Electron Devices*, Vol. 40, No. 1, January 1993, pp. 118–124.

[94] T. Sakurai, S. Kobayashi and M. Noda, "Simple expressions for interconnection delay," *IEEE International Symposium on Circuits and Systems (ISCAS)*, 1991, pp. 2375–2378.

[95] A. Kahng and S. Muddu, "An analytical delay model for RLC interconnects," *IEEE Transactions on Computer-Aided Design of Integrated Circuits and Systems*, Vol. 16, No. 12, December 1997, pp. 30–36.

[96] A. Kahng, K. Masuko and S. Muddu, "Delay models for MCM interconnects when response in non-monotone," *IEEE Multi-Chip Module Conference*, 1997, pp. 102–107.

[97] S. McCormick and J. Allen, "Waveform moment methods for improved interconnection analysis," *27th ACM/IEEE Design Automation Conference*, 1990, pp. 406–412.

[98] C. G. Lin-Hendel, "Accurate interconnect modeling for high frequency LSI/VLSI circuits and systems," in *Proceedings of IEEE International Conference on Computer Design (ICCD)*, 1990, 434–442.

[99] P. P. Sotiriadis and A. Chandrakasan, "Reducing bus delay in submicron technology using coding," in *Proceedings of the Conference on Asia South Pacific Design Automation (ASP-DAC)*, 2001, pp. 109–114.

[100] W. Elmore, "The transient response of damped linear network with particular regard to wide-band amplifier," *Journal of Applied Physics*, Vol. 19, 1948, pp. 55–63.

[101] B. Victor and K. Keutzer. "Bus encoding to prevent crosstalk delay," in *Proceedings of IEEE/ACM International Conference on Computer Aided Design (ICCAD)*, 2001, pp. 57–63.

[102] J. D. Z. Ma and L. He, "Formulae and applications of interconnect estimation considering shield insertion and net ordering," in *Proceedings of the 2001 IEEE/ACM International Conference on Computer-Aided Design (ICCAD)*, 2001, pp. 327–332.

[103] R. Arunachalam, E. Acar and S. R. Nassif, "Optimal shielding/spacing metrics for low power design," in *Proceedings of IEEE Computer Society Annual Symposium on VLSI*, 2003, pp. 167–172.

[104] C. Duan and S. P. Khatri, "Exploiting crosstalk to speed up on-chip buses," in *Proceedings of the Conference on Design, Automation, and Test in Europe (DATE)*, 2004.

[105] L. Li, N. Vijaykrishnan, M. Kandemir and M. J. Irwin, "A crosstalk aware interconnect with variable cycle transmission," in *Proceedings of the Conference on Design, Automation, and Test in Europe (DATE)*, 2004, pp. 102–107.

[106] R. Arunachalam, E. Acar and S. R. Nassif, "Optimal shielding/spacing metrics for low power design," in *Proceedings of IEEE Computer Society Annual Symposium on VLSI*, 2003, pp. 167–172.

[107] K. H. Baek, K. W. Kim and S. M. Kang, "A low energy encoding technique for reduction of coupling effects in SOC interconnects," in *Proceedings of 43rd IEEE Midwest Symp. Circuits and Systems*, Lansing, MI, August 2000, pp. 80–83.

[108] C.-G. Lyuh and T. Kim. "Low power bus encoding with crosstalk delay elimination," in *Proceedings of ASIC/SOC Conference*, 2002, pp. 93–100.

[109] P. Subrahmanya, R. Manimegalai and V. Kamakoti, "A bus encoding technique for power and crosstalk minimization," in *Proceedings of International Conference on VLSI Design (VLSID)*, 2004, pp. 443–448.

[110] Z. Khan, T. Arslan and A. T. Erdogan, "A novel bus encoding scheme from energy and crosstalk efficiency perspective for AMBA based generic SoC systems," in *Proceedings of the 18th International Conference on VLSI Design (VLSID)*, 2005, pp. 751–756.

[111] M. Mutyam, M. Eze, N. Vijaykrishnan and Y. Xie, "Delay and energy efficient data transmission for on-chip buses," in *Proceedings of the IEEE Computer Society Annual Symposium on Emerging VLSI Technologies and Architectures (ISVLSI)*, 2006.

[112] M. A. Elgamel and M. A. Bayoumi, "Interconnect noise analysis and optimization in deep submicron technology," *IEEE Circuits and Systems Magazine*, Vol. 3, No. 4, 2003, pp. 6–17.

[113] M. H. Chowdhury, Y. I. Ismail, C. V. Kashyap and B. L. Krauter, "Performance analysis of deep submicron VLSI circuits in the presence of self and mutual inductance," in *IEEE International Symposium Circuits and Systems*, Scottsdale, AZ, 2002, pp. 197–200.

[114] Y. I. Ismail, "On-chip inductance cons and pros," *IEEE Transactions on Very Large Scale Integration (VLSI) Systems*, Vol. 10, No. 6, December 2002, pp. 685–694.

[115] L. He and K. M. Lepak, "Simultaneous shield insertion and net ordering for capacitive and inductive coupling minimization," in *International Symposium on Physical Design*, San Diego, CA, 2000, pp. 55–60.

[116] C. Raghunandan, K. S. Sainarayanan and M. B. Srinivas, "Bus-encoding technique to reduce delay, power and simultaneous switching noise (SSN) in RLC interconnects," in

Proceedings of the 17th Great Lakes Symposium on Great Lakes Symposium on VLSI (GLS-VLSI), 2007, pp. 371–376.

[117] S.-W. Tu, Y.-W. Chang and J. Y. Jou, "RLC coupling-aware simulation and on-chip bus encoding for delay reduction," *IEEE Transactions on Computer-Aided Design of Integrated Circuits and Systems*, Vol. 25, No. 10, October 2006, pp. 2258–2264.

[118] B. J. LaMeres and S. P. Khatri, "Encoding-based minimization of inductive cross-talk for off-chip data transmission," in *Proceedings of the Conference on Design, Automation and Test in Europe (DATE)*, 2005, pp. 1318–1323.

[119] B. J. LaMeres and S. P. Khatri, "Bus stuttering: An encoding technique to reduce inductive noise in off-chip data transmission," in *Proceedings of the Conference on Design, Automation and Test in Europe (DATE)*, 2006.

[120] M. Lampropoulos, B. M. Al-Hashimi and P. Rosinger, "Minimization of crosstalk noise, delay and power using a modified bus invert technique," in *Proceedings of the Conference on Design, Automation and Test in Europe (DATE)*, 2004.

[121] K. Aingaran, F. Klass, C. M. Kim, C. Amir, J. Mitra, E. You, J. Mohd and S. K. Dong, "Coupling noise analysis for VLSI and ULSI circuits," in *Proceedings of International Symposium on Quality Electronic Design (ISQED)*, 2000, pp. 485–489.

[122] L. Anghel and M. Nicolaidis. "Cost reduction and evaluation of temporary faults detecting technique," in *Proceedings of the Conference on Design, Automation and Test in Europe (DATE)*, 2000, pp. 591–598.

[123] K. L. Shepard and V. Narayanan. "Noise in deep submicron digital design," in *Proceedings of IEEE/ACM International Conference on Computer-Aided Design (ICCAD)*, November 1996, pp. 524–531.

[124] A. Krstic, Y. M. Jiang and K. T. Cheng, "Pattern generation for delay testing and dynamic timing analysis considering power supply noise effects," *IEEE Transactions on CAD*, Vol. 20, No. 3, March 2001, pp. 416–425.

[125] D. Bertozzi, L. Benini and G. D. Micheli, "Low power error resilient encoding for on-chip data buses," in *Proceedings of the Conference on Design, Automation and Test in Europe (DATE)*, 2002, pp. 102–109.

[126] D. Bertozzi, L. Benini and G. De Micheli, "Error control schemes for on-chip communication links: the energy-reliability tradeoff," *IEEE Transactions on Computer-Aided Design of Integrated Circuits and Systems*, Vol. 24, No. 6, June 2005, pp. 818–831.

[127] D. Siewiorek and R. Swarz, *Reliable Computer Systems: Design and Evaluation*, A. K. Peters, Natick, MA, October 1998.

[128] P. Lala, *Self-Checking and Fault-Tolerant Digital Design*, Morgan Kaufmann, San Francisco, CA, 2000.

[129] C. Metra and M. Favalli, "Bus crosstalk fault-detection capabilities of error-detecting codes for on-line testing," *IEEE Transactions on Very Large Scale Integration (VLSI) Systems*, Vol. 7, No. 3, September 1999, pp. 392–396.

[130] D. Pradhan, *Fault Tolerant Computing: Theory and Techniques*, Prentice-Hall, Englewood Cliffs, NJ, 1986.

[131] D. Das and N. Touba, "Weight-based codes and their applications to concurrent error detection of multilevel circuits," in *Proceedings of IEEE VLSI Test Symposium*, 1999, pp. 370–376.

[132] M. Favalli and C. Metra, "Optimization of error detecting codes for the detection of crosstalk originated errors," in *Proceedings of Design, Automation and Test in Europe (DATE)*, March 2001, pp. 290–296.

[133] S. Lin and D. J. Costello, *Error Control Coding: Fundamentals and Applications*, Prentice-Hall, Englewood Cliffs, NJ, 1983.

[134] J. F. Wakerly, *Digital Design: Principles and Practices*, 3rd ed., Prentice-Hall, Englewood Cliffs, NJ, 2000.

[135] F. J. MacWilliams and N. J. A. Sloane, *The Theory of Error-Correcting Codes*, Amsterdam, The Netherlands: North-Holland, 1996.

[136] H. Zhang, V. George and J. M. Rabaey, "Low swing on-chip signaling techniques: Effectiveness and robustness," *IEEE Transactions on Very Large Scale Integration (VLSI) Systems*, Vol. 8, No. 3, August 2000, pp. 264–272.

[137] R. Hegde and N. R. Shanbhag, "Toward achieving energy efficiency in the presence of deep submicron noise," *IEEE Transactions on Very Large Scale Integration (VLSI) Systems*, Vol. 8, August 2000, pp. 379–391.

[138] R. Hegde and N. R. Shanbhag, "Lower bounds on energy dissipation and noise-tolerance for deep submicron VLSI," in *Proceedings of IEEE International Symposium on Circuits and Systems (ISCAS)*, 1999, pp. 334–337.

[139] D. Bertozzi, L. Benini and B. Ricco, "Energy-efficient and reliable low-swing signaling for on-chip buses based on redundant encoding," in *Proceedings of IEEE International Symposium on Circuits and Systems (ISCAS)*, I, May 2002, pp. 93–96.

[140] L. Lin, N. Vijaykrishnan, M. Kandemir and M. J. Irwin, "Adaptive error protection for energy efficiency," in *Proceedings of IEEE/ACM International Conference on Computer Aided Design (ICCAD)*, 2003, pp. 2–7.

[141] K. Patel and I. Markov, "Error-correction and crosstalk avoidance in DSM busses," in *Proceedings of International Workshop on System-Level Interconnect Prediction (SLIP)*, 2003, pp. 9–14.

[142] S. R. Sridhara and N. R. Shanbhag, "Coding for system-on-chip networks: A unified framework," *IEEE Transactions on Very Large Scale Integration (VLSI) Systems*, Vol. 13, No. 6, June 2005, pp. 655–667.

[143] S. Komatsu and M. Fujita, "An optimization of bus interconnects pitch for low-power and reliable bus encoding scheme," in *Proceedings of IEEE International Symposium on Circuits and Systems (ISCAS)*, 2006.

CHAPTER

8

Custom Bus-Based On-Chip Communication Architecture Design

In addition to the bus-based on-chip communication architecture standards described in Chapter 3, there has been a lot of research on developing custom bus architectures over the last several years. Such custom architectures attempt to address the shortcomings of standard on-chip communication architectures by utilizing new topologies and protocols to obtain improvements for common design goals, such as performance and power. These novel topologies and protocols are often customized to suit a particular application, and typically include optimizations to meet application-specific design goals. Figure 8.1 shows where the custom communication architecture selection process fits into a typical electronic system level (ESL) design flow. Once a custom on-chip communication architecture has been selected, the next step is usually to perform an exploration phase, to determine the protocol and topology parameters that can best meet the design goals. In this chapter, we present some of the more significant custom bus-based on-chip communication architectures that have been proposed over the past few years. In Section 8.1, we describe split bus architectures that are useful for reducing bus power consumption. Section 8.2 presents serial bus architectures that aim to reduce wire congestion and the area footprint of the on-chip communication architecture. Section 8.3 describes code division multiple access (CDMA) based bus architectures. Section 8.4 elaborates on asynchronous bus architectures that avoid a global clock signal for synchronization, in order to reduce power consumption. Finally, Section 8.5 presents reconfigurable bus architectures that allow dynamically changing the bus topology and/or parameters (such as arbitration schemes) in order to better adapt to changing traffic conditions during application execution and improve performance.

8.1 SPLIT BUS ARCHITECTURES

Single shared bus-based communication architectures have the advantage of a simple topology and low area cost. The disadvantages of shared-bus architectures

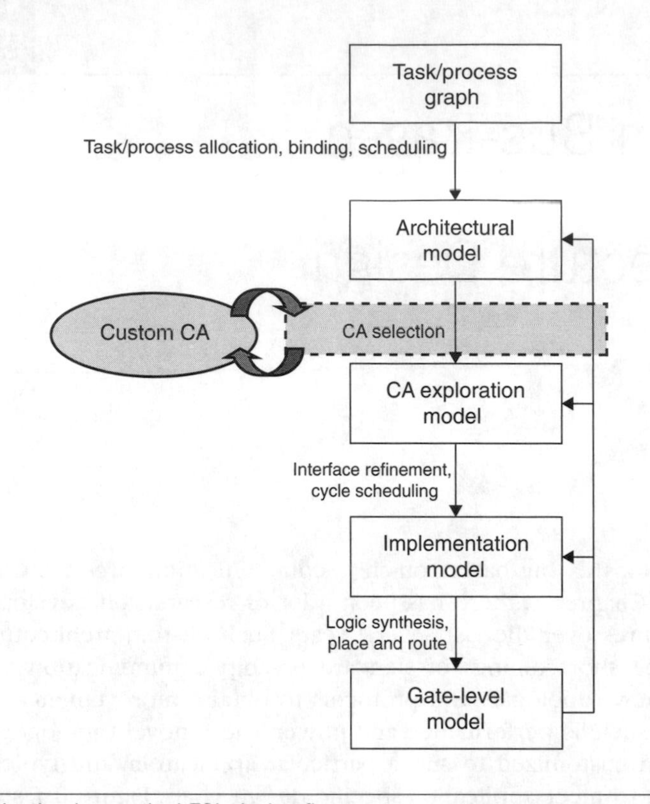

FIGURE 8.1

Custom CA selection in a typical ESL design flow

include larger wire and load capacitance, which results in a correspondingly larger power consumption as well as longer delay for data transfer (which leads to lower bandwidth). Split or segmented bus architectures attempt to overcome this drawback by splitting a shared bus into multiple segments. Split buses allow selective shutdown of unused bus segments, potentially saving energy. Furthermore, segmentation increases the parallelism by permitting parallel data transfers on different segments, which improves performance.

For instance, the segmented bus architecture [1] proposes the insertion of buffers to isolate the components on a bus. Figure 8.2 shows an example of the segmented bus architecture. An advantage of such an architecture is that only part of the bus is active at any given time. As an example, a transfer between the DSP and *X-mem* components in Fig. 8.2 causes only their corresponding segments and the buffers between them to be active, while the rest of the segments and buffers remain inactive. This reduces the effective load and wire capacitance of the active bus. Since the power consumption of the bus is proportional to the load capacitance and length of the active bus [2], the segmented bus results in power savings. Figure 8.3 compares the total bus power for the segmented bus [1] with a shared bus, for a set of benchmarks. It can be seen that a segmented bus dissipates much less power than a corresponding shared bus, since only part of the segmented bus is active at any given time, compared to a single shared bus, for which the

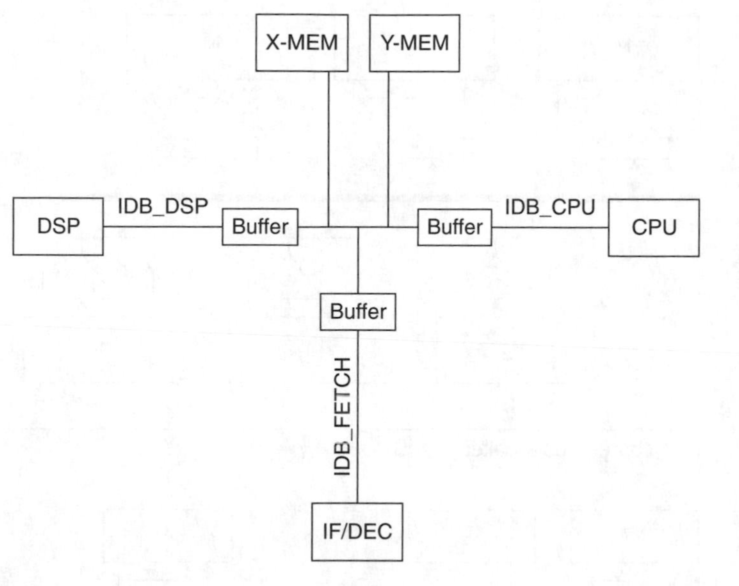

FIGURE 8.2

A segmented bus architecture [1]
© 1998 IEEE

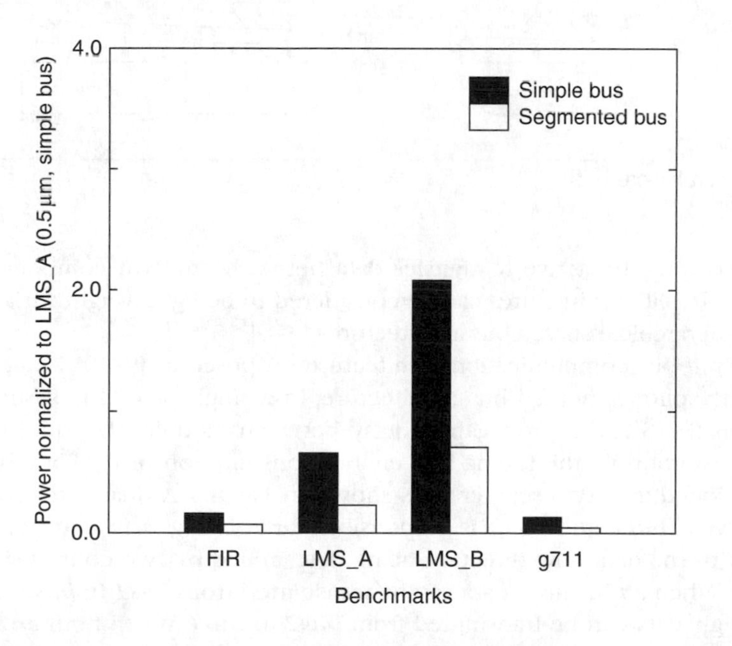

FIGURE 8.3

Total bus power comparison between segmented bus and shared bus architectures, for 0.35 micron technology [1]
© 1998 IEEE

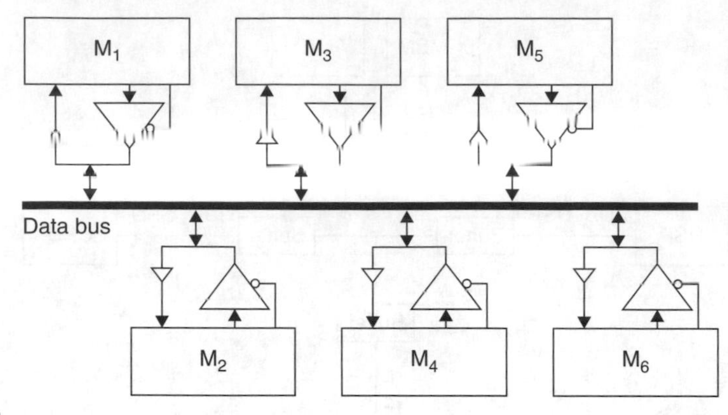

FIGURE 8.4

Monolithic single shared bus architecture [9]

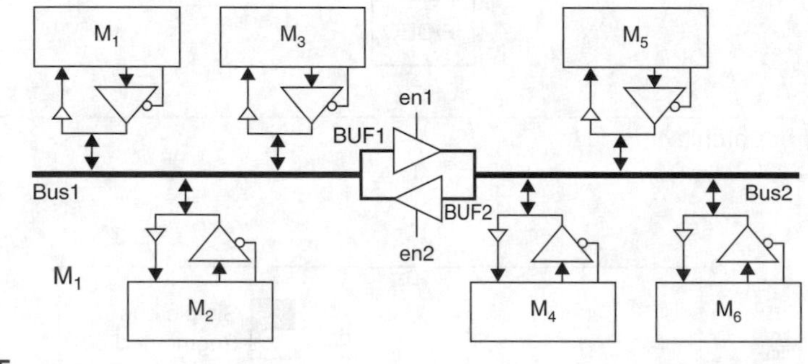

FIGURE 8.5

A split bus architecture [9]

entire bus must be active to transfer data between any two components. Such segmented/split architectures can be considered to be light-weight variants of traditional hierarchical shared bus architectures [3–8].

The split bus communication architecture proposed in [9, 10] is another variant of the split/segmented bus architecture. In a single shared bus, such as that shown in Fig. 8.4, the propagation delay between module *M1* and *M6* is large. In order to improve the timing and energy consumption of the long bus, it can be partitioned into two segments, as shown in Fig. 8.5. A dual-port driver at the boundary of *bus1* and *bus2* is responsible for relaying data between the two buses, with the data flow direction being determined by two control signals *en1* and *en2*. When *en1* is high, data can be transmitted from *bus1* to *bus2*, and when *en2* is high, data can be transmitted from *bus2* to *bus1*. When both *en1* and *en2* are low, the buses are isolated from each other.

Assuming that the internal energy of dual-port drivers is negligible and their intrinsic delay is smaller than the rest of the bus, and the energy dissipated to generate and connect control signals of the bus drivers is negligible, the split bus architecture results in energy savings compared to a simple shared bus architecture.

Table 8.1 Energy consumption of various bus architectures [9]

Architecture	Energy
$BUS = \{M_1, M_2, M_3, M_4\}$	1
$BUS1 = \{M_1, M_2\}$ $BUS2 = \{M_3, M_4\}$	0.75
$BUS1 = \{M_1, M_3\}$ $BUS2 = \{M_2, M_4\}$	0.875
$BUS1 = \{M_1, M_4\}$ $BUS2 = \{M_2, M_3\}$	0.875

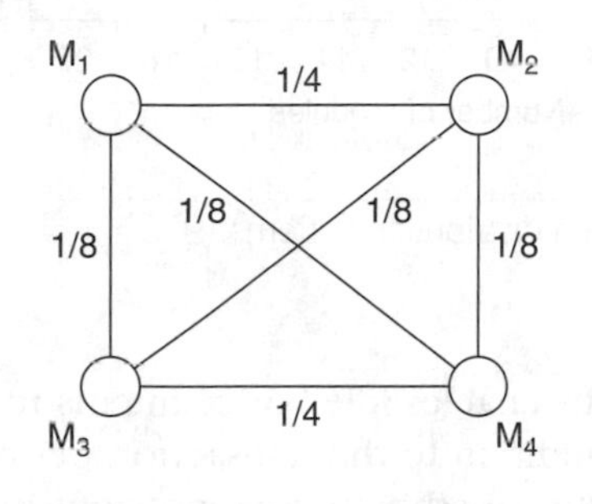

FIGURE 8.6

Data transfer probabilities between modules [9]

The energy savings will vary, depending on how the components are clustered on the different segments. Table 8.1 shows the energy consumption of different component allocations on the two segments, for a system with four modules—M_1, M_2, M_3, and M_4. The configuration in which modules M_1 and M_2 are on one segment, and modules M_3 and M_4 are on another segment results in the largest energy savings. Such a lowest energy consumption configuration can be derived from a data transfer probability graph, as shown in Fig. 8.6. The components having the highest probabilities of data transfer should be kept on the same segment, so that only that segment of the bus architecture is active during the transfer, which saves energy. This observation can be corroborated from the energy observations in Table 8.1 and the probability graph in Fig. 8.6.

Figure 8.7 shows the results of an experiment to determine energy savings of the split bus technique when compared to a shared bus architecture, where the data transfer probability between any two modules is assumed to be one of four probability distributions (*normal, exponential, delta function,* and *uniform*). The *x*-axis represents different system testbenches with varying number of modules. It can be seen that energy savings are obtained with all four data transfer probability functions when a split bus is used. Further reduction of bus energy can be obtained by using bus encoding and low-voltage swing signaling techniques [11, 12].

Another variant of the split bus was proposed by Lu and Koh [13]. The high performance bus architecture called SAMBA (single arbitration, multiple bus accesses) allows multiple masters to access the bus with only a single bus arbitration grant.

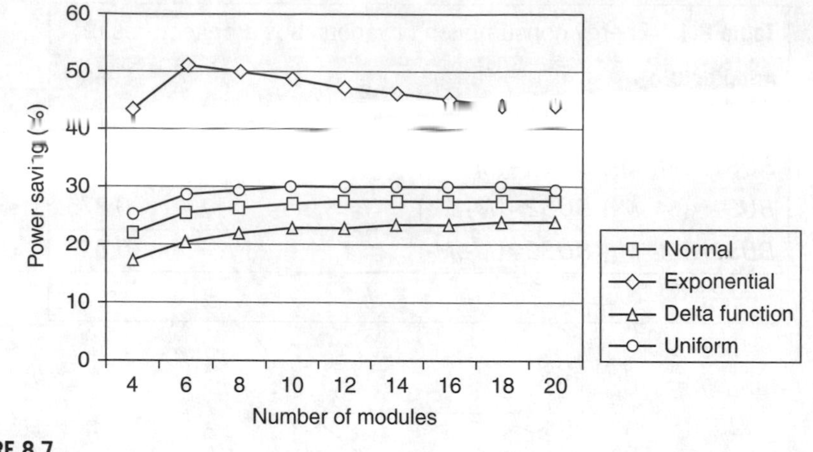

FIGURE 8.7

Power savings for different data distribution patterns [9]
© 2002 IEEE

The split nature of the bus enables idle bus segments to be used for other pending bus transactions (in addition to the transaction permitted by the arbitration), without introducing additional arbitration complexity. In traditional bus architectures, only one master can access the bus at any given time, after it is granted access to the bus by the arbiter. This leads to wastage of bus resources, which are completely consumed by a single transaction on the bus. Split [3] or out-of-order (OO) [4] transactions can overcome this bandwidth limitation, but at the cost of increased arbitration complexity, leading to an increase in arbitration delay. The SAMBA bus architecture attempts to overcome these drawbacks, and improve bus bandwidth and latency response.

Figure 8.8 shows the structure of the SAMBA bus architecture. It consists of two separate buses, each of which is used for data transfer in a forward or backward direction. The SAMBA bus architecture requires that the addresses of the modules be in an increasing or decreasing order from one end of the bus to the other. The bus transferring data from the lower address modules to the higher address modules is called the *forward* bus, while the bus transferring data from the higher address modules to the lower address modules is called the *backward bus*. A module is attached to the bus through an interface unit that can communicate with other interface units through both the forward and backward buses. Each bus operation consists of two phases: the request phase and the response phase. The communication request is sent by the initiator to the destination module via one of the buses in the request phase, and the response is sent by the destination on the other bus in the response phase. Before a transaction can begin on a bus, the interface unit of the initiator must decide whether to use the forward or backward bus, and then request the arbiter for access to that bus. The arbiter broadcasts the winner of the arbitration to all interface units, and it is possible that one module gains access to the forward bus, while another gets access to the backward bus. Multiplexers are used to combine all the signals, as shown in Fig. 8.8.

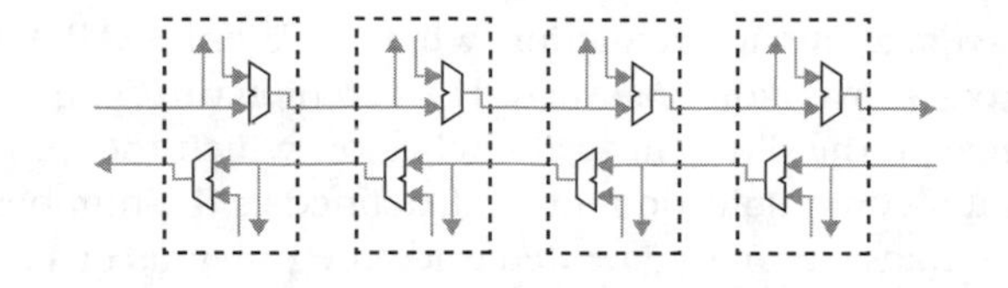

FIGURE 8.8

Structure of SAMBA bus architecture [13]
© 2003 IEEE

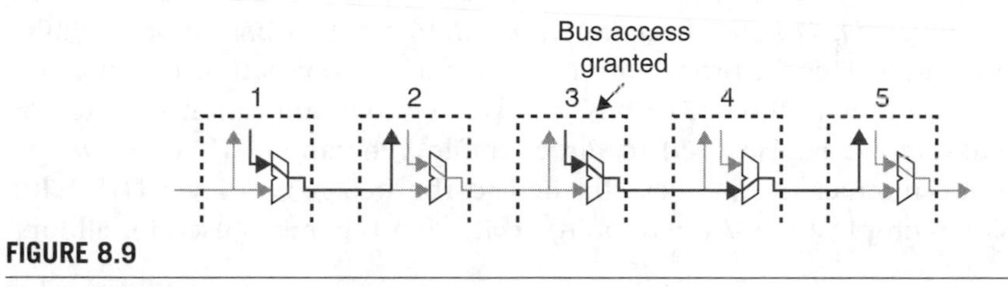

FIGURE 8.9

Multiple bus accesses with single arbitration [13]
© 2003 IEEE

Unlike a traditional bus architecture, a module connected to the SAMBA bus can access the bus even if it is not an arbitration winner. As long as there are no common bus segments in the paths of bus transactions, they can be performed simultaneously, allowing for more parallelism in data communication. If a module has a pending transaction for the forward bus, it can initiate the transaction if any of the following three conditions are met:

(*i*) The module wins the arbitration for the forward bus.

(*ii*) The bus transaction destination of the module is not after the arbitration winner (i.e., the address of the destination is lower than or equal to that of the arbitration winner), and no modules before this module are performing transactions with modules after it on the bus.

(*iii*) The module is after the arbitration winner and no module before it performs a bus transaction with a module after it.

Figure 8.9 depicts of a scenario where simultaneous multiple accesses to a bus occur after a single arbitration. Module 3 is the arbitration winner, and therefore the bus transaction from modules 3 to 5 is performed. Since the destination of the pending transaction at module 1 is module 2, which is before the arbitration winner, it satisfies condition (ii) above, and can therefore perform the transaction on the bus simultaneously. The ability to perform multiple transactions in a single cycle results in higher bandwidth and lower latency for the SAMBA bus. Communication latency is further reduced because transactions can be performed after automatic compatible transaction detection at the interface units without waiting for bus access from the arbiter. Figures 8.10 and 8.11 show the effective bandwidth and average latency reduction when different SAMBA bus configurations are compared with a traditional shared bus architecture. The *x*-axis shows the number of modules in the different testbenches on which the experiments were performed. The inter communication interval, defined as the number of bus cycles after which a new transaction is generated once the previous bus access has been granted, is randomly generated following a *Poisson* distribution. Three different SAMBA configurations are considered in the experiment, which are based on the communication distance distribution. The communication distance of a bus transaction is defined as the number of interface units between its initiator and destination. The three SAMBA configurations have different communication distance distributions: *uniform, Poisson, and exponential*. In the SAMBA-uniform configuration, a module has equal probability of generating a transaction for all other modules. In the other two distributions, an average communication distance of (*num_of_modules/4*) is also used to direct traffic generation, where *num_of_modules* is the number of modules attached to the bus. A two level TDMA/RR (time division multiple access/round-robin) arbitration scheme is used for all four

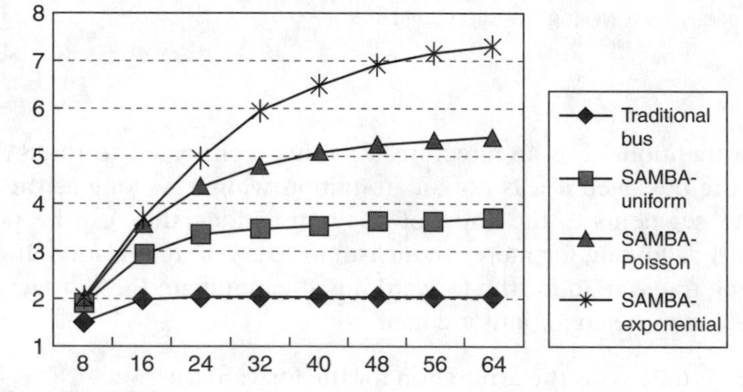

FIGURE 8.10

Effective bandwidth for buses with different number of modules [13]
© 2003 IEEE

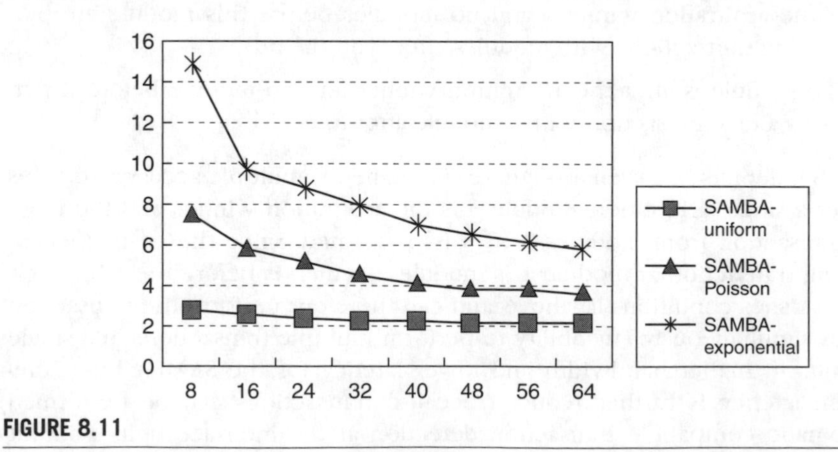

FIGURE 8.11

Average latency reduction for buses with different number of modules [13]
© 2003 IEEE

bus configurations that are compared, and an arbitration latency of one cycle is assumed. For fairness of comparison, it is assumed that the traditional bus architecture has two buses that can be used independently. Therefore, when the traffic on those buses is high enough, the effective bandwidth is two, as shown in Fig. 8.10. From Figs. 8.10 and 8.11 it can be seen that when the SAMBA bus is used, the effective bandwidth is improved by as much as 3.5 times, while the latency is reduced by up to 15 times.

8.2 SERIAL BUS ARCHITECTURES

Typical standard bus-based communication architectures [3–8, 14] have parallel-line buses with multiple signal lines for the data and address buses. In the deep submicron (DSM) era, coupling capacitance between adjacent signal lines leads to significant signal propagation delay and power consumption [15, 16]. Several techniques have been proposed to reduce this coupling capacitance, including increasing line-to-line spacing and non-uniform wire placement [17–20], bus ordering [21], bus swizzling [22], repeater staggering [23], and skewing signal transition timing of adjacent lines [24] (these techniques are explored in more detail in Chapters 6 and 11). One extremely effective way of reducing coupling capacitance is to reduce the number of physical signal lines, by coupling the data of m signal lines onto a single line, called a *serial link*. Such an approach was proposed by Ghoneima et al. [25] and Hatta et al. [26]. A parallel n line bus is converted into n/m serial links, as shown in Fig. 8.12. The reduction in the number of bus lines results in (i) a larger interconnect pitch, which reduces the coupling capacitance and (ii) a wider interconnect, which reduces the effective resistance,

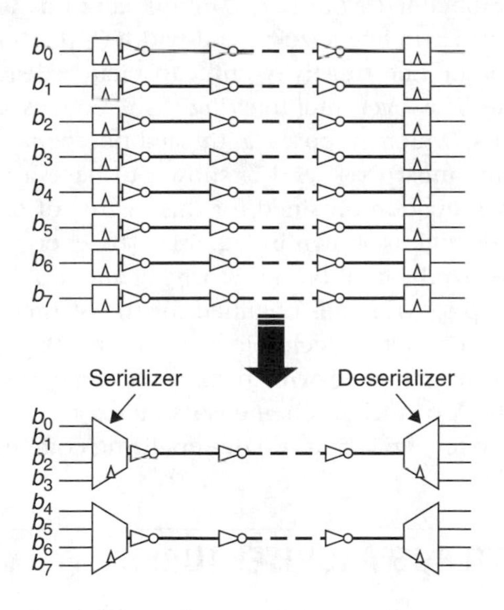

FIGURE 8.12

8-to-2 serial bus that converts an 8-bit parallel bus into 2 serial links ($n = 8$, $m = 4$) [25]

© 2005 IEEE

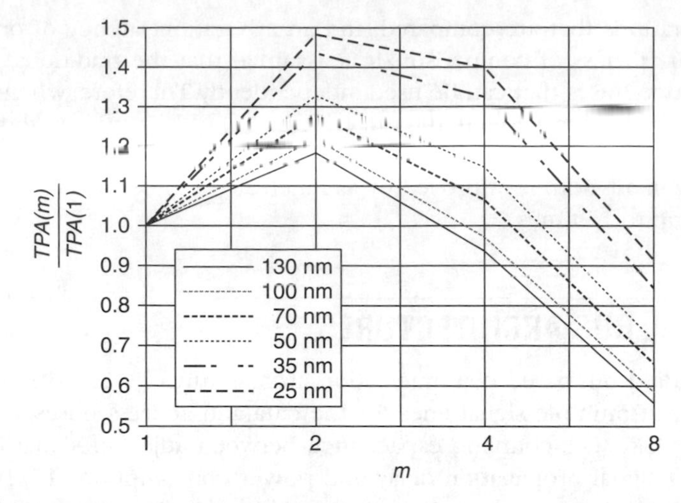

FIGURE 8.13

Relative throughput per unit area of a 64 to (64/m) serial link bus vs. degree of multiplexing *m* [25]

leading to a decrease in interconnect delay and energy consumption. A significant improvement in energy consumption can be obtained by carefully selecting the number of serial links.

Figure 8.13 shows the throughput per unit area of a 64 to $(64/m)$ serial link bus, *TPA(m)*, normalized to the throughput per unit area of a 64-bit bus, TPA(1), and plotted as a function of the degree of multiplexing m. The technology scaling parameters were obtained from [27]. Multiplexing bus lines to create serial links results in an increase in line spacing and width (if the bus area remains constant), which reduces the line resistivity and capacitance, leading to an improvement in line bit rate. However, multiplexing also reduces the number of bus lines transferring data, which imposes a throughput penalty. Thus an optimal degree of multiplexing must exist, and is shown to have a value of 2 from Fig. 8.13. The bus energy reduction obtained for this degree of multiplexing ($m = 2$) for a 64- to 32-bit serial link is shown in Fig. 8.14. The energy reduction numbers take into account the overhead of both the driving and multiplexing circuitry. An energy reduction of up to 31.42% is obtained for the 50-nm CMOS (complementary metal-oxide semiconductor) technology node and this energy reduction is expected to be more pronounced with further technology scaling, as shown in the figure. The drawback of such a scheme is its increased latency response and a performance overhead for serializing and de-serializing communication data.

8.3 CDMA-BASED BUS ARCHITECTURES

Typically, in standard bus-based communication architectures [3–8, 14], the physical interconnect resources are shared in the time domain. A bus inherently uses a

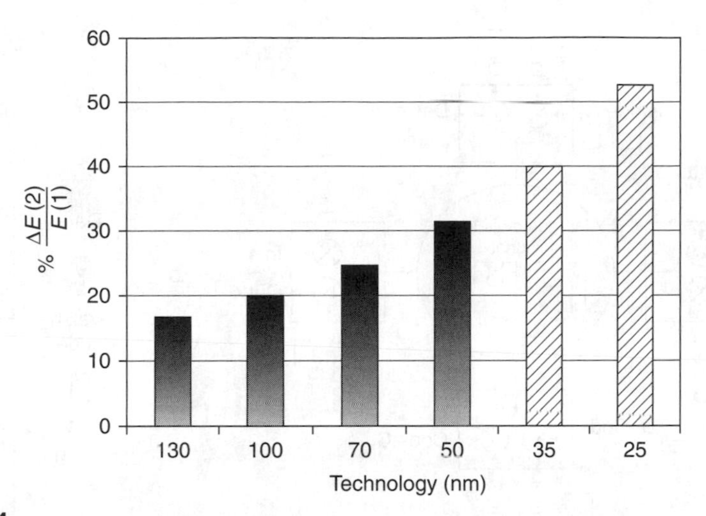

FIGURE 8.14

Bus energy reduction of a 64–32 serial link ($m = 2$) compared to a conventional 64-bit parallel line bus ($m = 1$) [25]

variant of TDMA to reuse expensive on-chip wires for multiple transactions that occur at different points of time. In contrast, CDMA is a spread spectrum technique that allows simultaneous use of the on-chip wires by multiple data transmission flows [28, 29]. It is based on the principle of codeword orthogonality, which avoids cross-correlation of codewords and allows perfect separation of data bits modulated with different codewords. In such a CDMA scheme, a transmitter on the bus modulates each of its transmitted bits with a spreading code before transmission on the bus, as shown in Fig. 8.15(a). On the bus, multiple transmissions can exist simultaneously as a multi-level signal. At the receiver's end, the signal is correlated with the same spreading code that was used by the transmitter. Since spreading codes are orthogonal, the original data that was transmitted can be retrieved. Figure 8.15(b) shows the *Walsh-Hadamard spreading* code [29] that is widely used in the communication domain. In CDMA, these spreading codes convert (or spread) each source data bit into *k* "chips" (which are user-specific fixed patterns), so that the spreading data rate (called *chip rate*) is *k* times the source bit rate. In a channel with two concurrent CDMA links, a 2-bit Walsh code is used and a data bit will be expanded into two "chips." Two clock cycles are needed to receive 1 bit of data if the receiver can receive one "chip" per clock cycle [30].

A mixed-mode bus architecture called CT-Bus was proposed by Lai et al. [30], which integrates CDMA and TDMA techniques in a hierarchical structure. The CT-Bus supports a fixed number of CDMA subchannels that are separated by different spreading codes. Two or more of these subchannels can be grouped into a subchannel group. Figure 8.16 shows the CT-Bus architecture having six subchannels, grouped into three different subchannel groups. *DF1* to *DF4* are data flows that will utilize the bus, and need to be assigned to different subchannel groups. Because of the channel isolation feature of the CDMA scheme, data flows in

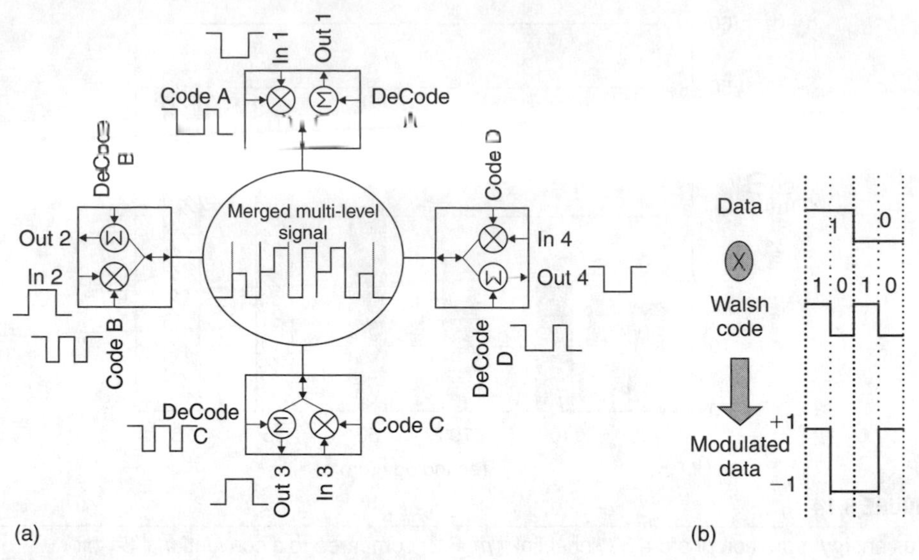

FIGURE 8.15

(a) CDMA-based interconnect (b) modulation of data and spreading code [30]
© 2004 IEEE

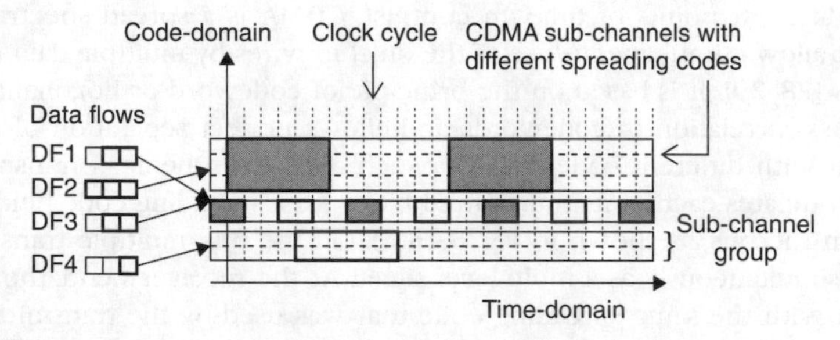

FIGURE 8.16

Architecture of CT-Bus (with three CDMA subchannel groups) [30]
© 2004 IEEE

different subchannel groups are isolated and do not impact each other. Within each CDMA subchannel group, data flows are assigned different TDMA timeslots. The subchannel groups can optionally use other arbitration schemes such as round-robin (*RR*) or fixed-priority (*FP*), which were described earlier in Chapter 2. The number of CDMA subchannels and subchannel groups are also configurable.

Figure 8.17 shows the latencies of two traffic flows, *DF3* and *DF4*, on a TDMA bus and on the CT-Bus, from a multimedia mobile phone case study that was simulated for a period of 1 second (*x*-axis). The *DF3* and *DF4* data flow latencies in the TDMA bus are impacted by the bursty traffic from other data flows. In contrast, due to the channel isolation feature of the CT-Bus, the latencies for the *DF3* and *DF4* data flows are very predictable and well controlled, after appropriate assignment to separate subchannel groups.

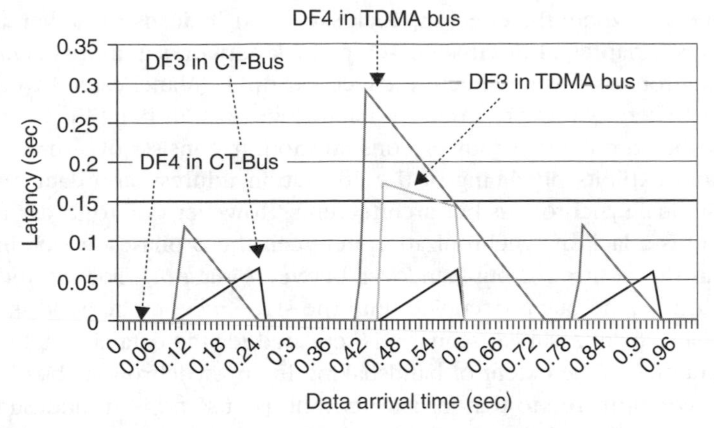

FIGURE 8.17

Latencies of DF3 and DF4 data traffic flows from multimedia mobile phone system case study [30]

An interesting discussion on CDMA and other advanced interconnects, such as frequency division multiple access (FDMA), was presented by Chang [31]. FDMA interconnects allow sharing of bus lines by assigning different data flows to different frequency channels. Data flows assigned to different frequency channels can communicate concurrently with virtually no interference, provided that undesired frequency channels are filtered out at the destination. The FDMA and CDMA can also be combined into a hybrid multi-carrier CDMA scheme, where concurrent data flows are possible by assigning appropriate codes and frequencies to each data flow pair. Such a multi-carrier CDMA scheme can have the highest aggregate data rate when compared to TDMA, CDMA, or FDMA interconnects, due to the increased bandwidth made available through the use of more than one frequency channel. However, the overhead of the modulation and demodulation transceivers at the interfaces of the communicating modules needs to be further analyzed, before CDMA-based interconnects can be considered commercially viable as on-chip communication architectures.

8.4 ASYNCHRONOUS BUS ARCHITECTURES

Asynchronous buses are implemented using clockless circuits, that is, they do not make use of a global clock signal for synchronization. Instead, synchronization occurs using additional handshake signals between transfer phases. Since the global clock generator and distributor contributes to a significant chunk of the on-chip bus power consumption, the clockless asynchronous buses have lower power consumption compared to traditional synchronous buses. Asynchronous buses also have another important advantage—resilience to clock skew even as the number of IPs (components) connected to the bus increases [32]. Performing long range communication across the chip using a widely distributed clock is difficult especially at high frequencies where the effect of clock skew usually leads to slower and wider interconnects (Chapter 11 discusses on-chip clock networks in more detail).

An asynchronous bus architecture called MARBLE (Manchester AsynchRonous Bus for Low Energy) was proposed by Bainbridge and Furber [33, 34] for reducing on-chip bus architecture power consumption. It consists of address and data channels, and exploits pipelining of the arbitration, address, and data phases, just like in traditional synchronous bus architectures. However, due to its asynchronous nature, there is a lack of synchronization between these phases, which introduces problems in the control of bus handover between initiators. For instance, during an address phase, the bus is occupied until the slave accepts the address and completes the handshake. A similar scenario is repeated for the data channel in the subsequent data phase. The extent of handshaking in an asynchronous bus is typically more involved compared to that for a synchronous bus and is an undesirable overhead. This overhead can be overcome to an extent by the introduction of latches at the component ports, as shown in Fig. 8.18. This allows, for instance, the address packet n to be held in the latch at the target, freeing up the address channel for the initiator, which can now send packet $n + 1$. These latches thus decouple the bus from the components, and free them up for subsequent transfers.

Two separate arbiters are used in MARBLE: one for the address channel and the other for the data channel. The bus utilizes centralized decoding and arbitration. It also supports spatial locality optimizations (implemented via dedicated signals)

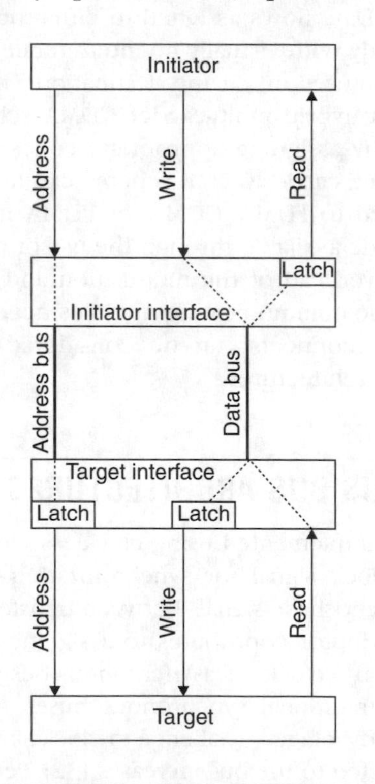

FIGURE 8.18

Decoupling the bus from components using latches in MARBLE [33] that indicate cache line fetches and DMA transfers. Finally, MARBLE supports burst mode and in-order split transactions. All these features, coupled with its inherent asynchronous behavior result in a bus architecture that has low latency and zero quiescent state power consumption. Two drawbacks of using an asynchronous bus like MARBLE are the additional hardware logic overhead at the module interfaces, and the additional bus lines needed for handshaking.

A high performance asynchronous bus, capable of multiple issue (i.e., allowing each master to issue another transaction before the response of a previous transaction arrives), and in-order as well as OO transaction completion was presented by Jung et al. [35]. A customized layered architecture is used for the proposed asynchronous bus in order to alleviate increasing system complexity and enable optimizations. This layered architecture is shown in Fig. 8.19. The *physical, data link*, and *transport* layers are mapped to the interfaces of the IPs connected to the bus. The physical layer handles data encoding, filtering, and driving functions. The data link layer is concerned with flow and access controls. The transport layer manages burst and split transactions, multiple issue, and in-order/out-of-order transaction completion. An asynchronous handshake protocol with two-phase signaling and data insensitive (DI) encoding is used for robust and high speed data transfers on the bus [35], while four-phase signaling and bundled data transfers [35] are used at the IP interfaces for high performance and low complexity. Two asynchronous buses are proposed: (i) MI-OCB, a multiple issue on-chip bus supporting in-order transaction completion and (ii) MO-OCB, a multiple issue on chip bus supporting OO transaction completion. Experimental results for the performance and power dissipation of these buses, when compared to a single issue on-chip asynchronous bus (SI-OCB) are shown in Figs. 8.20 and 8.21, respectively. The three asynchronous buses were implemented with a 0.25 μm CMOS process

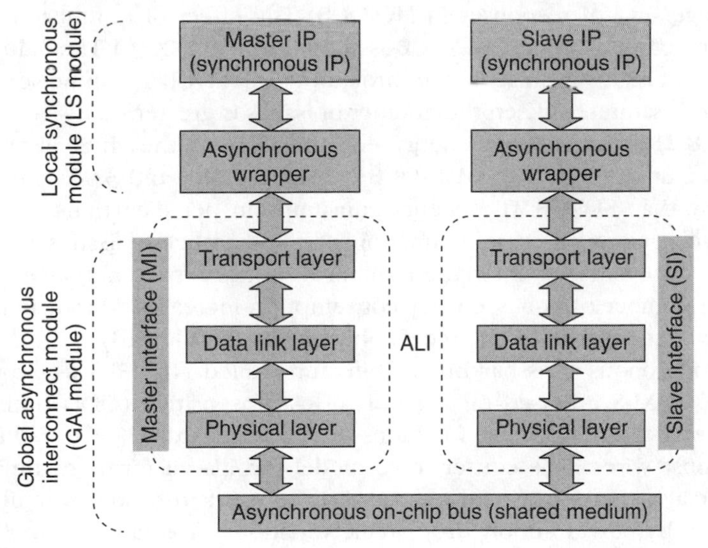

FIGURE 8.19

Layers of asynchronous bus architecture [35]
© 2005 ACM Press

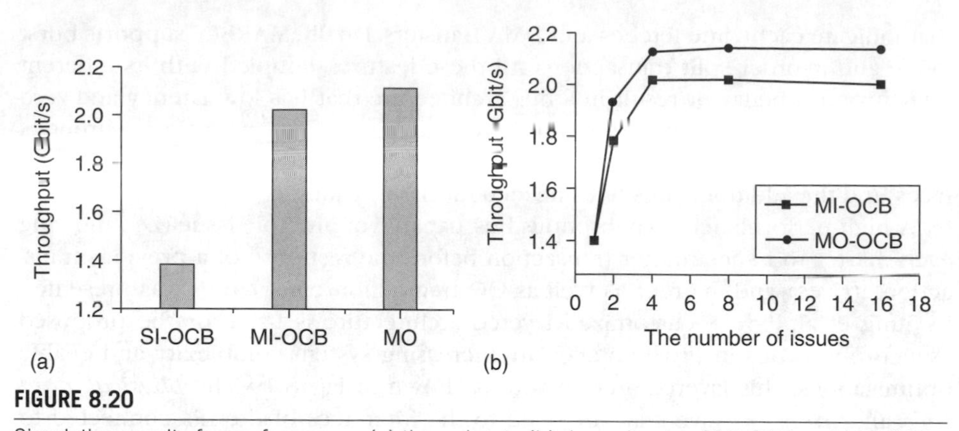

FIGURE 8.20

Simulation results for performance: (a) throughputs (b) throughputs of MI-OCB and MO-OCB as a function of the number of issues [35]
© 2005 ACM Press

and simulation results were obtained with Synopsys NanoSim [67] at the transistor level, using the pre-layout netlist. The testbench consists of 12 IPs (4 masters and 8 slaves), with each master communicating with the slaves with the same probability. The assumptions are that the delay of all the asynchronous modules is 0, and that the ratio of the non-bus transfer time to the total transfer time per synchronous IP is also 0. The testbench workload comprises of a total of 4800 transactions. From Fig. 8.20(a), it can be seen that the throughputs of MI-OCB and MO-OCB are 31.3% and 34.3% more than for SI-OCB. The throughput of MI-OCB is lower than that of MO-OCB because no reorder buffers and related hardware controllers are needed in MO-OCB (it is the responsibility of the master IP to handle rearrangement of responses in MO-OCB). The effect of number of issues on throughputs of MI-OCB and MO-OCB is shown in Fig. 8.20(b). The value of 1 for the number of issues represents the throughput of SI-OCB. It can be seen that the throughput is saturated when the number of issues is greater than 4.

Figure 8.21(a) shows the energy consumption of the three asynchronous bus architectures. MI-OCB and MO-OCB consume 6.76% and 3.98% more energy, respectively, than SI-OCB. The energy consumption per data transaction for the three asynchronous buses as a function of the number of issues is shown in Fig. 8.21(b). The value of 1 for the number of issues represents SI-OCB. With an increase in number of issues, energy consumption increases proportionally, since the hardware complexity of the reorder buffer increases linearly.

An asynchronous crossbar bus architecture called NEXUS was proposed by Lines [36]. NEXUS is based on the Quasi-Delay-Insensitive (QDI) timing model [37] requiring that the circuit functions correctly regardless of the delays of all gates or most wires. This conservative model forbids all forms of timing races, glitches, delay assumptions and clocks, and can work robustly over delay variations caused by power supply drop, in-die variations, crosstalk, and local heating. In such a QDI system, a separate wire cannot be used to indicate the validity of the data wire because one cannot make an assumption about the relative delay of the wires. Instead, the data and validity are mixed onto two wires. Together with

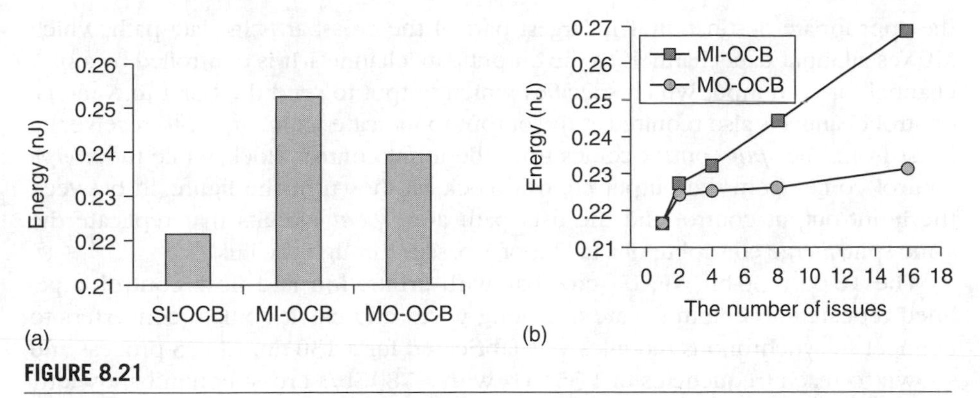

FIGURE 8.21

Simulation results for energy consumption: (a) energy consumption per data transaction (b) energy consumption per data transaction of MI-OCB and MO-OCB as a function of the number of issues [35]

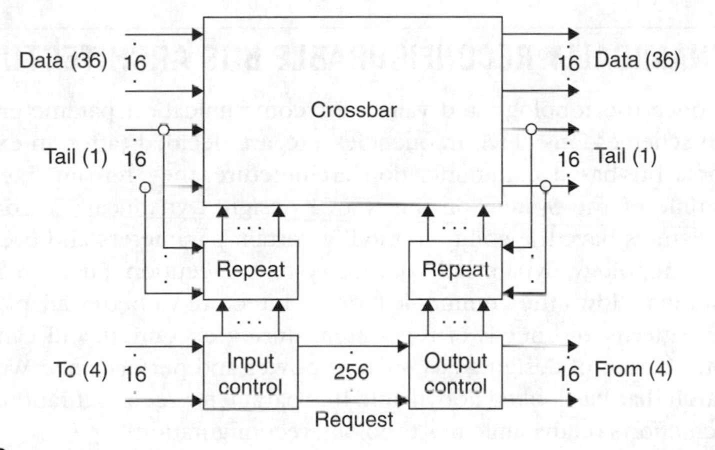

FIGURE 8.22

NEXUS crossbar decomposition [36]

a backward going acknowledge wire for flow control, these wires form an asynchronous channel. When both the data wires are 0, the channel is *neutral* and no data is present. When an initiator must send a bit, either the first or the second data wire is raised to send a logical 0 or 1. Once the receiver has received and stored the data, the receiver raises the acknowledge signal. Eventually, the sender puts the data wires back to *neutral*, after which the receiver lowers the acknowledge signal. This is called the four-phase dual rail handshake, and it is used in the asynchronous buses in NEXUS.

The NEXUS crossbar employs clock domain converters to bridge the asynchronous interconnect with the synchronous modules in the system. Figure 8.22 shows a decomposition of the NEXUS crossbar, which can support up to 16 modules. Data is transferred in bursts that cannot be fragmented, interleaved or dropped (i.e., *atomic* bursts). Each burst contains a variable number of words (NEXUS uses a 36-bit data path) terminated by a tail bit, and a 4-bit TO/FROM signaling to route the data to the appropriate destination. The largest part of the crossbar is its data path, which MUXes all input data channels to the output data channels. It is controlled by a *split* channel for each input which specifies which output to send the burst to. A merge control channel is also required at the output to indicate which input to receive the burst from. The *split* control comes from the input control block, while the *merge* control comes from the output control block, as shown in the figure. In between the input/output control and the data path are *repeat* circuits that replicate the same split/merge control until a tail bit of 1 passes through the link.

The 16-port, 36-bit NEXUS crossbar with arbitration and flow control, pipelined repeaters to communicate over long wires, and clock domain converters to connect to synchronous modules was fabricated for a 130-nm CMOS process and shown to reach frequencies of 1.35 GHz with a 780Gb/s cross-section bandwidth. The area footprint of a typical NEXUS system with all 16 ports used, and an average of two pipelined repeaters per link was reported to be 4.15 mm^2, which is a small and relatively reasonable fraction of the total chip area.

8.5 DYNAMICALLY RECONFIGURABLE BUS ARCHITECTURES

Typically, once the topology and values for communication parameters such as arbitration schemes, bus clock frequencies, etc. are decided (after an exploration phase) for a bus-based communication architecture, they remain fixed for the entire lifetime of the System on-chip (SoC) design. Dynamically reconfigurable bus architectures have the ability to modify certain parameters and even the bus architecture topology dynamically during system execution. Such an ability for reconfiguration allows the communication architecture to better adapt to changing traffic patterns and needs of the system during execution, and can result in better optimization of design goals, such as power and performance. We first present research that has looked at dynamic bus parameter reconfiguration, followed by research efforts on dynamic bus topology reconfiguration.

8.5.1 Dynamic Bus Architecture Parameter Reconfiguration

On-chip communication architecture standards such as AMBA [3, 14], IBM CoreConnect [4] and Sonics Smart Interconnect [6] provide limited support for dynamically reconfiguring their parameters, to adapt to changing application requirements at runtime [38]. For instance, Sonics Smart Interconnect allows software programmable arbitration and bandwidth allocation by dynamically varying TDMA slot allocation among components. Variable-length burst sizes and software programmable arbitration priorities are supported in AMBA and CoreConnect. An additional degree of configurability is provided in CoreConnect, by allowing each bus master to indicate a desired priority to the arbiter for each bus request.

8.5.1.1 *Communication Architecture Tuners*

Lahiri et al. proposed *communication architecture tuners* (CAT) [39, 40] to adapt on-chip communication architecture parameters (mainly arbitration priority) to the varying communication needs of the system and the varying nature of data being communicated. Figure 8.23 shows an example of a CAT-based communication

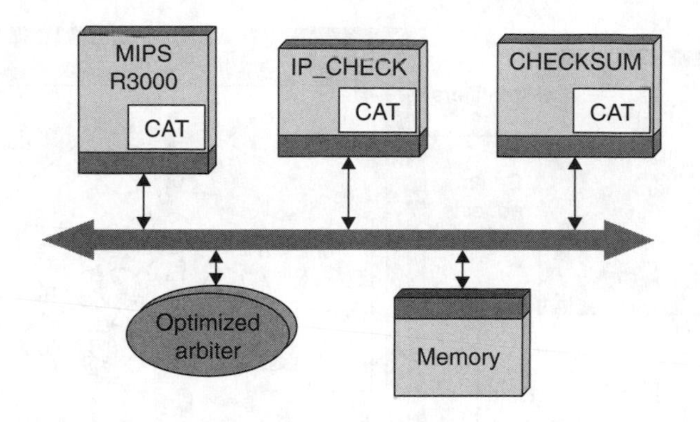

FIGURE 8.23

CAT-based communication architecture for a TCP system example [39, 40]
© 2004 IEEE

architecture for a TCP system from a network interface card. CATs are added to every master in the system and the arbiter and component interfaces are enhanced to handle CAT operation. Every time a communication request is generated by a component, its corresponding CAT is notified. The CAT also monitors the details of the data being communicated and the state of the component. For the system shown in Fig. 8.23, the CAT observes the packet size and deadline from the header of the packet being processed by the component, and groups communication requests based on the size and deadline of the packet being processed. The CAT then determines an appropriate arbitration priority value for the group of communication requests. Such a CAT-based communication architecture was able to dynamically adjust priorities and meet all the packet deadlines for the TCP example shown in Fig. 8.23. Whereas the traditional static priority assignment failed to meet the deadlines.

A more detailed view of the CAT module is shown in Fig. 8.24. CAT consists of a partition detector and a parameter generator circuit that generates arbitration priority values during system execution. A communication partition is a subset of the transactions generated by a component during execution. The partition detector circuit monitors and analyzes information generated by the component, such as transaction initiation requests, indications of importance of the data being processed, and tracer tokens. A component is enhanced to generate tracer tokens purely for the purpose of the CAT, to indicate specific operations to the CAT that the component is executing. The partition detector uses this information to identify the start and end of a sequence of consecutive communication transactions belonging to a partition. The parameter generator circuits generate values for communication arbitration priority based on the partition ID generated by the partition detector circuit (and other application specific data properties specified by the system designer). The generated value for the priority is sent to the arbiters and controllers in the communication architecture, to change its behavior.

Figure 8.25 shows the overall methodology for designing CAT-based communication architectures. In the first step, a performance analysis of the system execution traces for the partitioned/mapped system is performed, as described in [41], in order to obtain information and statistics for use in later steps. The trace-driven performance analysis technique used in [41] is considered comparable in accuracy to complete

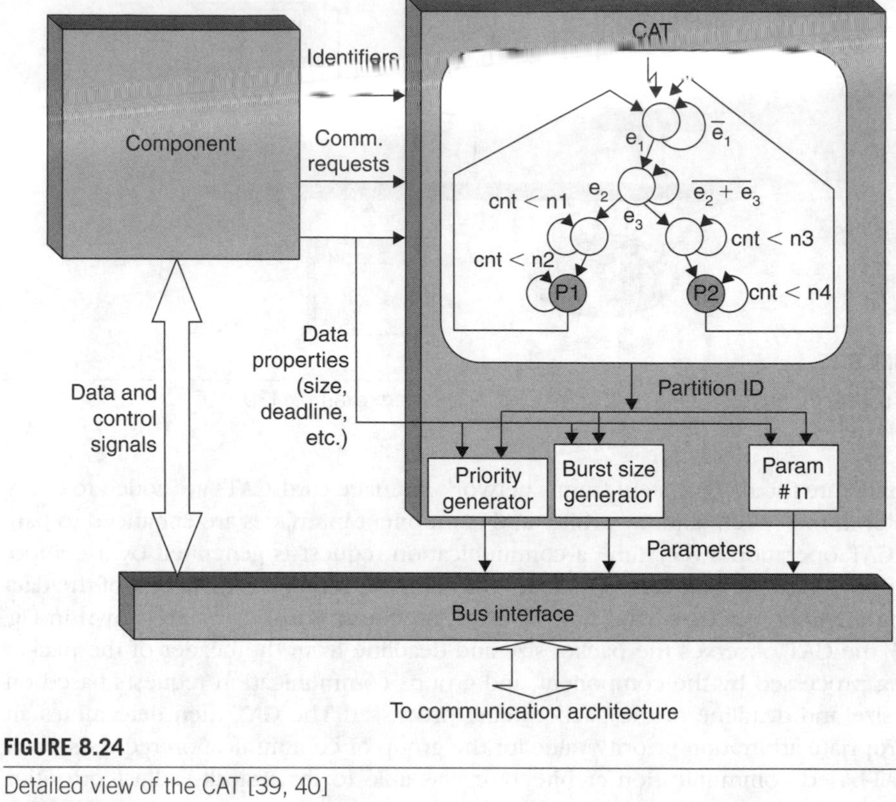

FIGURE 8.24

Detailed view of the CAT [39, 40]
© 2004 IEEE

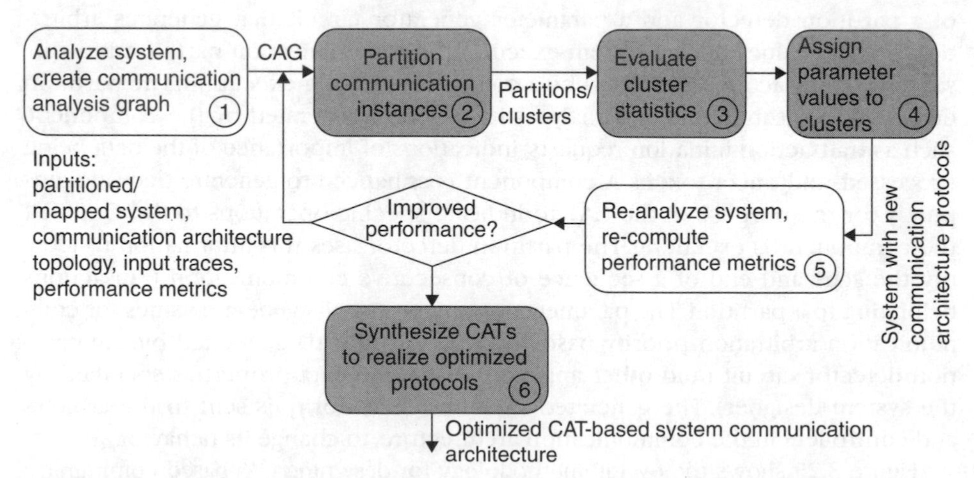

FIGURE 8.25

Methodology for designing CAT-based communication architectures [39, 40]
© 2004 IEEE

Table 8.2 Performance comparison of a CAT-based architecture with a conventional static priority-based architecture [39, 40]

Example system	Performance metric	Input trace	Static protocol-based architecture	CAT-based architecture	Improvement
TCP/IP	Missed deadlines	20 packets	10	0	–
SYS	Missed deadlines	573 transactions	413	17	24.3
ATM	Missed deadlines	169 packets	40	16	2.5
BRDG	Average cycles	10,000 cycles	304.72	254.1	1.2

system simulation, while being more efficient to employ in an iterative manner. The output of this analysis is a communication analysis graph (CAG), which is a compact representation of the system's execution for the given input traces. The vertices of this graph represent clusters of computation and abstract communications performed by the components during execution. The edges of the graph represent dependencies between the various computations and communications. The CAG can be quickly analyzed to obtain various performance statistics. In Step 2, the communication vertices in the CAG are grouped into a number of partitions. Each partition consists of events having similar communication requirements. In Step 3, various cluster statistics are evaluated, based on which arbitration priority values are assigned to the partitions in Step 4. Step 5 re-evaluates system performance for the new priority value assignments. If there is an improvement in performance, Steps 1–5 are repeated, till no further performance improvement can be achieved. Finally, Step 6 generates the CAT hardware to realize the optimized protocol generated in the previous steps.

Experimental studies were performed to compare an enhanced CAT-based communication architecture with conventional communication architectures, with static arbitration priority assignment. Four system testbenches were considered for the comparison study: (i) *TCP/IP*, a four component, single shared bus system shown in Fig. 8.23, (ii) SYS, a system with four components accessing a shared memory on a single shared bus, (iii) *ATM*, a packet forwarding unit of an ATM switch that consists of five components accessing a dual-port memory, over a single shared bus, and (iv) *BRDG*, a hierarchical shared bus-based system with two buses connected via a bridge, and six components, including two shared memories. Table 8.2 shows the performance benefits of using a CAT-based approach over a static arbitration priority-based conventional bus architecture. The performance objective for the *TCP/IP, SYS,* and *ATM* systems is to minimize the number of missed deadlines. For the *BRDG* system, each transaction is assigned a weight, and the overall performance of the system is measured using a weighted mean of the execution times of all the

Table 8.3 Effect of varying input traces (while maintaining comparable workloads) on the performance of CAT based architecture [39, 40]

	Deadlines Met (%)	
Input trace	**Static protocol-based architecture**	**CAT-based architecture**
T-6-0	13.06	94.62
T-6-1	12.86	93.47
T-6-2	12.06	93.47
T-6-3	11.9	94.1
T-6-4	10.64	95.48
T-6-5	11.62	94.08
T-6-6	11.24	96.89
T-6-7	13.3	95.07
T-6-8	12.17	94.47
T-6-9	14.76	94.55

bus transactions. The objective for this system is to minimize the weighted average processing time. Column 4 reports the performance of the static priority-based conventional bus architecture, while column 5 reports the results of the CAT-based communication architecture. The CAT-based architecture uses information such as weights on communication requests and deadlines to provide a more flexible and higher performance communication infrastructure.

Since this methodology is dependent on performance evaluation on execution traces, it is important to determine what influence the choice of input traces has on the performance of the CAT-based communication architecture. For this purpose, the performance of the CAT-based architecture and the conventional static priority bus architecture are compared for the *SYS* system testbench, for 10 different input traces (that present comparable workloads to the communication architecture), generated using random distributions for the timing, performance requirements, and size of the communication requests. Table 8.3 compares the fraction of deadlines met by the CAT-based architecture and the conventional static priority bus architecture. It can be seen that the performance gain for the CAT-based architecture over the conventional architecture is consistent across the traces, meeting 94.66% of the deadlines on an average. It is also clear that the performance advantage provided by CATs are relatively immune to the exact sequence and timing of input stimuli experienced by the system, because CATs are not tuned to the exact arrival times of communication requests, or packet sizes, and can therefore effectively track changes in deadlines and control-flow, which can impact performance.

Figure 8.26 shows an experiment comparing the performance of the CAT-based architecture with a static priority assignment based conventional bus architecture, for the *SYS* example, for 12 different input traces that present widely varying workloads to the communication architecture. Again, it can be seen from the figure

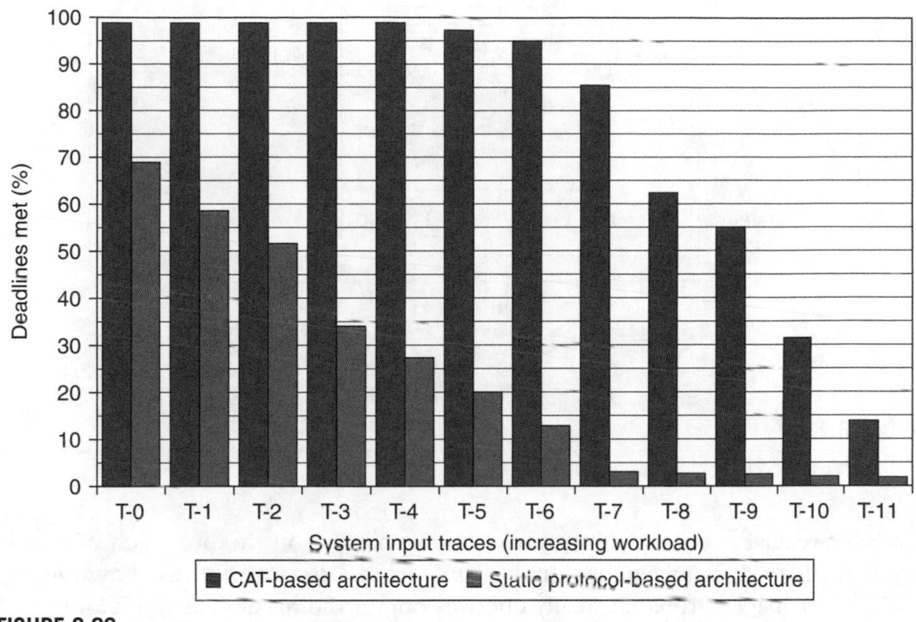

FIGURE 8.26

Performance comparison for CAT-based architecture for varying input traces (having widely different workloads) [39, 40]

© 2004 IEEE

that the CAT-based architecture provides better performance than the conventional bus architecture. The extent of the performance improvement varies, depending on the workload imposed by the trace. The benefit of an adaptive communication architecture like the CAT-based architecture is more pronounced for moderate to high workloads (e.g., *T-3* to *T-9*). For low workloads, both the conventional and CAT-based architectures are capable of meeting most of the deadlines, whereas for very high workloads, neither architecture can meet the deadlines. Consequently, for very low or very high workloads, the gains for the CAT-based architecture are comparatively smaller.

8.5.1.2 *LOTTERYBUS*

Lahiri et al. also proposed LOTTERYBUS [42, 43] to overcome the shortcomings of existing arbitration schemes, which can be inadequate under certain circumstances. For instance, the static priority scheme can lead to starvation of low priority masters under heavy traffic loads (i.e., the masters with low priority are rarely granted access to the bus, because of frequent high priority master transfers). On the other hand, the TDMA scheme provides a fairer distribution of bus bandwidth that can overcome starvation scenarios, but can lead to high transfer latencies due to the lack of flexibility in the static TDMA slot reservation. The LOTTERYBUS communication architecture attempts to provide effective bandwidth guarantees, while ensuring low latencies for bursty traffic with real-time latency constraints. LOTTERYBUS introduces a randomized arbitration algorithm implemented in a centralized *lottery manager* for each shared bus in an SoC. The lottery

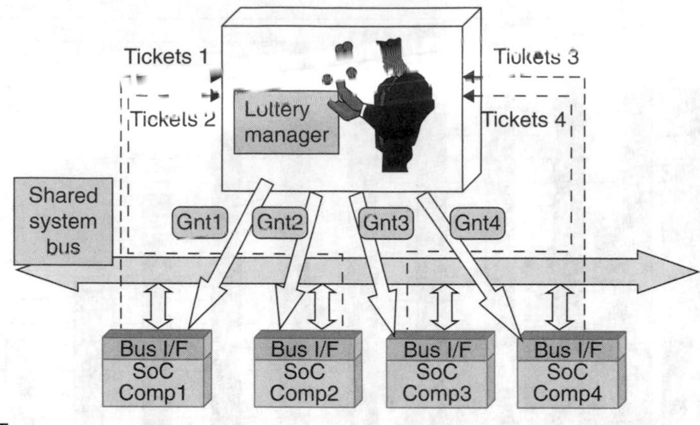

FIGURE 8.27

LOTTERYBUS communication architecture [42, 43]

manager receives requests from one or more masters on the bus, each of which is (statically or dynamically) assigned a number of *lottery tickets*, as shown in Fig. 8.27. The manager probabilistically chooses one of the masters as the winner and grants it access to the bus. While multiple word (burst) transfers are allowed, a maximum transfer size ensures that none of the masters monopolizes the bus for extended periods at a time.

The principle of the LOTTERYBUS operation can be explained as follows. Let $C_1, C_2, \ldots, C_n$ be the set of masters on the bus. Let the number of tickets held by a master be $t_1, t_2, \ldots, t_n$, and at any cycle. In addition, let the set of pending requests at any cycle be represented by $r_1, r_2, \ldots, r_n$, where $r_i = 1$ if master C_i has a pending request (and $r_i = 0$ otherwise). Then the probability that master C_i gets access to the bus is given by:

$$P(C_i) = \frac{r_i \cdot t_i}{\sum_{j=1}^{n} r_j \cdot t_j}$$

To decide on the arbitration winner, the lottery manager uses the notion of a lottery [68], and first examines the total number of tickets possessed by contending masters, given by $\sum_{j=1}^{n} r_j \cdot t_j$. It then generates a random number (or picks a winning lottery) from the range $[0, \sum_{j=1}^{n} r_j \cdot t_j)$[1] to determine which master to grant the bus to. If the number falls in the range $[0, r_1 \cdot t_1)$, the bus is granted to master C_1. If it falls in the range $[r_1 \cdot t_1, r_1 \cdot t_1 + r_2 \cdot t_2)$, it is granted to component C_2, and so on. For example, in Fig. 8.28, masters C_1, C_2, C_3, and C_4 are assigned 1, 2, 3, and 4 lottery tickets, respectively. In the bus cycle shown, only C_1, C_3, and C_4 have pending requests, and hence the number of current tickets is $\sum_{j=1}^{n} r_j \cdot t_j = 1 + 3 + 4 = 8$. The random number generator generates a number in the range [0,8) that happens to be 5, which lies between $r_1 \cdot t_1 + r_2 \cdot t_2 + r_3 \cdot t_3 = 4$,

[1] This set $[a, b)$ includes all the integers between a and b, inclusive of a but not b.

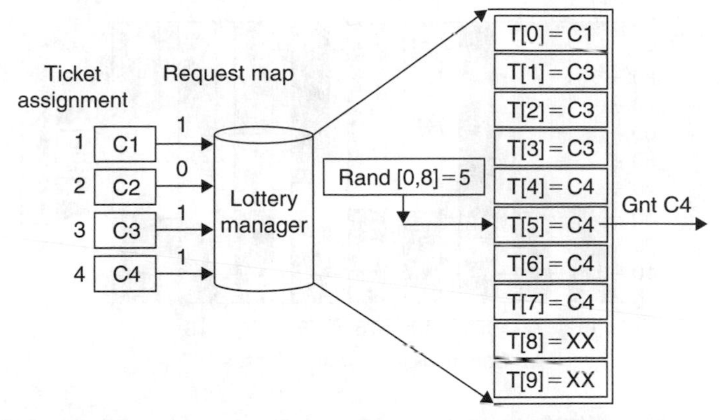

FIGURE 8.28

Example of lottery to determine arbitration winner in LOTTERYBUS [42, 43]
© 2006 IEEE

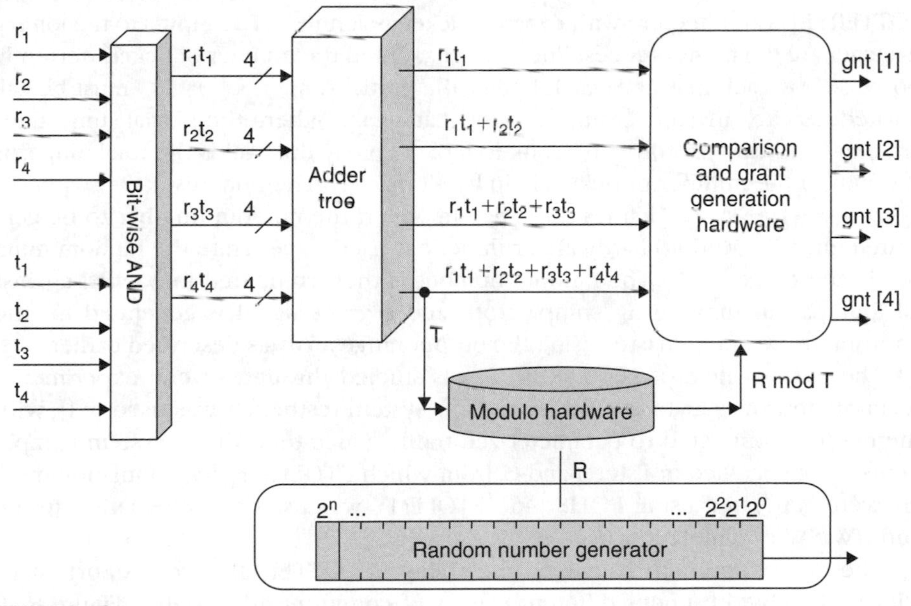

FIGURE 8.29

Lottery manager for dynamic LOTTERYBUS architecture [42, 43]
© 2006 IEEE

and $r_1 \cdot t_1 + r_2 \cdot t_2 + r_3 \cdot t_3 + r_4 \cdot t_4 = 8$. Therefore, the bus is granted to master C_4. LOTTERYBUS addresses the problem of a low priority master not being able to access the bus for extended periods of time, since the probability p that a component with t tickets is able to access the bus within n lottery drawings is given by the $1 - (t/T)^n$, which converges rapidly to 1, ensuring that no master is starved.

As mentioned earlier, tickets in the LOTTERYBUS architecture can be assigned to masters either statically or dynamically. The dynamic ticket assignment case is of particular interest, since it allows better adaptation to changing traffic and

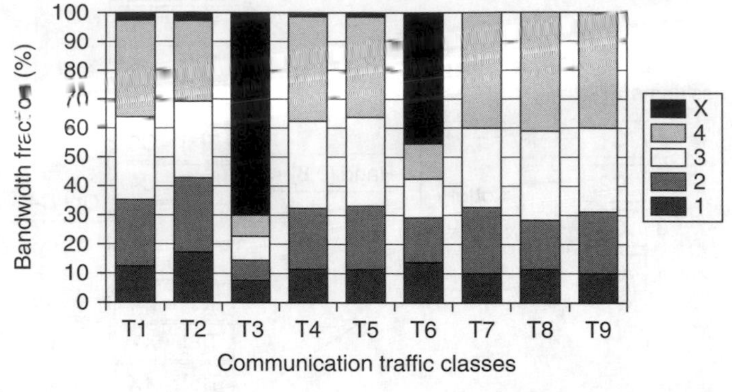

FIGURE 8.30

Bandwidth allocation of LOTTERYBUS for different communication traffic classes [42, 43]
© 2006 IEEE

performance requirements. Figure 8.29 shows the lottery manager for the LOTTERYBUS architecture with dynamic ticket assignment. The inputs to the lottery manager are the master request lines (r_1,r_2,r_3,r_4) and the number of tickets currently possessed by each master. At each lottery, the partial sum $\sum_{j=1}^{i} r_j \cdot t_j$ must be calculated for each master C_i (unlike in the static case, where the partial sum values are fixed and can be stored in a lookup table). For C_4, this yields the total range, or the sum of the number of tickets held by all masters. The final result, $T = r_1 \cdot t_1 + r_2 \cdot t_2 + r_3 \cdot t_3 + r_4 \cdot t_4$ defines the range in which the random number to be generated must lie. Modulo hardware arithmetic is used to generate the random number in the range $[0, T)$. The random number is then compared in parallel against all four partial sums using comparators, and a grant signal is generated for the appropriate winning master, using the output range analysis described earlier.

The performance of LOTTERYBUS was studied through several experiments. A simple four master, four slave shared bus system testbench was used [44], with the masters connected to parameterized traffic generators. All the system components were specified in Esterel and C, from which PTOLEMY [45] simulation models were generated using POLIS [46]. PTOLEMY was used for schematic capture and HW/SW co-simulation.

The first experiment examined the ability of LOTTERYBUS to proportionally allocate bandwidth under different classes of communication traffic. Figure 8.30 shows the results of this experiment, with the *x*-axis depicting nine different communication traffic classes and the *y*-axis depicting the fraction of the total bus bandwidth allocated to masters. It can be seen that for traffic classes with high bus utilization, the bandwidth allocated closely follows the assignment of lottery tickets. Tickets were assigned in the ratio 1:2:3:4 and for the traffic classes *T4, T5, T7, T8,* and *T9*, the bandwidth allocated is in the ratio of 1.15:2.09:2.96:3.83. However, for cases when the bus is partially un-utilized (e.g., *T3, T6*), the bandwidth allocation does not follow ticket assignment and is roughly the same for all components. This is because due to the sparse nature of communication in these classes, immediate grants are issued to most requests. These results show that LOTTERYBUS is capable of providing efficient control over bus bandwidth allocation for a variety of traffic classes and a varying level of bus utilization.

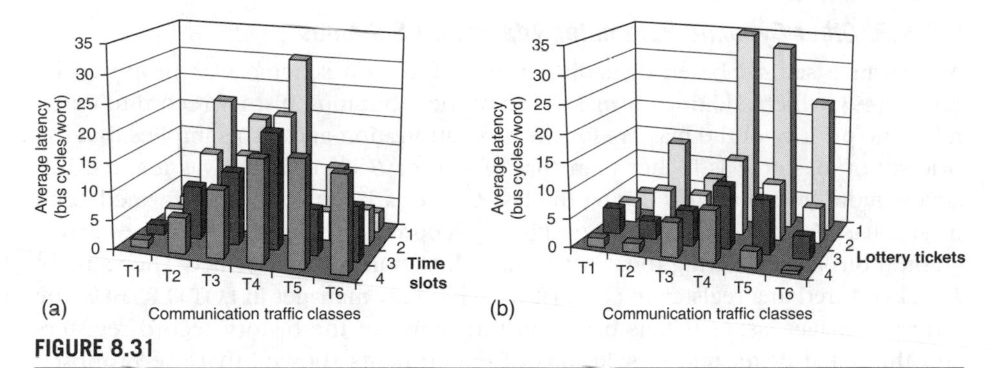

FIGURE 8.31

Communication latencies over different communication traffic classes for (a) TDMA and (b) LOTTERYBUS [42, 43]

Figure 8.31 compares the latency of the TDMA and LOTTERYBUS architectures over six different communication traffic classes. The *x*-axis depicts different traffic classes, while the *y*-axis depicts timeslots (Fig. 8.31(a)) and lottery tickets (Fig. 8.31(b)) assigned to the masters. The *z*-axis depicts the average communication latency per word. It can be seen from the figures that LOTTERYBUS exhibits better latency behavior than the TDMA architecture for a wide range of traffic classes. Most importantly, the communication latency for high priority masters varies significantly for the TDMA architecture (1.65 to 20.5 cycles per word), because the latency of communication in TDMA is highly sensitive to the timing wheel position (i.e., which master's slot currently has access to the bus) when the request arrives. The LOTTERYBUS architecture does not exhibit this phenomenon and ensures low latencies for high priority masters.

The experimental results thus show that LOTTEYRBUS is able to simultaneously provide low latencies for high priority traffic, while at the same time providing proportional bandwidth guarantees. The LOTTERYBUS architecture was implemented on top of the AMBA AHB architecture [3] and synthesized using the Synopsys Design Compiler [47] for a 0.15 μm CMOS cell library from NEC [48]. A communication architecture area increase of 16% for the static LOTTERYBUS and 24% for the dynamic LOTTERYBUS architecture was observed over the static priority-based communication architecture area. The critical path of the static priority-based architecture was observed to be 1.68 ns, enabling, at least theoretically, bus speeds up to 595 MHz. This critical path delay was unchanged for the static LOTTERYBUS architecture. The critical path delay for the arbiter in the dynamic LOTTERYBUS architecture was measured to be 1.92 ns, which is a 14% increase in the critical path of the overall communication architecture. With current technology scaling trends, such a logic delay will play a decreasing role compared to the global wire delay in determining overall communication architecture clock frequency [15], making the deployment of dynamic LOTTERYBUS architecture more feasible in future designs. The authors did not provide any information about the power dissipation overhead of the additional circuits, which is also a critical factor in considering such schemes for deployment.

8.5.1.3 *Other Dynamic Parameter Adaptation Schemes*

A derivative statistic-based dynamic lottery arbitration scheme was proposed by Zhang [49], which additionally makes use of the arbitration history record to determine the priority of the masters for the next arbitration grant. This implies that the priority of master *A* is higher than that of master *B* if master *A* was granted bus access more than master *B* during the last *L* times arbitration was performed. Each master has *M* registers to store their history record for the number of times it was granted bus access during the last *L* times arbitration was performed. The value of *L* is also stored in a register. In contrast to the lottery manager in LOTTERYBUS, the lottery manager issues tickets based on the values of the history record registers and the initial ticket registers. Results of experiments showed that the proposed statistic-based lottery scheme provided superior performance compared to the lottery-based scheme in LOTTERYBUS. However, no experiments were performed by the authors to determine the additional area overhead or timing impact of the history-based ticket generation. It was also not shown whether the statistic-based lottery scheme provides low latencies for high priority traffic, while at the same time providing proportional bandwidth guarantees, like LOTTERYBUS.

A *dynamic fraction control bus architecture* was proposed by Wang and Bayoumi [50] to provide similar benefits as LOTTERYBUS, but with lower system cost and design complexity. Additionally, since arbitration in LOTTERYBUS is based on probability, it becomes hard to implement accurate control over the bus bandwidth allocation for applications. In the proposed fraction control bus, bandwidth fractions are assigned to master components based on their communication requirements. The greater the fraction value, the greater the priority. Figure 8.32 shows the architecture of the fraction bus arbiter and decoder. The decoder is responsible for granting bus access to masters or produce chip select signals for comparators, based on master requests. The fraction calculator calculates the real-time bandwidth fraction for each master. The assigned fraction values for the masters are stored in a Lookup Table (LUT). Comparators perform fraction comparison and grant access to the master that satisfies arbitration conditions. The proposed fraction control bus can

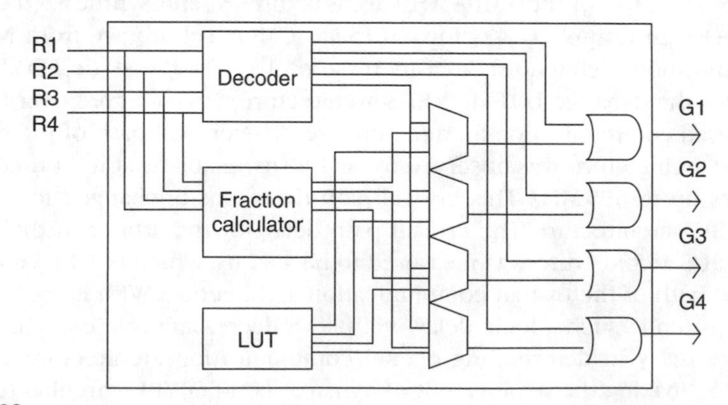

FIGURE 8.32

Architecture of the fraction bus arbiter and decoder [50]

be implemented statically or dynamically. In the *static fraction control bus* (SFCB), the fractions assigned to the masters are fixed. In the *dynamic fraction control bus* (DFCB), the fractions are initially set to a fixed value, and are thereafter continuously tuned to adapt to communication circumstances dynamically. The assigned bandwidth fraction for a master can be increased if it has pending requests more than a threshold value. In such a case, the master can *borrow* the bandwidth fractions from other masters temporarily, till the number of pending requests drops below the threshold value. Both the threshold value and the amount of bandwidth that can be borrowed are configurable by the designer.

Several experiments were performed to determine the effectiveness of the fraction control bus over LOTTERYBUS and conventional static priority bus architectures. Table 8.4 compares the gate count and achievable bus speed for the implementations of the static priority bus architecture, the LOTTERYBUS architecture and the static (SFCB) and dynamic (DFCB) fraction control bus architectures. The implementations were mapped onto the Xilinx Vertex2Pro FPGA. It can be seen from the table that the proposed fraction control bus architectures have lower area and delay compared to LOTTERYBUS. To evaluate and compare the performance of the fraction control bus, a four master subsystem for the cell forwarding unit of an ATM switch [51] was considered. Four versions of the system were implemented, with: (i) the static priority bus, with priorities 4, 3, 2, 1; (ii) LOTTERYBUS, with lottery numbers 1:1:4:6; and (iii), (iv) SFCB and DFCB architectures, with fractions 15%:15%:60%:10%. Table 8.5 shows the results of this experiment. For the static priority bus, the communication latency of the master with the highest priority is the lowest, but the bandwidth fraction for the masters with low priorities is extremely low, due to starvation. The fraction control buses (SFCB and DFCB) have lower system cost while

Table 8.4 Design complexity and achievable bus speed comparison [50]

	Gate counts	Delay (ns)	Max speed (MHz)
Priority	86	2.707	369.41
LotteryBus	152	3.276	305.72
SFCB	104	2.987	334.89
DFCB	134	3.113	321.23

Table 8.5 Performance comparison for ATM switch example [50]

	Port4 latency (cycles)	Port4 BW (%)	Port3 BW (%)	Port2 BW (%)	Port1 BW (%)
Priority	1.39	9.56	60.6	29.83	0.01
Lottery	1.4	9.32	63.6	15.15	11.93
SFCB	1.42	10.32	60.4	14.64	14.64
DFCB	1.46	9.74	58.25	16.34	15.67

maintaining comparable communication latencies, with LOTTERYBUS, SFCB and DFCB can also be seen to have a more accurate control over the allocation of bandwidth fractions, than other buses.

A time-division-based bus architecture which dynamically allocates TDMA timeslots (*dTDMA*), was proposed by Richardson et al. [52]. In dTDMA, the bus arbiter dynamically grows or shrinks the number of timeslots to match the number of active transmitters, as shown in Fig. 8.33. When a master needs to transmit data on the bus, it asserts its active signal to the arbiter, to request a timeslot. The arbiter uses a number of techniques to decide on a timeslot assignment for each master and produces a new configuration for each active transmitter and receiver before the beginning of the next clock cycle. On the next clock edge, the timeslot configuration data is loaded by the transmitters and receivers, and normal operation is continued. When a master finishes transmitting, it de-asserts its active signal, following which the arbiter de-allocates its timeslot in the same manner as it allocated it (Fig. 8.33). Such a dynamic timeslot assignment produces the most efficient timeslot allocation without any slot wastage. The only overhead is the one cycle initial communication delay when a timeslot is allocated. Various methods can be used to assign timeslot, including (but not limited to) methods based on the status of the transmit buffers or the length of the wait time.

The address mapped dTDMA bus architecture has several advantages over standard bus architectures such as AMBA [3]. Because of its memory-oriented design, AMBA imposes certain restrictions on the nature of addressing behavior. In addition to a 1kB address boundary on sequential transfers, a new transaction must be initiated if the next address is not an increment of the previous (i.e., a non-sequential access). These restrictions result in repeated arbitration overhead. For instance, a long transmission crossing the 1kB boundary must re-arbitrate, at the risk of losing bus ownership. Such re-arbitration can waste several cycles and add a significant overhead. To overcome these drawbacks during data streaming, the dTDMA bus architecture is transaction-less and address mapped, with each component on the bus being assigned a unique identifier. The dTDMA bus only requires re-arbitration when the destination of the data stream transfer changes. In contrast, for AMBA, transfer requests that are not sequential or that cross the 1kB boundary, require arbitration even if the destination of the transfer remains the same.

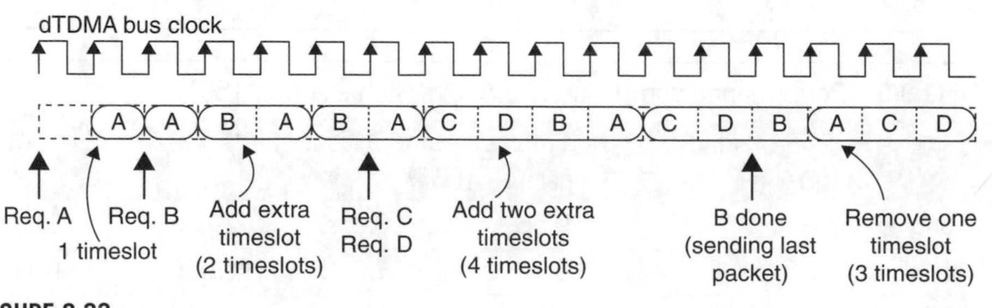

FIGURE 8.33

Dynamic timeslot allocation example [52]

A simple comparison between AMBA and dTDMA is shown in Fig. 8.34. In this example, master *A* requests access to the bus at clock *T1*, followed by a request from master *B* at clock *T2*. The AMBA arbiter does not grant access to master *A* before clock *T3*, whereas the dTDMA arbiter issues the new timeslot configuration before the end of clock *T1*. Data transmission for AMBA commences at *T5*, since *T4* is dedicated to the address phase of the transaction. In dTDMA, data transmission commences earlier, at *T2*. Master *B* has to wait five cycles from request to data transfer in AMBA, but only one cycle in dTDMA. In the example, the dTDMA bus architecture completes the transmission of two words from each master three cycles before the AMBA bus does.

An attractive quality of dTDMA is its predictable latencies, since a component is guaranteed to wait no longer than the number of active transmitters. In contrast, in the AMBA bus, a master may need to wait for an indeterminate amount of time before being granted access to the bus. The transaction-less, address-mapped dTDMA bus also requires fewer arbitrations compared to AMBA, and needs fewer cycles for the arbitration process, which improves overall performance.

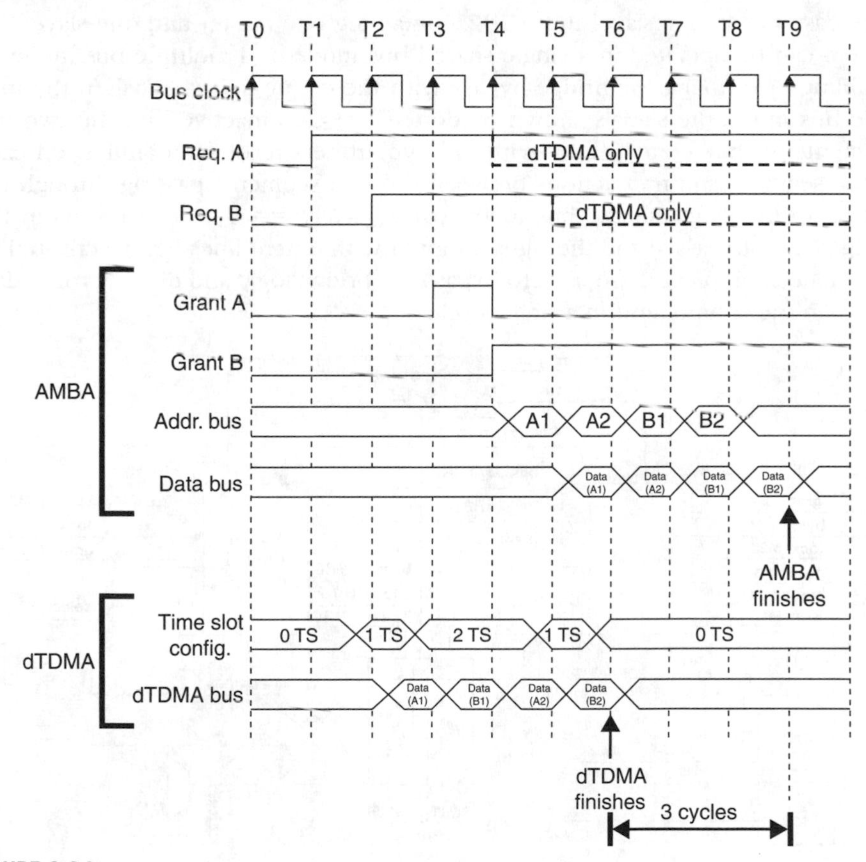

FIGURE 8.34

A simple two-component transfer on dTDMA and AMBA [52]

8.5.2 Dynamic Bus Architecture Topology Reconfiguration

In addition to dynamically configuring bus architecture protocol parameters such as arbitration schemes and burst sizes, it is also possible to change the topology of the bus architecture dynamically. Sekar et al. [51] proposed the *FLEXBUS* architecture, which is a high performance on-chip bus architecture with a dynamically configurable topology that can be implemented on top of an existing standard communication architecture such as AMBA AHB [3]. FLEXBUS is capable of detecting runtime variations in communication traffic, and adapting the topology of the communication architecture in two ways: (i) dynamic bridge bypass, which enables bus topology customizations via runtime fusing and splitting of bus segments and (ii) dynamic component re-mapping, which allows runtime switching of components from one bus segment to another. The key challenges of such an approach are maintaining compatibility with existing bus standards, minimizing the timing impact, minimizing the logic and wiring complexity, and providing low reconfiguration overhead.

The hardware required to support dynamic bridge bypass is shown in Fig. 8.35, for a system consisting of two AMBA AHB bus segments: *AHB1*, containing two masters and one slave; and *AHB2*, containing one master and one slave. This system can be operated in a single shared bus mode or a multiple bus mode by disabling or enabling the bridge bypass with the *config_select* signal. In the multiple bus mode, the signals shown by dotted lines are inactive, and the two bus segments operate concurrently, with the two arbiters resolving conflicts on each of the segment and transactions between the two segments passing through the bridge. In the single shared bus mode, *config_select == 1*, which results in the bridge being bypassed and the signals shown with dotted lines being activated. In this mode, multiplexers are used to bypass the bridge logic and directly route data between the components in a single cycle.

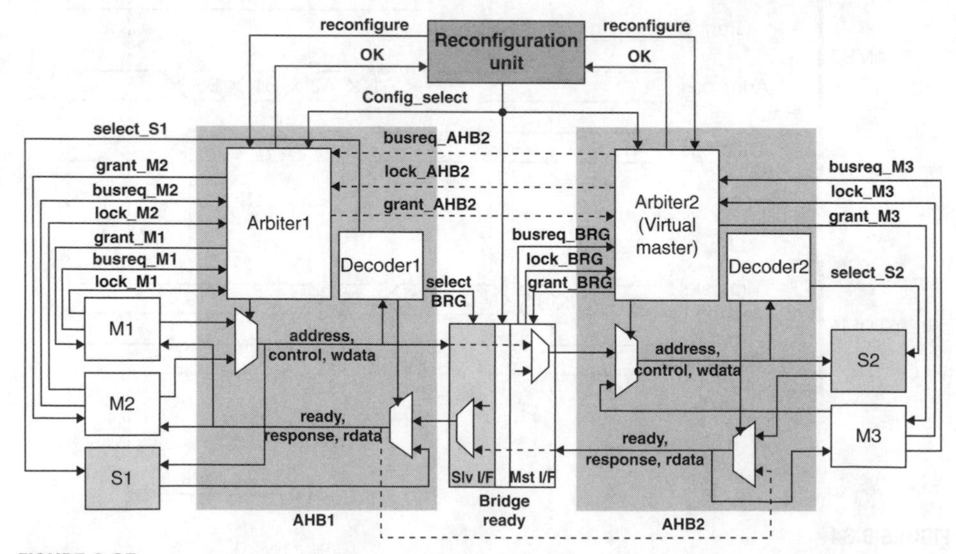

FIGURE 8.35

Dynamic bridge bypass hardware in FLEXBUS [51]

Arbitration in a single bus mode only grants one master access to the bus, whereas in multiple bus mode more than one master can have transactions executing in parallel. The arbitration mechanism of the multiple bus mode is adapted to meet the requirements of the single bus mode in FLEXBUS. This is done by using a distributed arbitration mechanism in the single bus mode, in which one of the arbiters acts as a virtual master that is regulated by the other arbiter. For example, in Fig. 8.35, if *Arbiter2* receives a transfer request from a master on *AHB2*, it immediately sends a request to *Arbiter1* using *busreq_AHB2* and *lock_AHB2* signals. *Arbiter1* arbitrates from the requests received on *AHB1*, as well as the requests from the virtual master. In parallel, to reduce arbitration latency, *Arbiter2* arbitrates among its received requests. However, it grants *AHB2* to its selected master only after receiving the *grant_AHB2* signal from *Arbiter1*, thus ensuring that only one master gets access to the bus in the single bus mode.

The reconfiguration unit (Fig. 8.35) selects the bus configuration at runtime, and ensures correct operation of the system when switching between the two configurations. The worst case overhead of bus reconfiguration for the two bus segment AMBA system is 17 clock cycles, assuming a single cycle slave response and that the bus is not locked. The runtime reconfiguration policy used in FLEXBUS can be described as follows: at runtime, the system observes the number of transactions on each local bus segment, as well as transactions between bus segments, for a time period *T*. Assuming that the average number of cycles required for a local transaction, and a cross-bridge transaction are known, the reconfiguration unit calculates the time required to process traffic for the single bus mode and the multiple bus mode. If the time taken to process the traffic for the single bus mode is less than the time for the multiple bus mode, the reconfiguration unit selects the single bus mode. Otherwise, the multiple bus mode is selected. The configuration time period *T* must be carefully set by the designer. A smaller value for *T* can result in a system more responsive to variations in traffic conditions, but if the traffic characteristics change rapidly, it can result in frequent switching between the configurations, and performance degradation due to the large reconfiguration overhead.

The hardware required to implement dynamic component re-mapping is shown in Fig. 8.36, which illustrates a two segment AMBA AHB architecture in which master *M2* and slave *S2* can be dynamically mapped to either *AHB1* or *AHB2*. The signals *config_select_M2* and *config_select_S2* are used to select the mapping of *M2* and *S2*, respectively. The signals of the re-mappable master or slave are connected to both buses, but the switch logic in *SWITCH_M* and *SWITCH_S* activates the signals for only one of the buses at a time, depending on the configuration chosen. The arbiters do not require any change when re-mapping components since master requests are only sent to the arbiter on the bus to which a master is connected. However, the decoders on the two bus segments need to be reconfigured to generate the correct signal for the re-mappable slave. The *Remap unit* (Fig. 8.36) is responsible for generating the signals to select the master and slave mapping configurations. Monitoring strategies, similar to the ones used for the dynamic bridge bypass can be used to determine when to remap the re-mappable masters and slaves.

Experimental studies were performed to evaluate the usefulness of the FLEXBUS approach. The AMBA AHB RTL description from the Synopsys

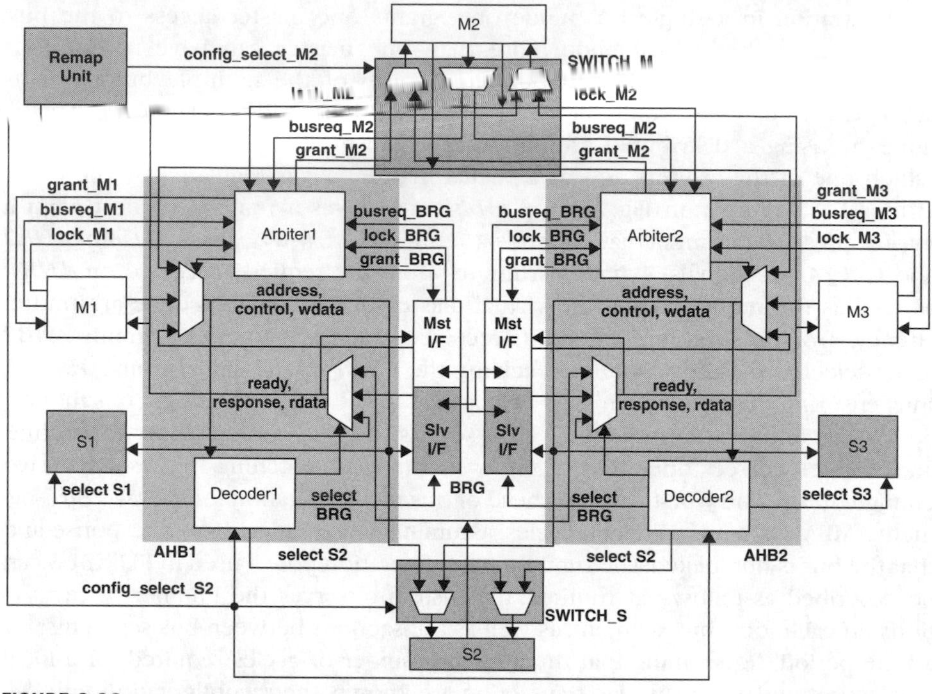

FIGURE 8.36

Dynamic component re-mapping hardware in FLEXBUS [51]
© 2005 IEEE

Designware [53] library was enhanced with the additional hardware descriptions needed to implement FLEXBUS. A system with eight masters (traffic generators) and eight slaves was used as a testbench. The first experiment explored the area and timing characteristics of the FLEXBUS implementation for a dynamic bridge bypass, and compared the results with those for a single shared bus architecture, and a multiple shared bus architecture. Table 8.6 shows the results from the experiment, for an implementation in the 0.13μm CMOS process technology [54]. FLEXBUS incurs a small delay penalty compared to statically configured architectures, due to the additional wiring and logic delay. For FLEXBUS, the critical path delay in the multiple bus mode is smaller than in the single bus mode since many of the long paths present in the single bus mode are not used in the multiple bus mode. The static multiple bus architecture has a smaller delay than the single shared bus due to less bus loading and shorter wire lengths.

Next, an experiment was performed to analyze the performance of FLEXBUS under a synthetic traffic profile [51]. Figure 8.37 plots the cumulative latency for the different architectures. It can be seen that FLEXBUS successfully adapts to frequent changes in traffic characteristics, to achieve performance improvements over the static single shared bus (21.3%) and multiple shared bus (17.5%) architectures. Finally, the performance of FLEXBUS and conventional architectures was compared for an IEEE 802.11 MAC (Message Authentication Code) processor-based SoC subsystem. All the buses were operated at 200 MHz. Table 8.7 shows the average time

Table 8.6 FLEXBUS hardware implementation results [51]

Bus architecture	Area (sq. mm)	Delay (ns)	Frequency (MHz)
Single shared bus	82.12	4.59	218
Multiple bus	84.27	3.79	264
FLEXBUS (single bus mode)	82.66	4.72	212
FLEXBUS (multiple bus mode)		3.93	254

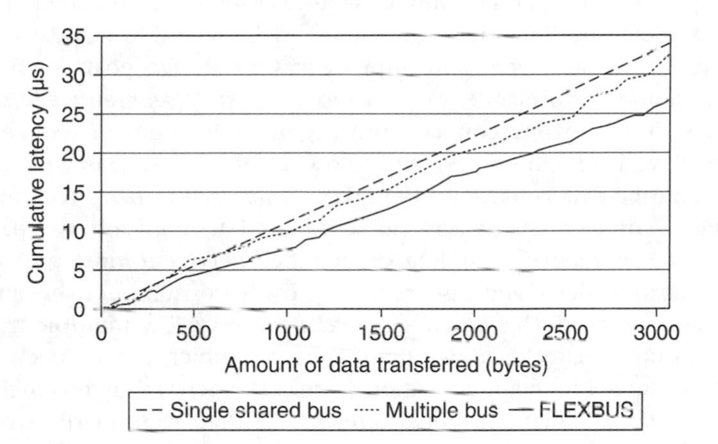

FIGURE 8.37

Cumulative frequency for different bus architecture, under synthetic traffic profiles [51]

Table 8.7 Performance of 802.11 MAC processor-based SoC subsystem for different communication architectures [51]

Bus architecture	Computation time (ns)	Data transfer time (ns)	Total time (ns)
Single shared bus	42,480	–	42,480
Multiple bus	26,905	12,800	39,705
FLEXBUS (bridge by-pass)	27,025	5,290	32,315
FLEXBUS (component re-mapping)	27,010	5,270	32,280
Ideally reconfigurable bus	26,905	5,120	32,025

taken to process a single frame of size 1kB, for different bus architectures. It can be seen that the times required by both variants of the FLEXBUS architecture are smaller compared to conventional architectures. FLEXBUS and its reconfiguration policies are also seen to perform close to the ideal case, which assumes no reconfiguration overhead, and an ideal reconfiguration policy having full knowledge of future bus traffic.

8.6 SUMMARY

In this chapter, we presented custom bus-based communication architectures that attempt to overcome limitations of commercially available standard bus-based communication architectures, and improve system design goals such as power consumption and performance. We first looked at *split/segmented bus architectures* that split a long interconnect into segments in order to reduce the wire delay, capacitive load and consequently power consumption, as well as increase parallelism during data communication. *Serial bus architectures* reduce the number of wires connecting components, compared to conventional parallel bus architectures. This reduces coupling capacitance between wires, which reduces signal propagation delay and also reduces power consumption because of the reduced capacitance. *CDMA-based bus architectures* allow multiple transmitters to send data on a shared medium simultaneously, which can reduce traffic conflicts and consequently communication latency. Asynchronous bus architectures, unlike conventional commercial bus architectures that are primarily synchronous, do not use a global clock signal for synchronization. Since the global clock consumes a significant amount of power, asynchronous bus architectures have lower power consumption. Finally, *dynamically reconfigurable bus architectures* allow the topology and/or protocol parameters such as arbitration schemes to dynamically change and adapt to changing traffic profiles and data characteristics, in order to improve performance.

FURTHER READING

Very early work in the area of custom bus architecture design resulted in a synchronous, high performance, split and burst transaction capable, pipelined, shared bus called the HiPi+ bus [55], which extended an even earlier work that proposed the high performance HiPi bus [56]. The concepts proposed in the work were a precursor to the enhancements that followed in several commercial on-chip bus architectures. Several works have proposed using custom circuit techniques [57–63] to improve the performance of hierarchical and crossbar bus-based conventional communication architectures such as AMBA AHB/AXI. A performance analysis of commonly used arbitration schemes was presented in [64], and custom variations on existing arbitration schemes, such as the direct mapped slot allocation TDMA [65], have been proposed to improve the performance of bus architectures. A custom wrapper-based bus (NECoBUS) [66] was proposed to reduce the latency of wrappers in wrapper-based bus architectures. NECoBUS employs several latency reduction techniques such as retry encapsulation, write-buffer switching, early bus requests, and converter-based multiple bit-width connections to remove the latency penalty induced in the conventional wrapper-based bus design.

REFERENCES

[1] Y. Zhang, W. Ye and M. J. Irwin, "An alternative architecture for on-chip global interconnect: Segmented bus power modeling," in *Proceedings of Thirty-Second Asilomar Conference on Signals, Systems & Computers*, 1998, pp. 1062–1065.

[2] Y. Zhang, R. Y. Chen, W. YE and M. J. Irwin, "System Level Interconnect Modeling," *Proceedings of the International ASIC Conference*, September 1998, pp. 289–293.

[3] ARM AMBA Specification and Multi layer AHB Specification (rev2.0), http://www.arm.com, 2001.

[4] IBM CoreConnect Specification, http://www.ibm.com/chips/techlib/techlib.nsf/product families/CoreConnect_Bus_Architecture.

[5] "STBus Communication System: Concepts and Definitions," *Reference Guide*, STMicro Electronics, May 2003.

[6] Sonics SMART Interconnect, http://www.sonicsinc.com.

[7] WISHBONE specification, http://www.opencores.org/wishbone.

[8] Altera AVALON Interface Specification, April 2006, http://www.altera.com/.

[9] R. Cheng-Ta Hsieh and M. Pedram, "Architectural energy optimization by bus splitting," in *Proceedings of IEEE Transactions on Computer-Aided Design of Integrated Circuits and Systems*, Vol. 21, No. 4, April 2002, pp. 408–414.

[10] R. Lu and C.-K. Koh, "A high performance bus communication architecture through bus splitting," in *Proceedings of the Asia and South Pacific Design Automation Conference (ASP-DAC)*, 2004, pp. 751–755.

[11] J. Rabaey and M. Pedram, *Low Power Design Methodologies*, Kluwer Academic Publishers, Norwell, MA, 1996.

[12] E. Macii, M. Pedram and F. Somenzi, "High-level power modeling, estimation and optimization," *IEEE Transactions on Computer Aided Design*, Vol. 17, Nov. 1998, pp. 1061–1079.

[13] R. Lu and C.-K. Koh, "SAMBA-bus: A high performance bus architecture for system-on-chips," in *Proceedings of International Conference on Computer Aided Design, (ICCAD)*, 2003, pp. 8–12.

[14] ARM AMBA 3.0 AXI Specification www.arm.com/armtech/AXI.

[15] R. Ho, K. W. Mai and M. A. Horowitz, "The Future of Wires," in *Proceedings of the IEEE*, Vol. 89, April 2001.

[16] Semiconductor Industry Association, International Technology Roadmap for Semi-conductors, 2003.

[17] P. Saxena and C. Liu, "A postprocessing algorithm for crosstalk-driven wire perturbation," *IEEE Transaction on Computer-Aided Design of Integrated Circuits and Systems*, April 2000, pp. 691–702.

[18] L. Macchiarulo, E. Macii and M. Poncino, "Wire placement for crosstalk energy minimization in address buses," in *Proceedings of Design, Automation and Test in Europe (DATE)*, March 2002, pp. 158–162.

[19] J. Cong, L. He, C.-K. Koh and Z. Pan, "Interconnect Sizing and Spacing with Consideration of Coupling Capacitance," *IEEE Transactions on Computer-Aided Design of Integrated Circuits and Systems*, Vol. 20, No. 9, September 2001, pp. 1161–1169.

[20] E. Macii, M. Poncino and S. Salerno, "Combining wire swapping and spacing for low-power deep-submicron buses," in *Proceedings of the IEEE Great Lakes Symposium on VLSI (GLS-VLSI)*, April 2003, pp. 198–202.

[21] P. Gupta and A. Kahng, "Wire swizzling to reduce delay uncertainty due to capacitive coupling," in *Proceedings of the International Conference on VLSI Design (VLSID)*, January 2004.

[22] Y. Shin and T. Sakurai, "Coupling-driven bus design for low-power application-specific systems," in *Proceedings of Annual ACM/IEEE Design Automation Conference (DAC)*, 2001, pp. 750–753.

[23] A. B. Kahng et al., "Interconnect Tuning Strategies for High Performance ICs," in *Proceedings of Design, Automation and Test in Europe (DATE)*, 1998, pp. 471–478.

[24] K. Hirose and H. Yasuura, "A bus delay reduction technique considering crosstalk," in *Proceedings of Design, Automation and Test in Europe (DATE)*, 2000, pp. 441–445.

[25] M. Ghoneima, Y. Ismail, M. Khellah, J. Tschanz and V. De, "Serial-link bus: a low-power on-chip bus architecture," *IEEE/ACM International Conference on Computer-Aided Design (ICCAD)*, 2005, pp. 541–546.

[26] N. Hatta, N. Demus Barli, C. Iwama, L. Dinh Hung, D. Tashiro, S. Sakai and H. Tanaka, "Bus serialization for reducing power consumption," *IPSJ Transactions on Advanced Computing Systems*, Vol. 47, No. 3, 2006, pp. 49–57.

[27] W. Steinhoegl et al., "Scaling laws for the resistivity increase of sub-100 nm interconnects," *International Conference on Simulation of Semiconductor Processes and Devices (SISPAD)*, September 2003, pp. 27–30.

[28] M. Frank Chang et al., "RF/wireless interconnect for inter- and intra-chip communications," in *Proceedings of the IEEE*, Vol. 89, No. 4, 2001, pp. 456–466.

[29] J. S. Lee and L. E. Miller, *CDMA Systems Engineering Handbook*, Artech House Publish, 1998. ISBN: 0-89006-990-5.

[30] B.-C. C. Lai, P. Schaumont and I. Verbauwhede, "CT-bus: a heterogeneous CDMA/TDMA bus for future SOC," in *Proceedings of the Thirty-Eighth Asilomar Conference on Signals, Systems and Computers*, 2004, pp. 1868–1872.

[31] M. F. Chang, "CDMA/FDMA-interconnects for future ULSI communications," *IEEE/ACM International Conference on Computer-Aided Design (ICCAD)*, 2005, pp. 975–978.

[32] E.-G. Jung, B.-S. Choi and D.-I. Lee, "High performance asynchronous bus for SoC," in *Proceedings IEEE International Symposium on Circuits and Systems (ISCAS)*, 2003, pp. 505–508.

[33] W. J. Bainbridge and S. B. Furber, "Asynchronous macrocell interconnect using MARBLE," in *Proceedings of Fourth International Symposium on Advanced Research in Asynchronous Circuits and Systems*, 1998, pp. 122–132.

[34] W. J. Bainbridge, "Asynchronous system-on-chip interconnect," Ph.D. Thesis, University of Manchester, March 2000.

[35] E. Jung, J. Lee, S. Kwak, K. Jhang, J. Lee and D. Har, "High performance asynchronous on-chip bus with multiple issue and out-of-order/in-order completion," in *Proceedings of the 15th ACM Great Lakes Symposium on VLSI (GLS-VSLI)*, 2005, pp. 152–155.

[36] A. Lines, "Nexus: An asynchronous crossbar interconnect for synchronous system-on-chip designs," in *Proceedings of 11th Symposium on High Performance Interconnects*, 2003, pp. 2–9.

[37] A. J. Martin, "The limitations to delay-insensitivity in asynchronous circuits," Sixth MIT Conference on Advanced Research in VLSI, MIT Press, 1990.

[38] K. Sekar, K. Lahiri and S. Dey, "Configurable platforms with dynamic platform management: An efficient alternative to application-specific system-on-chips," in *Proceedings of 17th International Conference on VLSI Design (VLSID)*, 2004, pp. 307–315.

[39] K. Lahiri, A. Raghunathan, G. Lakshminarayana and S. Dey, "Communication architecture tuners: a methodology for the design of high-performance communication architectures for systems-on-chips," in *Proceedings of the Conference on Design Automation (DAC)*, 2000, pp. 513–518.

[40] K. Lahiri, A. Raghunathan, G. Lakshminarayana and S. Dey, "Design of high-performance system-on-chips using communication architecture tuners," *IEEE Transactions on CAD of Integrated Circuits and Systems*, Vol. 23, No. 5, May 2004, pp. 620–636.

[41] K. Lahiri, A. Raghunathan and S. Dey, "Fast performance analysis of bus-based system-on-chip communication architectures," in *Proceedings of International Conference Computer-Aided Design (ICCAD)*, November 1999, pp. 566–572.

[42] K. Lahiri, A. Raghunathan and G. Lakshminarayana, "LOTTERYBUS: A new high-performance communication architecture for system-on-chip designs," in *Proceedings of the Conference on Design Automation (DAC)*, 2001, pp. 15–20.

[43] K. Lahiri, A. Raghunathan and G. Lakshminarayana, "The LOTTERYBUS on-chip communication architecture," *IEEE Transactions on Very Large Scale Integration (VLSI) Systems*, Vol. 14, No. 6, June 2006, pp. 596–608.

[44] K. Lahiri, A. Raghunathan and S. Dey, "Evaluation of the traffic performance characteristics of system-on-chip communication architectures," in *Proceedings of the International Conference VLSI Design (VLSID)*, January 2001, pp. 29–35.

[45] J. Buck, S. Ha, E. A. Lee and D. D. Masserchmitt, "Ptolemy: A framework for simulating and prototyping heterogeneous systems," *International Journal on Computer Simulation, Special Issue on Simulation Software Management*, Vol. 4, April 1994, pp. 155–182.

[46] F. Balarin, M. Chiodo, H. Hsieh, A. Jurcska, L. Lavagno, C. Passerone, A. Sangiovanni-Vincentelli, E. Sentovich, K. Suzuki and B. Tabbara, *Hardware-Software Co-Design of Embedded Systems: The POLIS Approach*, Kluwer Academic Publishers, Norwell, MA, 1997.

[47] Synopsys Inc., RTL Synthesis, Available http://www.synopsys.com/products/logic/.

[48] NEC Electronics, Cell Based IC CB-12 L/M/H Type Features/Basic Specifications. Available http://www.necel.com/cbic/en/cb12/cb12.html.

[49] Y. Zhang, "Architecture and performance comparison of a statistic-based lottery arbiter for shared bus on chip," in *Proceedings of the Conference on Asia South Pacific Design Automation (ASP-DAC)*, 2005, pp. 1313–1316.

[50] N. Wang, M. A. Bayoumi, "Dynamic fraction control bus: new SOC on-chip communication architecture design," in *Proceedings of SOC Conference*, 2005, pp. 199–202.

[51] K. Sekar, K. Lahiri, A. Raghunathan and S. Dey, "FLEXBUS: A high-performance system-on-chip communication architecture with a dynamically configurable topology," in *Proceedings of Design Automation Conference (DAC)*, 2005, pp. 571–574.

[52] T. D. Richardson, C. Nicopoulos, D. Park, V. Narayanan, Y. Xie, C. Das and V. Degalahal, "A hybrid SoC interconnect with dynamic TDMA-based transaction-less buses and on-chip networks," *19th International Conference on VLSI Design (VLSID)*, 2006, pp. 8–15.

[53] "Synopsys DesignWare Intellectual Property," http://www.synopsys.com/products/designware/designware.html.

[54] "CB-12" http://www.necel.com/cbic/en/cb12/cb12.html.

[55] W.-J. Hahn, A. Ki, K.-W. Rim and S.-W. Kim, "A multiprocessor server with a new highly pipelined bus," in *Proceedings of IPPS*, 1996, pp. 512–517.

[56] A. Ki, W. Sim, B. Park and Y. Yoon, "Highly pipelined bus: HiPi bus," *JTC-CSCC'91*, July 1991, pp. 528–533.

[57] J. P. Bissou, M. Dubois, Y. Savaria and G. Bois, "High-speed system bus for a SoC network processing platform," in *Proceedings of the 15th International Conference on Microelectronics (ICM)*, 2003, pp. 194–197.

[58] A. Landry, Y. Savaria and M. Nekili, "A Beyond-1 GHz AMBA High-Speed Bus for SoC DSP Platforms," *IEEE International Conference on Microelectronics (ICM)*, 2004, pp. 46–49.

[59] A. Landry, Y. Savaria and M. Nekili, "A novel 2 GHz multi-layer AMBA. High-speed bus interconnect matrix for SoC platforms," *IEEE International Symposium on Circuits and Systems, (ISCAS)*, 2005, pp. 3343–3346.

[60] A. Landry, Y. Savaria and M. Nekili, "Circuit techniques for a 2 GHz AMBA AHB bus," *The 3rd International IEEE-NEWCAS Conference*, 2005, pp. 311–314.

[61] M. Dubois, Y. Savaria and G. Bois, "A generic AHB bus for implementing high-speed locally synchronous islands," in *Proceedings of IEEE Southeast Conference*, 2005, pp. 11–16.

[62] S.-Y. Hwang and K.-S. Jhang, "An improved implementation method of AHB BusMatrix," in *Proceedings of IEEE International SOC Conference*, 2005, pp. 211–214.

[63] P. Wijetunga, "High-performance crossbar design for system-on-chip," *Proceedings of the 3rd IEEE International Workshop on System-on-Chip for Real-Time Applications*, 2003, pp. 138–143.

[64] F. Poletti, D. Bertozzi, B. Luca and A. Bogliolo, "Performance analysis of arbitration policies for SoC communication architectures," *Transactions on Design Automation for Embedded Systems*, Vol. 8, 2003, pp. 189–210.

[65] A. Olugbon, T. Arslan and I. Lindsay, "A formal approach to virtualisation and provisioning in AMBA AHB-based reconfigurable systems-on-chip," in *Proceedings of International Symposium on System-on-Chip*, 2005, pp. 175–178.

[66] K. Anjo, A. Okamura, T. Kajiwara, N. Mizushima, M. Omori and Y. Kuroda, "NECoBus: a high-end SOC bus with a portable and low-latency wrapper-based interface mechanism," in *Proceedings of the IEEE Custom Integrated Circuits Conference*, 2002, pp. 315–318.

[67] Synopsys NanoSim, http://www.synopsys.com/products/mixedsignal/nanosim/nanosim.html.

[68] A. C. Waldspurger and W. E. Weihl, "Lottery scheduling: Flexible proportional-share resource management," in *Proceedings of Symposium on Operating Systems Design and Implementation*, 1994, pp. 1–12.

CHAPTER

On-Chip Communication Architecture Refinement and Interface Synthesis

9

In a typical SoC design flow, several models of the system are created that capture different levels of detail, for different purposes. Figure 9.1 shows how communication architecture refinement and interface synthesis involve transformations between models with different levels of detail, in a typical design flow. *Functional* (or *task/process graph*) *level models* focus on capturing the functionality of the system, without any notion of hardware or software components that will ultimately implement the functionality. Such models are typically used as "golden reference" models to allow later stages of the design flow to check and validate the intended functionality of the system, as needed. *Architectural level models* on

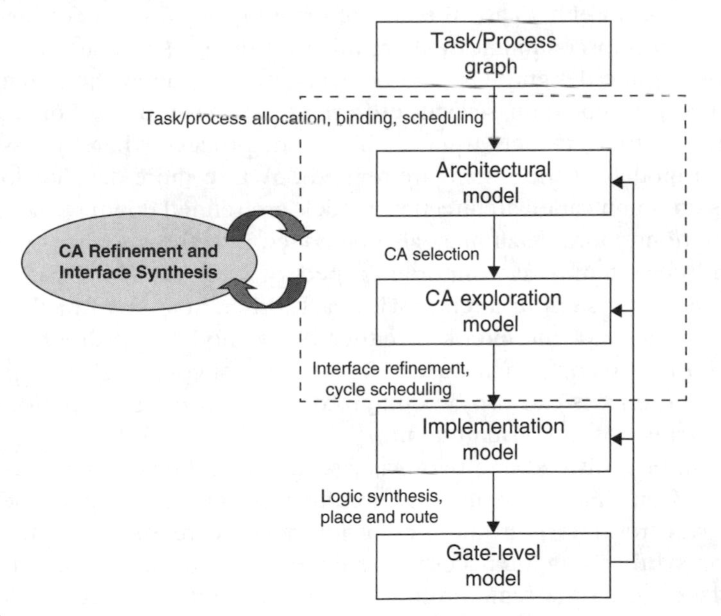

FIGURE 9.1

Communication architecture (CA) refinement and interface synthesis in a typical ESL design flow

the other hand are created after HW/SW partitioning (i.e., after mapping the various tasks to hardware components or software running on processors), and capture the behavior of system components communicating via abstract channels. Transaction level models (TLMs) described in detail in Chapter 4 are examples of architectural level models. Such models are intended for software developers (to initiate software development early in the design flow) and high level design space exploration. *Communication level models* capture the system communication infrastructure in much greater detail (typically cycle-accurate granularity) and are used for detailed system design space exploration. However, they do not necessarily capture the HW/SW component models in great detail. Finally, *implementation level models* capture the components and the communication infrastructure at the signal and cycle-accurate level of detail, and are typically used for detailed system verification and even more accurate analysis.

All of these models capture the communication between components in a system at different levels of detail—global variables (*functional* model), abstract transactions (*architectural* model), cycle-accurate transactions, possibly with some signals (*communication* model), and completely cycle and signal accurate (*implementation* model). Note that there can be many more models in a design flow that capture different parts of the same system with varying levels of detail. For instance, the processor memory subsystem in an SoC design is typically a critical factor in determining overall performance, and consequently may be modeled in a lot more detail than other subsystems consisting of less critical, low bandwidth peripherals. Conversely, designers can choose to merge some of these models (e.g., *architectural* and *communication* models), if adequate resources to create all these models during the different phases of a design cycle are not available. In either case, the task of creating all these models manually to explore the on-chip communication architecture design space at different levels of detail during the design phase can be very time consuming and cumbersome. To overcome this limitation, several research efforts have focused on automating the creation of these models, using a refinement process, whereby less detailed, higher-level models of the system are refined down to more detailed, lower-level models. As communication architecture models are refined down to more detailed implementations, more detail must also be added into the interfaces of the components in the system. A very important aspect of *interface synthesis* is to enable components in a system to interact with each other, in spite of having different interfaces (because of the interfaces either being modeled at different levels of details or if the two components use different types of communication protocols). Component interfaces can also be optimized to improve system performance, as well as reducing power consumption.

This chapter is divided into two major sections. Section 9.1 presents research in the area of on-chip communication architecture refinement, while Section 9.2 describes research efforts in the area of interface synthesis. The latter presents research on synthesizing interfaces to correct protocol mismatches, and to ensure that interfaces using the same protocol but having different level of details correctly interact with each other. This section also highlights research in the area of interface optimization to improve system performance characteristics.

9.1 ON-CHIP COMMUNICATION ARCHITECTURE REFINEMENT

Early work on interface-based design [1, 16] motivated the need for orthogonalizing communication from behavior to handle increasing SoC complexity, and using an iterative refinement methodology to create detailed implementation models from high level abstract models, for the on-chip communication architecture. While higher-level abstractions provide better simulation performance, it comes at the cost of accuracy and visibility into finer details, which can be realized only with lower abstraction levels. Much of the early research in this area [2–4] focused on obtaining signal-accurate implementation models from higher-level abstract message passing models. The approaches more or less refine the higher-level models by substituting abstract communication primitives with detailed implementations of the target protocol, from a library. A formal approach for such a step-wise refinement of communication abstractions was proposed by Plosila et al. [5]. The approach made use of *action systems* [6, 7], which is a state-based formalism for concurrent system specification, to refine an abstract formal specification into a concrete implementable form, such that the logical properties of the original description are preserved during the transformation. An abstract communication model, where communication between components is modeled using remote procedure calls (RPCs), is refined into a single variable signaling, where the communication occurs via a single control variable, and several data variables. Subsequently, communication is further refined to Boolean handshake signaling, which is directly transformable into actual wires.

9.1.1 COSY Methodology

Four different communication abstraction levels were proposed in [8, 9], as part of the *COSY methodology* [10], with a specific focus on exploring and designing signal processing applications. Figure 9.2 shows these models, with the higher-level models being successively refined to create the lower-level models. The application level model (*APP*) uses point-to-point FIFO channels of infinite size to communicate between functions. At this level, the application is modeled as a network of parallel processes that consist of sequential processes interconnected via unbounded point-to-point FIFOs (referred to as a Kahn Process Network [11] computation model). The *APP* model is refined into the system level model (*SYS*), where finite-sized FIFO channels facilitate communication between functions that have been mapped onto hardware and software components. These models are typically used for high level exploration of the design after annotating channels with communication delays. The channels are converted into more detailed, generic channels in the virtual component interface model (*VCI*). The VCI model uses the VCI interface proposed by the VSI Alliance [12], which does not commit to any particular bus protocol, and can be hooked onto specific bus protocols using simple wrappers. (Chapter 2 describes the VCI interface standard in more detail). At this level, data is split into chunks, from abstract data types used at the higher levels. Finally, at the physical-level model (*PHY*), a specific bus protocol is selected, and details of bus width, signaling, and arbitration protocol are

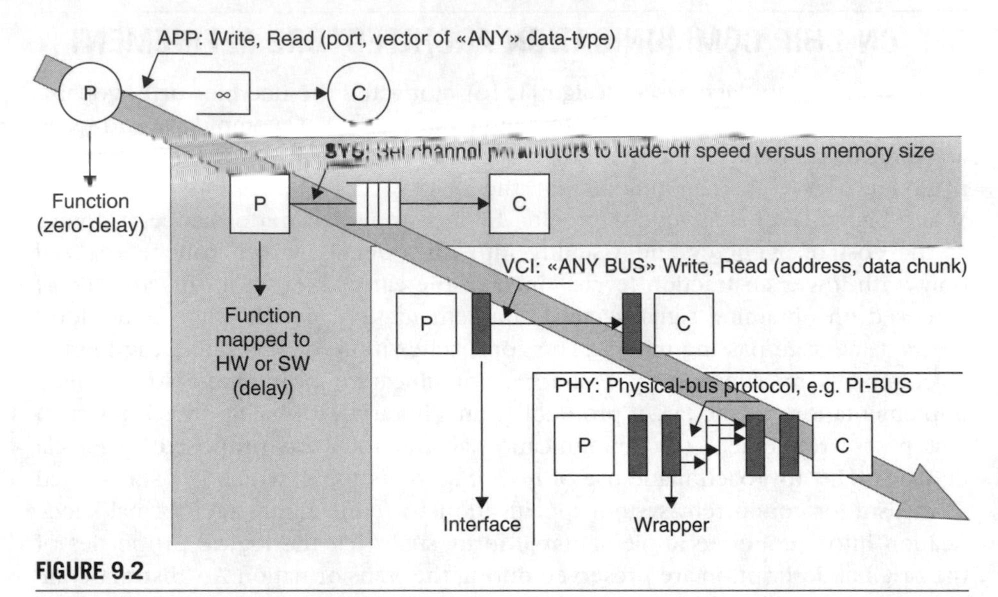

FIGURE 9.2

Communication architecture models [8, 9]
© 1999 IEEE

added. The process of conversion from the higher abstraction levels down to the lower levels (for instance the mapping of functional model to the architectural model, and then to the implementation model) is done in a partially automated manner using the VCC-COSY design flow developed by Philips and Cadence. This refinement-based methodology was used in the Nexperia digital video platform [13], developed by NXP (formerly Philips) Semiconductors. There have been a few other research efforts that have focused on automatically refining higher-level application models into more detailed architectural models for signal processing applications, using trace transformation techniques [14, 15].

9.1.2 SpecC Methodology

A more general, and automated on-chip communication architecture refinement methodology was proposed in [17, 18], as part of the SpecC [16] design methodology. Four model abstractions: *specification*, *architecture*, *communication*, and *implementation*, are defined in the refinement-based SpecC design methodology. The highest level of abstraction in the methodology is the *specification model*, in which the functionality of the system is captured using sequential or concurrent behaviors that communicate via global variables or abstract channels. No timing or structural information is captured at this level. In the next step, allocation, behavior partitioning, and scheduling is performed, to refine the specification model into an *architectural model*. The task of allocation involves (manually) selecting the processing elements (PEs), memories, and buses from a library to create the system target architecture. Partitioning maps the specification onto this architecture—behaviors are mapped to PEs, variables are mapped to memories, and channels are mapped to buses. Scheduling is needed to determine the order of behavior execution, when possibly concurrent behaviors are mapped to a sequential PE. The abstract communication channels between behaviors are mapped to the selected buses in the target architecture. The architecture model, however, is not yet accurate in terms of cycle-true timing and communication, since execution time and delays are estimated for all the behaviors in the PEs and all channels in the buses. In the next step, protocol insertion and inlining is performed to refine the architectural model to a *communication model*. During this phase, the virtual, abstract communication used in the architecture model is replaced with real communication protocols implemented on system buses. In particular, the virtual buses in the architecture model are refined into hierarchical channels that implement the required communication functionality over the actual wires of the bus. A protocol library is used to select and integrate the desired protocol, seamlessly replacing calls to abstract communication primitives in the architectural model with calls to the more detailed protocol implementation. The data exchanged between PEs is also translated from complex variables (that could be structures, integers, or multi-dimensional arrays) into bit streams that can be sent over the system buses, via a process called data slicing. The resulting communication model is a bus functional model, with the transactions on the buses represented in a signal and bit accurate manner, and with accurate timing in greater detail. The other components in the system are still represented at a higher abstraction level, allowing for fast simulation. Finally, the communication model is refined down to an *implementation model*. Software is compiled for target processors and behavior inside hardware PEs is scheduled in a cycle-accurate manner. The resulting implementation model is similar to a structural RTL level model that can be synthesized using logic synthesis tools to create a gate level netlist. Throughout this refinement methodology, design decisions are made manually by the designer (based on static estimation, or simulation-based exploration)—for instance, the choice of using particular PEs and bus protocols. However, once the design decision has been finalized, the refinement or translation process from higher abstraction levels down to the lower abstraction levels is automated.

9.1.3 Coral Framework

Bergamaschi et al. [19, 20] proposed the *CORAL framework* to automate refinement of designs at the *virtual* abstraction into real designs (at the structural RTL or signal-accurate abstraction). A virtual design abstraction is a higher-level structural and functional encapsulation of a real design, consisting of virtual components and interfaces. The communication architecture is represented as a set of virtual pins (or virtual wires). Each virtual pin represents potentially several pins in a real design. For instance, the PowerPC processor virtual component contains a single virtual pin representing all the pins that are responsible for interfacing between the internal data cache unit and the master side of the external processor bus. In fact, the virtual component for the PowerPC has only 10 virtual pins whereas the real PowerPC component has almost 160 pins. In order to refine the virtual communication model into a real one, properties are used to establish correspondence. Each real and virtual pin can have several functional and structural properties associated with it, such as bus type (type of bus the pin interfaces to, e.g., AMBA APB, or CoreConnect PLB), interface type (master or slave), function type (the function implemented by the pin, such as read, write, or interrupt) operation type (e.g., request, acknowledge), data type (e.g., address, data, or instruction), and pin group (to indicate grouping of pins in the same interface). A specialized language is used to specify these properties for the pins. By comparing the properties on pins, it can be determined whether the functionality of a real pin falls within the functionality of a virtual pin, during refinement.

To speed up this comparison process, two techniques are used: (i) property encoding using Binary Decision Diagrams (BDDs) [21] and (ii) property comparison and matching using logical operations on BDDs. Each property/value pair PV_i=<*Property type* T_i == *Property value* V_i> is mapped to a Boolean (e.g., BDD) variable. A group of these property/value pairs $PG = \{PV_1, PV_2, \ldots, PV_n\}$ is mapped to a set of BDD variables, denoted by $B(PG) = \{b_1, b_2, \ldots, b_n\}$. The BDD for the complete group is given by: $F(PG) = b_1 \wedge b_2 \wedge \ldots \wedge b_n$. When the property group PG is attached to a pin T, the complete BDD function $F(PG)$ is denoted as $F(T)|_{PG}$, or the property function F of pin T with respect to property group PG. During the virtual to real refinement, the compatible set of real pins needs to be determined for each virtual pin, with respect to their interconnection property group. This is done as follows. Let $F(V)|_{PG}$ and $F(R)|_{PG}$ be the Boolean functions representing the interconnection property groups of virtual pin V and real pin R, respectively. R is compatible with V *iff* $F(V)|_{PG} \supseteq F(R)|_{PG}$, that is, the property function for V contains the one for R. The containment operator $\supseteq$ is computed using BDD operations. In this manner, the virtual pins are refined down to the real pins, as shown in Fig. 9.3.

9.2 INTERFACE SYNTHESIS

As discussed earlier, during the process of communication refinement, and in the course of a typical system design flow, several models of the system are created that capture different levels of detail. Two major problems are encountered when interfacing components in these system models: (i) interfacing components modeled at different levels of detail and (ii) interfacing components that use different communication protocols. The first problem arises due to the need to quickly co-simulate systems at different points in a design flow, without spending time to ensure that all the components are modeled at the same abstraction, or level of detail. The second problem arises due to IP reuse in design flows today, where an IP (component) that interfaces with a particular on-chip communication protocol (e.g., IBM CoreConnect) is reused in another design that uses another on-chip communication protocol (e.g., ARM AMBA 2.0). Components are characterized by signals (or pins) at their interface to interact with other components. Since the signals for different protocols are usually different in number, as well as in terms of functionality, components supporting one on-chip communication protocol cannot be easily integrated into a system using another on-chip communication protocol. In each of these two cases, there is a need to modify or synthesize the interfaces for the components in the system, so that they can be seamlessly integrated

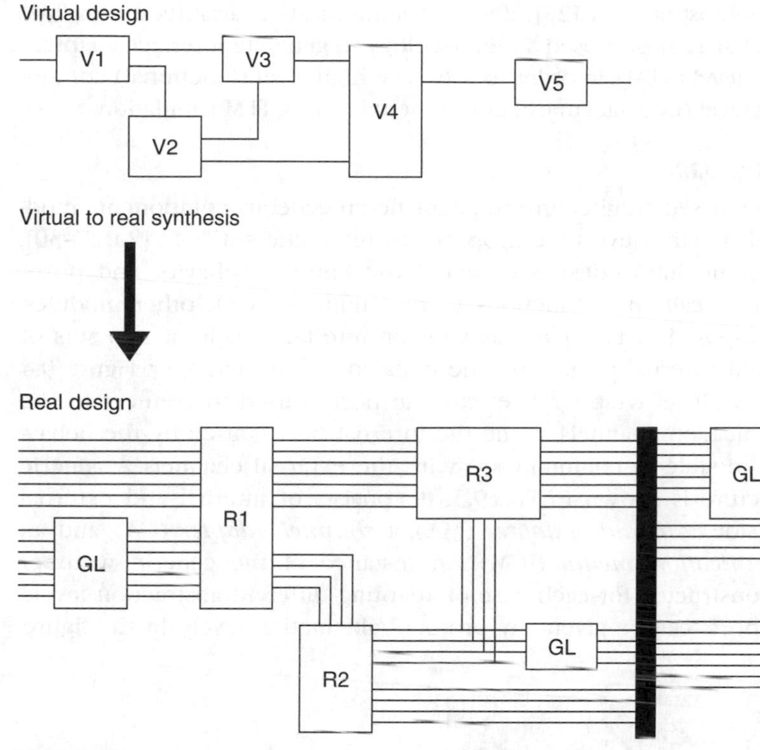

FIGURE 9.3

Refinement from virtual to real design (GL = glue logic) [19, 20]
© 2000 IEEE

with each other. This section will present research efforts to address the problems of interface synthesis, as well as efforts to optimize the interface design, in order to improve overall performance.

9.2.1 Connecting Components at Different Abstraction Levels

During communication refinement from a high abstraction level (e.g., functional) to a lower-level (e.g., cycle/pin accurate), mixed abstraction level co-simulation becomes necessary to validate if the refinement process correctly preserved the original functionality. To integrate heterogeneous components at different levels of abstraction (and thus with different interfaces), *wrappers* at the interfaces of components are a widely used solution. Bus functional models (BFM), for instance, encapsulate a high level and less detailed functional model of a component with a cycle-accurate interface wrapper for detailed low level simulation as part of methodologies proposed by Rowson and Sangiovanni-Vincentelli [1] and Semeria and Ghosh [22]. A bus cycle-accurate shell (BCASH) was used to adapt a remote procedure call (RPC) interface with a cycle-accurate one in Coware N2C [23]. Gerin et al. [24] and Nicolescu et al. [25] proposed using wrappers to interface between protocol-fixed and cycle-accurate abstractions [24], and between protocol-neutral and protocol-fixed abstractions [25], during communication architecture refinement. Wieferink et al. [26] proposed a methodology to generate interface adapters at the transaction level (TLM), in order to interface high level (functional) components with lower level (cycle-accurate) components, during TLM simulation.

9.2.1.1 *TIMA Approach*

A generic wrapper-based architecture to adapt heterogeneous components modeled at different abstraction levels was proposed by researchers at TIMA [24, 27–30]. A component (or module) consists of two basic entities: behavior and ports. The behavioral part calls port functions to communicate with other modules. A wrapper is composed of two parts as well: an interface made of two sets of ports (internal and external ports) and the behavior of the wrapper. Figure 9.4 shows a module with its wrapper. The external port is used to connect to the external communication channels, while the internal port is used by the behavioral part of the module to communicate with the external channels. A generic wrapper architecture is shown in Fig. 9.5. It consists of internal and external ports on either side, a *module adapter* (MA), a *channel adapter* (CA), and an *internal communication media* (ICM). An instance of the generic wrapper architecture is constructed for each case of adapting different abstraction levels. In the instance, ports can be given any protocol/abstraction levels. In the figure,

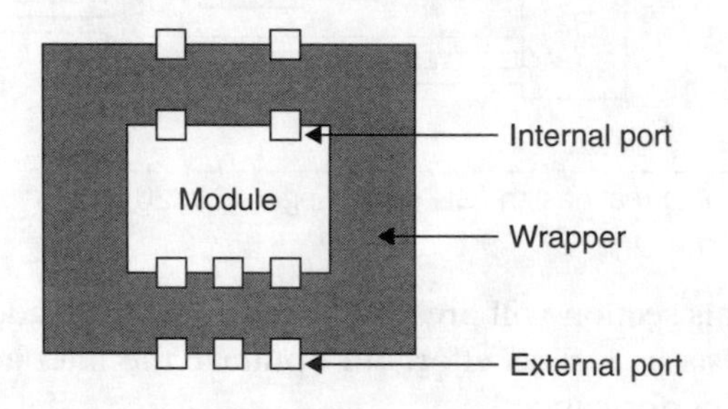

FIGURE 9.4

Module with a wrapper [27]
© 2001 IEEE

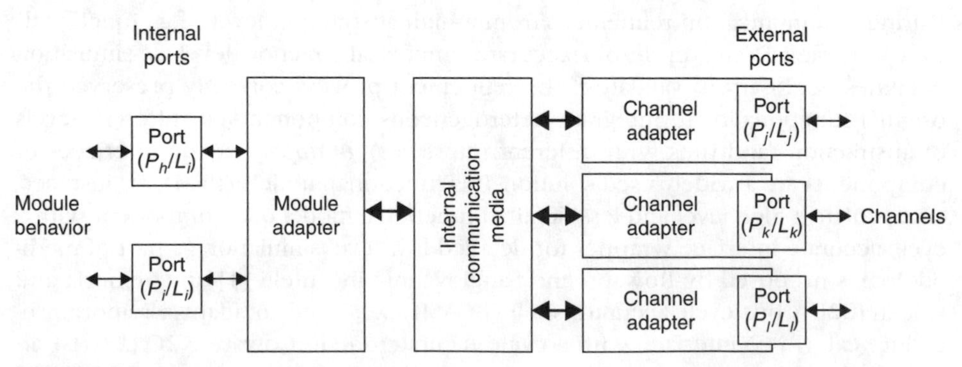

FIGURE 9.5

Generic wrapper architecture [27]
© 2001 IEEE

different protocol/abstraction levels are represented as P_h/L_i, P_i/L_i, P_j/L_j, P_k/L_k, and P_l/L_l. Data is transferred between the MAs and CAs via the ICM. Depending on the abstraction level of the module, the ICM can be either a function call (at the system or architecture abstraction) or an internal bus (at the RTL abstraction).

Compared to previously proposed techniques for mixed level co-simulation, such as [31-33], the generic wrapper architecture enables many-to-many correspondence between internal and external ports in the conversion of abstraction levels and communication protocols, while one-to-one correspondence is assumed in [31-33]. Figure 9.6 shows the communication refinement process, where the wrappers and communication channels are implemented from an abstract architecture, down to a physical architecture through the process of communication refinement. In the physical architecture, the wrappers are converted to an implementation in either hardware, as a communication coprocessor (CC), and/or as software, as the operating system (OS), depending on whether the module is mapped onto hardware, or a processor running software. During refinement, communication can be represented at the system level (SL), architecture level (AL), and the register transfer level (RTL). Because of the adaptation provided by the wrapper architecture, modules and channels can be refined independent to each other. At the system level, modules communicate with each other using message passing, over communication protocol-independent generic channels. At the architecture level, each channel is assigned a protocol (e.g., handshaking, FIFO, etc.) and parameters (e.g., FIFO size). Abstract modules from the system level are mapped to a hardware block or a software running processor at the architecture level. At the register transfer level, the modules are connected together via their communication coprocessors and a physical communication network.

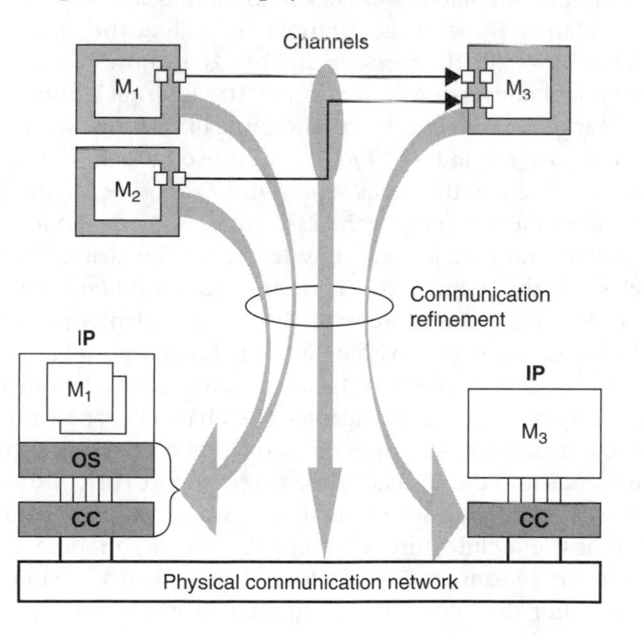

FIGURE 9.6

Communication refinement [27]
© 2001 IEEE

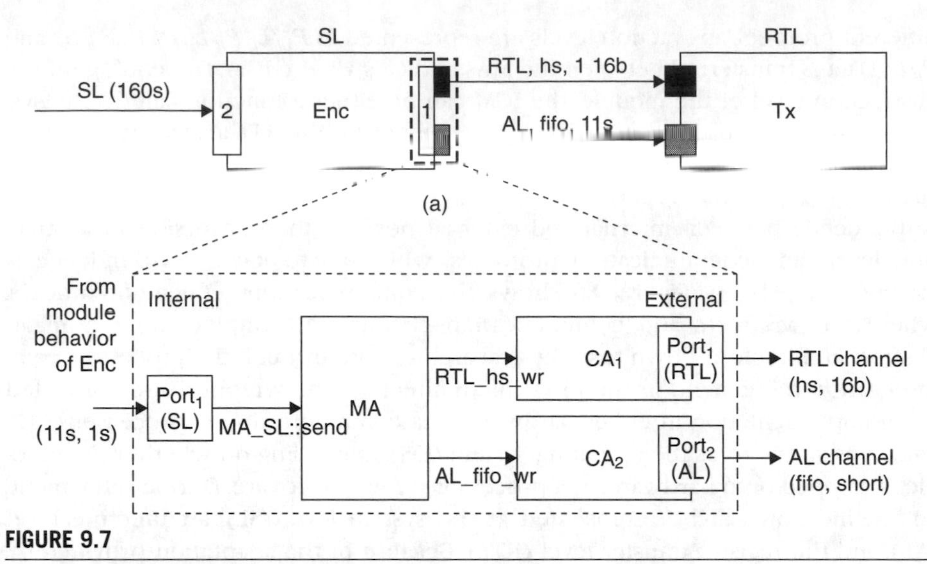

FIGURE 9.7

Example of wrapper for interfacing module at system level (SL) [27]
© 2001 IEEE

Figure 9.7 shows an example of a wrapper, when a module (*Enc*) is at the system level (*SL*), another module (*Tx*) is at register transfer level (*RTL*), and the external ports are at different abstraction levels. There are two channels between the modules: one at an architectural level (FIFO channel with 11 data items of short type) and the other at an RT level (handshake channel with 1 data item of 16-bit logic vector). The *Enc* module is shown sending a message via internal port $Port_1$ to the module adapter (*MA*) in the wrapper, by calling the SL channel function *MA_SL::send*. The *MA* splits the message and sends it to the two channel adapters, where data/protocol conversion is performed. The internal communication media (*ICM*) in the wrapper architecture is a function call relation (with the *MA* calling 2 functions—*RTL_ hs_wr* and *AL_ fifo_wr*) because module *Enc* is at the system level abstraction. Subsequently, data is sent to the *Tx* module via the external ports.

Figure 9.8 shows the wrapper, for the case where the *Enc* module is at the architecture level, and the *Tx* module is at the system level. The channel between the *Enc* and *Tx* modules is at the system level. The *ICM* is again a function call relation, since the *Enc* module is at the architecture level. The module adapter provides three internal ports at the architecture level, with *AL* channel functions for FIFO and handshake protocols. When sending data, the *MA* merges data received from the two input ports and passes it onto the channel adapter. The channel adapter CA_1 receives data from the *MA*, constructs the messages (i.e., converts received data to generic message type), and sends it to the SL channel via external port $Port_1$ at the system level.

Figure 9.9 shows the case where modules *Tx* and *Enc* are at the RT level, the *CaP* module is at the architecture level and the user *i/f* module is at the system level (*SL*). There are 4 channels (2 at RT level, 1 at AL, and 1 at SL) connecting the *Enc* module to other modules. The *Enc* module is implemented on a processor, with a wrapper consisting of a processor adapter (similar to a module adapter or *MA*) and channel adapters (at 3 different abstraction levels). Since *Enc* is at the

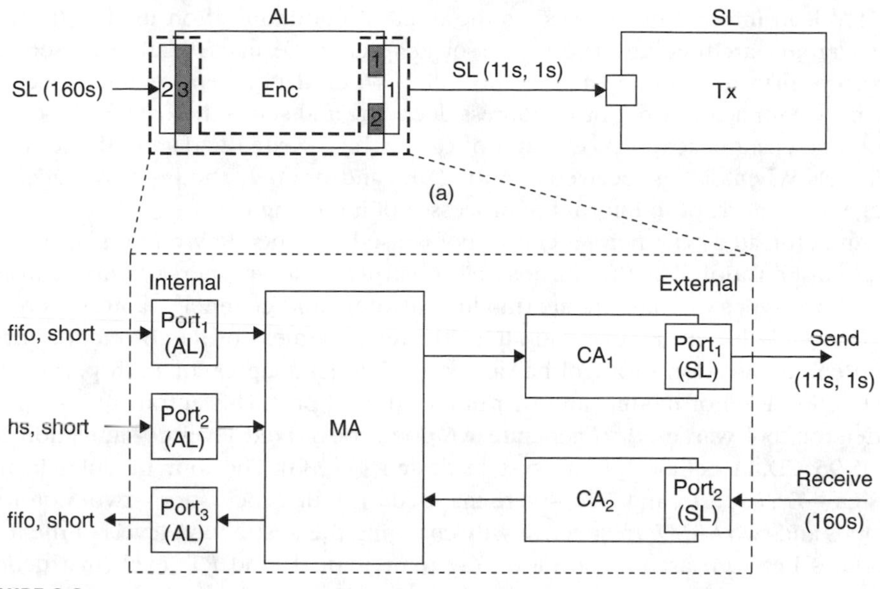

FIGURE 9.8

Example of wrapper for interfacing module at architecture level (AL) [27]
© 2000 IEEE

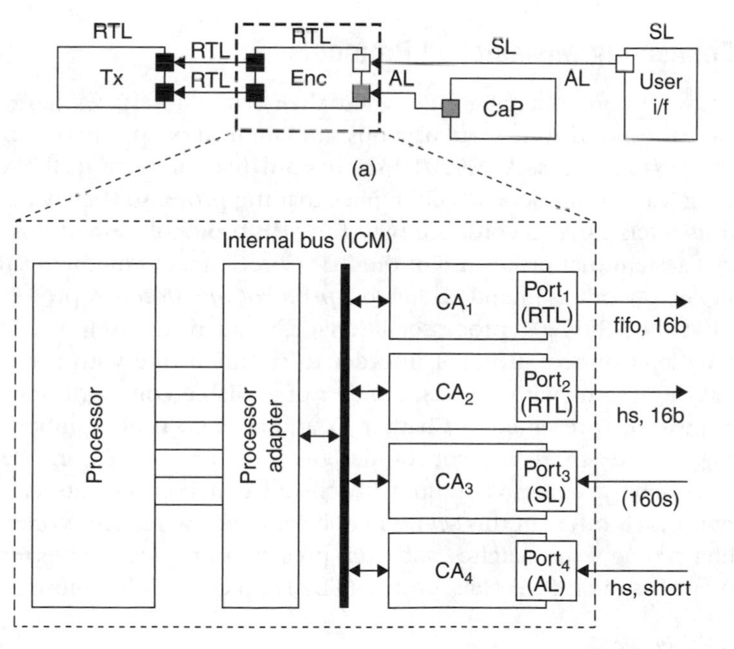

FIGURE 9.9

Example of wrapper for interfacing modules at register transfer level (RTL) [27]
© 2001 IEEE

RT level, an internal bus is used as the internal communication media (*ICM*) of the wrapper architecture. The processor adapter performs several tasks such as interrupt management and address decoding. When data is sent by the processor, the processor adapter performs address decoding and sends the data to the corresponding channel adapters (CA_1 and/or CA_2), which write the data to the RT level channels. When data is received at ports $Port_3$ and/or $Port_4$, the processor adapter triggers an interrupt to inform the processor of incoming data.

An automated wrapper generator tool is used to generate wrappers for mixed level co-simulation. For the purpose of constructing a wrapper, a library consisting of two types of components (module adapters and channel adapters) is used. Each module has at least one module adapter regardless of its abstraction level. Each communication protocol has also one channel adapter for each pair of the abstraction level of module and that of its external port. This automated wrapper generator tool was used to generate wrappers for mixed level co-simulation for an IS-95 CDMA cellular phone system design [34, 83]. The four modules in the design—*Tx*, *Rx*, *Enc*, and *Dec*—were mapped on four processors - two Motorola 68000s and two ARM7s, respectively. By changing the abstraction levels of the four modules between architecture level (SystemC model) and RT level (Instruction Set Simulator or ISS of 68000 and ARM7), the simulation speeds were found to vary from 380 KHz (entire design at SystemC architectural level), 1 KHz (single ISS, all other models at architecture level), 500 Hz (two ISSs), to 250 Hz (four ISSs). Thus, the proposed wrapper-based mixed co-simulation approach enables a system designer to trade-off simulation time with accuracy.

9.2.2 Connecting Mismatched Protocols

When hardware components are designed, their native interfaces have signals or ports that correspond to a particular bus communication protocol. For instance, an ARM processor, such as ARM920T [35], has a native interface to the AMBA AHB bus communication protocol, which implies that the processor behavior drives the signals at its interface in accordance with the AHB protocol. Now if this processor is used in a system that makes use of the IBM CoreConnect communication protocol, it will fail to work as intended, due to a *protocol mismatch*. A protocol adapter is needed at the ARM920T processor interface, to adapt the AMBA AHB protocol to the IBM CoreConnect protocol, in order to communicate with the other components in the system. With the vast number of available communication protocol standards [36–40] (presented in Chapter 3), and an increasing number of components being reused across different SoC designs, these protocol (or interface) adapters are becoming a common solution to ensure that components can correctly interact with each other. In this section we briefly review related work in the area of handling protocol mismatches, and then present a protocol converter synthesis methodology based on Petri Nets, proposed by Filho et al. [74] in more detail.

9.2.2.1 *Early Work*

Early work in this area was concerned with synthesizing interfaces between simple protocols (such as serial, handshake, etc.). Borriello and Katz [41] proposed an approach for the synthesis of interface transducers (i.e., glue logic that connects two circuits) between custom chips and buses. Timing diagrams of the two incompatible interfaces were specified as inputs, which were used to derive event graphs for the interfaces. These event graphs were subsequently merged into a single graph, with the help of the designer who manually specified merge labels to combine corresponding event graphs. The combined graph was used to generate a transducer logic circuit. Any timing violations or race conditions were corrected by adding appropriate logic circuitry in the transducer. Data width mismatches at the interfaces were, however, not handled. Akella and McMillan [42] presented an approach to synthesize protocol converters, by matching control signals of the two incompatible protocols. The protocols being interfaced are specified using finite state machines in Verilog, and it is assumed that the data path of the converter is given. A cross product of the two state machines is obtained, and optimized to obtain a state machine description of the converter. A third finite state machine (called the *C machine*) must be specified by the designer to describe the legal sequence of operations between the two FSMs, representing the protocols. This synthesis approach can however lead to a very large number of states in the protocol converter. Data width mismatches at the interfaces are also not handled. Sun and Brodersen [43] proposed a method for the synthesis of system interface modules, as part of the SIERA design environment [87]. A module library is created and used to obtain event graphs for the two incompatible protocols, which are subsequently interconnected based on data dependencies. From the resulting event graph, a protocol controller is synthesized to respond to the control signals of the two module protocols. Additional details of the interface module, such as the sequence of data transfers and interconnection of data ports, must be specified by the designer.

Another approach to synthesize an interface process between incompatible protocols was proposed by Narayan and Gajski [44]. The inputs are hardware description language (HDL) descriptions of the two protocols, detailing the number of control and data lines, and the sequence of data transfers over those lines. The protocol specification is reduced to a combination of five basic operations—data read, data write, control read, control write, and time delay. The protocol description is then broken down into blocks (referred to as a group of relations), whose execution is guarded by a condition on one of the control wires, or a time delay. Finally, the relations of the two protocols are matched into sets that transfer the same amount of data. The output of the synthesis procedure is an interface process, and information regarding the interconnection of ports of the two protocols. This approach is able to handle data width mismatches, as well as cases where the mismatched interfaces are driven by different clocks. However, the procedural specification of the protocols makes it difficult to adapt to situations with different data sequencing. Smith and Micheli [45] presented an approach to generate a cycle-accurate synchronous interface between two hardware subsystems, given an HDL model of each subsystem. The mismatched protocols are mapped to a standard communication scheme, which is then implemented as a state-machine-based interface architecture. The synchronous component interfaces can handle frequency mismatches, as well as both unidirectional and bidirectional buses. This work was extended by Shin and Gajski [46], by using protocol flow graphs to synthesize interfaces that use queues and internal control logic to regulate buffering.

A more formal approach for addressing protocol mismatch was proposed by Passerone et al. [47, 48]. The two mismatched protocols are described using regular expressions and translated into corresponding deterministic automata. Then the product of the two automata is taken using a production computation algorithm, so that only the legal sequence of operations is retained. Any resulting non-determinism in the product FSM is resolved by using a simple set of rules or arbitrary choices based on the order of the states, to remove the states and transitions that result in a violation of either of the two protocols. The resulting product FSM is used to create the interface logic to handle mismatch between the protocols. In this approach, the correspondence between data on the two sides is automatically resolved, and the interface translates different sequences of data. However, the two communicating entities must be driven by the same clock, which is a limitation. This limitation is overcome by Park et al. [49], by inserting additional states and edges in the FSM, and introducing a queue in the protocol converter. Passerone et al. [50] subsequently extended the work by Park et al. [49] and gave a more rigorous formalism to create interface protocol converters using game-theoretic algorithms that are extended from [51]. Unlike [49], the converter in [50] is derived from a starting specification of what constitutes an acceptable protocol conversion. The protocol conversion problem is set up and solved for send–receive protocols, where the sender and receiver are specified as automata. A third automaton, called the requirement automaton, specifies constraints on the converter, such as buffer size and the possibility of message loss. The solution consists of a winning strategy that complies with both the receive protocol and the requirement in a game against a sender, which must be allowed to follow any sequence of actions permitted by the send protocol. Under this game-theoretic formulation, if a winning strategy is found to exist, then the two protocols are convertible, and a protocol converter is derived, as an automaton.

9.2.2.2 *Addressing Protocol Mismatches for Interface Standards*

More recent work in the area of interface synthesis to address protocol mismatches has been concerned with generating bridge converters and interface adapters for emerging standard bus communication protocols such as AMBA [36], CoreConnect [39], STBus [38], AVALON [52], and interface standards such as OCP [53] and VCI [54] (all of which are described in detail in Chapter 3).

Bridge Protocol Converters

Bridge protocol converters enable data transfers across buses with different protocols. Instead of using protocol converter interface wrappers for every component (i.e., distributed protocol conversion) which can increase chip area, a bridge converter is an external component that encapsulates protocol conversion functionality, performing the required protocol conversion in a centralized manner. Zhonghai et al. [55] presented the design of a bridge (at the RTL level) to connect the AMBA AHB bus with the off-chip PCI [56] bus. The AHB/PCI bridge enables transfer of address, data, and control commands between components on the two types of buses. Choi and Kang [57] presented a bridge to interface an advanced, high speed protocol (AHB) with a simpler, low speed protocol (ISA). The proposed bridge overcomes frequency mismatches between protocols. Zitouni et al. [58] presented a framework for creating bridges based on the amalgamation of FSMs of the two mismatched protocols, to create a bridge FSM. FIFO-based AMBA/PI-BUS and AMBA/ PCI bridges were generated, based on the proposed approach. A bridge to interface between the AMBA and TMS DSP [59] buses was described by Cho et al. [60].

Component Interface Protocol Converters

Several approaches have proposed protocol converters at component interfaces for standard on-chip communication protocols. Strano et al. [61] presented a wrapper to interface OCP [53] and STBus [38] protocols. The components in the system are assumed to have a native OCP interface, and the wrapper is used to convert transactions originating from the OCP interface into STBus transactions. A similar wrapper is developed for converting between VCI [54] and PI-BUS [62] and between VCI and AMBA AHB [63]. Kim and Lee [64] proposed a methodology for synthesizing wrappers for interfacing components running at different clock frequencies compared to the bus. Traditionally used FIFO-based wrappers across different clock domains [65] incur a large latency. The proposed approach uses combinational logic inside wrappers to reduce latency. Several interfaces were described for the AMBA AHB and APB buses for slow/fast masters/slaves. Androutsopoulos et al. [66] proposed an automated protocol converter approach by extending the product FSM construction approach from previous work [42, 47, 48], where the two mismatched protocols are modeled as FSMs and the converter FSM is derived from the product of the two FSMs. Unlike the previous work, where only acyclic FSMs are considered, the proposed approach allows arbitrary cyclic FSMs, which are necessary to compute the bandwidth that is achieved by the protocol converter, allowing bandwidth (rather than latency) to be optimized during converter synthesis. The proposed approach was used to generate protocol converters to manage transfers between the AMBA AHB and APB protocols.

Another approach to automate protocol converter synthesis using synchronous finite state automata was proposed by D'silva et al. [67, 68]. In this approach, a designer provides FSM style protocol descriptions and a mapping between the data buses of the protocols, which are combined to create a protocol converter FSM. The technique was used to generate protocol converters between commonly used SoC protocols such as AMBA AHB, IBM CoreConnect PLB and OCP. The protocol converters are able to handle mismatched data widths, types, clock speeds, and pipelining. Another approach to automatic protocol converter synthesis proposed by Roychoudhury et al. [69] makes use of the visual formalism of message sequence charts (MSCs) for the specification and realization of the converters. An MSC is viewed as an atomic unit of interaction between two or more components. An interaction pattern will consist of a (concatenated) sequence of MSCs, and the set of all such patterns of interest is captured by a high level MSC (HMSC) [70]. The HSMC, which is basically a finite state automaton with an MSC associated with each state, is required as an input to the converter generator. This input is used to generate a SystemC implementation of the converter, when one exists (since it is possible

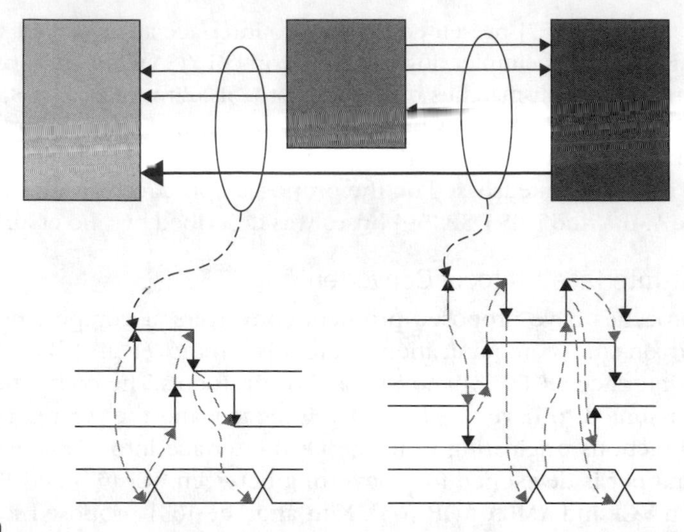

FIGURE 9.10

Protocol descriptions as marked timing diagrams [74]
© 2003 IEEE

that a converter cannot be generated between two protocols due to extreme mismatches). A protocol converter was generated, using the proposed approach, to interconnect a master and a slave. The converter supports features from several common bus communication standards such as AMBA and CoreConnect.

An Example of Protocol Conversion Using Petri Nets

A protocol converter synthesis methodology based on Petri Nets [71–73] was proposed by Filho et al. [74]. Initially, timing diagrams describing both the mismatched protocols are marked to establish causal relationships between signal transitions, as shown in Fig. 9.10. A marked arrow that goes from a signal transition to another creates a precedence order between signal events described along the protocol. An event refers to an instance of a signal action, which can be either: (i) transition from high to low, denoted as $x-$, (ii) transition from low to high, denoted as $x+$, (iii) a data assertion ($x\#$), or (iv) a data de-assertion ($x*$) where the data bus becomes valid or invalid, respectively. These marked timing diagrams are translated into a class of Petri Net models called Signal Transition Graphs (STGs) [75]. Figure 9.11 shows marked timing diagrams and STGs for a VME peripheral interface protocol and the AVALON [52] bus, for the case of a single read operation on the peripheral. The goal of the synthesis procedure is to obtain one unique Petri Net that describes the interface process behavior. As shown in Fig. 9.11, the timing diagrams of each protocol are annotated and translated to STG format, followed by safeness, liveness, validity, and output persistency checking to ensure that an interface can actually be synthesized for the mismatched protocols.

The interface synthesis proceeds by applying successive transformations over the derived STGs. First, the STG whose module is responsible for controlling the data assertion should be modified. This is because when the data becomes available for the transfer, this module indicates it to the environment through the change

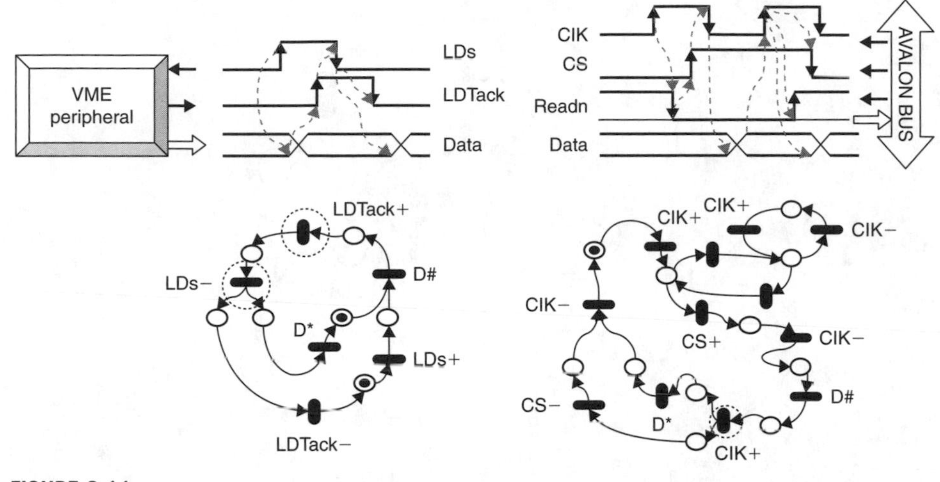

FIGURE 9.11

VME peripheral single read operation to AVALON bus: timing diagram and STG [74]
© 2003 IEEE

of one or more output signals. The STG transitions related to those signals will then be used to synchronize the data assertion during transfers. In the example shown in Fig. 9.11, when a read operation takes place, data assertion by the VME compatible module is followed by a rising edge on the *LDTack* signal. The transition labeled with *LDTack+* is then used to synchronize the validation of the data between the two protocol descriptions. Similarly, one or more input signals from the environment indicate to the module that the data can be de-asserted and so the STG transitions related to those signals can be used to synchronize the data de-assertion phase. In this case, the *LDs* falling edge indicates for the module that the data can be de-asserted, though the transition marked with the label *LDs−* will be used to synchronize the moment when data may become invalid. Using a similar procedure, the synchronization transitions are identified on the AVALON bus STG. The transition labeled as *Clk+* that succeeds the data assertion transition is both the assertion and the de-assertion synchronization event. Figure 9.12 shows the transformations that are subsequently performed on the STGs. The transitions *ta*, *td* and the places *pa*, *pd* are created in the protocol responsible for the data assertion. Arcs are added from the transition assigned to synchronize data assertion to *pa*, and from *pa* to *ta*. Similarly, arcs are added from *td* to *pd* and from *pd* to the transition assigned to synchronize data de-assertion. The respective chosen transitions are temporarily re-labeled as *ta* and *td*. This procedure thus selects the synchronization points between the protocols.

In the next step, a mirror operation is performed on both the interface protocols, as shown in Fig. 9.13. This results in the input signals associated with a net being turned into output signals, and vice versa. This is done because when a protocol converter is inserted between the two modules, the input signals on the modules must be output signals on the protocol converter, and similarly, output signals on the modules must be input signals on the protocol converter.

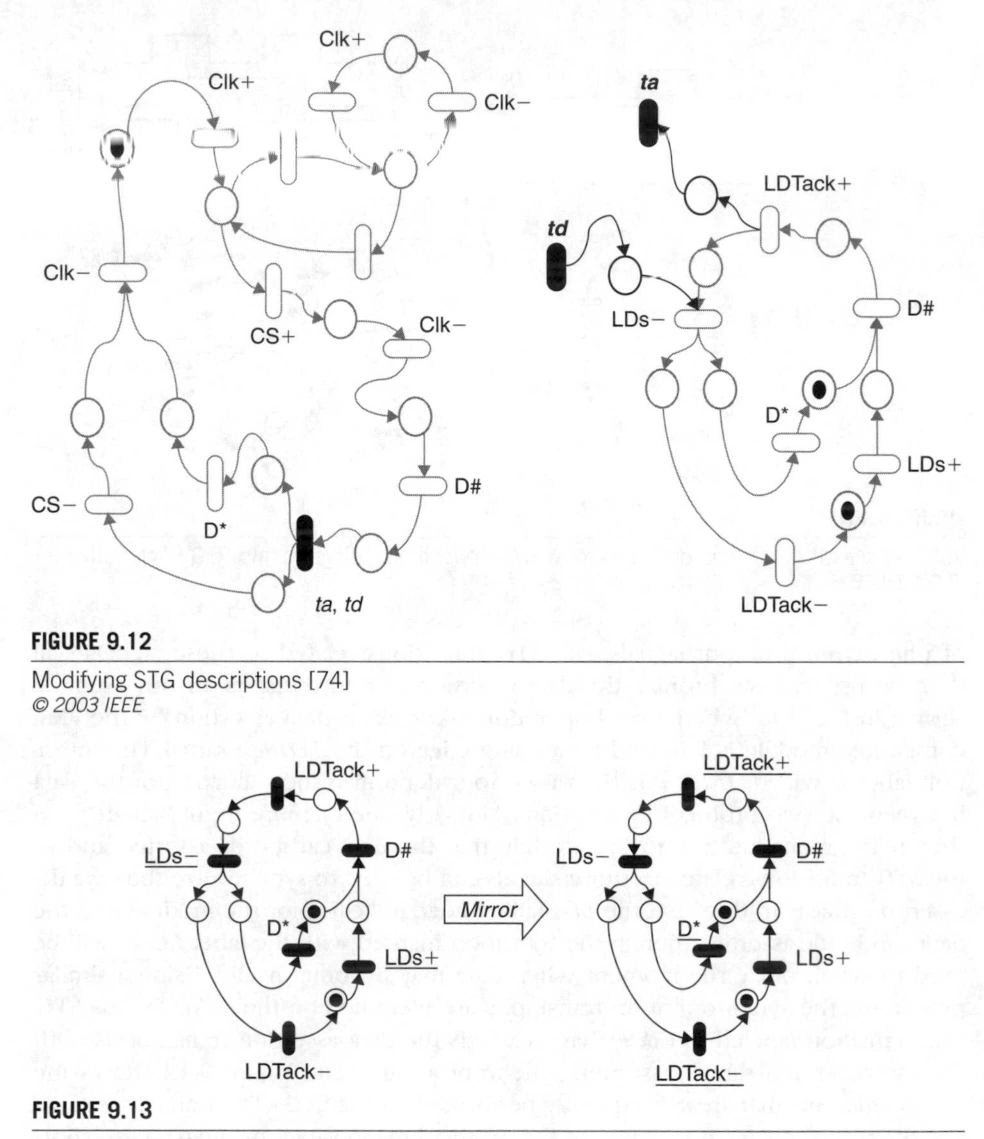

FIGURE 9.12

Modifying STG descriptions [74]
© 2003 IEEE

FIGURE 9.13

Mirroring STG descriptions [74]
© 2003 IEEE

The third and last step is to compose the protocol converter description through a parallel composition over the two modified and mirrored interface protocols. This is done by merging the transitions created in a previous step, in such a way that the protocol behaviors are synchronized, as shown in Fig. 9.14. The resulting STG no longer describes the interface protocols, but rather the protocol converter behavior. An important issue in this procedure is to ensure that safeness and liveness properties are preserved after the transformations. Conditions for implementability must also be verified, and a complete state encoding must be guaranteed to ensure that VHDL code for the protocol converter can be generated.

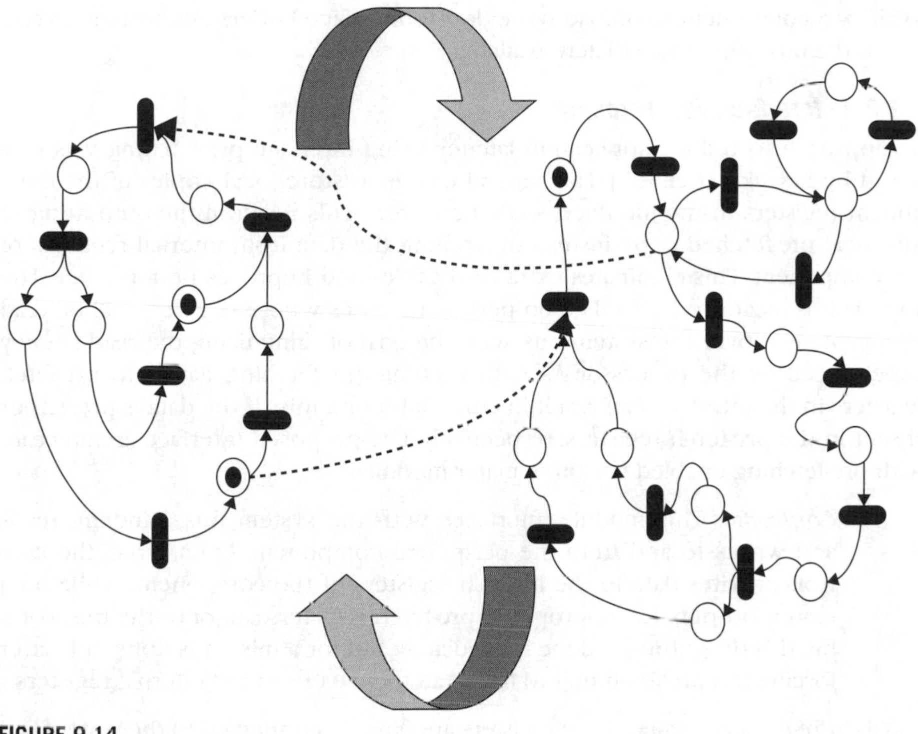

FIGURE 9.14

Interface converter synthesis [74]
© 2003 IEEE

This is possible if for any given actual state of the system, there is only one corresponding marking on the Petri Net, and vice versa. CSC property verification [76, 77] can be used for this purpose, and has been extensively addressed in literature for state graphs generated from different classes of Petri Nets using a region-based approach. Finally, protocol converter behavior consistency is verified by checking prohibitive submarkings reachability. The approach described above was used to automate the synthesis of a protocol converter between the AMBA APB and AVALON bus protocols, considering single read and write data transfer operations.

9.2.3 Interface Optimization

In addition to overcoming protocol mismatches, interfaces can also be used to optimize power or performance during data transfers. A large body of work exists to optimize power consumption, reliability, and latency using encoding techniques at interfaces. These are presented in great detail in Chapter 7. Here we present some *interface synthesis approaches* to optimize communication performance. There has been some research in optimizing the design of component interfaces during interface synthesis, to improve communication performance. For instance, a few approaches have proposed improving application performance by embedding DMA logic in the interface [78], and by implementing write buffers in the interface to allow a component to offload write data to interface buffers and resume execution, if the bus is not immediately available [79].

9.2.3.1 *Interface Pre-Fetching*

An approach to reduce application latency using interface pre-fetching was proposed by Lysecky et al. [80]. The general idea is to store local copies of the component registers in the interface, so that register reads result in the outputting of this local pre-fetched copy, instead of fetching the data from internal registers of the component. This eliminates extra read cycles and improves performance. The approach is meant to be applied on peripheral cores whose registers will be read by a processor over the system bus, with the goal of minimizing the read latency experienced by the processor. As with caching, pre-fetching data into pre-fetch registers in the interface can result in either a hit or a miss if the data is present or absent in the pre-fetch registers, respectively. The proposed interface architecture with pre-fetching enabled has three major modules:

(i) *Controller*: This module interfaces with the system bus, handling reads and writes to and from the peripheral component. On a write, the controller writes data to the internal registers of the component, while on a read it outputs the appropriate pre-fetch register data onto the bus. For a hit, this outputting is done immediately, but for a miss, it is done only after forcing the pre-fetch unit to first read the data from the internal registers.

(ii) *Pre-fetch registers*: These registers are directly connected to the system bus for fast output. Any output to the bus must pass through these registers.

(iii) *Pre-fetch unit (PFU)*: This module implements pre-fetch heuristics, and is responsible for reading data from the internal registers of the component. The pre-fetch heuristics used include updating pre-fetch registers (in addition to the internal component registers) on a write to the component and repeatedly reading any internal register values that change, due to an external write or due to a change in the execution status, into the pre-fetch registers.

To evaluate the usefulness of the interface pre-fetching mechanism, it was applied to a digital camera subsystem case study [81]. The digital camera consists of a CCD preprocessor component for capturing images, a CODEC component to compress and decompress picture frames, and other components, including a processor, BIOS, and memory. Figure 9.15 shows the result of the experiment, when the CCD and CODEC components in the system were implemented with and without pre-fetching. It can be clearly seen that pre-fetching improves performance when reading registers from the CCD and CODEC components. The pre-fetching performance increase of the digital camera is directly related to the ratio of I/O access to processor computation. Since the digital camera system spends 88% of execution time performing computation and only 12% performing I/O access, pre-fetching did not have a large impact on overall performance. However, the improvement in performance for peripheral I/O access was 25%. Therefore, for a design that is more I/O intensive, a greater overall performance improvement can be achieved. The energy and power overhead of pre-fetching modules was found to be fairly negligible.

	Reads	Cycles without pre-fetching	Cycles with pre-fetching
CCD – Status	3	12	6
CCD – Data	256	1024	512
CODEC – Status	256	1024	512
CODEC – Data	257	1028	514
Total for two cores	772	3088	1544
Digital camera		48,616	47,072
Digital camera peripheral I/O access		6,224	4,680
Digital camera processor execution		42,392	42,392

FIGURE 9.15

Impact of pre-fetching on digital camera performance [80]
© 1999 IEEE

A technique to automatically generate the pre-fetch unit to satisfy user-imposed register constraints was proposed by the authors in an extension to this work in [82]. The proposed approach performs pre-fetch scheduling to trade-off performance with power consumption—updating pre-fetch registers frequently increases power consumption but potentially improves performance due to greater number of hits, whereas less frequent pre-fetch register updates save power, but do not provide as much performance gain. Two major techniques, based on real-time scheduling, and Petri Net dependency models were proposed to create a balanced schedule for pre-fetch register updates.

9.3 DISCUSSION: INTERFACE SYNTHESIS

Interface synthesis has been the focus of much research in the software engineering domain, and a lot of work has been done on object-oriented (OO) design and component-based design [84–86]. For SoC designs, a lot of the same principles and concepts can be applied. Like in OO/component-based software design, SoC interface design must address issues of reusability, component inter-compatibility, and performance overheads. Additionally, unlike OO/component-based software design, SoC interface design needs to deal with compatibility between various hardware abstractions and mismatching hardware protocols, as well as satisfying application constraints such as power and area.

9.4 SUMMARY

In this chapter, we presented research efforts in the related areas of communication architecture refinement and interface synthesis. Communication architecture refinement is an essential part of any SoC design flow, as high level models are iteratively refined with more detail being added, till an implementation model (typically at the synthesizable RTL abstraction) is obtained. Models at different levels of abstraction are needed because each abstraction level possesses its own unique advantages and disadvantages. High level models enable faster simulation but less accuracy, while lower level models have greater accuracy, but suffer from slow simulation speeds due to the large amount of captured detail. Research efforts were presented that propose using different abstraction levels during a design flow, and automating the task of communication architecture refinement, to reduce their modeling time. Because of the various abstraction levels of models used in a typical design flow, as well as the different protocols used by components that are reused in designs, there is a need to generate interfaces that can translate between models at different levels of detail, as well as between different protocols. Several research efforts were presented that facilitate the task of creating interface logic, for interfacing components at different abstraction levels and for rectifying mismatches in protocols used by two or more components, so that the system works correctly, as intended. Finally, we looked at some work in the area of interface optimization, to improve application performance by using latency reducing techniques during the design of component interfaces.

REFERENCES

[1] J. A. Rowson and A. L. Sangiovanni-Vincentelli, "Interface based design," in *Proceedings of the 34th Design Automation Conference (DAC)*, 1997, pp. 178–183.

[2] F. Vahid and L. Tauro, "An object-oriented communication library for hardware/software codesign," in *Proceedings of the International Workshop on Hardware Software Codesign (CODES)*, March 1997, pp. 81–86.

[3] J.-M. Daveau, G. F. Marchioro, T. Ben-Ismail and A. A. Jerraya, "Protocol Selection and Interface Generation for HW–SW Codesign," *IEEE Transactions on VLSI Systems*, Vol. 5, No. 1, March 1997, pp. 136–144.

[4] K. Hines and G. Borriello, "Dynamic communication models in embedded system co-simulation," in *Proceedings of the Design Automation Conference (DAC)*, June 1997, pp. 395–400.

[5] J. Plosila, P. Liljeberg and J. Isoaho, "Refinement of on-chip communication channels," *Proceedings of the International Symposium on System-on-Chip*, 2004, pp. 197–200.

[6] R. Back and K. Sere, "From action systems to modular systems," *Software—Concepts and Tools*, Vol. 17, Springer 1996, pp. 26–39.

[7] J. Plosila, "Self-timed circuit design—the action system approach," Ph.D. Thesis, University of Turkq, September 1999.

[8] J.-Y. Brunel, E. A. de Kock, W. M. Kruijtzer, H. J. Kenter and W. J. Smits, "Communication refinement in video systems on chip," in *Proceedings of the 7th International Workshop on Hardware/Software Codesign (CODES)*, 1999, pp. 142–146.

[9] J.-Y. Brunel, W. M. Kruijtzer, H. J. H. N. Kenter, F. Pétrot, L. Pasquier, E. A. de Kock and W. J. M. Smits, "COSY communication IP's," in *Proceedings of the 37th Conference on Design Automation (DAC)*, 2000, pp. 406–409.

[10] J. Y. Brunel et al., "COSY: a methodology for system design based on reusable hardware and software IP's," *Technologies for the Information Society*, J.-Y. Roger (Ed.), IOS Press, 1998, pp. 709–716.

[11] G. Kahn, "The semantics of a simple language for parallel programming," *Information Processing*, 1974, pp. 471–474.

[12] D. Fairbank et al., "The VSI Alliance: journey from vision to production," *Electronic Design*, Vol. 46, No. 1, January 12, 1998, pp. 86–92.

[13] P. Clarke, "Philips extends TriMedia reuse into Nexperia cores," *EE Times*, August 30, 1999.

[14] P. Lieverse, P. van der Wolf and E. F. Deprettere, "A trace transformation technique for communication refinement," in *Proceedings of the 9th International Symposium on Hardware/Software Codesign (CODES)*, April 2001, pp. 134–139.

[15] A. D. Pimentel and C. Erbas, "An IDF-based trace transformation method for communication refinement," in *Proceedings of the ACM/IEEE Design Automation Conference (DAC)*, June 2003, pp. 402–407.

[16] D. D. Gajski, J. Zhu, R. Dömer, A. Gerstlauer and S. Zhao, *"SpecC: Specification Language and Methodology"*, Kluwer Academic Publishers, Boston, MA, March 2000. ISBN 0-7923-7822-9, 336 p.

[17] A. Gerstlauer, R. Domer, J. Peng and D. Gajski, *"System Design: A Practical Guide with SpecC,"* Kluwer Academic Publishers, May 2001.

[18] A. Gerstlauer, D. Shin, J. Peng, R. Dömer and D. D. Gajski, "Automatic, layer-based generation of system-on-chip bus communication models," *IEEE Transactions on Computer-Aided Design of Integrated Circuits and Systems (TCAD)*, Vol. 26, No. 9, September 2007, pp. 1676–1687.

[19] R. A. Bergamaschi and William R. Lee, "Designing systems-on-chip using cores," in *Proceedings of Design Automation Conference (DAC)*, 2000, pp. 420–425.

[20] R. Bergamaschi, W. R. Lee, D. Richardson, S. Bhattacharya, M. Muhlada, R. Wagner, A. Weiner and F. White, "Coral-automating the design of systems-on-chip using cores," in *Proceedings of the IEEE Custom Integrated Circuits Conference, (CICC)*, 2000, pp. 109–112.

[21] R. E. Bryant, "Graph based algorithms for Boolean function manipulation," *IEEE Transactions on Computers*, Vol. 35, No. 8, August 1986, pp. 677–691.

[22] L. Semeria and A. Ghosh. "Methodology for hardware/software co-verification in C/C++," in *Proceedings of the Asia South Pacific Design Automation Conference (ASP-DAC)*, January 2000.

[23] Coware Inc., "N2C," available at http://www.coware.com.

[24] P. Gerin, S. Yoo, G. Nicolescu and A. A. Jerraya, "Scalable and flexible cosimulation of SoC designs with heterogeneous multi-processor target architectures," in *Proceedings of the Asia South Pacific Design Automation Conference (ASP-DAC)*, 2001.

[25] G. Nicolescu, S. Yoo and A. A. Jerraya, "Mixed-level cosimulation for fine gradual refinement of communication in SoC design," in *Proceedings of the Design Automation and Test in Europe (DATE)*, 2001.

[26] A. Wieferink, R. Leupers, G. Ascheid, H. Meyr, T. Michiels, A. Nohl and T. Kogel, "Retargetable generation of TLM bus interfaces for MP-SoC platforms," in *Proceedings of the 3rd IEEE/ACM/IFIP International Conference on Hardware/Software Codesign and System Synthesis (CODES+ISSS)*, 2005, pp. 249–254.

[27] S. Yoo, G. Nicolescu, D. Lyonnard, A. Baghdadi and A. A. Jerraya, "A generic wrapper architecture for multi-processor SoC cosimulation and design," in *Proceedings of the International Workshop on Hardware-Software Codesign*, 2001, pp. 195–200.

[28] D. Lyonnard, S. Yoo, A. Baghdadi and A. A. Jerraya, "Automatic generation of application-specific architectures for heterogeneous multiprocessor system-on-chip," in *Proceedings of the Design Automation Conference (DAC)*, June 2001, pp. 518–523.

[29] W. Cesário, D. Lyonnard, G. Nicolescu, Y. Paviot, S. Yoo, A. A. Jerraya, L. Gauthier and M. Diaz-Nava, "Multiprocessor SoC platforms: a component-based design approach," *IEEE Design and Test of Computers*, Vol. 19, No. 6, November 2002, pp. 52–63.

[30] M. Dziri, W. Cesário, F. R. Wagner and A. A. Jerraya, "Unified component integration flow for multi-processor SoC design and validation," in *Proceedings of the Conference on Design, Automation and Test in Europe (DATE)*, 2004.

[31] K. Hines and G. Borriello, "Optimizing communication in embedded system co-simulation," in *Proceedings of the International Workshop on Hardware-Software Codesign*, March 1997, pp. 121–125.

[32] K. Hines and G. Borriello, "Dynamic communication models in embedded system co-simulation," in *Proceedings of the Design Automation Conference (DAC)*, June 1997, pp. 395–400.

[33] K. Hines and G. Borriello, "A geographically distributed framework for embedded system design and validation," in *Proceedings of the Design Automation Conference (DAC)*, June 1998, pp. 140–145.

[34] TIA/EIA-95A, "Mobile Station-Base Station Compatibility Standard for Dual-Mode Wideband Spread Spectrum Cellular Systems," 1995.

[35] ARM 9 Family, http://www.arm.com/products/CPUs/families/ARM9Family.html.

[36] ARM AMBA Specification and Multilayer AHB Specification, (rev2.0), http://www.arm.com, 2001.

[37] ARM AMBA AXI Specification, www.arm.com/armtech/AXI.

[38] "STBus Communication System: Concepts and Definitions," Reference Guide, STMicroelectronics, May 2003.

[39] IBM CoreConnect Specification, http://www.ibm.com/chips/techlib/techlib.nsf/product families/CoreConnect_Bus_Architecture.

[40] Sonics, *Sonics μNetworks Technical Overview*, Sonics Inc, June 2000.

[41] G. Borriello and R. Katz, "Synthesis and optimization of interface transducer logic," in *Proceedings of the International Conference on Computer Aided Design (ICCAD)*, 1987, pp. 274–277.

[42] J. Akella and K. McMillan, "Synthesizing converters between finite state protocols," in *Proceedings of the IEEE International Conference on Computer Design (ICCD)*, 1991, pp. 410–413.

[43] J. Sun and R. Brodersen, "Design of system interface modules," in *Proceedings of the International Conference on Computer Aided Design (ICCAD)*, 1992, pp. 478–481.

[44] S. Narayan and D. D. Gajski, "Interfacing incompatible protocols using interface process generation," in *Proceedings of the Design Automation Conference (DAC)*, June 1995, pp. 468–473.

[45] J. Smith and G. D. Micheli, "Automated composition of hardware components," in *Proceedings of the 35th Design Automation Conference (DAC)*, June 1998, pp. 14–19.

[46] D. Shin and D. D. Gajski, "Interface synthesis from protocol specification," CECS Technical Report 02-13, April 2002.

[47] R. Passerone, J. A. Rowson and A. L. Sangiovanni-Vincentelli, "Automatic synthesis of interfaces between incompatible protocols," in *Proceedings of the Design Automation Conference (DAC)*, 1998, pp. 8–13.

[48] R. Passerone, "Automatic synthesis of interfaces between incompatible protocols," M.S. Thesis, University of California, Berkeley, 1997.

[49] B.-H. Park, H. Choi, I-C. Park and C.-M. Kyung, "Synthesis and optimization of interface hardware between ip's operating at different clock frequencies," in *Proceedings of the International Conference on Computer Design (ICCD)*, June 2000, pp. 519–524.

[50] R. Passerone, L. de Alfaro, T. A. Henzinger and A. L. Sangiovanni-Vincentelli, "Convertibility verification and converter synthesis: two faces of the same coin," in *Proceedings of the International Conference on Computer-Aided Design (ICCAD)*, November 2002.

[51] L. de Alfaro and T. A. Henzinger, "Interface automata," in *Proceedings of the 9th Annual Symposium on Foundations of Software Engineering*, ACM Press, 2001, pp. 109–120.

[52] Altera AVALON Interface Specification, April 2006, http://www.altera.com/.

[53] Open Core Protocol International Partnership (OCP-IP) OCP datasheet, Release version 2.0/2.1, http://www.ocpip.org.

[54] VSI Alliance™ On-Chip Bus Development Working Group, Virtual Component Interface (VCI) Standard Version 2 (OCB 2 2.0), April 2001.

[55] W. Zhonghai, Y. Yizheng, W. Jinxiang and Y. Mingyan, "Designing AHB/PCI bridge," in *4th International Conference on ASIC*, 2001, pp. 578–580.

[56] PCI Special Interest Group, PCI Local Bus Specification Revision 2.2 (PCI SIG, 1998), p. 1.

[57] S. Choi and S. Kang, "Implementation of an on-chip bus bridge between heterogeneous buses with different clock frequencies," *5th International Workshop on System-on-Chip for Real-Time Applications*, 2005, pp. 530–534.

[58] A. Zitouni, S. Badrouchi and R. Tourki, "Communication architecture synthesis for multi-bus SoC," *Journal of Computer Science*, Vol. 2, No. 1, 2006, pp. 63–71.

[59] Texas Instruments Inc. TMS32C5x User's Guide. Available at http://focus.ti.com/.

[60] H. Cho, S. Abdi and D. Gajski, "Design and implementation of transducer for ARM-TMS communication," in *Proceedings of the 2006 Conference on Asia South Pacific Design Automation (ASP-DAC)*, 2006, pp. 126–127.

[61] G. Strano, S. Tiralongo and C. Pistritto, "OCP STBUS plug-in methodology," in *The International Embedded Solutions Event*, 2004.

[62] D. Hommais, F. Pétrot and I. Augé, "A practical tool box for system level communication synthesis," in *Proceedings of the 9th International Symposium on Hardware/Software Codesign (CODES)*, 2001, pp. 48–53.

[63] Y. Ming-Yan, Z. Qing-Li, W. Jin-xiang, Y. Yi-zheng and L. Feng-chang, "The design of AMBA AHB/VCI wrapper," *5th International Conference on ASIC*, 2003, pp. 438–442.

[64] N.-J. Kim and H.-J. Lee, "Design of AMBA™ wrappers for multiple-clock operations," in *International Conference on Communications, Circuits and Systems*, 2004, pp. 1438–1442.

[65] T. Chelcea and S. M. Nowick, "A low-latency FIFO for mixed-clock systems," in *Proceedings of the IEEE Computer Science Workshop on VLSI*, 2000, pp. 119–126.

[66] V. Androutsopoulos, D. M. Brookes and T. J. W. Clarke, "Protocol converter synthesis," *IEE Proceedings—Computers and Digital Techniques*, Vol. 151, No. 6, November 18, 2004, pp. 391–401.

[67] V. D'silva, S. Ramesh and A. Sowmya, "Bridge over troubled wrappers: automated interface synthesis," in *Proceedings of the 17th International Conference on VLSI Design (VLSID)*, 2004.

[68] V. D'silva, S. Ramesh and A. Sowmya, "Synchronous protocol automata: a framework for modelling and verification of SoC communication architectures," in *Proceedings of the Conference on Design, Automation and Test in Europe (DATE)*, 2004.

[69] A. Roychoudhury, P. S. Thiagarajan, T. Tran and V. A. Zvereva, "Automatic generation of protocol converters from scenario-based specifications," in *Proceedings of the 25th IEEE International Real-Time Systems Symposium RTSS 2004*, pp. 447–458.

[70] Z.120. Message Sequence Charts (MSC96), 1996.

[71] T. Murata, "Petri nets: properties, analysis and applications," *Proceedings of the IEEE* 77, Vol. 44 (April), 1989, pp. 541–580.

[72] M. Kishinevsky, A. Kondratyev, A. Taubin and V. Varshavsky, "Concurrent hardware: the theory and practice of self-timed design," *Series in Parallel Computing* John Wiley & Sons, 1994.

[73] J. Desel and J. Esparza, *Free Choice Petri Nets*, Cambridge University Press, Cambridge, Great Britain, 1995.

[74] J. A. O. Filho, M. E. Lima and P. R. Maciel, "Petri net based interface analysis for fast IP-core integration," in *Proceedings of the 1st ACM and IEEE International Conference on Formal Methods and Models for Co-design (MEMOCODE)*, 2003.

[75] T. A. Chu, "Synthesis of self-timed VLSI circuits from graph theoretic specifications," Ph.D. Thesis, Department of EECS, Massachusetts Institute of Technology, September 1987.

[76] J. Cortadella, M. Kishinevsky, A. Kondratyev, L. Lavagno and A. Yakovlev, "A region-based theory for state assignment in speed-independent circuits," *IEEE Transactions on Computer-Aided Design*, Vol. 16, No. 8, August 1997, pp. 793–812.

[77] E. Pastor, J. Cortadella, A. Kondratyev and O. Roig, "Cover approximations for the synthesis of speed independent circuits," *Workshop on Logic and Architecture Synthesis*, December 1995, pp. 150–159.

[78] A. Olugbon, S. Khawam, T. Arslan, I. Nousias and I. Lindsay, "An AMBA AHB-based reconfigurable SOC architecture using multiplicity of dedicated flyby DMA blocks," in *Proceedings of the 2005 Conference on Asia South Pacific Design Automation (ASP-DAC)*, 2005, pp. 1256–1259.

[79] K. Anjo, A. Okamura and M. Motomura, "Wrapper-based bus implementation techniques for performance improvement and cost reduction," *IEEE Journal of Solid-State Circuits*, Vol. 39, No. 5, May 2004, pp. 804–817.

[80] R. Lysecky, F. Vahid, T. Givargis and R. Patel, "Pre-fetching for improved core interfacing," *International Symposium on System Synthesis (ISSS)*, 1999, pp. 51–55.

[81] F. Vahid and T. Givargis, "The case for a configure-and-execute paradigm," *International Workshop on Hardware/Software Codesign (CODES)*, 1999, pp. 59–63.

[82] R. Lysecky, F. Vahid and T. Givargis, "Techniques for reducing read latency of core bus wrappers," in *Proceedings of the Design Automation and Test in Europe (DATE)*, 2000, pp. 84–91.

[83] S. Yoo, J. Lee, J. Jung, K. Rha, Y. Cho and K. Choi, "Fast prototyping of an IS-95 CDMA cellular phone: a case study," in *Proceedings of the 6th Conference of Asia Pacific Chip Design Languages*, October 1999, pp. 61–66.

[84] I. Jacobson, *"Object-Oriented Software Engineering: A Use Case-Driven Approach,"* Addison-Wesley Professional, 1992.

[85] M. Abadi and L. Cardelli, *A Theory of Objects*, Springer-Verlag, 1996.

[86] J. Rumbaugh, M. Blaha, W. Premerlani, F. Eddy and W. Lorensen, *Object-Oriented Modeling and Design*, Prentice Hall, 1990.

[87] J. S. Sun et al., "SIERA: a CAD environment for real-time systems," *3rd Physical Design Workshop*, May 1991.

CHAPTER

10

Verification and Security Issues in On-Chip Communication Architecture Design

With the growing complexity of on-chip communication protocols, it is becoming essential to verify the correctness of these protocols connecting intellectual property (IP) blocks (components) in a system-on-chip (SoC), to ensure freedom from bugs, and prevent deadlock or livelock conditions from arising during application execution. The problem of on-chip communication protocol verification is exacerbated by the observation that almost every communication protocol specification is documented informally, using natural languages (e.g., the 230 page AMBA 2.0 [1] specification document). It is quite common for these bulky specification documents to be misinterpreted by designers due to their vagueness. It is also quite common for these documents to have inconsistencies that a human reader will not notice or have missing rules that may be inferred automatically by a human reader, but which due to their inherent incompleteness and inconsistency can cause bugs in an implementation. Formal modeling of on-chip communication protocols can help address this problem by ensuring the *verification* of the properties and constraints inherent to a communication protocol. Another challenge for verification in SoC designs is to ensure correctness during IP block integration. Increasingly, designers are reusing existing IP blocks (that may have been used in older SoC designs) in an effort to speed up the design process by eliminating the need to develop existing functionality from scratch. These IP blocks are either developed in-house or acquired from third party IP vendors. While it is important for the reused IP blocks to be functionally verified by the IP designer to ensure correct operation, it is equally important for an IP integrator to ensure that IP blocks obtained from external vendors and integrated into an SoC design can interface with one another correctly. In other words, IP integrators must verify that the interfaces of the IP blocks interact properly with the rest of the system. The interactions between the various IP blocks in an SoC typically occur over on-chip communication architectures. Due to the variety and complexity of on-chip communication protocol standards being used today, the task of verifying whether an IP block behaves correctly in an SoC environment has become non-trivial and time consuming.

Designing *secure on-chip communication architectures* is emerging as another major challenge for designers. SoC designs used in PDAs (personal digital assistants), smartphones, remote cameras, sensors, etc. are increasingly capturing, storing, and manipulating sensitive data. Security in such systems is an important concern, especially as these systems become accessible to outsiders in a highly networked environment. These networked connections expose applications to intrusions and malicious attacks that can compromise the security of a system. For instance, there has been a significant increase in viruses targeting mobile smartphones lately that can harm the system in many ways such as crashing the system ROM (read-only memory), disabling phone features, or installing malware to covertly track phone usage. These viruses spread via unsecured Bluetooth connections, MMS (multimedia messaging service) messages, or Internet downloads on the phones. The cost of potential insecurities in systems can have grave consequences—for instance it has been estimated that the "I Love You" virus that attacked Windows PCs caused a loss of nearly one billion dollars in revenues worldwide [90]. Security attacks on SoCs can take many forms such as physical, side channel, or software attacks. It is thus not enough to secure a system from network-based intrusions—malicious entities can logically or physically gain access to the system and compromise the integrity and confidentiality of system data as well (e.g., gaining access to device and reverse engineering/probing it to gain access to stored on-chip cryptographic keys or proprietary firmware). Secure system design can thus no more be left as an afterthought, but must be tightly integrated with the system implementation stage. The communication architecture in an SoC design facilitates communication between all the components in a system, and has a global view of the system. As such, it can be exploited to detect and prevent malicious attacks on the integrity of system data.

Figure 10.1 shows where the verification of communication protocols and secure communication architecture design effort fits into a typical electronic system level (ESL) design flow. In this chapter, we focus on research in the areas of verification and security in on-chip communication architecture design. The first half of the chapter presents issues and research aimed at the verification of on-chip communication architectures. Section 10.1 presents research in the area of verification of on-chip communication protocols and communication architecture logic components, such as arbiters. Section 10.2 describes research that addresses the verification of IP blocks being integrated into an SoC, to ensure that they are compliant with the on-chip communication protocol. The second half of the chapter discusses details of secure on-chip communication architecture design. The security features that a communication infrastructure offers can be categorized into two classes: (i) features that are supported natively by the standard communication architecture and (ii) features that enhance existing standard architecture specifications by adding additional components, for instance modules to encrypt communicated data. In Section 10.3, we discuss the basic requirements and concepts related to SoC security. Section 10.4 then presents the first class of security features that are built into bus-based communication architecture standards. Finally, Section 10.5 describes research efforts that fall into the second class of security features that modify existing standards to enhance system security. These research efforts propose novel techniques to prevent malicious attacks on SoC designs.

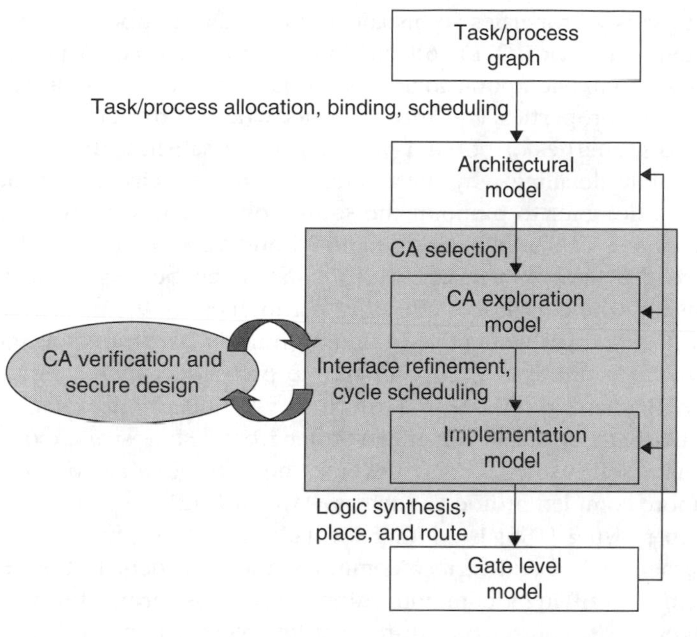

FIGURE 10.1

Communication architecture verification and secure design in a typical ESL design flow

10.1 VERIFICATION OF ON-CHIP COMMUNICATION PROTOCOLS

On-chip communication architecture standards that are widely used today, such as AMBA 2.0/3.0 [1, 2], STBus [3], IBM CoreConnect [4], and Sonics Smart Interconnect [5], are primarily described using informal specification documents that combine timing diagrams with English language descriptions. Such informal specifications can be the source of many bugs when an on-chip communication architecture is being implemented, due to misinterpretation of dense information, inconsistencies in the specification, or incorrect assumptions for incompletely specified aspects of the protocols. In an attempt to avoid these informal specification-related bugs, several research efforts have proposed using formal specification and verification of on-chip communication protocols. Formal verification approaches typically develop formal descriptions of communication protocols, such as finite state machine (FSM)-based representations, and then use techniques to formally verify these representations. There are two main approaches to formal verification, namely *model checking* and *theorem proving*.

Model checking [6, 7] is a commonly used automated formal verification technique that enables checking of functionality properties of finite state concurrent systems, and verifies properties such as *safety* and *liveness* that are specified in temporal logic. Model checking tools such as SMV [37] or SPIN [134] take two main inputs when verifying communication protocols: (i) a model of the communication protocol (typically a variant of an FSM is used, although other models can also be used) and (ii) properties (or specifications) that the protocol is expected to satisfy. Properties are usually expressed as temporal logic statements. Computation tree logic (CTL) [6] and linear time logic (LTL) [135] are examples of two popular temporal logic notations. A model checking tool analyzes a model and its properties, and then outputs whether the given model satisfies the specified properties or not. If a property is not satisfied, the tool generates a counter-example detailing why the model does not specify the property, which then allows a designer to pinpoint the source of error in the model. Reasonably large state spaces can be efficiently handled and verified by model checking—for instance, FSM models with up to 10^{30} states can be verified in a matter of minutes, using symbolic model checking techniques [8, 9]. Model checking has been shown to be highly effective in finding bugs in hardware specifications, for instance in the Futurebus+ cache coherence protocol, which was adopted as a standard by IEEE and the US Navy [10]. However, since model checking is based on exhaustive state space search, it can suffer from a state space explosion problem (too many states to effectively traverse and verify in a reasonable amount of time) for more complex protocols such as AMBA AXI [2].

Theorem proving [136] is another formal verification technique that can be used to formally define and verify a communication protocol. In theorem proving, the problem of verifying a communication protocol is formulated formally using axioms, hypothesis, and a conjecture, usually expressed in classical first order logic (or possibly higher order logic). Automated theorem proving tools such as ACL2 [122] and PVS [20] generate proofs to show why a statement (conjecture) is a logical consequence of a set of other statements (the axioms and hypothesis). Theorem proving is useful when the protocol being verified is not finite state (i.e., involving a potentially infinite state space) and cannot be fully analyzed by model checking. But unlike model checking, theorem proving is not truly automated, but rather semi-automated (the operation of theorem proving tools is usually guided interactively by an expert to get a solution in a reasonable amount of time) and the verification effort involved can be substantial [137]. It is, however, sometimes possible to reduce the verification time and effort by combining finite state (model checking) and theorem proving verification techniques [43, 44].

10.1.1 Early Work with PCI Bus Protocol Verification

Early work [11–17] in the area of formal verification of communication protocols focused on the formal specification and verification of the off-chip PCI local bus protocol [18, 19]. Mokkedem et al. [14] proposed using theorem proving techniques to formalize a significant portion of the PCI v2.1 [18] specification. The PVS [20] theorem prover was used to verify a proposed solution to fix previously detected bugs [21–23] (arising due to inconsistency and deadlock issues) in the PCI v2.1 protocol specification. Ivanov and Nunna [16] made use of a particular type of partial ordering technique [24–27] called series-parallel posets [28–30], for formal verification of the PCI local bus protocol. Partial orders are useful for modeling and verifying system behavior because they avoid the need for studying all possible interleavings of events during system execution, and as such reduce the time complexity of verifying properties from exponential to polynomial time in many cases. The series-parallel posets in [16] were used to verify ordering and independence properties for events specified in the PCI data transfer protocol.

Chauhan et al. [13] proposed using symbolic model checking to verify the correctness of the PCI v2.2 [19] local bus protocol. The process of symbolic model checking is broken down into the three phases of modeling, specification, and verification. In the modeling phase, an SMV [32] temporal logic model checker based on binary decision diagrams (BDDs) [31] is used. The SMV checker has a built-in data flow oriented hardware description language (HDL) for modeling and accepting specifications expressed in CTL [6] notation. SMV extracts a finite state model as a state transition graph from an SMV program, and BDDs are used to represent and manipulate transition relations. In the specification phase, the properties are written in the CTL [6] branching time temporal logic notation. Formulas in CTL are built from three components: atomic propositions, Boolean connectives, and temporal operators. The final task of verification is an automatic process, with every CTL formula having a fixed point characterization that can be used to find the set of states satisfying the formula [33]. Because PCI is a high performance bus, there are strict requirements on various events, such as number of wait states, arbitration latency, latency for the target to assert its select signal, etc. The PCI bridges also have some restrictions on transaction posting and must obey PCI ordering rules. Chauhan et al. [13] created a simple model of a system with PCI bus, master, slave, arbiter, and bridge components to verify bus protocol properties about transaction termination, arbitration, and latency requirements, as well as the PCI bridge ordering rules. The verification process found two potential bugs in the standard PCI specification, arising due to inconsistencies between the specification document and the state machines given in the specification Appendix [19].

Formal verification techniques such as bounded model checking that use satisfiability (SAT) solvers, to improve upon BDD-based approaches (such as the one used in [13]), have also been proposed [34, 35]. BDDs are based on canonical representations, and require a complete representation of a function before checking its satisfiability. They have the disadvantage of being too large in size to fit into available computer memory, for certain large functions. In contrast, SAT-based approaches intelligently sample the variable space and are usually more effective because they are less likely to fail due to exponential space blowup. Aloul and Sakallah [12] proposed a Boolean SAT-based approach to verify the PCI local bus. Rules from the PCI specification document were extracted and written as a set of logical expressions in conjunctive normal form (CNF). CTL formulas to analyze dead state and property checking were converted into CNF. A simple model of a PCI-based system containing interconnected master, slave, and arbiter components was created (the bridge component was not considered). Several properties related to bus arbitration, master-target terminations, read/write transactions, and back-to-back transactions were verified using the GRASP SAT solver [36], which was able to detect several bugs in the PCI specification, similar to those detected in [13] and [15]. The SAT-based approach took significantly less memory and time to verify the protocol, compared to the BDD-based approach in [13].

Campos et al. [11] used formal verification techniques to gain insight into the performance of the PCI local bus. The PCI local bus was modeled with a state transition graph, and represented symbolically using BDDs [31]. The graph was subsequently traversed using algorithms based on symbolic model checking techniques. Issues such as transaction completion and arbitration fairness were verified by the approach, and response time bounds (i.e., specification of minimum and maximum times) for transactions were calculated to estimate latency during the various phases within a transaction. The proposed approach thus computed quantitative timing information about the model, to analyze and identify factors (such as arbitration schemes) that influenced the performance of the PCI bus.

While formal techniques are able to eliminate problems encountered in interpreting informal specifications, in practice it is often hard to write these formal specifications. A new specification style based on writing the protocol specification as *monitors* was proposed by Shimuzu et al. [15]. The bus protocol in such a specification style is described via an observer (or monitor) that raises error indications when the protocol is violated. The observer can also detect the component responsible for the error. In addition to finding specification errors, the monitor-based specification style can also aid in the development of a runtime monitor that checks protocol implementation. The simplicity of writing these monitors allows them to be used in existing HDL such as Verilog and VHDL. Shimuzu et al. [15] modeled the PCI v2.2 specification using monitor-style specification, to cover almost all the signal requirements, and used the Cadence SMV [37] model checking tool to verify the formal PCI specification.

10.1.2 Verifying On-Chip Communication Architectures

Several approaches [38–42] have looked at formally specifying on-chip communication protocols in order to detect ambiguities and inconsistencies in their specification documents. A formal specification of the AMBA 2.0 [1] protocol was proposed by Roychoudhury et al. [40]. The bus protocol is defined as a synchronous composition of several FSMs. The finite state description of the protocol can then be formally verified (for properties such as safety and liveness) using a state space search technique, such as model checking. Roychoudhury et al. [40] made use of the Cadence SMV [37] model checker to automatically verify safety properties in the ARM microcontroller bus architecture (AMBA) protocol. The goal of their work was to detect hard-to-find bugs in the informal AMBA specification document. The various components of an AMBA-based system, such as the AHB (advanced high performance bus) bus protocol, the APB (advanced peripheral bus) bus protocol, bridges, arbiters, and address decoders, together with masters and slaves, were first specified by the designer in the form of FSMs. From these descriptions, SMV was used to create a global state transition graph of the entire system. The transition relation and set of states were represented compactly by using BDDs [31].

The SMV model checker enables specification of the properties of the protocol in CTL notation [6]. CTL is a branching time logic in which time is represented by a tree-like structure and the future is not determined—there are possibly several paths in the future, any one of which might be the actual path that is realized. Assume that ξ is a property of states in the transition system we are verifying (i.e., a state formula). Then "ξ is an invariant" is stated as AGξ in CTL. "ξ is true in all the next states (at least one next state)" is stated as AXξ (EXξ). "ξ is eventually true in all outgoing paths (at least one outgoing path)" is stated as AFξ (EFξ). Since in CTL the operators can be nested, therefore the property AG($\xi \Rightarrow$ AFψ) means that in all reachable states of the system, either (i) ξ is not true, or (ii) ξ holds, and ψ holds along all outgoing paths eventually. This type of a property is very useful in specifying guaranteed response, or lack of starvation. For instance, in AMBA, a master *m* would request bus access by asserting the *HBUSREQm* signal and the arbiter would grant bus access to *m* by asserting the *HGRANTm* signal. Then one can specify and check the property AG(*HBUSREQm* ⇒ *AFHGRANTm*), which states that whenever master *m* requests bus access, it is always eventually granted bus access by the arbiter. Verifying this property is very important in a pipelined protocol such as AMBA, since there might be corner cases that can prevent a master from getting access to the bus, especially after being SPLIT [1]. Roychoudhury et al. [40] used the SMV model checker to find a violation of this no-starvation property as described below.

Consider a simple system with two masters and one slave. Master *m1* transfers two data items in a burst in cycles 1 and 2, as shown in Fig. 10.2. Note that the *HMASTER* signal specifies which master currently has bus access. Master *m2* gains access in cycle 3 and starts a burst. Due to the pipelined nature of AMBA, the response for the address transmitted in cycle 2 by master *m1* arrives in cycle 3. If this is a SPLIT response (as shown in Fig. 10.2), then the fact that master *m1* has been split is recorded in the slave. The arbiter in the meantime snoops on the split response sent by the slave in cycle 3. In cycle 4, the arbiter masks master *m2* instead of master *m1*, because *HMASTER* changed to *m2* in cycle 3. Because the arbiter detected a split in cycle 3, and noted the value of *HMASTER* as *m2* in the same cycle, it erroneously assumes that *m2* has been SPLIT, instead of *m1*. Subsequently, the slave becomes ready to service *m1* (which is indicated by the slave to the arbiter by asserting *HSPLIT1*), and *m1* is allowed to access the bus. But the arbiter never grants access to *m2*, even if *m2* requests bus access again, since it has been masked out, due to the erroneous assumption that *m2* has been SPLIT. Since the slave has no record of *m2* being split, it never asserts *HSPLIT2* to inform the arbiter to service *m2*. This leads to a starvation of master *m2*, and a violation of the property AG(*HBUSREQm* ⇒ AF*HGRANTm*). This behavior is a bug arising due to the incompleteness of the AMBA specification document. To avoid the above scenario, the arbiter needs to keep track of the *HMASTER* value for not just the current cycle, but also for the previous cycle, due to the pipelined nature of the transfers. This needs to be explicitly clarified in the AMBA specification document, in order to avoid buggy implementations.

Madl et al. [38] proposed a formal approach based on model checking to verify the AMBA AHB bus protocol, as part of their framework to evaluate the performance of real-time applications. An open-source NuSMV model checker [88] was used to verify CTL [6] properties on the FSM models of the AMBA AHB protocol. Using this approach, the authors were able to uncover an ambiguity in the AMBA AHB specification that might lead to a flawed implementation. Consider a simple SoC with two masters (*master1*, *master2*) and a slave connected to an AMBA AHB bus. The arbiter has to keep track of the masters' state in order to manage SPLIT transfers. This could be implemented by providing dedicated wires

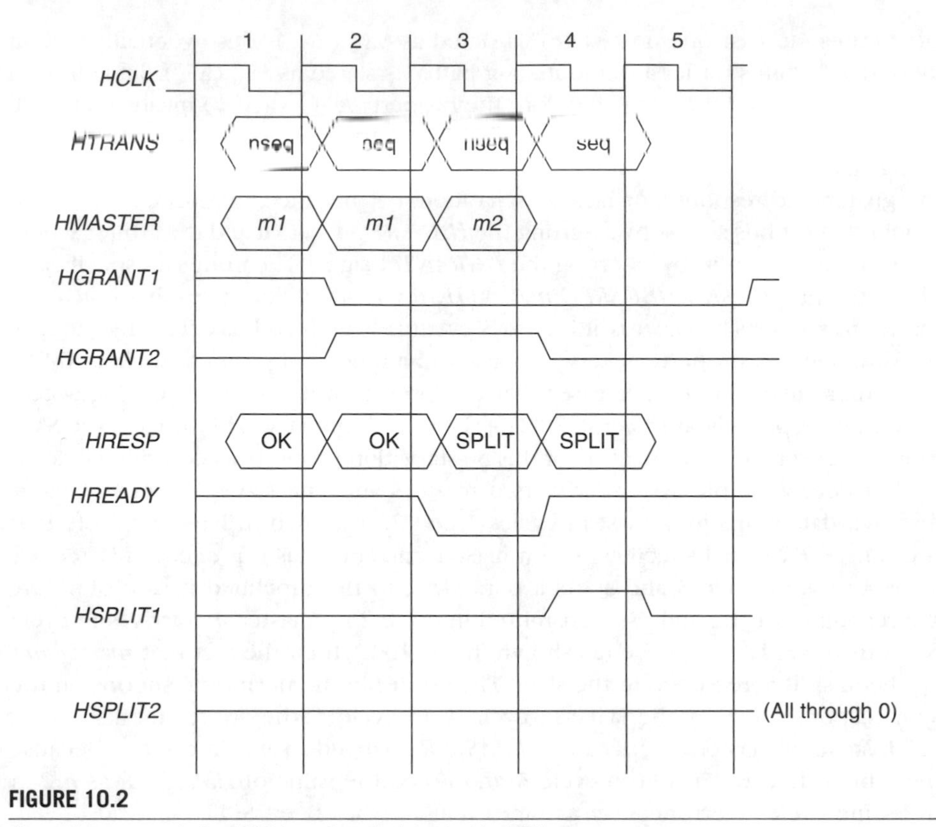

FIGURE 10.2

Starvation scenario in AMBA protocol, detected in [40]
© 2003 IEEE

between the masters and the arbiter. However, this is impractical in most cases as it requires extra computation and hardware. An alternative method is to monitor the bus traffic to obtain the master and slave states. The arbiter may use the *HTRANS* signal to check whether the master is idle or transmitting (NONSEQ, SEQ), the *HBURST* signal to predict the remaining cycles from the transfer, and the *HRESP* signal to monitor whether the active master and slave has to step back to repeat a transaction. Now consider the case where the slave has previously SPLIT *master1* (*master1* is masked by the arbiter), and is involved in a transaction with *master2*. The slave can unmask a master by issuing an *HSPLITx* signal using the masked master's address to the arbiter. Consider the scenario where the slave tries to unmask *master1* by setting *HSPLITx* when it issues a RETRY response intended for *master2*. The AMBA specification is ambiguous on what the arbiter should do in this case. The specification says that a master has to repeat the last transaction when it receives the RETRY response. If the arbiter monitors the bus signals to keep track of the masters' states it will try to go back to its previous state to keep synchronized with the master and the slave. However, if the arbiter implements this behavior it will not unmask *master1* as requested by the slave. Since there is no acknowledgement for *HSPLITx* signals, the slave will assume that *master1* is already unmasked, and won't request that the arbiter unmasks it again. This may result in deadlock as *master1* never gets access to the bus again. The AMBA specification states that, "A slave which issues RETRY responses must only be accessed by one master at a time." The authors concluded that there was an ambiguity in this statement, and it was not clear whether it meant that the slave can be accessed by one master at a time while having SPLIT another master, or if the slave can be accessed by only one master at any time with no SPLITs being allowed for other masters.

Schmaltz and Borrione [41] made use of theorem proving to verify properties of the AMBA AHB bus. The ACL2 theorem prover [122] was used by the authors to prove arbitration and coherence properties for arbitrary numbers of masters and slaves. However, time was abstracted away and complex functionality such as bursts, wait states, splits, and retries were not handled in the approach. This is because theorem provers are generally better suited for attacking data path properties at a high level of abstraction, without the complexity of handling cycle level control signals. Amjad [43, 44] made use of a combination of theorem proving and model checking to verify the control and data path properties of the AMBA bus protocol. The verification was conducted using HOLCHECK, a model checker for the propositional μ-calculus L_μ [45], which is part of the HOL theorem prover [46]. This enabled results from the model checker to be represented as HOL theorems, for full compositionality with more abstract theorems proved in HOL using a formal model theory of L_μ that the authors developed. The AMBA specification document was converted into a formal model in HOL. The AHB and APB protocols were represented as state machines, with signals modeled as Boolean variables, and a state being represented as a tuple of these variables. Properties of the AMBA AHB and APB protocols to be checked were expressed as CTL properties and translated to L_μ via HOL. Model checking was then applied to check for the properties of latency (checking if bus becomes available within a given number of cycles), coherence (checking if registers are updated correctly at the end of a data transfer), and deadlock freedom (checking for abnormal termination or freezing of the system) for the APB protocol. For the AHB protocol, in addition to latency, coherence and deadlock freedom, another property related to arbitration (checking for fairness of bus access) was also verified. Subsequently a compositional proof was constructed in the theorem prover to verify the entire AMBA communication architecture specification (i.e., a combination of AHB and APB protocols).

An approach to formally verify the IBM CoreConnect PLB (processor local bus) [4] arbiter module was proposed by Goel and Lee [42]. The authors used symbolic model checking, and the RuleBase SMV model checker [47] to verify a Verilog RTL (register transfer level) implementation of a PLB arbiter core. RuleBase is an extension of the SMV model checker developed at IBM. A model in RuleBase is specified in the environment description language (EDL), which is a dialect of the SMV description language. Notably, RuleBase allows verification of designs in traditional hardware description languages such as VHDL and Verilog. Rules are written in IBM's Sugar [48], which adds additional properties on top of CTL, to make it easier to specify properties for hardware designs, including the ability to enter regular expression like sequences to describe computation paths.

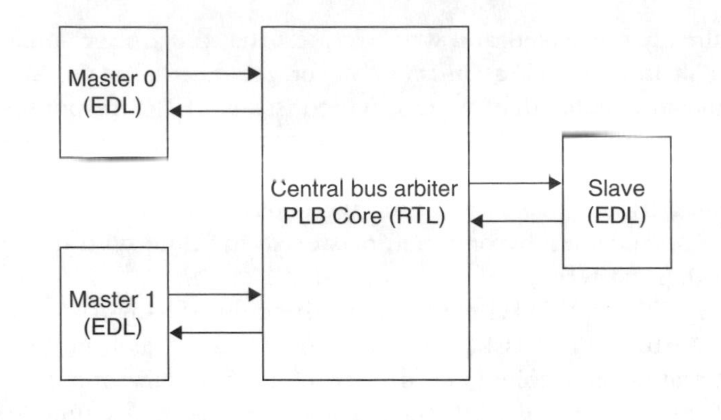

FIGURE 10.3

PLB arbiter verification setup [42]
© 2000 IEEE

Figure 10.3 shows the verification setup for the arbiter core, which included master and slave modules written in the RuleBase EDL. Properties of the arbiter such as arbitration, bus-locking, timeout, and pipeline transfers were checked for compliance with the specification document. Several bugs were found in the arbiter core as a result of the verification effort. Most of these bugs were traced back to problems in the specification document, which were classified as:

(i) *Ambiguity*: Some statements could be interpreted in different ways.
(ii) *Incompleteness*: The behavior of some signals was specified incompletely.
(iii) *Redundancy*: Information useful for designers, but not essential for implementation was included.
(iv) *Inconsistency*: Separate statements make different requirements that might be inconsistent.

While the verification for the setup shown in Fig. 10.3 took a few minutes to verify a set of properties, it did not scale well when the number of masters was increased to eight, in which case it took more than a day to verify a single property. This shows one of the limitations of formal verification techniques, which can consume an exorbitant amount of time due to the phenomenon of state space explosion [52].

10.2 COMPLIANCE VERIFICATION FOR IP BLOCK INTEGRATION

In addition to verifying the correctness of the on-chip communication components and protocols used in SoC designs, it is essential for the IP cores in the design to have the same interpretation of (and adhere to) the communication protocol specification. IP cores today are typically reused across many designs, to reduce design effort and time-to-market. But different SoC designs may use completely different on-chip communication standards. To ensure that IP cores can be reused across different designs seamlessly, the core functional logic of an IP core is separated from its interface description. Thus, if the IP core is ported from an

AMBA-based system to a CoreConnect-based one, only the interface logic of the IP core needs to be changed. However, even for interface logic that is designed for a particular communication protocol, it is necessary to verify compliance with the protocol, due to the vast number of valid configurations and options available in today's communication architecture standards. The IP core interfaces thus need to be verified for compliance with the on-chip communication architecture. Generally speaking, there are two ways to verify IP core compliance:

(i) *Dynamic simulation-based techniques*: Simulation-based techniques have been around for a long time. They usually make use of hardware monitors [15, 49, 50], which are circuits that monitor the interface between the IP core and the rest of the environment, and flag any violations of the interconnection protocol during simulation. Creating a monitor for a complex protocol, however, can be a challenging task. Simulation-based approaches also suffer from slow simulation speeds for large systems, and inadequate coverage during system simulation.

(ii) *Formal (static) verification techniques*: Formal verification techniques specify properties of the communication protocol in terms of a temporal logic formula such as CTL [6] or LTL [135]. Subsequently, existing formal verification tools such as SMV [37] and VIS [51] are used to verify the IP core under test against these properties. However, extracting properties from informal specification written in a natural language can be tedious. There is a high probability that certain properties implied by the specification will be either ignored or misinterpreted. Formal model checking techniques also suffer from massive memory requirements and long runtimes for more complex systems, due to the state space explosion problem [52].

Commercially, several vendors have released compliance toolkits, referred to as "Verification IP" (VIP) solutions that enable an IP core to be tested for compliance with a particular communication protocol. At the time of this book's publication, some commercially available solutions include Incisive VIP from Cadence [53], VCS VIP from Synopsys [54], SolidPC VIP from Averant [55], PureSpec VIP from Denali [56], eVc from Verisity (now acquired by Cadence) [57], and nVS VIP from nSys [58]. These "Verification IP" solutions are typically comprised of numerous components such as protocol checkers, testbench generators, test environments, coverage monitors, etc. Increasingly, assertion-based verification (ABV) is becoming popular for testing IP core compliance. Historically, hardware designers have used assertions in VHDL and Verilog code to flag errors during simulation and improve bug observability. These interface-checking assertions can be as simple as "signals X and Y should never be active (logic 1) at the same time," or express more complex temporal relationships such as "once the request signal goes high, the acknowledge signal should become active within 5–8 clock cycles." Such assertions have also been used as properties in formal verification efforts, to statically verify IP core interface compliance. Several property specification languages such as PSL [59], OVL [60], OVA [61], and SVA [62] have been developed (as an alternative to CTL and LTL), to ease the specification of more complex assertions. These emerging languages are generally easier to understand at the semantic and syntactic levels than CTL or LTL.

Most of the commercial VIP solutions provide support for (static) formal ABV and (dynamic) simulation-based ABV. This section presents an overview of the research that has been done to ease the compliance verification of IP blocks.

10.2.1 Dynamic Simulation-Based Techniques

Nightingale and Goodenough [63] proposed an AMBA compliance testing toolkit for IP cores. Many standard on-chip communication architectures such as AMBA require compliance testing of an IP core to demonstrate that it has achieved a pre-defined level of quality, which gives the IP integrator confidence that the IP core will function correctly when integrated into an SoC design built around the AMBA communication architecture. When AMBA compliance is achieved, the IP core is certified and distributed as *AMBA system architecture compliant*. AMBA compliance testing is performed using a simulation environment in which the IP core is exercised. The simulation environment consists of, among other things, a protocol checker and a coverage monitor. The protocol checker performs cycle by cycle checks to ensure that the IP core (also referred to as the device under test or DUT) obeys the rules within the AMBA specification. The protocol checks are essentially a fixed set of rules, such as "In the second cycle of a split or retry response, a master must drive *HTRANS* to IDLE," that must not be violated. The coverage monitor ensures that coverage points (a list of bus transactions) must be observed within the test environment for compliance. Coverage points ensure that the tests have sufficiently exercised the IP core under test. An example of a coverage point is "A bus slave must be accessed by a read transfer followed immediately by a write transfer to the same address."

Two types of AMBA compliance test (ACT) environments are presented in [63], and are available for AMBA compliance testing for any IP core. In the first type of environment (*active mode*), shown in Fig. 10.4(a), the DUT is connected to a standard

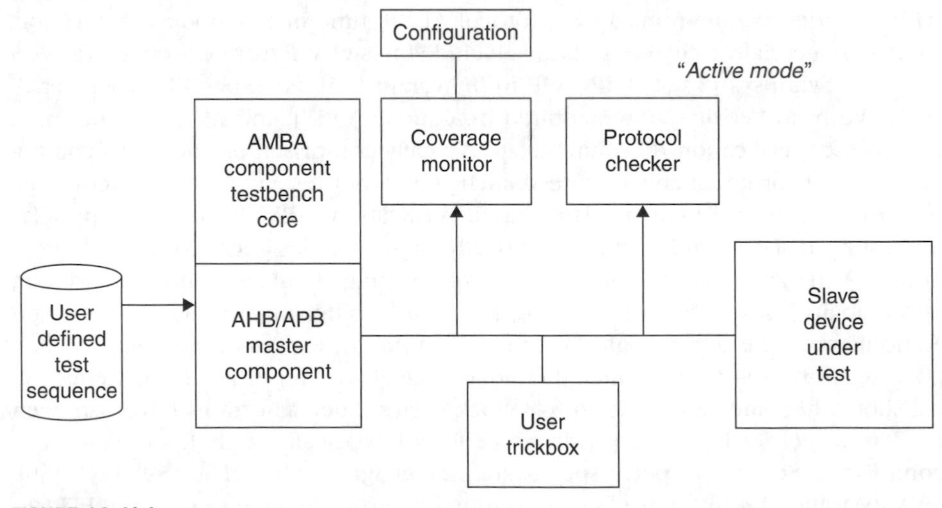

FIGURE 10.4(a)

Compliance test environments: (a) active mode with slave DUT
© 2001 IEEE

AMBA testbench, which reads stimulus generation information and response checking information from a user defined test sequence file. The user provides an RTL implementation of the DUT, the *configuration* and the *user defined test sequence*. The *configuration* allows coverage points to be configured, to allow for the fact that some IP cores will not utilize all the features of the AMBA bus. As a basic rule, all AMBA compliant IP cores must be able to accept all possible input combinations, but do not have to generate all possible output combinations. The *trickbox* is an RTL component that provides examples for interfacing with the DUT. It has an AHB/APB slave interface that can be controlled by the testbench, using test vectors that access the *trickbox*, instead of the DUT. The *trickbox* allows non-AMBA signals of the DUT to be controlled and observed. Also, while the DUT shown in Fig. 10.4(a) is a slave, masters are also supported, and the environment appears as shown in Fig. 10.4(b). In this case, the environment created around the master DUT is treated as a slave and drives back directed responses. Additional features such as a behavioral memory (slave) can also be enabled during master DUT compliance testing. In the second type of environment (*passive mode*), shown in Fig. 10.4(c), a *system simulation environment* is used to provide the stimulus and response checking for the DUT. The user provides an RTL implementation of the DUT, the *configuration*, and the *system simulation environment*. It is more challenging to achieve full coverage in this mode, since the stimulus source to the DUT may have a limited transaction or scenario generation capability. Protocol checking in both types of ACT environments is done with the help of temporal expression specification using the Verisity Specman "e" language [64]. Temporal expressions capture

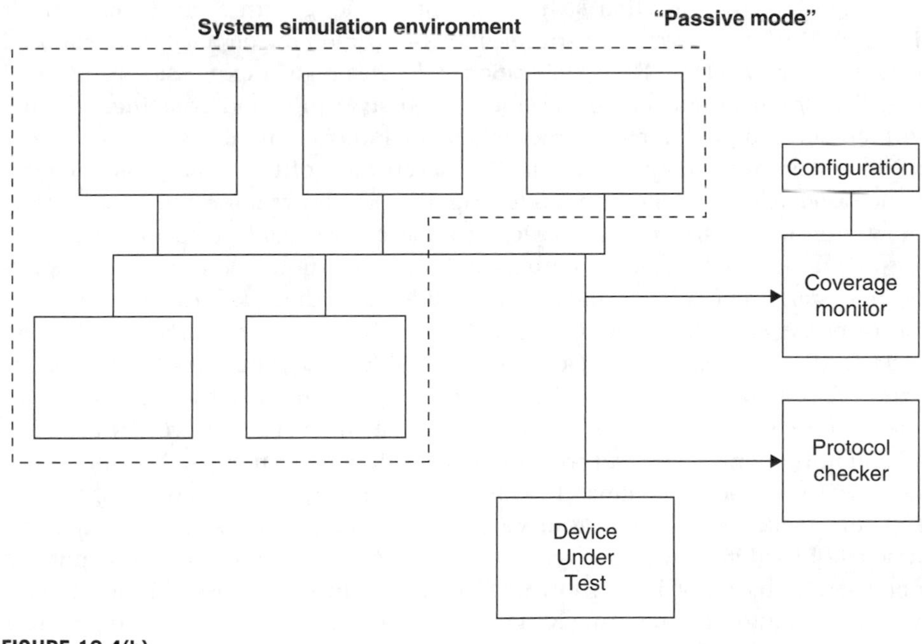

FIGURE 10.4(b)

Active mode with master DUT

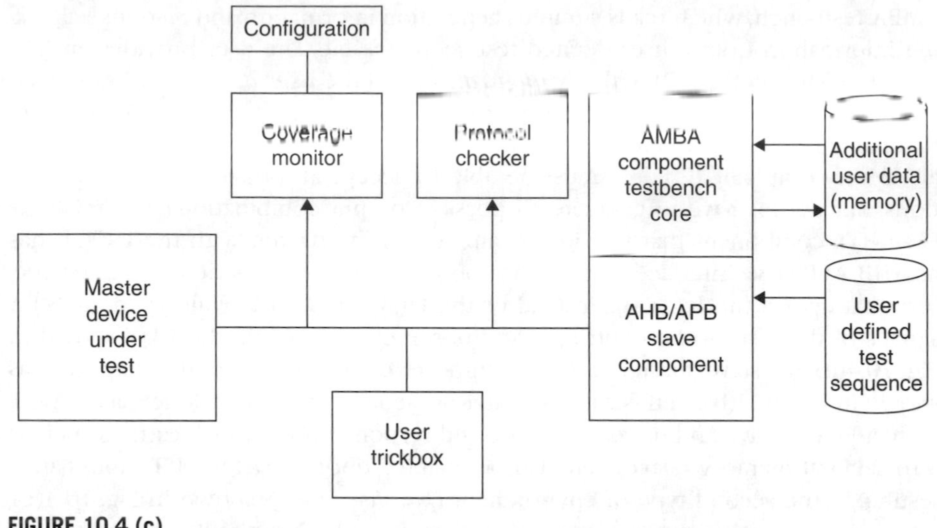

FIGURE 10.4 (c)

Passive mode [63]
© 2001 IEEE

temporal relationships between events, values of fields, variables or other items during compliance testing, and flag errors if temporal conditions (e.g.,"wait for one cycle before asserting a signal") are violated by the DUT. Formal tools for property checking can also be used in place of the temporal expressions.

A high level specification style for monitors, along with a linear time, linear size algorithm to translate the specifications into monitor circuits was proposed by Oliveira and Hu [65]. The specification style uses regular expressions enhanced with the production-based specification (PBS) style [66], and constructs to support storage and pipelining, in order to specify interface monitors. A specification in PBS is compact, compared to an HDL description of the same protocol, making it easier to read, write, and modify. Figure 10.5 shows an example of a monitor written in this specification style, for a basic open core protocol (OCP) [89] master that can only handle one outstanding transaction at one time. The specification starts by declaring the interface wires, and then defines the command, and response encodings. The first production defines the monitor's behavior and declares that the master interface should exhibit a sequence of idle commands or transfers. A transfer can be either a read or a write. A write transfer consists of zero or more states waiting for a slave to accept a command, followed by the slave accepting the write. The read transfer starts with zero or more states waiting for the slave to accept the command, followed by either an instantaneous response or a response with zero or more wait states. For simplicity, checks to ensure that the master holds address and data values (if the slave does not accept the command immediately) have not been shown. Oliveira and Hu [65] translated the monitor specification into a circuit using insights from previous work in converting regular expressions into circuits [67, 68], and extended it to support handling of storage and pipelining. The size of the generated circuit and the runtime of the translation algorithm were both linear in the size of the specification, because the translation

```
/* Inputs and Outputs of Master */
/* The monitor treats these as inputs. */
input SCmdAccept, SResp[1:0], SData[31:0];
output MAddr[31:0], MCmd[2:0], MData[31:0];


/* Response codes defined in standard. */
/* NULL, Data VAlid, and ERRor */
define null_resp = !SResp[0] & !SResp[1];
define dva_resp  =  SResp[0] & !SResp[1];
define err_resp  =  SResp[0] &  SResp[1];


/* Commands defined in standard. */
define cmd_idle  = !MCmd[0] & !MCmd[1] & !MCmd[2];
define cmd_write =  MCmd[0] & !MCmd[1] & !MCmd[2];
define cmd_read  = !MCmd[0] &  MCmd[1] & !MCmd[2];


master -> (cmd_idle || transfer)*;


transfer -> write transfer || read_transfer;


/* SCmdAccept indicates that the slave
   has accepted the command. */


write_transfer ->
     (cmd_write & !SCmdAccept)*, (cmd_write & SCmdAccept);


read_transfer ->
     (cmd_read & !SCmdAccept)*,
     (wait_state_resp || instant resp);


wait_state_resp ->
     (cmd_read & SCmdAccept & null_resp),
     null_resp* , response;


instant_resp -> (cmd_read & SCmdAccept & dva_resp) ||
                (cmd_read & SCmdAccept & err_resp);


response -> dva_resp || err_resp;
```

FIGURE 10.5

Complete interface monitor specification for a basic OCP master that can only handle one outstanding transaction at a time [65]

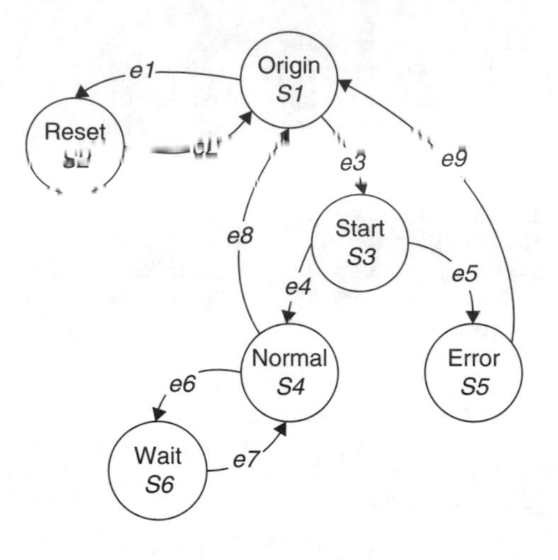

FIGURE 10.6

An example of an FSM model [69]
© 2004 IEEE

created a constant size circuit for each operator in the specification. The output of the translation algorithm was a monitor circuit in Verilog (or VHDL) code that can be used for IP compliance testing.

In the monitor-based verification methodology proposed by Lin et al. [69], the on-chip communication protocols are modeled as FSMs. The FSM is defined as a 5-tuple $\{Q, I, S, q0, \delta\}$, where Q is the finite set of states, I is the set of bus signals that need to be considered, S is a finite set of statuses describing the values of bus signals for each state, $q0 \in Q$ and is the initial state, and $Q \times 2^I \rightarrow Q$ is the set of state transition functions. The combination of a state and a transition can indicate one or more properties, and a transition between two states indicates the relationship of the two states. The status of each state describes the correct signal values, which can also be used to represent the properties. An example of the FSM model is shown in Fig. 10.6, where *Origin*, *Start*, *Normal*, *Wait*, *Reset*, and *Error* are the states, and *S1–S6* are the statuses. Such FSM models were created for the Wishbone [70] and AMBA AHB [1] communication protocols from their specification documents. These FSM models were used in monitors during simulation to verify IP core compliance for three designs. Experiments were performed by injecting errors in correct designs to verify compliance for several IP cores. Figure 10.7 shows the three designs and the states in which error was detected successfully by the monitor in each of them.

A methodology to synthesize monitors from visual specifications given in CESC (clocked event sequence charts) was proposed by Gadkari and Ramesh [75]. CESC [71] is a visual language designed for the purpose of specifying system level interactions between single and multiple clock domains. CESC overcomes the limitations of capturing high level assertions (properties) with specification languages such as Sugar/PSL [59] or temporal language, which can become complex for long event sequences [72]. CESC is also more scalable and intuitive than a manual construction of assertion monitors using native languages, which is error-pone and

Design	Inserted error	Results
PWM/Timer/Counter (WISHBONE)	ACK and ERR are asserted simultaneously	Test fails in the Normal state: ERR!=0
MAC (AMBA AHB)	The ERROR response is given with a single cycle	Test fails in the ERROR state: ERROR response needs two cycles
Convolution (AMBA AHB)	Provide a wait state to IDLE transfer	Test fails in the "Normal with IDLE/BUSY" state: HREADY!=1 Test fails in the "WAIT from Master" state: READY!=1

FIGURE 10.7

Experimental results for erroneous designs [69]
© 2004 IEEE

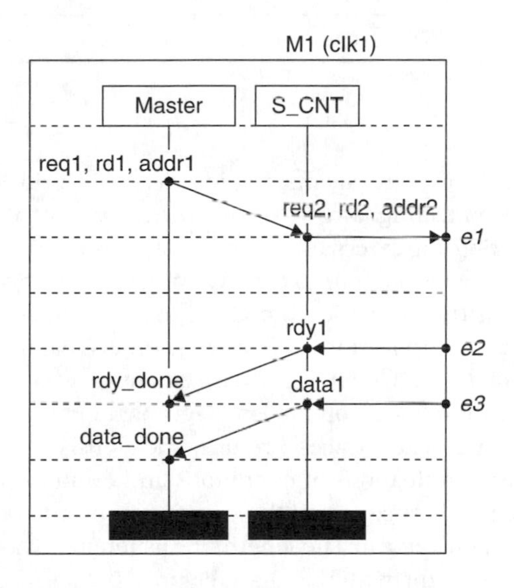

FIGURE 10.8

SCESC for typical read operation (single clock domain) [75]
© 2005 IEEE

does not scale well. A few approaches [73, 74] have attempted to automatically create monitors and protocol checkers from temporal language specifications, involving the generation of automata equivalent to the specified property in temporal logic. The approach by Gadkari and Ramesh [75] attempts to take this one step further and synthesize assertion monitors directly from CESC visual specifications. CESC also improves upon efforts to formalize message sequence chart (MSC) like notations for system level protocol compliance verification [76, 77] by offering features such as synchronizing clocks, structural constructs, and multiple clocks.

The simplest CESC is referred to as SCESC (single clocked event sequence chart). It represents a finite duration event sequence or interaction scenario. Figure 10.8 shows an SCESC corresponding to a typical read operation within a single clock

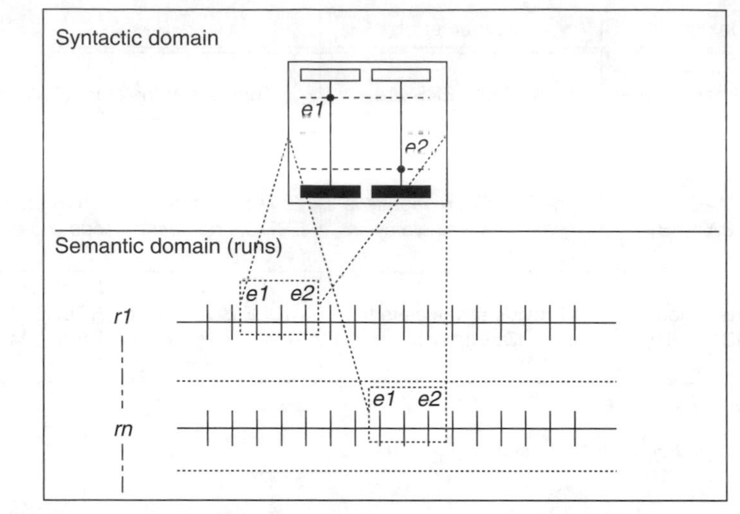

FIGURE 10.9

Semantic mapping for SCESC [75]
© 2005 IEEE

domain. Vertical lines represent agents, while horizontal grid lines represent synchronizing clocks. Connecting arrows show the causality relationship between events. Events occurring on external agents are shown on the frame of the chart, and are referred to as environment events. Complex interaction scenarios can be represented using constructs provided in CESC for sequential and parallel composition, loop, alternative, implication, and asynchronous parallel composition (for interactions across clock domains). A formal semantics was defined for CESC by the authors, based on the synchronous language paradigm [78, 79]. The semantic domain was based on clocked traces or runs, and consisted of all possible runs. Each run described the evaluations of conditions and events along the sequence of clock ticks. Each SCESC chart specified the constraints on the occurrence of events within a finite interval on any run. The length of this interval was determined by the number of grid lines present within a chart. Figure 10.9 shows how for every run associated with the SCESC, there is a finite interval in which events occur according to the ordering specified by the SCESC.

Subsequently, monitors were synthesized from the CESC specifications. These monitors were essentially FSMs that could detect the traces (runs) which exhibited event sequence behavior specified in the CESC. The monitor automaton was an extension of the string-matching automaton described in [80], and operated on clocked event traces. The monitor automaton made use of a dynamic scoreboard to store information regarding event occurrences, which was useful for implementing checks related to causality relationships between events during a run. The synthesis algorithm constructed localized monitors for every SCESC, and then combined them using various composition operations. Monitors communicated and synchronized across clock domains using the dynamic scoreboard. CESC was used to synthesize monitors for the OCP [89] and AMBA AHB [1] communication

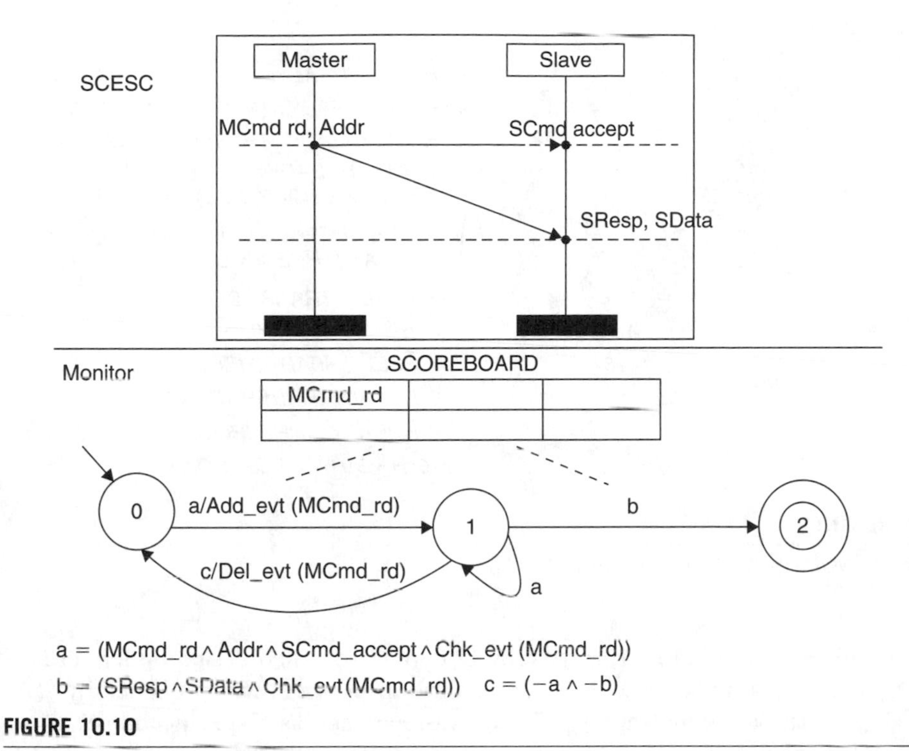

FIGURE 10.10

Simple read operation in OCP, represented as SCESC, and the synthesized monitor [75]
© 2005 IEEE

protocols. Figure 10.10 shows the SCESC corresponding to a simple read transaction in the OCP, along with the synthesized monitor that can detect the depicted scenario. A composition of such monitors can be used to verify IP core compliance for complex, multi-clock domain SoC designs.

10.2.2 Static Formal Verification-Based Techniques

A few approaches [81, 82] have proposed creating a verification platform to formally perform compliance verification. Karlsson et al. [83] made use of a Petri-net-based representation, PRES+ [84], in order to represent the IP core, its interface and the communication protocol. Timed computation tree logic, T(CTL) [123], was used to represent properties that needed to be satisfied by the IP core, for compliance. Subsequently, the PRES+ models were translated into Timed Automata [85], and model checking was performed with the UPPAAL environment [86], to determine if the properties were satisfied or not. If the properties were not satisfied, then the model checker produced a diagnostic trace to indicate which properties were violated. Experiments were performed with an AMBA-based mobile telephone SoC design, to check if the components in the system are compliant with the AMBA protocol. Using the formal verification approach, several bugs in the interfaces of IP cores were detected.

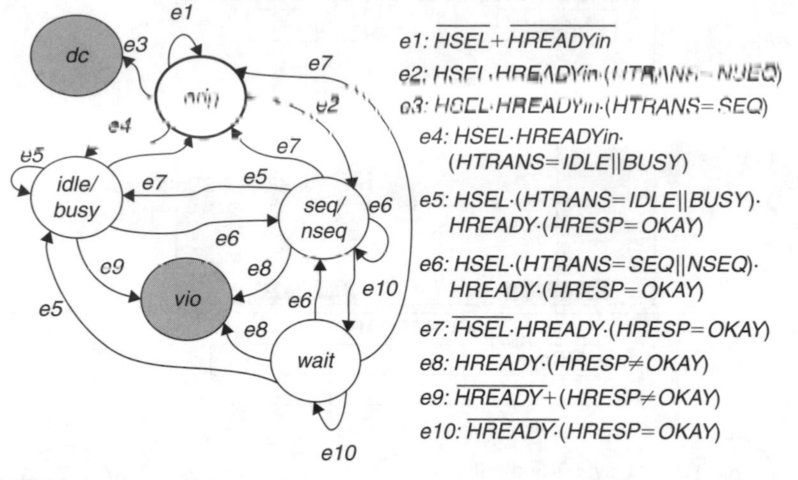

FIGURE 10.11

Spec FSM for simplified AHB slave protocol [87]
© 2005 IEEE

Another approach to formally verify IP core (also called design under verification or DUV) compliance with a communication protocol was proposed by Yang et al. [87]. The communication protocol was specified as a specification FSM (or spec FSM), which acted as a functional monitor for the DUV. The possible DUV behaviors were classified as: (i) *don't care*—the behavior is not defined since the input sequence is not supposed to appear, (ii) *legal*—the output sequence is allowed by the protocol under a valid input sequence, and (iii) *illegal*—the output sequence is prohibited by the protocol under a valid input sequence. Two special states were added to every spec FSM—*vio* and *dc*. The spec FSM moved to the state *dc* if a don't-care input sequence was applied to the DUV. If the DUV behaved illegally for a valid input sequence, it moved to the *vio* state, to indicate a protocol violation. If the DUV behaved legally for a valid input sequence, then the spec FSM moved among the normal states of the FSM. Figure 10.11 shows the spec FSM for a simplified AMBA AHB slave protocol. In the idle/busy state, if *HREADY* is not asserted or *HRESP* is not set to OKAY, the spec FSM moves to the *vio* (*e*9) state. This implies that a slave can only respond with an OKAY to an IDLE or BUSY transfer, which is explicitly defined in the AMBA specification. Additionally, in the *orig* state, if a transfer is initiated by asserting the signal *HSEL* and *HREADYin* as well as setting *HTRANS* to SEQ, the spec FSM moves to the state *dc* (*e*3). This implies that the master should never set *HTRANS* to SEQ for the first transfer, which is an input constraint to the slave. These inputs can be treated as don't-cares since they are not supposed to appear. A DUV is compliant with a protocol specification if and only if there exists no valid input sequence (of any arbitrary length) along with the corresponding DUV output sequence that can drive the spec FSM into its *vio* state.

Figure 10.12 shows the FSM for an AHB slave interface design. The outputs of the interface are *HREADY*, *HRESP*[1], and *HRESP*[0]. When a write request is detected from a master, it moves to the *write* state and responds with OKAY to

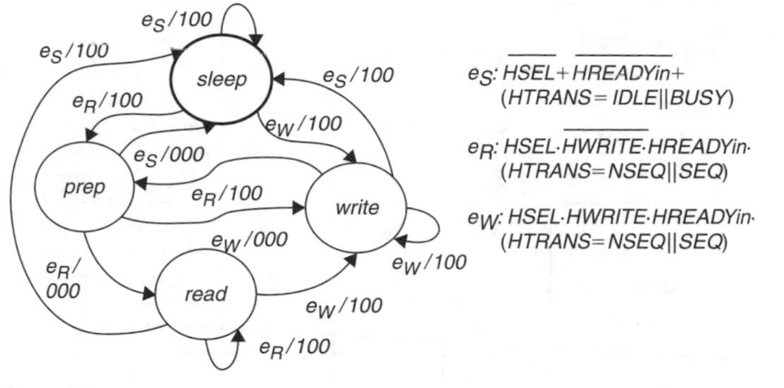

FIGURE 10.12

FSM for AHB slave interface design [87]
© 2005 IEEE

indicate the completion of the write operation. When a read request is detected from the master, it first moves to the *prep* state to insert a wait cycle, and then moves to the *read* state, responding with an OKAY to indicate that the read operation is done on the next cycle. In the absence of any requests from the master or in case of an IDLE or BUSY transfer request, it goes to the *sleep* state. The problem of interface compliance verification is essentially the problem of compliance verification between the protocol FSM (Fig. 10.11) and the DUV FSM (Fig. 10.12). This comparison between the two FSMs is not simply equivalence checking, since they are intrinsically different. But the states in the two FSMs have some sort of corresponding relations. For instance, when the DUV FSM is in the *sleep* state, the protocol spec FSM may be in the *orig* state, because both states imply that the slave is not requested. The two states are referred to as a state pair. The corresponding relation between states does not have to be 1-to-1, and can also be *n*-to-1 or 1-to-*n*. For instance, the state *orig* and the *idle/busy* states in the protocol spec FSM are both able to correspond to the *sleep* state in the DUV FSM, since the DUV responds identically if it is not requested, or receives an IDLE or BUSY transfer. Thus the DUV FSM is not simply a subset or superset of the protocol spec FSM. The compliance verification problem (i.e., proving that the DUV is compliant) can be solved by examining all the state pairs, and ensuring that none of them includes the *vio* state. In order to find all the possible state pairs for the given FSMs, a branch-and-bound algorithm is used to create a search tree whose nodes are state pairs. The root node consists of the initial two states. If there exist I/O values that make the two FSMs move to state *A* and *B*, respectively, then node (*A, B*) is added as a child of the root node. In this way, the process exhaustively grows children and grandchildren from the root node, to create a search tree with all possible state pairs. To ensure that such a tree does not have an infinite depth, a bounding function is used to remove duplicate state pairs, which results in a pruning of the search tree. The algorithm was implemented in the *C* language, and experiments were performed on a set of designs to test for compliance with AMBA AHB [1] and Wishbone [70] communication protocols. Figure 10.13 shows the results of these experiments. To check if the proposed approach could detect design flaws, some

I/F protocol type	DUV	Result	Iteration count	$(\lvert Q_s \rvert \times \lvert Q_d \rvert)^2$
WISHBONE slave	*spi*	Pass	180	$(7 \times 3)^2 = 442$
	ac3 ctrl	Pass	221	$(7 \times 5)^2 = 1225$
AMBA AHB slave	*con7*	Pass	204	$(7 \times 4)^2 = 784$
	mac	Pass	191	$(7 \times 6)^2 = 1764$
	remap	Pass	136	$(7 \times 6)^2 = 1764$
	mac_err	Fail	57	$(7 \times 6)^2 = 1764$

FIGURE 10.13

Results of verifying real designs [87]

errors were intentionally injected into the *mac* design, to create the *mac_err* benchmark. The results indicate that the actual (tree-growing) iteration count is far less than the worst case, presented in the last column of Fig. 10.13.

10.3 BASIC CONCEPTS OF SoC SECURITY

Having discussed verification issues in the previous sections, we now focus on the challenges associated with designing secure SoCs. The concept of information security was first explored in the context of communication systems [91]. When sensitive data is exchanged between two entities using public networks, it is essential to provide security services such as *data integrity*, *data confidentiality*, and *peer authentication*. Data integrity ensures that the information has not been altered by malicious entities. Data confidentiality ensures that no outsider can snoop on sensitive information. Finally, peer authentication ensures that the information is exchanged between appropriate parties, rather than someone impersonating a trusted party. The set of integrity, confidentiality, and peer authentication are the basic security services required of SoC designs that are used in a wide range of applications today. In addition to these basic security services, SoC designs have several additional security requirements. The *availability* of the SoC design is one such security service, which is needed to prevent malicious entities from launching denial of service (DoS) attacks or degrading system performance. *Secure storage* is another service that is essential for securing sensitive information in the internal and external storage devices associated with an SoC design. The sensitive content used in a system can have content security or digital rights management (DRM) requirements that must be guaranteed. Finally, *tamper resistance* is required to ensure that even when the system does fall into the hands of malicious hackers, physical or logical probing does not compromise sensitive information.

The security requirements for SoC designs are often achieved using three main classes of cryptographic algorithms [92,93]:

(i) *Symmetric algorithms*: These algorithms involve using a single key that is shared between the sender and the receiver. The sender encrypts the data (called *plaintext*) using the key and send the encrypted data (called *ciphertext*) to the receiver. When the ciphertext reaches its destination, the receiver uses the same key to decrypt it into plaintext. Examples of symmetric algorithms include AES [124], DES [125], 3DES [126], and RC4 [127].

(ii) *Asymmetric algorithms*: These algorithms involve a pair of keys, one of which is called the *public key*, while the other is called the *private key*. The sender uses the public key to encrypt data and send it to the receiver. This public key is available for use to the whole world. However, decryption of the encrypted data is only possible using the receiver's private key. The private key is not known to anyone except the receiver, which thus ensures secure data communication. It is also possible to use the private key to encrypt the data, and then have possibly several receivers use the public key to decrypt the data. Such a scenario is used for digital signatures, where the content provider "signs" its content, and the users verify the authenticity of the content source by decrypting the signature with their public key. Examples of asymmetric algorithms include RSA [128], DSS [129], and Diffie and Hellman [130].

(iii) *Cryptographic hash algorithms*: These algorithms convert arbitrary length input messages into fixed length output values that are sometimes termed as a *message digest* or a *digital fingerprint*. Examples of cryptographic hash algorithms include SHA-1 [131], MD5 [132], and RIPEMD-160 [133]. These hash algorithms are also used to calculate a message authentication code (MAC) with the help of a secret key. The resulting keyed-hash message authentication code (HMAC) is useful for preventing tampering of hashed data. Any cryptographic hash function such as SHA-1 or MD5 can be used to calculate HMAC—the resulting algorithm is termed as HMAC-SHA-A and HMAC-MD5, respectively.

A combination of these cryptographic algorithms is typically used in security protocols, to provide various security services. For instance, in a typical networked SoC, DRM protocols such as MOSES [94], OpenIPMP [95], and ISMA [96] are used to protect content from unauthorized use, while security protocols such as SSL [97] and IPSec [98] provide secure communication between the SoC and the outside world, and are widely used for secure web transactions, and virtual private networks (VPNs), respectively.

The existence of well-designed cryptographic algorithms alone does not prevent attacks on SoCs from hackers. This is because hackers, instead of attacking these well-designed algorithms, typically exploit weaknesses in the hardware and software components in an SoC implementation. Figure 10.14 shows a classification of attacks that are possible on embedded SoC designs [99]. At the top level, attacks can be classified into three categories: (i) *privacy attacks*, which attempt to gain access to sensitive information stored or manipulated within an SoC;

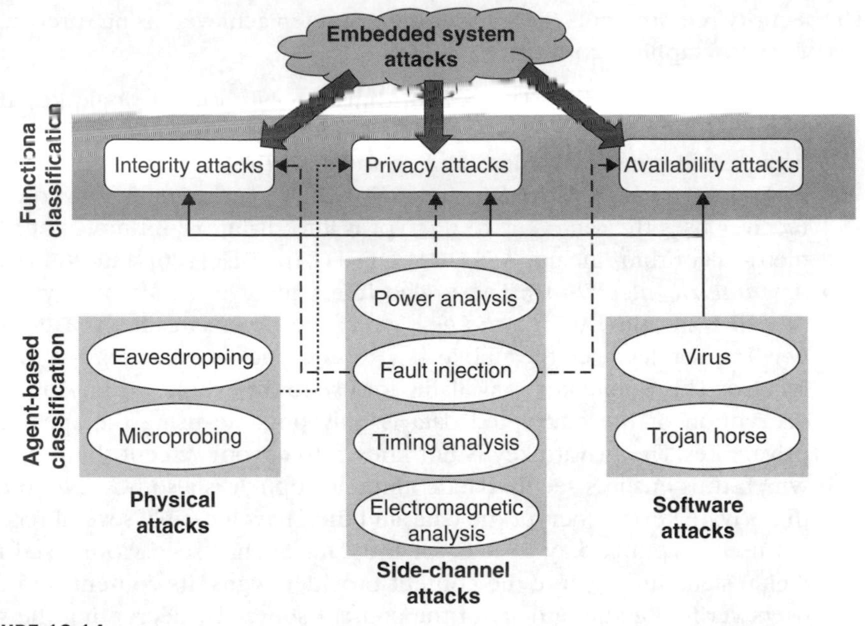

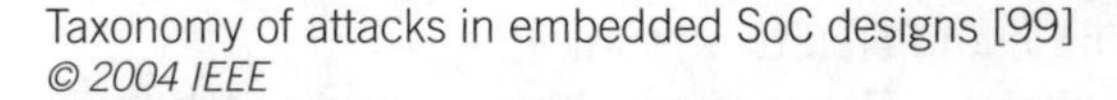

FIGURE 10.14

Taxonomy of attacks in embedded SoC designs [99]

(ii) *integrity attacks*, which attempt to change the data (or code) associated with an SoC; and (iii) *availability attacks*, which attempt to utilize the SoC resources maliciously, in order to disrupt the normal functioning of the system.

At a second level, attacks can be classified according to the means of launching the attacks. There are three categories at this level, as shown in Fig. 10.14:

(i) *Software attacks*: These are attacks launched through software agents such as viruses, worms, Trojans, etc. Due to the increasing software complexity, possible extensible nature of software code (that allows malicious updates) and the ubiquity of networking, these types of attacks are possibly the most common, and will continue to pose a major risk in the future [100–102].

(ii) *Physical attacks*: These are attacks involving physical intrusion into the SoC at some level after de-packaging the chip [103, 104]. These attacks can typically involve eavesdropping on inter-component communication on the chip, but require highly sophisticated probing techniques, such as e-beam microscopy [105]. Such attacks are therefore more expensive and less frequently used compared to other low overhead attacks, at the SoC level.

(iii) *Side-channel attacks*: These are attacks that involve observing the state and properties of the system while it processes information, such as when performing cryptographic operations. Examples of observed system properties include execution time, electromagnetic analysis, power consumption, or behavior in the presence of fault injection [91, 106–108].

Here, we survey research efforts that attempt to improve security in SoC designs through enhancements to on-chip communication architectures.

10.4 SECURITY SUPPORT IN STANDARD BUS PROTOCOLS

The AMBA AHB [1] and AXI [2] protocols have bus lines reserved as protection/security signals. These user configurable signals are intended for use by components that need to implement some level of security to prevent software attacks stemming from illegal instruction execution. They can be used by a master to indicate whether a particular transfer is a normal/user mode, or a privileged mode transfer, which determines if the transfer can access non-secure or secure regions (e.g., memory spaces) in the system. ARM uses these protection/security signals as part of its TrustZone [121] security initiative, which enables the protection of on- and off-chip memory and peripherals from software attack. The primary component of this initiative is a software-controlled AMBA APB peripheral called the TrustZone protection controller (TZPC) that determines which other peripherals can be sent secure accesses and which peripherals can only handle non-secure accesses. The TZPC looks at each transaction being requested on the bus and aborts those which it determines to be illegal. The TZPC registers are memory mapped into a secure area of memory and are generally only configured at boot-up.

Sonics SLX and SMX protocols [5] support an access protection mechanism via dedicated in-band signals, whereby slaves can have two protection regions: secure and non-secure. Two types of checks are performed for any transfer intended for such a slave. First, the type of transfer (read or write) is checked to ensure that it is allowed according to the read or write permissions. Secondly, the source of the transfer is checked (by checking ID signals) to ensure that the master issuing the transfer is allowed to access the requested region within the slave. The protection mechanism also allows groups of masters to be defined as secure, essentially partitioning the entire system into secure and non-secure regions.

10.5 COMMUNICATION ARCHITECTURE ENHANCEMENTS FOR IMPROVING SoC SECURITY

A few research efforts have attempted to extend the limited support for security provided by standard bus-based communication architectures, presented in the previous section. A security-enhanced communication architecture (SECA) was proposed by Coburn et al. [109] to prevent software type attacks on an SoC that either exploit weaknesses in the installed software, or execute malicious code on the system. Such attacks are by far the most common type of SoC attacks, because of their ease of design and deployment. SECA extends the AMBA architecture, with a centralized security enforcement module (SEM), and enhancements to the bus interfaces of components in the system, to implement three types of security services: (i) *address-based services*, which limit and regulate access privileges to certain regions of the memory and peripheral space for certain components;

(ii) *data-based services*, which provide fine grained control and restrictions on data values that can be assumed by certain memory and peripheral locations, and (iii) *sequence-based services*, that perform complex checks on sequences of transfers on the communication architecture. A simple instance of SECA, involving the address-based protection scheme was implemented in NEC's MP211 mobile phone SoC application [109].

Figure 10.15 illustrates the implementation of SECA on an AMBA on-chip communication architecture [109]. The SECA configuration consists of two main components: a single security enforcement module (SEM) and a security enforcement interface (SEI) for each slave. The SEM monitors system communication and enforces

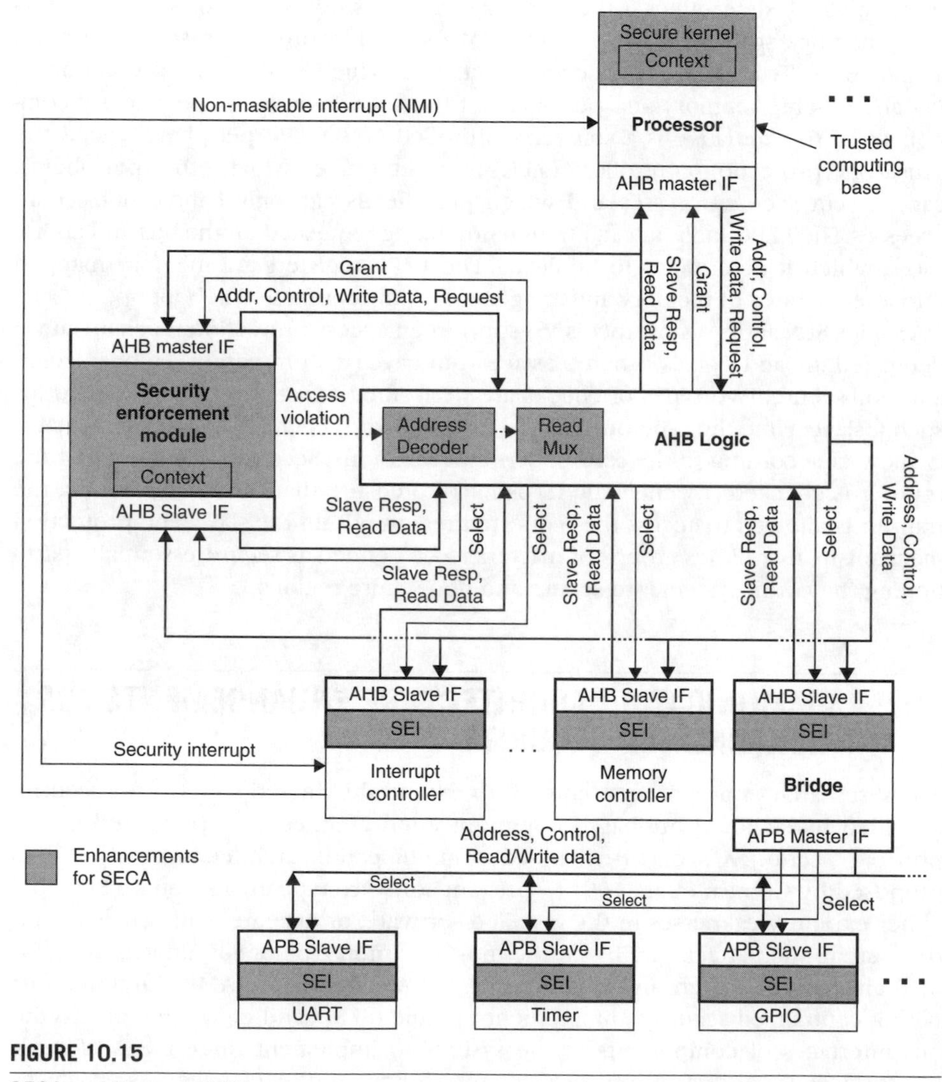

FIGURE 10.15

SECA architecture [109]

© 2005 ACM Press

security policies that can detect intrusions and provide access control services. It is possible to program the SEM, through its slave interface, to enforce a wide range of security policies having different granularities (and overheads). The SEM uses the SEI to filter data values reaching the internal memory and registers of peripherals, and ensures that access privileges are respected by the data reaching the slave.

The SEM is the central architectural component responsible for monitoring communication, and is shown in Fig. 10.16. It consists of three security modules:

(i) *Address-based protection unit (APU)*: This enforces access control rules that specify how a master accesses a component that is, whether it is allowed read-only, read-write, write-only, or no access. A lookup table is used to associate permissions with regions of the address space that are accessible (readable, writeable, or both).

(ii) *Data-based protection unit (DPU)*: This is responsible for configuring SEI at each peripheral for data-based protection. The DPU has memory to store *access level* values for a peripheral, and the address of each peripheral's configuration register. An access level represents a set of valid operations. The SEI of each peripheral has a configuration register that stores the current access level. There can be several of such access levels for a peripheral in a system. For example, access level 0 can only allow a

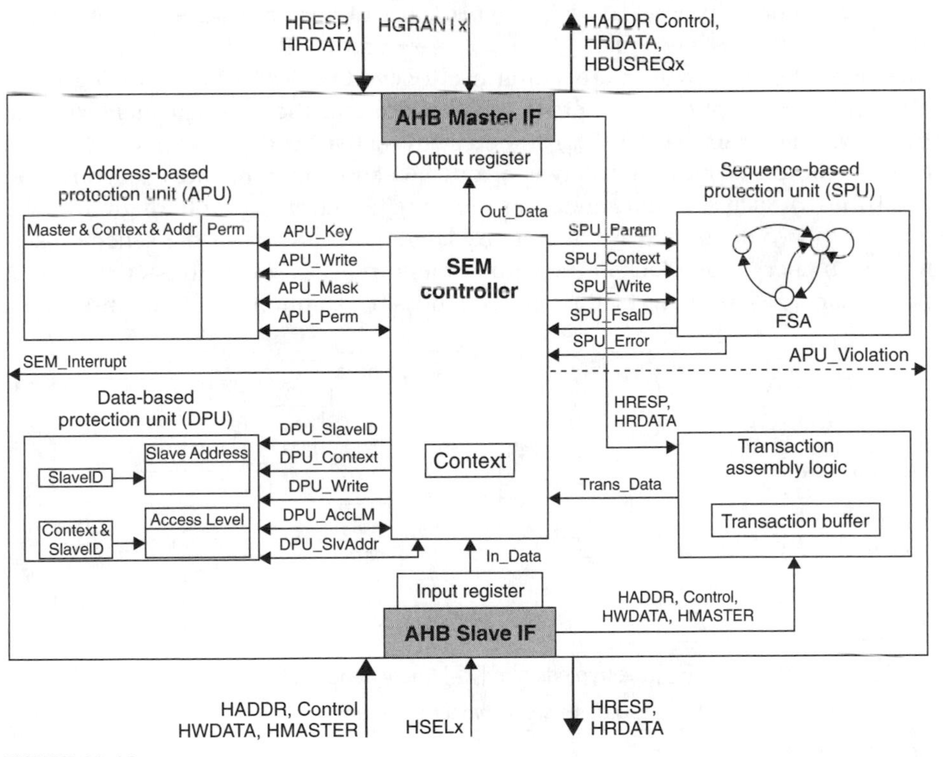

FIGURE 10.16

SEM architecture from [109]

© 2005 ACM Press

peripheral to be disabled by writes to the appropriate registers; access level 1 can allow the peripheral to be configured to be partially operational; while access level 2 enables the peripheral to be configured for full functionality. These access levels can be customized for each peripheral in a system using a lookup table in the SEI, which limits the accessible registers at each access level.

(iii) *Sequence-based protection unit (SPU)*: This enables monitoring of a sequence of bus transactions to implement application specific security features. It makes use of security automata [119] to enforce security policies. For example, the DRM security policy of "play content at most x times" can be enforced by using two security automata—one that monitors the bus to detect the case when the content is played, and then signals the second automata, which monitors and enforces whether the content has been played less than x times.

A survey of existing techniques for bus encryption was presented by Elbaz et al. [110]. Encryption of data communicated on the bus is essential to prevent probe-based physical and side-channel attacks on the processor–memory bus. While the work is primarily concerned with the external bus connecting the on-chip processor and external memory, the proposed concepts can be easily extended to on-chip buses as well. The principle of bus encryption is quite old—and was first introduced more than 25 years ago by Best [111–113]. Some of the rules proposed by Best are still in use today [114–117, 120], including using a hardware encryption unit between the cache and the external memory controller, as shown in Fig. 10.17(a). Another alternative to this placement [110] is to have the encryption unit between the CPU and cache memory, as shown in Fig. 10.17(b). Such a scheme enhances the system security by ciphering the cache at the expense of an increase in memory latency. Regardless of the scheme used, having an encryption block in the path to memory has a severe impact on system performance. Elbaz et al. [110] proposed a possible solution to this, by inserting a

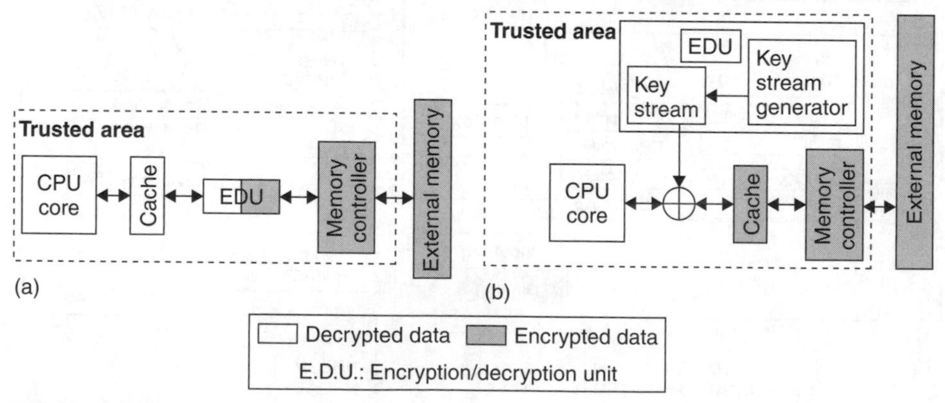

FIGURE 10.17 (a,b)

Bus encryption schemes from [110]

compression step in addition to the encryption step in the scheme shown in Fig. 10.17(a). The compression step occurs before the encryption stage, and improves the encryption performance by decreasing the size of data to cipher and decipher. Compression also results in a decrease in memory footprint, for the encrypted and compressed data. Additionally, compression increases message entropy and thus improves the efficiency of the encryption algorithm. The authors proposed using the IBM CodePack tool [118] for code compression, which claims to increase memory density by as much as 35% for a performance impact of +/− 10% (the exact value depends on the type of memory used).

10.6 SUMMARY

In this chapter, we presented research in the areas of verification and secure design of on-chip communication architectures. For on-chip communication architecture verification, we focused on two major aspects: (i) verifying the on-chip communication architecture specification and implementations, and (ii) verifying if IP cores are compliant with the on-chip communication protocol, to which they will be connected, in an SoC design. In the first case, research efforts that have formally modeled several on-chip communication architecture specification from informally specified, natural language documents, found a few cases where the specification document was ambiguous, inconsistent, and incomplete. This motivates a more formal description of on-chip communication architectures, in order to ensure that designers following the specification do not misinterpret it, or make incorrect assumptions during implementation that can cause bugs. In the second case, several research efforts have proposed techniques to test if IP cores are compliant with on-chip communication protocols that are used in an SoC integration environment. This compliance testing can either be done using dynamic simulation-based techniques, or static formal verification-based approaches. Simulation-based approaches that monitor interactions between the IP core and the on-chip communication protocol are quite commonly used, but cannot guarantee 100% coverage, due to the limited test cases that can be simulated in a reasonable amount of time. Formal verification techniques can provide this guarantee, but the formal modeling process can be quite complex. Formal verification techniques can also suffer from state space explosion for large designs. Several commercial products exist for both simulation-based and formal verification-based compliance verification. But a lot more research is needed to overcome the inherent drawbacks of the two techniques, especially in the face of increasing SoC complexity.

We also presented the motivation and fundamental concepts for secure SoC architectures. Malevolent entities typically resort to software, physical or side-channel attacks on SoC designs, in order to gain access to sensitive information. The on-chip communication architecture has a global view of the entire system and can therefore serve as an efficient means to manage and implement security policies that can keep malevolent entities at bay. Standard bus-based communication architectures such as AMBA and Sonics Interconnect support basic features in the form of dedicated, customizable signals intended to improve SoC security by allowing components to define secure access regions. To extend this basic security feature set, research efforts have looked at enhancing existing communication architectures by adding additional security enforcement modules, more complex component interfaces that define secure access regions, and encryption engines. These efforts can thwart off a wider army of complex software, physical, and side-channel attacks. However, the field of secure SoC design is still in its infancy, and much research still remains to be done. One glaring area of concern is the overhead of using security schemes in SoC designs. Typically, there is an unfortunate trade-off between system security and the traditional design goals of performance, power, area, cost, etc. that makes it harder for designers to dedicate resources toward making a system secure and hacker-proof. With a rising penetration of SoC designs in almost all walks of life, and their increasing use in large networked environments, research efforts must focus on building lightweight, yet robust and security architectures that can meet the desired levels of security.

REFERENCES

[1] ARM AMBA Specification and Multilayer AHB Specification (rev2.0), http://www.arm.com, 2001.

[2] ARM AMBA AXI Specification, www.arm.com/armtech/AXI.

[3] "STBus Communication System: Concepts and Definitions," *Reference Guide*, STMicroelectronics, May 2003.

[4] IBM CoreConnect Specification, http://www.ibm.com/chips/techlib/techlib.nsf/product families/CoreConnect_Bus_Architecture.

[5] Sonics SMART Interconnect, http://www.sonicsinc.com.

[6] E. Clarke, E. Emerson and A. Sistla, "Automatic verification of finite-state concurrent systems using temporal logic specifications," *ACM Transactions on Programming Languages and Systems*, Vol. 8, No. 2, 1986.

[7] E. M. Clarke and E. A. Emerson, "Synthesis of synchronization skeletons for branching time temporal logic," Logic of Programs: Workshop Lecture Notes in Computer Science, London, Springer-Verlag, Vol. 131, May 1981.

[8] J. R. Burch, E. M. Clarke, K. L. McMillan, D. L. Dill and J. Hwang, "Symbolic model checking: 10^{20} states and beyond," *Symposium on Logic in Computer Science*, 1990, pp. 1–33.

[9] K. L. McMillan, "Symbolic model checking—An approach to the state explosion problem," Ph.D. Thesis, SCS, Carnegie Mellon University, 1992.

[10] E. M. Clarke, O. Grumberg, H. Hiraishi, S. Jha, D. E. Long, K. L. McMillan and L. A. Ness, "Verification of the Futurebus+ cache coherence protocol," in *Proceedings of International Symposium on Computer Hardware Description Languages and Their Applications*, North-Holland, April 1993, pp. 5–20.

[11] S. Campos, E. Clarke, W. Marrero and M. Minea, "Verifying the performance of the PCI local bus using symbolic techniques," in *Proceedings. of International Conference on Computer Design (ICCD)*, 1995, pp. 1–15.

[12] F. Aloul and K. Sakallah, "Efficient verification of the PCI local bus using boolean satisfiability," in *Proceedings of International Workshop on Logic Synthesis (IWLS)*, 2000, pp. 131–135.

[13] P. Chauhan, E. Clarke, Y. Lu and D. Wang, "Verifying IP core based system-on-chip designs," in *Proceedings of IEEE ASIC SOC Conference*, 1999, pp. 1–5.

[14] A. Mokkedem, R. Hosabettu, M. Jones and G. Gopalakrishnan, "Formalization and analysis of a solution to the PCI 2.1 bus transaction ordering problem," *Formal Methods in System Design*, Vol. 16, 2000, pp. 1-30.

[15] K. Shimuzu, D. Dill and A. Hu, "Monitor-based formal specification of PCI," in *Proceedings of International Conference on Formal Methods in Computer-Aided Design (FMCAD)*, 2000, pp. 335-353.

[16] L. Ivanov and R. Nunna, "Specification and formal verification of interconnect bus protocols," in *Proceedings of the 43rd IEEE Midwest Symposium on Circuits and Systems*, 2000, pp. 378-382.

[17] K. Oumalou, A. Habibi and S. Tahar, "Design for verification of a PCI bus in SystemC," in *Proceedings of International Symposium on System-on-Chip*, 2004, pp. 201-204.

[18] PCISIG, "PCI Special Interest Group—PCI Local Bus Specification, Revision 2.1," 1995.

[19] PCI Special Interest Group, PCI Local Bus Specification Revision 2.2, December 1998.

[20] S. Owre, J. Rushby, N. Shankar and F. von Henke, "Formal verification for fault-tolerant architectures: Prolegomena to the design of PVS," *IEEE Transactions on Software Engineering*, Vol. 21, No. 2, 1995, pp. 107-125.

[21] F. Corella, "Proposal to fix ordering problem in PCI 2.1," http://www.pcisig.com/reflector/thrd8.htm1#00706, 1996.

[22] F. Corella, "Verifying memory ordering model of I/O system," *Computer Hardware Description Languages (CHDL)*, Toledo, Spain, 1997. (Invited Talk)

[23] F. Corella, R. Shaw and C. Zhang, "A formal proof of absence of dead-lock for any acyclic network of PCI buses," in *Proceedings of Computer Hardware Description Languages (CHDL)*, 1997, pp. 134-156.

[24] V. Pratt, "Modeling concurrency with partial orders," *International Journal of Parallel Programming*, Vol. 15, No. 1, November 1986, pp. 33-71.

[25] P. Godefroid, "Partial order methods for the verification of concurrent systems: An approach to the state explosion problem," Doctoral Dissertation, University of Liege, 1995.

[26] D. Peled, "Combining partial order reductions with on-the-fly model checking," *Journal of Formal Methods in Systems Design*, Vol. 8, No. 1, 1996, pp. 39-64.

[27] R. Nalumasu and G. Gopalakrishnan, "A new partial order reduction algorithm for concurrent system verification," in *Proceedings of IFIP*, 1996, pp. 1-10.

[28] L. Ivanov, R. Nunna and S. Bloom, "Modeling and analysis of non-iterated systems: An approach based on series-parallel posets," in *Proceedings IEEE International Symposium on Circuits and Systems (ISCAS)*, 1999, pp. 404-406.

[29] L. Ivanov and R. Nunna, "Formal verification with series-parallel posets of globally-iterated/locally-non-iterated systems," in *Proceedings of the MWSCAS*, 1999.

[30] L. Ivanov and R. Nunna, "Formal verification: A new partial order approach," *Proceedings of ASIC/SOC*, Washington, DC, 1999, pp. 32-36.

[31] R. E. Bryant, "Graph-based algorithms for boolean function manipulation," *IEEE Transactions on Computers*, Vol. C-35, No. 8, 1986, pp. 677-691.

[32] K. McMillan, *Symbolic Model Checking*, Kluwer Academic Publishers, 1993.

[33] J. R. Burch, E. M. Clarke, K. L. McMillan and D. L. Dill, "Sequential circuit verification using symbolic model checking," in *Proceedings of the 27th ACM/IEEE Design Automation Conference (DAC)*, June 1990, pp. 46-51.

[34] A. Biere, A. Cimatti, E. Clarke, M. Fujita and Y. Zhu, "Symbolic model checking using SAT procedures instead of BDDs," in *Proceedings of the Design Automation Conference (DAC)*, 1999, pp. 317-320.

[35] A. Biere, A. Cimatti, E. Clarke and Y. Zhu, "Symbolic model checking without BDDs," in *Proceedings of TACAS*, 1999, pp. 193–207.

[36] J. Silva and K. Sakallah, "GRASP—A new search algorithm for satisfiability," in *Proceedings of IEEE/ACM International Conference on Computer Aided Design (ICCAD)*, 1996, pp. 220–227.

[37] K. McMillan, "The SMV model checker," 1999, http://www-cad.eecs.berkeley.edu/~kenmcmil/smv/.

[38] G. Madl, S. Pasricha, Q. Zhu, L. Bathen and N. Dutt, "Formal performance evaluation of AMBA-based system-on-chip designs," *6th Annual ACM Conference on Embedded Software (EMSOFT)*, Seoul, Korea, October 2006, pp. 311–320.

[39] V. D'silva, S. Ramesh and A. Sowmya, "Synchronous protocol automata: A framework for modelling and verification of SoC communication architectures," in *Proceedings of the Conference on Design, Automation and Test in Europe (DATE)*, 2004.

[40] A. Roychoudhury, T. Mitra and S. R. Karri, "Using formal techniques to debug the AMBA system-on-chip bus protocol," in *Proceedings of Design, Automation and Test in Europe Conference and Exhibition (DATE)*, 2003, pp. 828–833.

[41] J. Schmaltz and D. Borrione, "Validation of a parameterized bus architecture using ACL2," in *Proceedings of Fourth International Workshop on the ACL2 Theorem Prover and Its Applications*, Boulder, CO, 2003, pp. 1–20.

[42] A. Goel and W. R. Lee, "Formal verification of an IBM CoreConnect processor local bus arbiter core," in *Proceedings of the 37th Conference on Design Automation (DAC)*, 2000, pp. 196–200.

[43] H. Amjad, "Verification of AMBA using a combination of model checking and theorem proving," *Automated Verification of Critical Systems, Volume 145 of Electronic Notes in Theoretical Computer Science*, 2005, pp. 45–61.

[44] H. Amjad, "Model checking the AMBA protocol in HOL," Technical Report, University of Cambridge, UCAM-CL-TR-602, September 2004.

[45] D. Kozen, "Results on the propositional mu-calculus," *Theoretical Computer Science*, Vol. 27, 1983, pp. 333–354.

[46] H. Amjad, "Programming a symbolic model checker in a fully expansive theorem prover," in *Proceedings of the 16th International Conference on Theorem Proving in Higher Order Logics, Lecture Notes in Computer Science*, 2003, pp. 171–187.

[47] I. Beer, S. Ben-David, C. Eisner and A. Landver, "Rulebase: An industry-oriented formal verification tool," in *Proceedings of ACM/IEEE Design Automation Conference (DAC)*, 1996, pp. 655–660.

[48] I. Beer, S. Ben-David, C. Eisner, D. Fisman, A. Gringauze and Y. Rodeh, "The temporal logic sugar," *13th International Conference on Computer-Aided Verification*, LNCS 2102, Springer, 2001, pp. 363–367.

[49] M. S. Jahanpour and E. Cerny, "Compositional verification of an ATM switch module using interface recognizer/suppliers (IRS)," *IEEE International High-Level Design, Validation, and Test Workshop*, 2000, pp. 71–76.

[50] M. Kaufmann, A. Martin and C. Pixley, "Design constraints in symbolic model checking," *10th International Conference on Computer-Aided Verification*, Springer, 1998, pp. 477–487.

[51] R. K. Brayton et al., "VIS: A system for verification and synthesis," in *Proceedings of Computer Aided Verification, Lecture Notes in Computer Science*, Vol. 1102, 1996, Springer, Berlin Heidelberg, New York, pp. 428–432.

[52] O. Kupferman and M. Y. Vardi, "Module checking," *8th International Conference on Computer-Aided Verification*, LNCS 1102, Springer, 1996, pp. 75–86.

[53] Cadence Incisive Verification IP, http://www.cadence.com/products/functional_ver/verification_ip/index.aspx.

[54] Synopsys VCS Verification IP, http://www.synopsys.com/products/designware/vcs_verification_library.html.

[55] Averant SolidPC Verification IP, www.averant.com/storage/documents/SolidPC.pdf.

[56] Denali Purespec Verification IP, https://www.denali.com/products/purespec.html.

[57] Verisity eVc Verification IP, http://www.verisity.com/products/evc.html.

[58] nSys nVS Verification IP, http://www.nsysinc.com/products.html.

[59] PSL, http://www.eda.org/vfv/docs/psl_lrm-1.01.pdf/.

[60] OVL, http://www.verificationlib.org/.

[61] OVA, http://www.opervera.org/.

[62] SVA, http://www.systemverilog.org/.

[63] A. Nightingale and J, Goodenough, "Testing for AMBATM compliance," in *Proceedings of 14th Annual IEEE International ASIC/SOC Conference*, 2001, pp. 301–305.

[64] Verisity Specman "e" language, http://www.verisity.com/products/specman.html.

[65] M. T. Oliveira and A. J. Hu, "High-level specification and automatic generation of IP interface monitors," in *Proceedings of the 39th Conference on Design Automation (DAC)*, 2002, pp. 129–134.

[66] A. Seawright and F. Brewer, "High-level symbolic construction techniques for high performance sequential synthesis," in *Proceedings of 30th Design Automation Conference (DAC)*, 1993, pp. 424–428.

[67] P. Raymond, "Recognizing regular expressions by means of dataflow networks," *23rd International Colloquium on Automata, Languages, and Programming*, Springer, 1996, pp. 336–347.

[68] A. Seawright and F. Brewer, "High-level symbolic construction techniques for high performance sequential synthesis," in *Proceedings of ACM/IEEE 30th Design Automation Conference (DAC)*, 1993, pp. 424–428.

[69] H. Lin, C. Yen, C. Shih and J. Jou, "On compliance test of on-chip bus for SOC," in *Proceedings of the Asia South Pacific Design Automation Conference (ASPDAC)*, 2004, pp. 328–333.

[70] Wishbone, Revision B.3 Specification, http://www.opencores.org/wishbone/.

[71] A. Gadkari, S. Ramesh and R. Parekhji, "CESC: A visual formalism for specification and verification of SoCs," in *Proceedings of the IEEE Great Lakes Symposium on VLSI (GLS-VLSI)*, 2004, pp. 354–357.

[72] K. Fisler, "Towards diagrammability and efficiency in event sequence languages," in *Proceedings of CHARME*, 2003, pp. 431–447.

[73] M. C. W. Geilen, "On construction of monitors for temporal logic properties," in *Proceedings of Runtime Verification (RV)*, 2001, pp. 1–19.

[74] Y. Abarbanel et al., "FoCs: Automatic generation of simulation checkers from formal specifications," in *Proceedings of CAV*, 2000, pp. 538–542.

[75] A. A. Gadkari and S. Ramesh, "Automated synthesis of assertion monitors using visual specifications," in *Proceedings of Design, Automation and Test in Europe (DATE)*, 2005, pp. 390–395.

[76] A. Bunker and G. Gopalakrishnan, "Using live sequence charts for hardware protocol specification and compliance verification," in *Proceedings of International Workshop HLDVT'01*, 2001, pp. 95–100.

[77] I. Krüger, "Distributed system design with message sequence charts," Dissertation, Technische Universität München, 2000.

[78] G. Berry, "The Esterel V5 Language Primer," Version 5.91, 2000.

[79] N. Halbwachs, *Synchronous Programming of Reactive Systems*, Kluwer Academic Publishers, 1993.

[80] T. Cormen, C. Leiserson and R. Rivest, *Introduction to Algorithms*, MIT Press, 1990.

[81] K. W. Susanto and T. Melham, "AMBA-ARM7 formal verification platform," in *Proceedings of International Conference of Formal Engineering Methods*, 2003, pp. 48–67.

[82] W. S. Liao and P.-A. Hsiung, "Creating a formal verification platform for IBM CoreConnect-based SoC," in *Proceedings of ATVA*, 2003.

[83] D. Karlsson, P. Eles and Z. Peng, "A formal verification methodology for IP-based designs," in *Proceedings of the Digital System Design (DSD)*, 2004, pp. 372–379.

[84] L. A. Cortés, P. Eles and Z. Peng, "Verification of embedded systems using a petri net based representation," in *Proceedings of International Symposium on System Synthesis (ISSS)*, 2000, pp. 149–155.

[85] E. M. Clarke Jr, O. Grumberg and D. A. Peled, *Model Checking*, MIT Press, 1999.

[86] UPPAAL homepage, http://www.uppaal.com.

[87] Y.-C. Yang, J.-D. Huang, C.-C. Yen, C.-H. Shih and J.-Y. Jou, "Formal compliance verification of interface protocols," in *Proceedings of the International Conference on VLSI Design (VLSID)*, 2005, pp. 12–15.

[88] A. Cimatti, E. Clarke, E. Giunchiglia, F. Giunchiglia, M. Pistore, M. Roveri, R. Sebastiani and A. Tacchella, "NuSMV 2: An openSource tool for symbolic model checking," in *Proceedings of the 14th International Conference on Computer-Aided Verification (CAV)*, 2002.

[89] Open Core Protocol International Partnership (OCP-IP). OCP Datasheet, Release Version 1.0, http://www.ocpip.org.

[90] Counterpane Internet Security, Inc., http://www.counterpane.com.

[91] P. Kocher, R. Lee, G. McGraw and A. Raghunathan, "Security as a new dimension in embedded system design," in *Proceedings of Design Automation Conference (DAC)*, 2004, pp. 753–760.

[92] W. Stallings, *Cryptography and Network Security: Principles and Practice*, Prentice Hall, 1998.

[93] B. Schneier, *Applied Cryptography: Protocols, Algorithms and Source Code in C*, John Wiley and Sons, 1996.

[94] MPEG Open Security for Embedded Systems (MOSES), http://www.crl.co.uk/projects/moses/.

[95] OpenIPMP, http://www.openipmp.org.

[96] Internet Streaming Media Alliance (ISMA), http://www.isma.tv/home.

[97] SSL 3.0 Specification, http://wp.netscape.com/eng/ssl3/.

[98] IPSec Working Group, http://www.ietf.org/html.charters/ipsec-charter.html.

[99] S. Ravi, A. Raghunathan and S. Chakradhar, "Tamper resistance mechanisms for secure embedded systems," *International Conference on VLSI Design*, January 2004.

[100] G. Hoglund and G. McGraw, *Exploiting Software: How to Break Code*, Addison-Wesley, 2004, http://www.exploitingsoftware.com.

[101] J. Viega and G. McGraw, *Building Secure Software*, Addison-Wesley, 2001, http://www.buildingsecuresoftware.com.

[102] G. McGraw, "Software security," *IEEE Security and Privacy*, Vol. 2, March–April 2004, pp. 80–83.

[103] R. Anderson and M. Kuhn, "Tamper resistance—A cautionary note," 1996.

[104] R. Anderson and M. Kuhn, "Low cost attacks on tamper resistant devices," in *IWSP: International. Workshop on Security Protocols, Lecture Notes on Computer Science*, 1997, pp. 125–136.

[105] O. Kommerling and M. G. Kuhn, "Design principles for tamper-resistant smartcard processors," In *Proceedings of the USENIX Workshop on Smartcard Technology (Smartcard)*, May 1999, pp. 9–20.

[106] E. Hess, N. Janssen, B. Meyer and T. Schutze, "Information leakage attacks against smart card implementations of cryptographic algorithms and countermeasures," in *Proceedings of EUROSMART Security Conference*, June 2000, pp. 55–64.

[107] J. J. Quisquater and D. Samyde, "Side channel cryptanalysis," in *Proceedings of the SECI*, 2002, pp. 179–184.

[108] J. Kelsey, B. Schneier, D. Wagner and C. Hall, "Side channel cryptanalysis of product ciphers," in *Proceedings of ESORICS'98*, September 1998, pp. 97–110.

[109] J. Coburn, S. Ravi, A. Raghunathan and S. T. Chakradhar, "SECA: Security-enhanced communication architecture," in *Proceedings of CASES*, 2005, pp. 78–89.

[110] R. Elbaz, L. Torres, G. Sassatelli, P. Guillemin, C. Anguille, C. Buatois and J. B. Rigaud, "Hardware engines for bus encryption: A survey of existing techniques," in *Proceedings of Design, Automation and Test in Europe (DATE)*, 2005, pp. 40–45.

[111] R. M. Best, "Microprocessor for executing enciphered programs," US Patent No. 4 168 396, September 18, 1979.

[112] R. M. Best, "Crypto microprocessor for executing enciphered programs," US Patent No. 4 278 837, July 14, 1981.

[113] R. M. Best, "Crypto microprocessor that executes enciphered programs," US Patent No. 4465 901, August 14, 1984.

[114] R. Elbaz, L. Torres, G. Sassatelli, P. Guillemin, M. Bardouillet and A. Martinez, "A parallelized way to provide data encryption and integrity checking on a processor-memory bus," in *Proceedings of Design Automation Conference (DAC)*, 2006, pp. 506–509.

[115] R. Takahashi and D. N. Heer, "Secure memory management unit for microprocessor," US Patent (from VLSI Technology, Inc.) No. 5 825 878, October 20, 1998.

[116] B. Candelore and E. Sprunk, "Secure processor with external memory using block chaining and block reordering," US Patent (from General Instrument Corporation) No. 6 061 449, May 9, 2000.

[117] T. Gilmont, J.-D. Legat and J. J. Quisquater, "Enhancing security in the memory management unit," in *Proceedings of 25th Euromicro Conference*, September 1999, pp. 449–456.

[118] IBM code pack, http://www.ibm.com/us/.

[119] F. Schneider, "Enforceable security policies," *ACM Transactions on Information and System Security*, Vol. 3, February 2000, pp. 30–50.

[120] G. E. Suh, D. Clarke, B. Gassend, M. van Dijk and S. Devadas, "AEGIS: Architecture for tamper-evident and tamper-resistant processing," in *Proceedings of the International Conference on Supercomputing (ICS)*, 2003, pp. 160–171.

[121] ARM TrustZone, http://www.arm.com/products/esd/trustzone_home.html.

[122] ACL2 Theorem Prover, http://www.cs.utexas.edu/users/moore/acl2/.

[123] R. Alur, C. Courcoubetis and D. Dill, "Model-checking for real-time systems," in *Proceedings of the 5th Annual Symposium on Logic in Computer Science*, 1990, pp. 414–425.

[124] Advanced Encryption Standard (AES), http://www.csrc.nist.gov/publications/fips/fips197/fips-197.pdf.

[125] Data Encryption Standard (DES), http://csrc.nist.gov/publications/fips/fips46-3/fips46-3.pdf.

[126] R. Merkle and M. Hellman, "On the security of multiple encryption," *Communications of the ACM*, Vol. 24, No. 7, July 1981, pp. 465–467.

[127] S. Mister and S. E. Tavares, "Cryptanalysis of RC4-like ciphers," *Selected Areas in Cryptography*, London, Springer-Verlag, 1998, pp. 131–143.

[128] A. Menezes, P. C. Oorschot and S. A. Vanstone, *Handbook of Applied Cryptography*, CRC Press, October 1996.

[129] Digital Signature Standard (DSS), http://www.itl.nist.gov/fipspubs/fip186.htm.

[130] W. Diffie and M. E. Hellman, "New directions in cryptography," *IEEE Transactions on Information Theory*, Vol. IT-22, November 1976, pp. 644–654.

[131] US Secure Hash Algorithm 1 (SHA-1), http://tools.ietf.org/html/rfc3174.

[132] The MD5 Message-Digest Algorithm, http://tools.ietf.org/html/rfc1321.

[133] RIPEMD-160 (RACE Integrity Primitives Evaluation Message Digest) Algorithm, http://www.esat.kuleuven.ac.be/~bosselae/ripemd160.html.

[134] G. J. Holzmann, "The model checker spin," *IEEE Transactions on Software Engineering*, Vol. 23, No. 5, May 1997, pp. 279–295.

[135] A. Pnueli, "A temporal logic of concurrent programs," *Theoretical Computer Science*, Vol. 13, 1981, pp. 45–60.

[136] D. Cyrluk, S. Rajan, N. Shankar and M. K. Srivas, "Effective theorem proving for hardware verification," in *Theorem Provers in Circuit Design, TPCD '94*, September 1994, pp. 203–222.

[137] K. Havelund and N. Shankar, "Experiments in theorem proving and model checking for protocol verification" in *Proceedings of International Symposium of Formal Methods Europe (FME)*, 1996, pp. 662–681.

CHAPTER

Physical Design Trends for Interconnects

11

Ioannis Savidis and Eby G. Friedman

Over the past 10 years, the source of the critical signal delays has undergone a major transition. With the scaling of active device feature sizes into the deep submicrometer (DSM) regime, the on-chip interconnect has become the primary bottleneck in signal flow within high complexity, high speed integrated circuits (ICs). The smaller feature size in DSM technology nodes reduces the delay of the active devices; however, the effect on delay due to the passive interconnects has increased rapidly, as described by the 2005 International Technology Roadmap for Semiconductors (ITRS) [1]. The transition from an IC dominated by gate delays for feature sizes greater than 250μm to where the interconnects are the primary source of delay is graphically illustrated in Fig. 11.1. As noted in the figure, the disparity between the relative delay of the interconnects and the active devices is exacerbated in each successive technology node [1]. The local wire delay decreases with feature size due to a reduction in the distance among the active devices. Special attention must, however, be placed on the global lines, since the overall speed of current ICs is most often limited by the long distance global interconnects [2–6].

Low power dissipation has become a critical design criterion. With shrinking feature size and larger chip die dimensions, the sheer number of interconnects has increased exponentially [1, 7]. Interconnect capacitance often dominates the total gate load [8]; therefore, a large portion of the total transient power is dissipated by these on-chip lines. This characteristic is particularly true for those long interconnects that distribute the clock signals, where as much as 40–50% of the total power of an IC can be dissipated [1]. The gains achieved in performance, however, are often accompanied by an increase in power dissipation. As an example, additional interconnect layers enhance circuit speed at the expense of higher power consumption due to the larger interconnect capacitance. Considering both power consumption and propagation delay, interconnect design has become a dominant issue in high speed ICs.

In addition to an increase in interconnect power consumption, the design complexity of the various interconnect networks is continuing to present a significant

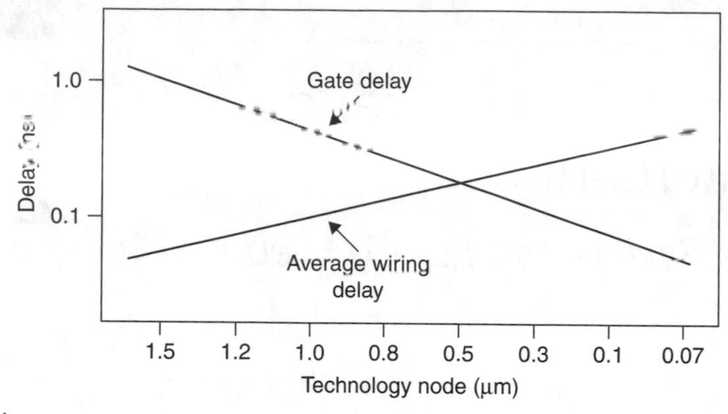

FIGURE 11.1

Comparison of interconnect (wiring) delay to gate delay

challenge. Semiconductor companies are currently building microprocessors with over 500 million transistors, and this number is increasing [9, 10]. In addition to the interconnects among the various on-chip devices, the clock and power distribution networks both require significant metal resources. Accurately modeling the clock, power, and signal nets is a difficult task; optimally allocating metal to properly design these networks presents an even greater challenge.

A significant factor that contributes to the complexity of interconnect modeling is the increase in the sheer number of lines. *RC* interconnect models are not sufficiently accurate to properly capture signal propagation in lines with fast transition times. Including inductance (*L*) in the *RC* model has become a necessary modification. *RLC* models are therefore becoming increasingly common at the expense of greater computational cost. On-chip clock rates also contribute to the complexity of the interconnect modeling process. Until recently, the semiconductor industry had been focusing primarily on faster clock rates. However, over the past couple of years, there has been a shift in this industrial paradigm toward multi-core processing. With this shift, operational frequencies have been somewhat reduced while throughput improved at the expense of increased die area. Inductance may be ignored at these lower operating frequencies under certain conditions. The length of the line, the cross-sectional area of the line, the signal waveform properties, and the available current return paths must all be considered in determining whether to include inductance in the interconnect model when operating at mid-range frequencies ranging from 1 to 3 Gigahertz [13]. For these reasons, interconnect modeling and metal allocation have become a complex design problem.

The intention of this chapter is to provide insight into the complexity of the on-chip interconnect design process, particularly in high performance applications. The effects of scaling interconnect in the DSM regime is discussed in Section 11.1. Low power, high speed circuit design techniques in support of global signaling are presented in Section 11.2. Once an understanding of current interconnect models and analysis techniques has been established, application of these models and techniques to global power and clock distribution networks is examined in

Sections 11.3 and 11.4, respectively. A brief introduction to 3-D interconnects is provided in Section 11.5 as a glimpse into future interconnect technologies. Finally, some concluding remarks are offered in Section 11.6.

11.1 DSM INTERCONNECT DESIGN

The steady decrease in the feature size of semiconductor devices has enhanced circuit complexity and performance. Scaling of the lateral dimensions in planar devices, such as MOS transistors, has produced improvements in the area, power, and speed of the devices [12]. The specific characteristics of interconnects, particularly the global lines, are degraded by current scaling trends. The power consumption and signal propagation delays of these long resistive lines have increased. Inductive effects must now be considered for possible inclusion in the interconnect models. For these reasons, accurate on-chip interconnect models are required to determine the signal characteristics and design requirements of high speed DSM interconnect.

Accurate interconnect models are used in both the design and analysis of ICs. Local interconnects can be neglected if the line capacitance is much smaller than the load capacitance. As the line capacitance becomes comparable to the load capacitance, the local line can be modeled as a single lumped capacitor, as depicted in Fig. 11.2(a) [13]. Since these lines are short, the signal propagation delay is negligible as compared to the gate delay [14]. In addition, these short interconnects have a negligible line resistance, minimally degrading the signal propagation characteristics.

Long interconnects suffer from resistive effects that impede signal propagation. The signal delay through these lines can be comparable to or exceed the gate delay. A lumped capacitor model of a global line is typically highly inaccurate with errors often exceeding 30%. A more elaborate model that includes the resistive effects of long lines is therefore better suited to represent global interconnects. As shown in Fig. 11.2(b), the simplest *RC* model is a lumped model, which does not consider the distributed nature of the impedance of a long global line. *RC* lines are often divided into sections of distributed impedances, capturing the distributed nature of the line impedance [15–17]. Each subsection is modeled as an equivalent *RC* circuit. Two common circuits often used to model long interconnect lines are T and Π circuits, depicted in Fig. 11.3(a) and 11.3(b), respectively. The accuracy of these models depends upon the number of sections used to represent the interconnect lines, with five or six sections typically being more than sufficient to accurately model a line. A T or Π equivalent circuit can model an *RC* line with less than 3% relative error in the delay, even in the case where only three ladder stages are used [15]. An *RC* model is usually adequate for low to medium operating frequencies; however, at frequencies exceeding a GHz, an *RC* model is often inadequate to accurately characterize the waveform properties along a wide interconnect line. An *RLC* model is often necessary to accurately characterize these interconnects.

Long global lines are usually much wider than local lines, exhibiting a lower resistance per unit length. With the reduction in line resistance and the higher clock frequencies, the line inductance contributes to the signal propagation

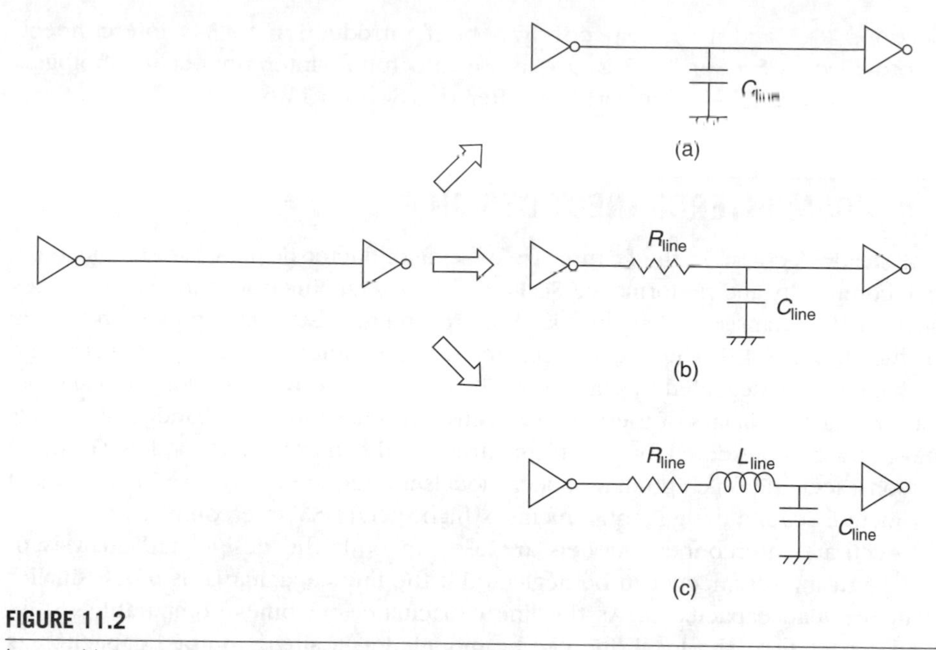

FIGURE 11.2

Lumped interconnect models: (a) *C* model, (b) *RC* model, (c) *RLC* model [13]

Figure reused with kind permission from Springer Science and Business Media

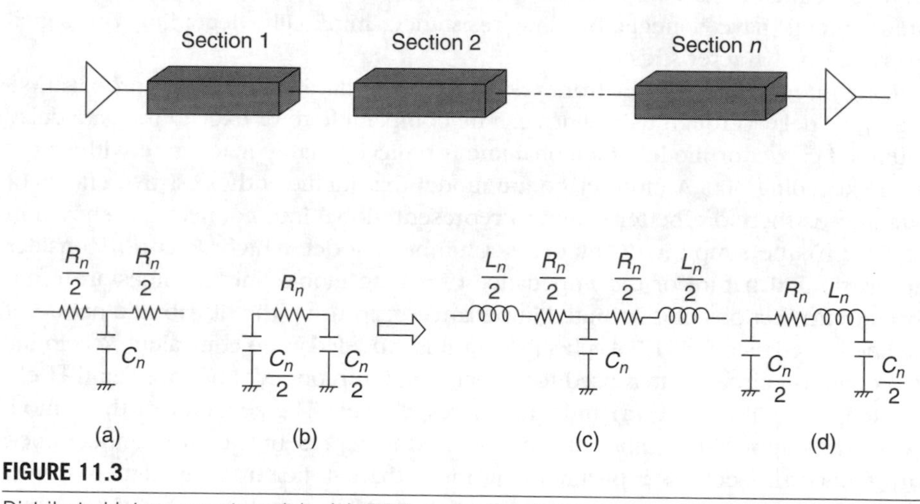

FIGURE 11.3

Distributed interconnect models: (a) *RC* T model, (b) *RC* Π model, (c) *RLC* T model, (d) *RLC* Π model [13]

Figure reused with kind permission from Springer Science and Business Media

characteristics [11]. A first order approximation of an inductive interconnect is shown in Fig. 1.2(c). Distributed T and Π equivalent models for an *RLC* line are shown in Fig. 11.3(c) and (d), respectively [18, 19].

The conditions necessary for the line inductance to be included in the interconnect model have been examined in [20–22]. In high speed digital circuits,

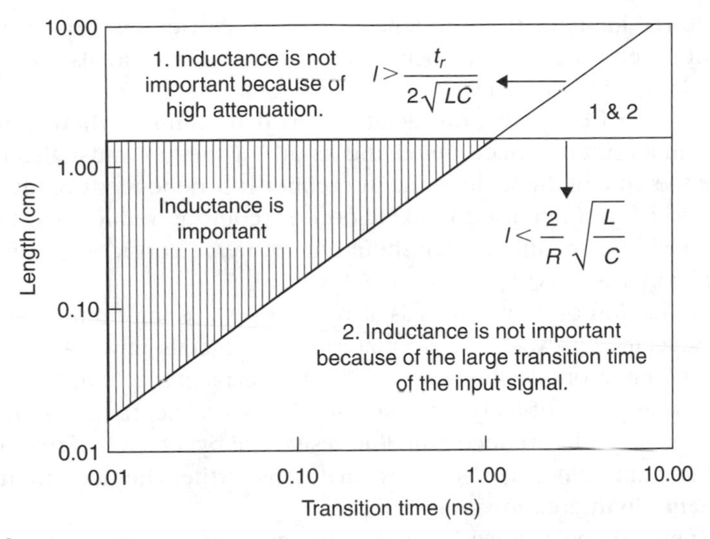

FIGURE 11.4

Transition time (t_r) versus the length of the interconnect line (l). The crosshatched area denotes the region where inductance is important ($L = 10^{-8}$ H/cm, $R = 400\Omega$/cm, and $C = 10^{-12}$ F/cm) [11]

limits to include the line inductance in the interconnect model based on the line impedance and transition time are described [11]. A signal propagating in an underdriven uniform lossy transmission line exhibits significant inductive effects if the line length l satisfies the following condition [11, 12],

$$\frac{t_r}{2\sqrt{LC}} < l < \frac{2}{R}\sqrt{\frac{L}{C}} \tag{11.1}$$

where R, L, and C are the per unit length resistance, inductance, and capacitance, respectively, and t_r is the rise time of the signal waveform. A graphical depiction of the length and transition time that requires the inclusion of inductance for arbitrary values of line resistance, capacitance, and inductance is depicted in Fig. 11.4 [11]. In addition, a criterion is presented in Fig. 11.4, consistent with (11.1), that graphically illustrates the conditions in which the inductance can be omitted from an interconnect model. The first condition states that as the attenuation of the line increases and the magnitude of the subsequent reflections from the load decreases, ringing on the line can be eliminated, permitting the inductance to be ignored. The second condition is that if the transition time of the output signal from the CMOS gate driving a transmission line is greater than twice the time of flight of the signal propagating across the line, the inductance can again be ignored. It is important to note that increasing signal frequencies typically require faster signal transition times. The resulting effect of a decreased signal transition time is a lower limit on the line length, making shorter on-chip interconnects behave inductively. As a consequence, medium length lines can also behave inductively at high signal frequencies.

In addition to frequency considerations in determining whether to include inductance in an interconnect model, the length of the line must also be considered. Since the time of flight along the interconnect is dependent on the length of the line, longer lines increase the likelihood of requiring inductance in an interconnect model. The line inductance should therefore be considered in high speed, high complexity ICs [23–31].

The introduction of new materials also increases the importance of the interconnect inductance. New dielectric materials and metals have been introduced to reduce the interconnect impedance. The line capacitance can be reduced by half of the capacitance of SiO_2 with the use of low k dielectrics [13]. In addition, copper interconnect has reduced the line resistance by a factor of two to three as compared to aluminum [13]. These new materials further the need to include the line inductance in interconnect models.

Design methodologies for driving global interconnects have been proposed to reduce the propagation delay of long resistive lines; however, these techniques have also ignored the inductance of the line. Under certain conditions, ignoring the line inductance may lead to high area and power inefficient circuits. Novel techniques for designing both low power and high speed circuits to drive both *RC* and *RLC* lines are presented in the next section.

11.2 LOW POWER, HIGH SPEED CIRCUIT DESIGN TECHNIQUES

Power and noise are important characteristics when considering design techniques to optimize circuit performance in low power, high speed circuits. Noise from both inter- and intra-layer capacitive and inductive interconnect coupling, as illustrated in Fig. 11.5, affects the delay, degrades the waveform shape, and most importantly, creates the possibility of an erroneous interpretation of the digital signals [11, 32–34]. In addition, faster clock rates create higher slew rates, further increasing the on-chip noise. A variety of design techniques have been developed to mitigate deleterious on-chip noise. Lowering the power consumed by circuits, however, requires a variety of different design techniques that target a combination of static, dynamic, and short-circuit power. Power dissipation in CMOS circuits is therefore reviewed in Subsection 11.2.1. Once an understanding of the basic power dissipation principles has been established, wire sizing is introduced in Subsection 11.2.2 as a useful technique to improve circuit performance. In Subsection 11.2.3, a driver sizing optimization procedure is presented as another technique to enhance system performance. Tapered buffers are introduced in Subsection 11.2.4. Subsection 11.2.5 focuses on repeater insertion as a means to partition long interconnects into smaller segments, thereby reducing the capacitive load of each section driven by an individual repeater (as compared to the total load driven by a single large inverter). And finally, some summarizing remarks on low power, high speed circuit design techniques are presented in Subsection 11.2.6.

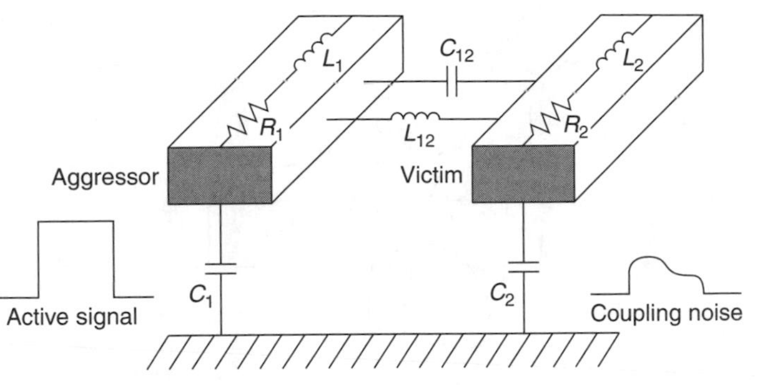

FIGURE 11.5

Cross-coupled interconnect noise in ICs. An active transition on an aggressor line can induce a coupling noise voltage on a victim line [6]

Figure reused with kind permission from Springer Science and Business Media

11.2.1 CMOS Power Dissipation

There are three primary components of power dissipation in CMOS circuits. The three components are

$$P_{\text{total}} = P_{\text{dynamic}} + P_{\text{SC}} + P_{\text{leakage}} \tag{11.2}$$

where the dynamic and short-circuit power are components of the transient power, and the leakage power is a component of the static power. The dynamic, short-circuit, and leakage components can be further expanded as

$$P_{\text{total}} = CV_{\text{DD}}^2 f_{\text{switch}} + \frac{1}{2} I_{\text{peak}} t_{\text{base}} V_{\text{DD}} f_{\text{swith}} + I_{\text{leakage}} V_{\text{DD}} \tag{11.3}$$

The dynamic power P_{dynamic} accounts for the energy dissipated in charging and discharging the nodal capacitances. When a nodal capacitance C is charged, $\frac{1}{2}CV_{\text{DD}}^2$ joules of energy is stored on the capacitor, and an equal amount is dissipated in the transistors. In the discharge phase, the remaining $\frac{1}{2}CV_{\text{DD}}^2$ joules of energy stored on the capacitance is dissipated by the transistors through the discharge path, as shown in Fig. 11.6. Thus, the total energy expended in the charge and discharge cycle is CV_{DD}^2. The average dynamic power consumed is the product of CV_{DD}^2 over the frequency of the charge and discharge cycle f_{switch}, producing the well known expression for dynamic power in CMOS circuits, $CV_{\text{DD}}^2 f_{\text{switch}}$ [6].

Short-circuit current flows in a static CMOS gate when a conductive path exists from the power rail to the ground rail. A path exists when a signal transitions at the input, passing through intermediate voltage levels [35–38]. For a static CMOS inverter, this voltage range is from the n-type transistor threshold voltage V_{Tn}, the voltage at which the n-type transistor turns on, to $V_{\text{DD}} + V_{\text{Tp}}$, the voltage at which the p-type transistor turns off. Within this voltage range, both the pull-up and pull-down networks conduct DC current, producing short-circuit current, as illustrated in Fig. 11.7. The period of time when this conductive path exists is denoted as t_{base} in Eq. (11.3) [6].

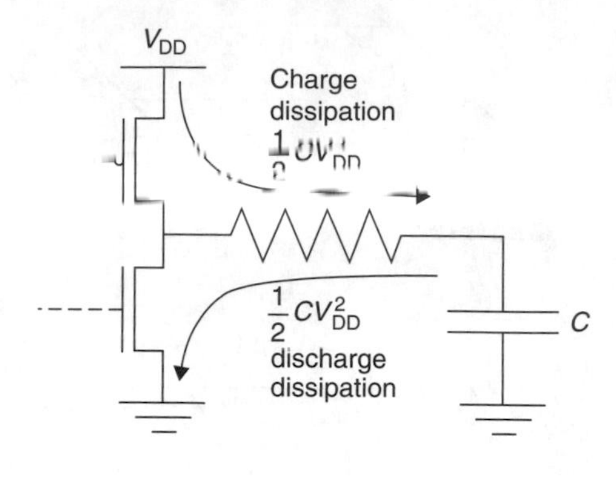

FIGURE 11.6

Energy dissipation during the charge/discharge cycle [6]
Figure reused with kind permission from Springer Science and Business Media

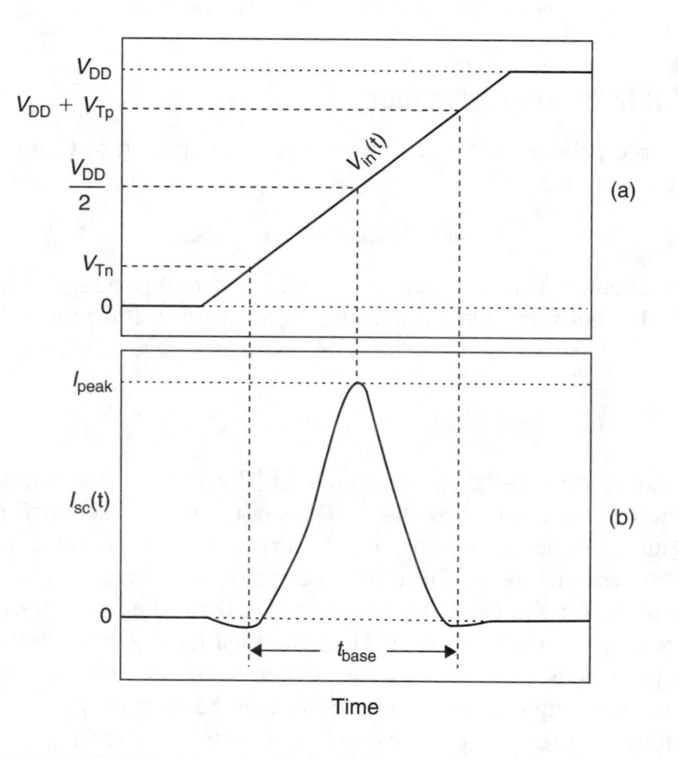

FIGURE 11.7

Short-circuit current waveform of a CMOS gate: (a) ramp-shaped input waveform, (b) short-circuit current waveform [6]
Figure reused with kind permission from Springer Science and Business Media

The transistor leakage current $I_{leakage}$ is the current that flows between the power terminals in the absence of any switching, giving rise to a leakage power component $P_{leakage}$. Typically, 50–70% of the total power dissipation is contributed by the dynamic power component [6]. Therefore, an effective strategy for reducing the total transient power consumption is to reduce the dynamic dissipation by lowering V_{DD}, operating at a lower frequency, or reducing nodal capacitances.

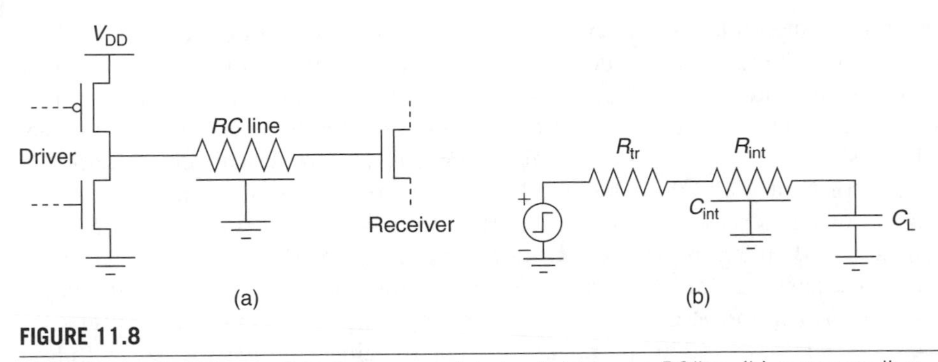

FIGURE 11.8

A CMOS circuit driving an *RC* interconnect: (a) circuit driving an *RC* line, (b) corresponding model [6]

Figure reused with kind permission from Springer Science and Business Media

Static power due to leakage current, however, is expected to grow significantly in the near future, soon exceeding 50% of the total power dissipated on-chip [39–41].

11.2.2 Wire Sizing

The width of an interconnect affects the power characteristics and propagation delay of an IC. As wiring becomes longer and the interconnect cross-sectional area smaller, it results in an increase in the *RC* interconnect impedances, thereby degrading the delay of the gates. Consider a CMOS inverter driving an *RC* interconnect line, as illustrated in Fig. 11.8. A simple first-order model of the delay of this circuit is [6]

$$T_{50\%} = 0.4R_{\text{int}}C_{\text{int}} + 0.7(R_{\text{tr}}C_{\text{int}} + R_{\text{tr}}C_{\text{L}} + R_{\text{int}}C_{\text{L}}) \tag{11.4}$$

If the driver load is effectively capacitive, where the interconnect resistance R_{int} is much less than the effective driver resistance R_{tr}, the interconnect capacitance can be combined with the input capacitance of the terminating gate to model the load as a lumped capacitance, permitting the circuit delay to be characterized by an *RC* circuit delay, $0.7R_{\text{tr}}(C_{\text{int}} + C_{\text{L}})$. Increasing the driver transistor width reduces R_{tr}, decreasing the circuit delay, thereby trading off circuit power and area for higher speed. This behavior, however, changes when R_{int} becomes comparable to R_{tr}. The delay cannot be reduced below $R_{\text{int}}(0.4C_{\text{int}} + 0.7C_{\text{L}})$. Note that the interconnect component, $0.4R_{\text{int}}C_{\text{int}}$, increases quadratically with interconnect length since both R_{int} and C_{int} are proportional to the length of the line. Increasing the width of the interconnect to reduce R_{int} does not significantly reduce the delay caused by the *RC* interconnect impedance since this decrease in wire resistance is offset by an increase in the wire capacitance. Many algorithms have been proposed to determine the optimum wire size that minimizes a target cost function. Some of these algorithms address reliability issues by reducing clock skew [42], while most of these algorithms focus on minimizing delay [43–47]. The results described in [48–50] consider simultaneous driver and wire sizing based on the Elmore delay model [51] with capacitance, resistance, and power models.

Additionally, tradeoffs exist between the dynamic and short-circuit power characteristics as there is a dependence of the power dissipation on the interconnect width, as illustrated in Fig. 11.9. As the line inductance-to-resistance ratio increases with wider lines, the short-circuit power decreases due to a reduction in the signal transition time. For an *RC* line, the short-circuit power will remain approximately constant with increasing width as the decrease in interconnect resistance is offset by an increase in capacitance, maintaining a relatively unchanged *RC* time constant, and therefore signal transition time. If the width of the interconnect exceeds a specific limit (shown in Fig. 11.9), the short-circuit power increases for both an *RC* and *RLC* line due to the change in the matching characteristics between the driver and interconnect [52]. The dynamic power increases with line width since the line capacitance is greater. As shown in Fig. 11.9, an optimum interconnect width exists at which the total transient power is a minimum if the line exhibits inductive behavior.

11.2.3 Driver Sizing

Transistor sizing is another design approach, producing tradeoffs at the circuit level in CMOS logic families [54–63]. Wider transistors produce more current; however, the physical area and gate capacitance also increase linearly with width, increasing the circuit area and power. Thus, optimal transistor sizing is strongly dependent on the design tradeoffs among area, power, and speed.

A common objective of transistor sizing is lower delay. Consider a CMOS circuit with the output load dominated by the input capacitance of the following stage. The charge time monotonically decreases with increasing driver width. The input load of the transistor, however, also increases linearly with the driver width, loading the preceding gate. The net result is that the total delay of a data path with additional stages can be smaller. Similarly, a uniform increase in all of the transistors does not substantially change the propagation delay of a circuit where the output loads are dominated by the input capacitance of the fanout. The current

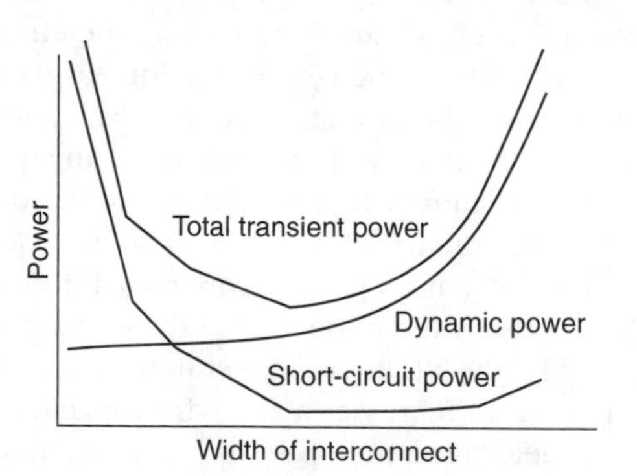

FIGURE 11.9

Dynamic, short-circuit, and total transient power as a function of interconnect line width assuming an inductive line [53]

Figure reused with kind permission from Springer Science and Business Media

drive of the gates I_{out} will increase which is offset, however, by an increase in the output capacitive load C_L. The I_{out}/C_L ratio remains essentially constant [6]. A careful balance of the current drive and output load is therefore necessary to enhance circuit performance.

Aside from improving circuit performance by increasing the current drive, transistor sizing also affects the power characteristics of a circuit. A simple approximation treats the circuit power as linearly proportional to the total active area A of a driver, that is, $P = CV^2f$, where $C = C_{ox}A$, and the gate oxide capacitance per unit area C_{ox} is constant for a given technology. An increase in the transistor width increases the area of the driver, which dissipates more power. Using the product of the power consumed and the delay of the driver as a figure of merit for optimizing the transistor size, the power-delay product is minimum when the gate output capacitance equals the sum of the interconnect and load capacitances [56, 64]. The power optimal transistor size is smaller than the power-delay optimal transistor size. An efficient tradeoff between power and delay is needed, however, with intermediate sized transistors. Non-optimal tradeoffs beyond the power-delay optimal size can be pursued in performance aggressive circuits [6].

11.2.4 Tapered Buffers

An important example of transistor sizing to drive large capacitive loads is tapered buffers [6]. The intermediate buffers are used to drive the intermediate capacitive loads. An inverter appropriately scaled for the capacitive load, as shown in Fig. 11.10(a), reduces the delay; however, the large input capacitance of the inverter loads the previous logic stage. A similar argument can be made when inserting another inverter sufficiently large to drive the inverter driving the load. This process continues until the initial input inverter of the buffer is sufficiently small to be driven by a logic gate at an acceptable speed. Thus, a tapered buffer consists of a chain of inverters of gradually increasing size, as illustrated in Fig. 11.10(b). The ratio of the size of an inverter to the size of the preceding inverter is the tapering factor β. Under the assumption that a stage load is proportional to the size of the next stage,

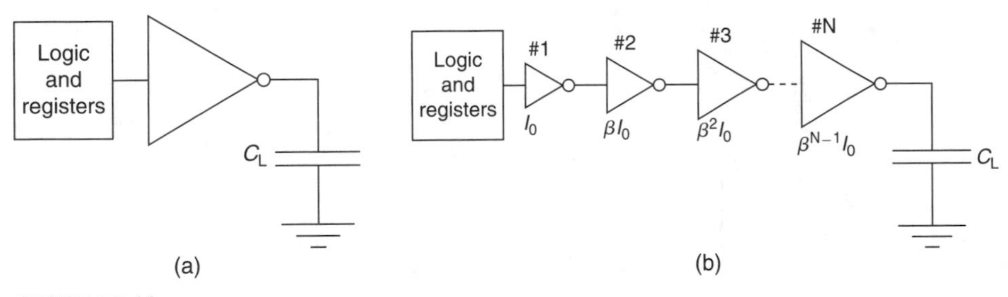

FIGURE 11.10

Circuit techniques to drive a large capacitive load: (a) a large single inverter, (b) A system of tapered buffers. The delay of the tapered buffer system is often less than the delay of a single large inverter [6]

Figure reused with kind permission from Springer Science and Business Media

thereby neglecting the interconnect and output capacitance of the gate, the delay of a tapered buffer is minimum at a constant exponential tapering factor $\beta_{opt} = e$, the base of the natural logarithm [66]. This constant tapering factor also corresponds to an optimal number of stages $N_{opt} = \ln M$, where $M = C_L/C_o$ is the ratio of the load capacitance C_L to the input capacitance C_o of the initial inverter in the chain [65, 66]. Note that since the number of buffer stages N is an integer, this condition cannot in general be satisfied precisely. Therefore, one of the two integers closest to $\ln M$ is chosen, where β is selected to satisfy $\beta^N = M$. More accurate delay models [67–69] and capacitance models [70, 71] have been employed to include the intrinsic drain and source capacitance, a ramp input, and short-circuit current. The delay optimal tapering factor increases with the ratio of the intrinsic output capacitance (which includes the diffusion and gate overlap capacitance) to the input gate capacitance [67, 68]. Further enhancements to the model, such as the effects of a finite slew rate, producing short-circuit current, have also been incorporated [70, 71].

Tradeoffs among area, power, and delay have also been considered [72–75]. For a specific load, the dependence of the buffer delay on the tapering factor is relatively flat around β_{opt}, as depicted in Fig. 11.11. The total area of the buffer is also a relatively strong function of β_{opt}. Thus, an effective tradeoff among the delay, area, and power is possible. For example, if a buffer with an optimum number of stages is implemented with both four stages and three stages, the buffer delay rises by 3% and 22% but the area shrinks by 35% and 54%, respectively. Tapered buffers with a fixed tapering factor have been compared with a geometrically increasing tapered buffer system [75]. The minimum delay of a variable-taper buffer can be reduced to within a few percent of the delay of a fixed-taper buffer by implementing the first few stages with a fixed-taper factor [6]. Optimal area-delay tradeoffs are therefore achieved in a fixed-taper buffer system with the final one to two stages utilizing a larger tapering factor.

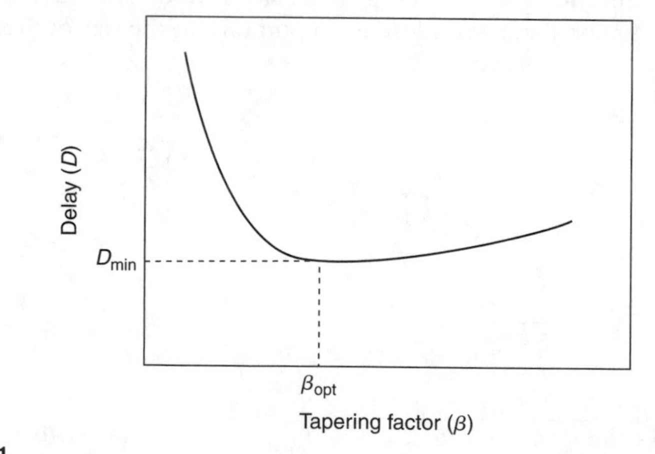

FIGURE 11.11

Dependence of the tapered buffer propagation delay on the tapering factor [6]
Figure reused with kind permission from Springer Science and Business Media

11.2.5 Repeater Insertion

Widening a uniform line has a marginal impact on the overall wire delay. A more effective strategy for reducing the delay of a long interconnect is to strategically insert buffers along a line. These buffers are typically called repeaters and the process is called repeater insertion [14]. Repeaters circumvent the quadratic increase in interconnect delay by partitioning the line into smaller and approximately equal sections, as shown in Fig. 11.12. The sum of the section delays is smaller than the delay of the original path since the delay of each section is reduced. The decreased interconnect delay is partially offset by the additional delay of the inserted repeaters. The optimal number of repeaters is determined by considering the delay of each individual repeater added to the repeater system, and determining the number of repeaters at which the increase in the repeater delay outweighs the lower interconnect delay. The optimal number of repeaters k_{opt} and the optimal size of the repeaters h_{opt} as compared to a minimum sized repeater h are

$$k_{opt} = \sqrt{\frac{a_1 R_t C_t}{a_2 R_0 C_0}} \tag{11.5}$$

$$h_{opt} = \sqrt{\frac{R_0 C_t}{R_t C_{g0}}} \tag{11.6}$$

respectively, where R_t and C_t are the total interconnect resistance and capacitance, respectively, R_0 and C_0 are the input and output repeater resistance and capacitance, respectively, and C_{g0} is the input capacitance of the repeater. The two fitting parameters, a_1 and a_2, account for the rise and fall time of the propagating signal [76].

A number of repeater insertion methods have been proposed [77–82]. Bakoglu presents a method based on characterizing the repeaters by the input capacitance and the effective output resistance of each repeater [14, 83]. The minimum delay of the resulting *RC* circuit is achieved when the delay of the repeater section equals the wire segment delay.

Techniques to improve interconnect performance vary depending upon the electrical characteristics of the line. For an *RC* line, repeater insertion techniques outperform wire sizing [84]. Unlike an *RC* line, the minimum signal propagation delay always decreases with increasing line width for *RLC* lines if an optimum repeater system is used [85, 87]. In *RLC* lines, wire sizing outperforms repeater insertion as the minimum signal propagation delay with no repeaters is smaller

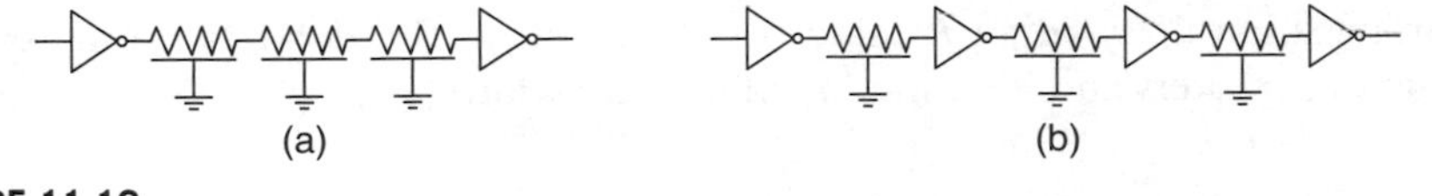

FIGURE 11.12

Repeater insertion: (a) original interconnect line, (b) interconnect line with inserted repeaters [6]
Figure reused with kind permission from Springer Science and Business Media

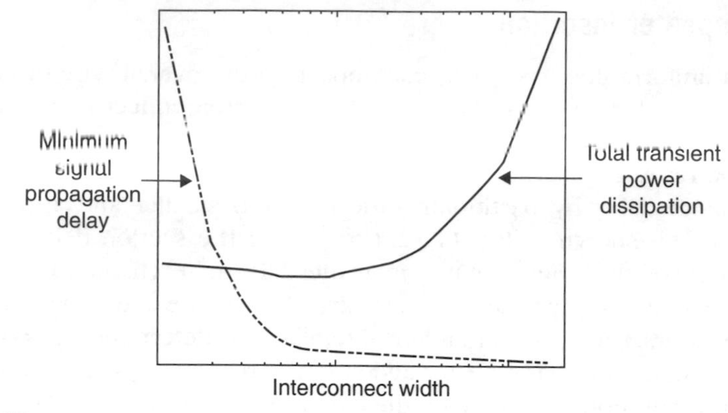

FIGURE 11.13

Minimum signal propagation delay and transient power dissipation as a function of line width for a repeater system [13]

Figure reused with kind permission from Springer Science and Business Media

than the minimum signal propagation delay using any number of repeaters. For an *RLC* line, the minimum signal propagation delay always decreases with wider lines until the number of repeaters equals zero. As shown in Fig. 11.13, the minimum propagation delay decreases while the power dissipation increases for wider interconnect, delineating the tradeoff between minimum delay and total power dissipation [13].

The interconnect resistance decreases with wider lines, increasing the ratio between the line inductance and resistance L/R, and decreasing the number of inserted repeaters to achieve the minimum propagation delay. The minimum delay produced by an optimum repeater system decreases with increasing line width as the total gate delay decreases. For an inductive interconnect line, the total signal propagation delay is [13]

$$t_{\text{pd-total}} = k_{\text{opt-}RLC} t_{\text{pd-section}} \quad (11.7)$$

where $t_{\text{pd-section}}$ is the signal delay of each *RLC* section [86] and $k_{\text{opt-}RLC}$ is the optimum number of repeaters. As shown in Fig. 11.14, for different line lengths l, the optimum number of repeaters $k_{\text{opt-}RLC}$ which minimizes the signal propagation delay decreases with increasing line width for all line lengths until the number of repeaters reaches zero, the point at which only a single driver at the beginning of the line is effective. *RC* lines, however, require repeaters, increasing $k_{\text{opt-}RC}$, as a wider line increases the line capacitance driven by each repeater. The propagation delay of an *RLC* line is therefore a decreasing function of the line width, whereas the propagation delay of an *RC* line is dependent on the delay of an increasing number of repeaters and is a function of the line width [13].

11.2.6 Summary

In this section, an introduction to several low power, high speed circuit design techniques has been presented. Wire sizing is shown to produce nominal improvements

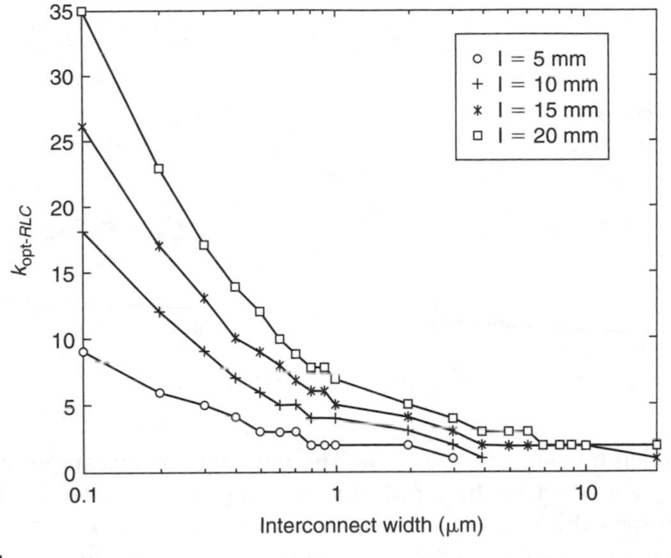

FIGURE 11.14

Optimum number of repeaters for minimum propagation delay for different line widths [13]
Figure reused with kind permission from Springer Science and Business Media

in circuit speed. Cascaded buffers have been suggested to drive large capacitive loads. Repeater insertion techniques have also been introduced to improve signal propagation delay in long resistive interconnects. The remaining sections of Chapter 11 focus on applying the basic principles presented in this chapter to power and clock distribution networks, as well as 3-D interconnect technologies.

11.3 GLOBAL POWER DISTRIBUTION NETWORKS

Distributing power in high speed, high complexity ICs has become a challenging task. This section provides insight and intuition into the behavior and design of power distribution networks. An overview of noise issues related to the power distribution network is presented. Within this noise framework, the basic model of a power grid is discussed. Decoupling capacitors are also introduced as a means to temporarily provide charge from within the power network. After a basic understanding of the models, noise issues, and use of decoupling capacitors, two important characteristics of power networks are examined in this section. The first phenomenon, multi-path current redistribution, is directly related to the frequency dependent impedance variation amongst the different levels of the on-chip metallization layers. The second topic, electromigration, is a consequence of the increasing current densities encountered in high complexity circuits.

11.3.1 Noise in Power Distribution Networks

Noise in power distribution networks is depicted by the power delivery system illustrated in Fig. 11.15. The power grid consists of a supply, load, and interconnect

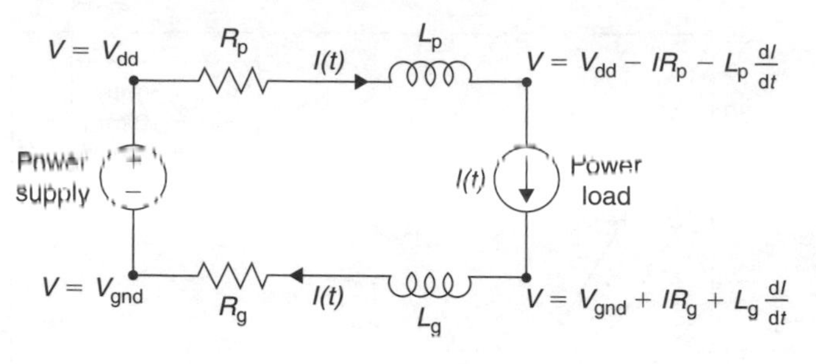

FIGURE 11.15

Power delivery system consisting of the power supply, power load, and non-ideal interconnect [12]

Figure reused with kind permission from Springer Science and Business Media

lines connecting the supply to the load. The nominal power and ground voltages, V_{dd} and V_{gnd}, are provided by an ideal power supply. A variable current source $I(t)$, which is typically a transistor or group of transistors, models the power load. The interconnect lines connecting the power supply to the load are considered non-ideal with a finite resistance and inductance, R_p, L_p and R_g, L_g, for the power and ground lines, respectively. Resistive voltage drops $\Delta V_R = IR$ and inductive voltage drops $\Delta V_L = L\,(dI/dt)$ develop across the parasitic interconnect impedances as the load draws current from the power network. The voltage levels at the load terminal change from the nominal levels provided by the supply, decreasing to $V_{dd} - IR_p - L_p\,(dI/dt)$ at the power terminal and rising to $V_{gnd} + IR_g + L_g\,(dI/dt)$ at the ground terminal. This change in supply voltages is referred to as power supply noise [12].

Power supply noise can adversely affect circuit operation. One major consequence is an increase in signal delay uncertainty. When power supply variations reduce the rail-to-rail power voltage, the gate-to-source voltage across both the NMOS and PMOS transistors also decreases, thereby lowering the output drive current of these devices. The signal delay increases as compared to the delay under a nominal power supply voltage. Conversely, a higher power voltage and a lower ground voltage shorten the propagation delay. The net effect of power noise on propagating clock and data signals is an increase in both delay and delay uncertainty within the data paths [88, 89]. Consequently, power supply noise can severely limit the maximum operating frequency of an IC [90, 91].

The power distribution network should exhibit a small impedance at the terminals of the load to ensure a small variation in the power supply voltage. Decoupling capacitors ensure correct and reliable operation of an IC in the frequency range from DC to some target operating frequency f_o by ensuring that the impedance of the power network is maintained below a specified upper bound within this target frequency range. The function of a decoupling capacitor is to provide charge when transient current demands on the power grid are high. These decoupling capacitors are distributed across a system, placed at the board, package, and on-chip levels (see Fig. 11.16). Each decoupling capacitor provides transient current to the load, effectively reducing the local transient noise. The

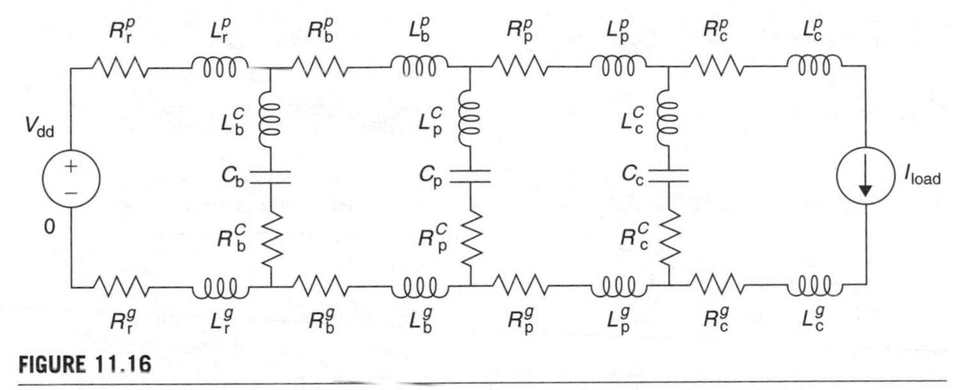

FIGURE 11.16

Power network with board, package, and on-chip decoupling capacitances [12]
Figure reused with kind permission from Springer Science and Business Media

low frequency power noise is lowered by those decoupling capacitors farthest from the load, whereas the high frequency transients require a fast injection of charge which is provided by those capacitors closest to the load.

11.3.2 Multi-Path Current Redistribution

Multi-path current redistribution is an extension of current redistribution within a single conductor to return current flowing among several parallel conductors. Adjacent signal lines, power networks, and the substrate can provide a variety of potential current return paths. Significant redistribution of the return current among these return paths can occur as signal frequencies increase. At low frequencies, the line impedance $Z(\omega) = R(\omega) + j\omega L(\omega)$ is dominated by the interconnect resistance. In this case, the path resistance determines the distribution of the return current among the available return paths, as shown in Fig. 11.17(a). At high frequencies, the line impedance $Z(\omega) = R(\omega) + j\omega L(\omega)$ is dominated by the reactive component $j\omega L(\omega)$. The minimum impedance path is primarily determined by the least inductive $L(\omega)$ path, as shown in Fig. 11.17(b). In power grids, both the forward and return currents undergo multi-path redistribution as both the forward and return paths can change with frequency since the paths consist of multiple conductors connected in parallel [12]. Current redistribution can lead to excessive current densities along certain paths. Under these conditions, a phenomenon known as electromigration must be considered.

11.3.3 Electromigration

On-chip current densities can reach several hundred thousand amperes per square centimeter, making electromigration a significant issue. Electromigration is the transport of metal atoms under the force of an electron flux. The significance of electromigration has been established early in the development of ICs [93, 94]. The depletion and accumulation of metal material resulting from atomic flow

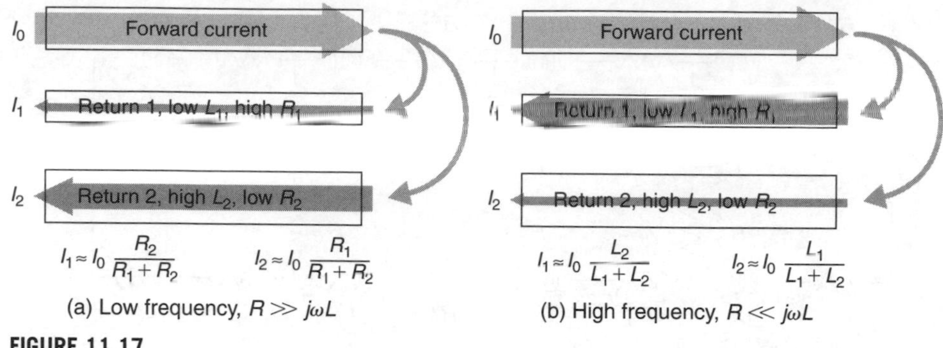

FIGURE 11.17

Current loop with two alternative current return paths. The forward current I_0 returns both through return path one with resistance R_1 and inductance L_1, and return path two with resistance R_2 and inductance L_2. In this structure, $L_1 < L_2$ and $R_1 > R_2$: (a) At low frequencies, the path impedance is dominated by the line resistance. (b) At high frequencies, the path impedance is dominated by the line inductance [12]

Figure reused with kind permission from Springer Science and Business Media

can lead to the formation of extrusions and voids in the metal structures. These extrusions and voids can lead to short circuits and open circuit faults, respectively, degrading the reliability of an IC [92]. The mass transport of metal ions through diffusion under an electrical driving force F is

$$J_a = C_a \mu F \tag{11.8}$$

where C_a is the atomic concentration and μ is the mobility of the atoms.

Two forces, an electric field force and an electron wind force, act on the metal ions. The electric field force is proportional to the electric field E and acts in the direction of the field. Conduction electrons accelerate in the direction opposite to the electric field, transferring momentum to the metal ions in the course of scattering. The force exerted by these electrons is also in the direction opposite to the field E, and is commonly referred to as the electron wind force. In metals of interest, such as aluminum and copper, the electron wind force dominates and the net force acts in the direction opposite to the electric current. The resulting atomic flux is therefore in the opposite direction of the electric current j, as shown in Fig. 11.18.

11.3.4 Summary

An overview of noise modeling is presented in this section as a means to introduce basic concepts behind the design of power distribution networks. Decoupling capacitors have been introduced as a means to reduce the noise within the power grid. Additionally, an examination of both multi-path current redistribution and electromigration provides insight into two important characteristics of interconnect, in general, and power grids, in particular. A more detailed discussion of power distribution networks can be found in [12].

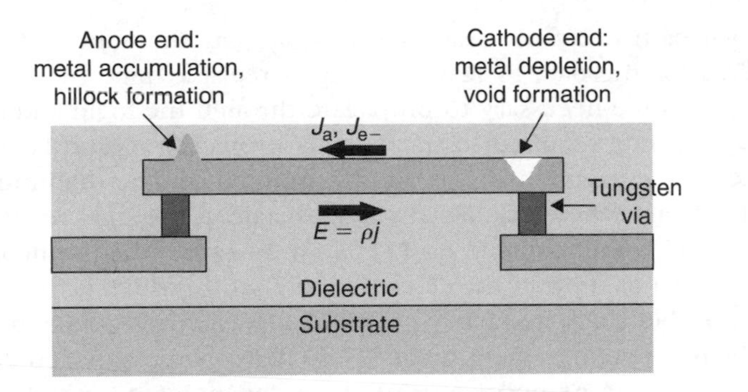

FIGURE 11.18

Electromigration mass transport in an interconnect line. An electron flux J_{e-} flowing in the opposite direction of the electric field $E = \rho j$ induces an atomic flow J_a in the direction of the electron flow [12]

Figure reused with kind permission from Springer Science and Business Media

11.4 CLOCK DISTRIBUTION NETWORKS

In a synchronous digital system, the clock signal provides a time reference for the movement of data within that system. Clock signals are typically loaded with the greatest fanout, travel over the longest distances, and operate at the highest speeds of any signal, either control or data, within the entire system. The control of any differences in the delay of the clock signals can severely limit the maximum performance of the entire system as well as create catastrophic race conditions in which an incorrect data signal may latch within a register. Therefore, understanding the basic principles governing clock distribution networks is of primary importance. An overview of the primary timing relationships is therefore provided in Subsection 11.4.1 to introduce the important concepts of local data paths and clock skew. Timing constraints and relationships are followed by a general introduction to clock topologies in Subsection 11.4.2, with more specific asymmetric and symmetric topologies presented in Subsections 11.4.3 and 11.4.4, respectively. Low power clock design is reviewed in Subsection 11.4.5, which is followed by a short summary in Subsection 11.4.6.

11.4.1 Timing Relationships

General synchronous systems are composed of the following three delay components: (i) memory storage elements; (ii) logic elements; and (iii) clocking circuitry and distribution networks [95]. The minimum allowable clock period $T_{CP(min)}$ between any two registers in a sequential data path is

$$\frac{1}{f_{clk,MAX}} = T_{CP(MIN)} = T_{PD(MAX)} + T_{Skew} \tag{11.9}$$

where

$$T_{PD(MAX)} = T_{C\text{-}Q} + T_{Logic} + T_{Int} + T_{Set\text{-}up} + T_{Hold} \tag{11.10}$$

and the total path delay of a data path $T_{PD(MAX)}$ is the sum of the maximum time required for the data to leave the initial register once the clock signal C_i arrives T_{C-Q}, the time necessary to propagate through the logic and interconnect $T_{Logic} + T_{Int}$, the time required to successfully propagate to and latch within the final register of the data path $T_{Set\text{-}up}$, and the amount of time the input data signal must be stable once the clock signal changes state at the final register T_{Hold}. The sum of the delay components in Eq. (11.10) must satisfy the timing constraint of Eq. (11.9) in order to support the clock period $T_{CP(MIN)}$, which is the inverse of the maximum possible clock frequency $f_{clk,MAX}$. The waveforms depicted in Fig. 11.19 illustrate the timing requirement of Eq. (11.9) being barely satisfied. Note that the clock skew T_{skewij} can be positive or negative depending on whether C_f leads or lags C_i, respectively.

Clock distribution networks are based on equipotential clocking, where the entire network is considered a temporal surface which must be maintained at a specific voltage at each half of the clock cycle. Ideally, clocking events occur simultaneously at all registers. Given this global clocking strategy, the clock signal arrival times at each register are defined with respect to a universal time reference. The difference in the clock signal arrival time between two sequentially adjacent registers is the clock skew T_{Skew}. Zero clock skew occurs if the clock signals C_i and C_f are in complete synchronism. The clock skew between two sequentially adjacent registers, R_i and R_j, and an equipotential clock distribution network is defined as

$$T_{Skewij} \equiv T_{Ci} - T_{Cj} \tag{11.11}$$

where T_{Ci} and T_{Cj} are the clock delay from the clock source to the registers R_i and R_j, respectively. Note that system-wide or chip-wide clock skew between two non-sequentially adjacent registers, from an analysis viewpoint, has no effect on the performance and reliability of a synchronous system and is essentially meaningless. System-wide global clock skew only places constraints on the permissible local clock skew. The clock skew between any two registers in a global data path which are not necessarily sequentially adjacent is the sum of the clock skew between each pair of registers along the global data path between those same two registers.

Depending upon whether C_i leads or lags C_f and upon the magnitude of T_{Skew} with respect to T_{PD}, system performance and reliability can either be degraded or enhanced. If the time of arrival of the clock signal at the final register of a data path T_{Cf} leads that of the time of arrival of the clock signal at the initial register of the same sequential data path T_{Ci}, as depicted in Fig. 11.20(a), the clock skew is referred to as positive clock skew and, under this condition, the maximum attainable operating frequency is decreased. Positive clock skew is the additional amount of time which must be added to the minimum clock period to reliably apply a new clock signal at the final register. Also note that positive clock skew only affects the maximum frequency of a system and cannot produce a race condition. If the clock signal arrives at R_i before the signal reaches R_f, as shown in Fig. 11.20(b), the clock skew is defined as being negative.

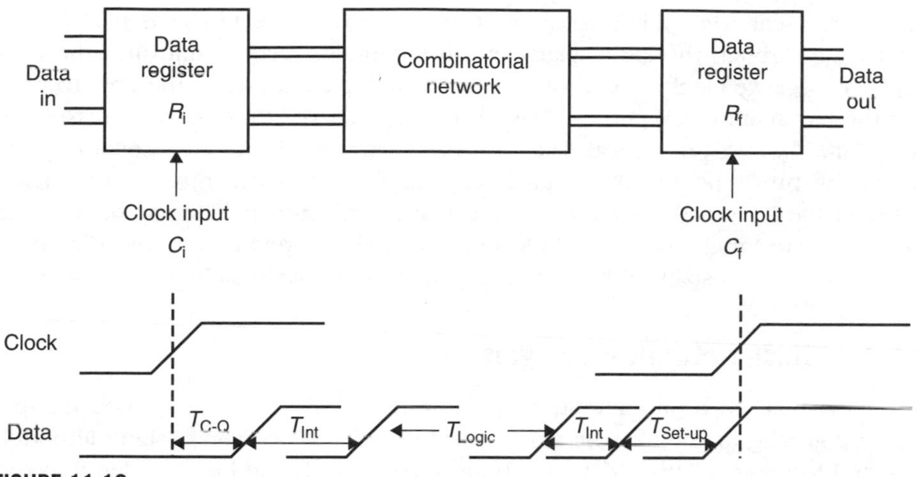

FIGURE 11.19

Timing diagram of clocked data path [95]
© 2001 IEEE

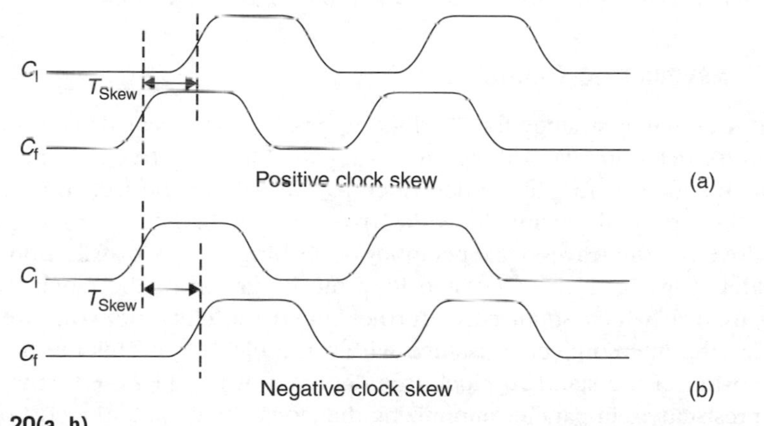

FIGURE 11.20(a, b)

Clock timing diagrams [95]
© 2001 IEEE

Negative clock skew can be used to improve the maximum performance of a synchronous system by decreasing the delay of a critical path; however, a potential minimum constraint can occur, creating a race condition [104–108]. In this case, when C_f lags C_i, the clock skew must be less than the time required for the data signal to leave the initial register, propagate through the interconnect and combinatorial logic, and successfully set up in the final register. If this condition is not met, the data stored in register R_f is overwritten by the data that had been stored in register R_i and has propagated through the combinatorial logic. By forcing C_i to lead C_f at each critical local data path, excess time is shifted from the neighboring less critical local data paths to the critical local data paths. This negative clock skew represents the additional amount of time that the data signal at R_i has to propagate through the logic stages and interconnect sections and into the final register. Negative clock skew subtracts from the logic path delay, thereby decreasing the minimum clock period. Thus, applying negative clock skew increases the total time that a given critical data path has to accomplish its functional requirements by providing the data signal released from R_i extra time to propagate through the logic and interconnect stages and latch into R_f. The use of negative clock skew in a random path i results in positive clock skew in the preceding path $i - 1$, which may establish the new upper limit on the system clock frequency.

11.4.2 Clock Network Topologies

Tradeoffs that exist among system speed, physical die area, and power dissipation are greatly affected by the clock distribution network. The design methodology and topology of the clock distribution network should be considered in the development of the structure of the network for distributing the clock signals. The most common and general approach to equipotential clock distribution is the use of buffered trees. In contrast to these highly asymmetric structures, symmetric trees, such as H-trees, are also used to distribute high speed clock signals [95].

11.4.3 Asymmetric Topologies

The most common strategy for distributing on-chip clock signals is to insert buffers at the clock source and along the clock path, forming a tree structure [96, 97]. The clock source is frequently described as the root of the tree, the initial portion of the tree as the trunk, individual paths as the branches, and the registers being driven as the leaves. This metaphor describing a clock distribution network is illustrated in Fig. 11.21. Occasionally, a mesh version of the clock tree structure is used in which shunt paths further down the clock network are used to minimize the interconnect resistance within the clock tree. This mesh structure, an extension of the standard clock tree depicted in Fig. 11.21, effectively places branch resistances in parallel, minimizing the clock skew but at the cost of greater power dissipation. If the interconnect resistance of the buffer at the clock source is small as compared to the buffer output resistance, a single buffer is often used to drive the entire clock distribution network. The primary requirement of a single buffer system is that the buffer should provide sufficient current to drive the network capacitance, which includes both interconnect and fanout, while maintaining high quality waveform shapes. Additionally, a single buffer can be used if the output resistance of the buffer is much greater than the resistance of the interconnect section being driven. Alternatively, buffers, acting as repeaters, distributed throughout the clock network may be used in place of a single buffer. This approach requires additional area but greatly improves the precision and control of the clock signal waveforms and is necessary if the resistance of the interconnect lines is non-negligible. The distributed buffers serve the double function of amplifying the clock signals degraded by the distributed interconnect impedances while isolating the local clock nets from the upstream load impedances. Note that

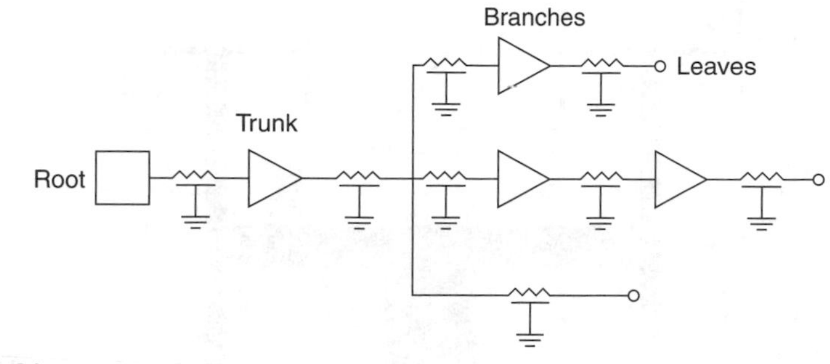

FIGURE 11.21

Tree structure of clock distribution network [95]
© 2001 IEEE

the buffers are a primary source of clock skew within a well-balanced clock distribution network since the active device characteristics vary much more greatly than the passive interconnect characteristics.

11.4.4 Symmetric Topologies

Another approach for distributing clock signals, an extension of the distributed buffer approach depicted in Fig. 11.21, utilizes a hierarchy of planar symmetric H-tree or X-tree structures [14, 98, 99] to ensure zero clock skew by maintaining identical distributed interconnect and buffer paths from the clock signal source to the clocked registers. This approach ensures that each clock path from the source to each register has practically the same delay. The primary delay difference among the clock signal paths is due to variations in process parameters that affect the interconnect impedance and, in particular, any active distributed amplifying buffers. The clock skew within an H-tree clock network is dependent upon the physical size, the semiconductor process, and the number of active buffers and clocked latches distributed within the H-tree structure. The conductor widths in H-tree structures are designed to progressively decrease as the signal propagates to lower levels of the hierarchy, ensuring that reflections are minimized at the branch points. Specifically, the impedance of the conductor leaving each branch point Z_{K+1} must be twice the impedance of the conductor providing the signal at that branch point Z_K for an H-tree structure and four times the impedance for an X-tree structure. Therefore, for the tapered H-tree structure illustrated in Fig. 11.22, $Z_K = Z_{K+1}/2$. A drawback to H-tree structures as compared to standard clock trees is that the interconnect capacitance and therefore the power dissipation is much greater since the total wire length tends to be much longer. Symmetric clock structures such as H-trees are used to minimize clock skew; however, an increase in the clock signal delay is incurred. The increase in clock delay must be considered when choosing between buffered asymmetric and symmetric clock networks. Additionally, H- and X-tree distribution networks are difficult to implement in high complexity integrated systems which are typically irregular in nature [95].

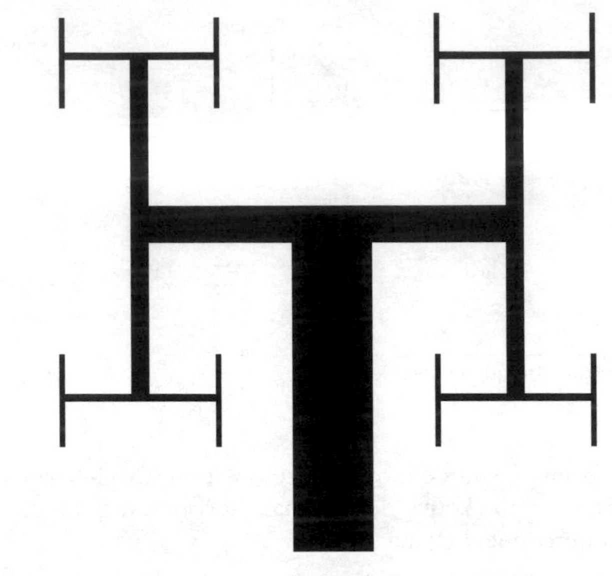

FIGURE 11.22

Tapered H-tree clock distribution network

11.4.5 Power Considerations

In modern integrated systems, the clock signal may drive many tens of thousands of registers, placing a large capacitive load on the network. The combination of a large capacitive load and a demand for higher clock frequencies has led to an increasingly larger proportion of the total power of a system dissipated within the clock network, in some applications much greater than 40% of the total power [100–102]. The primary component of power dissipation in most CMOS-based digital circuits is dynamic power. It is possible to reduce CV^2f dynamic power by lowering the clock frequency, the power supply, and/or the capacitive load of the clock distribution network. Multi-core processors target dynamic power reduction by maintaining the same logical throughput but at a reduced frequency. Each core can operate at a lower frequency while maintaining the same total workload as a single processor operating at a much higher frequency. Additionally, decreasing the total effective capacitance required to implement a clock tree can also reduce power consumption. Reductions of 10–25% in power dissipated within the clock tree have been reported with no degradation in clock frequency. Targeting the quadratic voltage term of the dynamic power has the greatest potential for power savings. A technique has been described for designing clock buffers and pipeline registers such that the clock distribution network operates at half the power supply swing, reducing the power dissipated in the clock tree by 60% without compromising the clock frequency [103]. The degradation in system speed is minimal since the data signals operate over the full power supply rail. The voltage is therefore only reduced in the clocking circuitry, resulting in significantly lower power with a minimal degradation in system speed [95].

11.4.6 Summary

In this section, clock distribution timing constraints are introduced and an overview of clock skew in terms of the local data path delay is provided. Once an understanding of the basic timing relationships is established, several asymmetric and symmetric clock topologies are presented, and the benefits and drawbacks of these networks are discussed. Finally, methods to minimize the power consumed within the clock distribution network are addressed as the clock network can consume more than 40% of the total power dissipated on-chip.

11.5 3-D INTERCONNECTS

As previously mentioned, two major effects of device scaling on interconnects are the increase in the number of metal layers necessary to achieve a higher integration density and an increase in the number and length of the global lines. These effects, caused by device scaling, increase both the line delay and capacitive crosstalk. 3-D interconnects have been proposed as a possible solution to address these issues. The introduction of a third dimension significantly alters the distribution of the interconnect length in ICs. As the number of planes is increased, the length and number of the global interconnects decrease as depicted in Fig. 11.23. Since the total number of IC interconnects are the same, the number of short interconnects increases [109]. Various characteristics, including the power dissipation and area allocated for metallization, can be improved with 3-D interconnects.

The corner-to-corner interconnect length is one such characteristic that benefits directly from 3-D integration. Since it is feasible to partition a 2-D IC into multiple subsections and stack these sections in the vertical dimension, the corner-to-corner interconnect length significantly decreases. As a result, for a constant clock frequency, several global interconnects in the upper metallization levels can be transferred to local, smaller aspect ratio metal layers. This strategy, in turn, implies a reduction in the total number of metal levels within a 3-D circuit. Alternatively, assuming a constant number of registers along a sequential data path and number of metal layers, an increase in the clock frequency is possible as the worst case data path delay can be reduced (decreasing T_{int} in Eq. (11.10)). For a constant clock frequency, the wiring pitch can be reduced, lowering the interconnect area. In addition, 3-D ICs with smaller interconnect lengths consume less power as compared to 2-D ICs as a consequence of the reduced capacitive load of the global interconnect lines. A graphical depiction of the variation in gate pitch, interconnect length, and power consumption as a function of the number of planes for two different values of Rent's exponent is illustrated in Fig. 11.24 [110]. Rent's exponent is a component of Rent's rule, a rule that correlates the number of I/O terminals with the number of circuit elements. Rent's rule describes the increase in the number of interconnects due to scaling similar to how Moore's Law describes the increase in device density due to scaling. Note that Rent's exponent increases as the parallelism of a system increases [111, 112].

Increasing the number of planes that can be integrated into a single 3-D system requires interplane interconnects that connect signals between vertically stacked

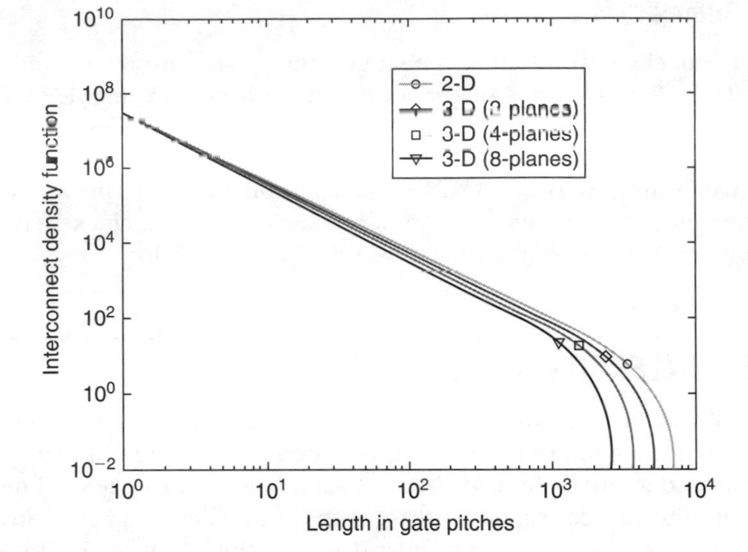

FIGURE 11.23

Interconnect length distribution for 2-D and 3-D ICs

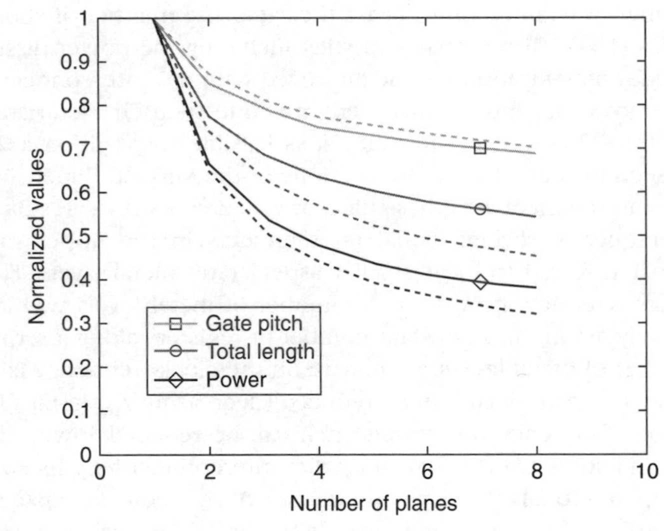

FIGURE 11.24

Variation of gate pitch, total interconnect length, and interconnect power as a function of the number of planes

devices. Interplane interconnects implemented as through silicon vias (TSV) or 3-D vias can produce the shortest path within a 3-D system, as compared to wire bonding, peripheral vertical interconnects, and solder ball arrays. Fabrication processes utilizing 3-D vias should be reliable and inexpensive, exhibit low

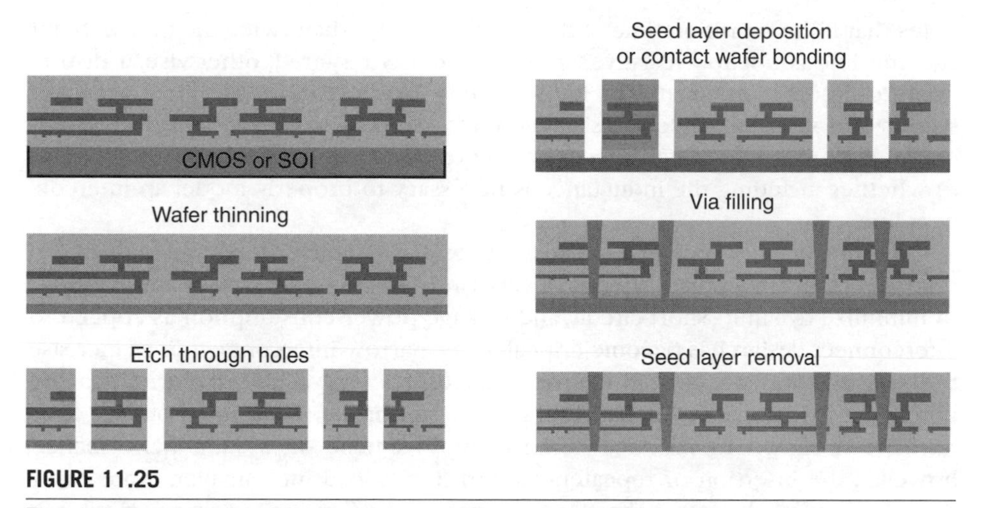

FIGURE 11.25

The via last approach to TSV fabrication and filling after wafer thinning

impedance characteristics, and have a negligible effect on the performance and reliability of the nearby active devices. There are two industrial fabrication approaches to produce TSVs. In the via first approach, the 3 D vias are formed after fabricating the active devices on each plane and prior to wafer thinning. A major disadvantage of this approach is the degradation in the reliability of the TSV due to wafer thinning and bonding. The advantages of this technique include simpler wafer handling, and compatibility with existing process flows [113]. In the via last approach, fabrication of the 3-D vias is performed after wafer thinning, as depicted in Fig. 11.25 [114–117]. The reliability of the TSVs is much improved as compared to the via first approach; however, the via last approach requires handling the thin wafers for several processing steps. Currently, the most popular method of TSV fabrication is via first.

3-D integration is a novel technology of growing importance that has the potential to offer significant performance and functional benefits as compared to 2-D ICs. Much work, however, is needed to properly characterize and model the interplane TSV, which is the primary technological innovation required to exploit the benefits of 3-D integration.

11.6 SUMMARY AND CONCLUDING REMARKS

The complexity of properly designing interconnects in the DSM regime increases with each successive technology generation. Modeling, following the same trend as design, has also increased in complexity. Many factors contribute to the complexity of properly modeling the interconnect. Deciding between modeling the interconnect as a distributed *RC* or *RLC* is dependent on such factors as the signal waveform characteristics and the length of the line. A relevant criterion notes that if the transition time of the signal is shorter than twice the time of flight and the line is not too resistive, an *RLC* model is required; otherwise, a distributed *RC* model is sufficient. The inductance is therefore more prevalent at higher speeds and with longer, low resistance interconnects. However, with the advent of parallel processing, clock speeds can be reduced, complicating the decision as to whether including the inductance is necessary to properly model an interconnect line.

Low power, high speed circuit techniques are essential to expand battery lifetime and maintain ambient thermal levels. Understanding the causes and methods to minimize dynamic, short-circuit, and leakage power consumption as applied to interconnect design has become critical. Long, narrow interconnect lines increase the resistive and possibly inductive characteristics of a line, requiring unique design methodologies to enhance system performance. Wide interconnect lines have been shown to produce nominal improvements in system performance; however, the insertion of repeaters to partition a load into smaller, more easily driven segments decreases the signal propagation delay. Proper sizing of tapered buffers, while considering tradeoffs among area, power, and speed, is an effective technique to drive nodes with large capacitive loads.

Clock and power distribution are important applications of the general interconnect design problem. Power distribution networks are greatly affected by *IR* and $L(di/dt)$ noise. Special treatment of power lines is required to assure that these noise sources do not adversely affect both the delay and the delay uncertainty of the clock and data signals. Decoupling capacitors are added between the source and sink nodes of a power network to provide charge at the terminals of the load. The large currents that are often present on the power grid can cause electromigration, the mass transport of metal ions through diffusion under an electrical driving force. These extrusions and voids can cause circuit failures in the form of short or open circuits. Clock distribution networks affect circuit behavior in other ways. Negative clock skew along a sequential data path can create race conditions although negative skew can also be used to improve the performance of a path, whereas positive clock skew degrades the maximum operating frequency. Topologies that distribute clock signals can be either asymmetric or symmetric, with buffers to maintain proper waveform characteristics. Symmetric clock distribution schemes, such as an H-tree structure, have an additional benefit of minimizing clock skew.

Novel techniques in interconnect design can also help alleviate deleterious trends that include longer line lengths, greater line impedances, and increased propagation delays. 3-D interconnects is one such technique that can help reduce the line length, and therefore the line impedance. Additional research is required to fully exploit the benefits of 3-D integration, including design, analysis, modeling, and process manufacturing technologies. As interconnect continues to become more complex, requiring greater resources, an understanding of the basic principles, as presented in this chapter, is essential to the development of new process technologies and design techniques while ensuring that electrical interconnect remains a viable means of signal transmission within the foreseeable future.

REFERENCES

[1] International Technology Roadmap for Semiconductors: Semiconductor Industry Association, 2005.

[2] K. Lee, "On-chip interconnects—Gigahertz and beyond," in *Proceedings of the IEEE International Interconnect Technology Conference*, December 1998, pp. 15–17.

[3] C. S. Chang, "Interconnection challenges and the national technology roadmap for semiconductors," in *Proceedings of the IEEE International Interconnect Technology Conference*, December 1998, pp. 3–6.

[4] M.T. Bohr, "Interconnect scaling—The real limiter to high performance ULSI," in *Proceedings of the IEEE International Electron Devices Meeting*, December 1995, pp. 241–244.

[5] F. Caignet, S. Delmas-Bendhia and E. Sicard, "The challenge of signal integrity in deep-submicrometer CMOS technology," in *Proceedings of the IEEE*, Vol. 89, No. 4, April 2001, pp. 556–573.

[6] A. V. Mezhiba and E. G. Friedman, "Trade-offs in CMOS VLSI circuits," *Trade-offs in Analog Circuit Design: The Designer's Companion*, C. Toumazou, G. Moschytz and B. Gilbert (Eds.), Kluwer Academic Publishers, Dordrecht, The Netherlands, 2002, pp. 75–114.

[7] R. Ho, K. W. Mai and M. A. Horowitz, "The future of wires," in *Proceedings of the IEEE*, Vol. 89, No. 4, April 2001, pp. 490–504.

[8] Y. Shin and T. Sakurai, "Power distribution analysis of VLSI interconnects using model order reduction," *IEEE Transactions on Computer-Aided Design of Integrated Circuits and Systems*, Vol. 21, No. 6, June 2002, pp. 739–745.

[9] G. E. Moore, "Cramming more components onto integrated circuits," *Electronics*, April 1965, pp. 114–117.

[10] S. Borkar, "Obeying Moore's law beyond 0.18 micron," in *Proceedings of the ASIC/SOC Conference*, September 2000, pp. 26–31.

[11] Y. I. Ismail, E. G. Friedman and J. L. Neves, "Figures of merit to characterize the importance of on-chip inductance," *IEEE Transactions on Very Large Scale Integration (VLSI) Systems*, Vol. 7, No. 4, December 1999, pp. 442–449.

[12] M. Popovich, A. V. Mezhiba and E. G. Friedman, *Power Distribution Networks with On-Chip Decoupling Capacitors,* Springer Science+ Business Media, New York, 2008.

[13] M. A. El-Moursy and E. G. Friedman, "Design methodologies for on-chip inductive interconnect," *Interconnect-Centric Design for Advanced SoC and NoC*, J. Nurmi, H. Tenhunen, J. Isoaho and A. Jantsch (Eds.), Kluwer Academic Publishers, Boston, 2004.

[14] H. B. Bakoglu, *Circuits, Interconnects, and Packaging for VLSI, Reading*, Addison-Wesley Publishing Company, MA, 1990.

[15] T. Sakurai, "Approximation of wiring delay in MOSFET LSI," *IEEE Journal of Solid-State Circuits*, Vol. 18, No. 4, August 1983, pp. 418–426.

[16] R. Antinone and G. W. Brown, "The modeling of resistive interconnects for integrated circuits," *IEEE Journal of Solid-State Circuit*, Vol. 18, No. 2, April 1983, pp. 200–203.

[17] G. Y. Yacoub, H. Pham, M. Ma and E. G. Friedman, "A system for critical path analysis based on back annotation and distributed interconnect impedance models," *Microelectronics Journal*, Vol. 19, No. 3, May/June 1988, pp. 21–30.

[18] N. Gopal, E. Tuncer, D. P. Neikirk and L. T. Pillage, "Non-uniform lumped models for transmission line analysis," in *Proceedings of the IEEE Topical Meeting on Electrical Performance of Electronic Packaging,* April 1992, pp. 119–121.

[19] T. Dhaene and D. D. Zutter, "Selection of lumper element models for coupled lossy transmission lines," *IEEE Transactions on Computer-Aided Design of Integrated Circuits and Systems*, Vol. 11, No. 7, July 1992, pp. 805–815.

[20] L. Chang, K. Chang and R. Mathews, "When should on-chip inductance modeling become necessary for VLSI timing analysis?," in *Proceedings of the IEEE International Technology Conference*, 2000, pp. 170–172.

[21] A. Deutsch et al., "When are transmission-line effects important for on-chip interconnections?," *IEEE Transactions on Microwave Theory and Techniques*, Vol. 45, No. 10, October 1997, pp. 1836–1846.

[22] B. Krauter, S. Mehrotra and V. Chandramouli, "Including inductive effects in interconnect timing analysis," in *Proceedings of the IEEE Custom Integrated Circuits Conference*, May 1999, pp. 445–452.

[23] A. Deutsch et al., "Electrical characteristics of interconnections for high-performance systems," in *Proceedings of the IEEE*, Vol. 86, No. 2, February 1998, pp. 313–355.

[24] P. J. Restle, A. E. Ruehli, S. G. Walker and G. Papadopoulos, "Full-wave PEEC time-domain method for the modeling of on-chip interconnects," *IEEE Transactions on Computer-Aided Design of Integrated Circuits and Systems*, Vol. 20, No. 7, July 2001, pp. 877–887.

[25] J. A. Davis and J. D. Meindl, "Compact distributed RLC interconnect models—Part I: Single line transient, time delay, and overshoot expressions," *IEEE Transactions on Electron Devices*, Vol. 47, No. 11, November 2000, pp. 2068–2077.

[26] J. A. Davis and J. D. Meindl, "Compact distributed RLC interconnect models—Part II: Coupled line transient expressions and peak crosstalk in multilevel networks," *IEEE Transactions on Electron Devices*, Vol. 47, No. 11, November 2000, pp. 2078–2087.

[27] Y. I. Ismail and E. G. Friedman, *On-chip Inductance in High Speed Integrated Circuits*, Kluwer Academic Publishers, Boston, 2001.

[28] A. Deutsch et al., "Functional high-speed characterization and modeling of a six-layer copper wiring structure and performance comparison with aluminum on-chip interconnections, in *Proceedings of the IEEE International Electron Devices Meeting*, December 1998, pp. 295–298.

[29] A. Deutsch et al., "Design guidelines for short, medium, and long on-chip interconnections," in *Proceedings of the IEEE Topical Meeting on Electrical Performance of Electronic Packaging*, October 1996, pp. 30–32.

[30] S. V. Morton, "On-chip inductance issues in multiconductor systems," in *Proceedings of the IEEE/ACM Design Automation Conference*, October 1999, pp. 921, 926.

[31] A. Deutsch et al., "The importance of inductance and inductive coupling for on-chip design guidelines for short, medium, and long on-chip interconnections," in *Proceedings of the IEEE Topical Meeting on Electrical Performance of Electronic Packaging*, October 1996, pp. 30–32.

[32] I. Catt, "Crosstalk (noise) in digital systems," *IEEE Transactions on Electronic Computers*, Vol. 16, No. 6, December 1967, pp. 743–763.

[33] T. Sakurai, "Closed-form expressions for interconnection delay, coupling, and crosstalk in VLSI's," *IEEE Transactions on Electron Devices*, Vol. 40, No. 1, January 1993, pp. 118–124.

[34] K. T. Tang and E. G. Friedman, "Interconnect coupling noise in CMOS VLSI circuits," in *Proceedings of the ACM/IEEE International Symposium on Physical Design*, April 1999, pp. 48–53.

[35] L. Bisduonis, S. Nikolaidis, O. Koufopavlou and C. E. Goutis, "Modeling the CMOS short-circuit power dissipation," in *Proceedings of the IEEE International Symposium on Circuits and Systems*, Vol. 4, May 1996, pp. 469–472.

[36] H. J. M. Veendrick, "Short-circuit dissipation of static CMOS circuitry and its impact on the design of buffer circuits," *IEEE Journal of Solid-State Circuits*, Vol. 19, No. 4, August 1984, pp. 468–473.

[37] S. R. Vemuru and N. Scheinberg, "Short-circuit power dissipation estimation for CMOS logic gates," *IEEE Transactions on Circuits and Systems I: Fundamental Theory and Application*, Vol. 41, No. 11, November 1994, pp. 762–766.

[38] A. M. Hill and S.-M. Kang, "Statistical estimation of short-circuit power in VLSI design," in *Proceedings of the IEEE International Symposium on Circuits and Systems*, Vol. 4, May 1996, pp. 105–108.

[39] A. Chanrakasan and R. W. Brodersen, "Minimizing power consumption in digital CMOS circuits," in *Proceedings of the IEEE*, Vol. 83, No. 4, April 1995.

[40] J. Kao, S. Narendra and A. Chandrakasan, "Subthreshold leakage modeling and reduction techniques," in *Proceedings of the IEEE/ACM International Conference on Computer-Aided Design*, November 2002, pp.141–148.

[41] R. Rao, A. Srivastava, D. Blaauw and D. Sylvester, "Statistical estimation of leakage current considering inter- and intra-die process variation," in *Proceedings of the International Symposium on Low Power Electronics and Design*, August 2003, pp. 84–89.

[42] S. Pullela, N. Menezes and L. T. Pillage, "Reliable non-zero skew clock trees using wire width optimization," in *Proceedings of the IEEE/ACM Design Automation Conference*, June 1998, pp. 165–170.

[43] S. Pullela, N. Menezes and L. T. Pillage, "Moment-sensitivity-based wire sizing for skew reduction in on-chip clock nets," *IEEE Transactions on Computer-Aided Design of Integrated Circuits and Systems*, Vol. 16, No. 2, February 1997, pp. 210–215.

[44] T. D. Hodes, B. A. McCoy and G. Robins, Dynamically wire-sized elmore-based routing constructions," in *Proceedings of the IEEE International Symposium on Circuits and Systems*, Vol. 1, May 1994, pp. 463–466.

[45] M. Edahiro, "Delay minimization for zero-skew routing," in *Proceedings of the IEEE International Conference on Computer-Aided Design*, November 1993, pp. 563, 566.

[46] S. S. Sapatnekar, "RC interconnect optimization under the elmore delay model," in *Proceedings of the IEEE/ACM Design Automation Conference*, June 1994, pp. 387, 391.

[47] J. J. Cong and K. Leung, "Optimal wiresizing under elmore delay model," *IEEE Transactions on Computer-Aided Design of Integrated Circuits and Systems*, Vol. 14, No. 3, March 1995, pp. 321–336.

[48] S. Pullela, N. Menezes, L. T. Pillage, "Simultaneous gate and interconnect sizing for circuit-level delay optimization," in *Proceedings of the IEEE/ACM Design Automation Conference*, June 1995, pp. 690–695.

[49] J. J. Cong and C.-K. Koh, "Simultaneous driver and wire sizing for performance and power optimization," *IEEE Transactions of Very Large Scale Integration (VLSI) Systems*, Vol. 2, No. 4, December 1994, pp. 408–425.

[50] C. P. Chen and N. Menezes, "Spec-based repeater insertion and wire sizing for on-chip interconnect," in *Proceedings of the IEEE International Conference on VLSI Design*, January 1999, pp. 476-483.

[51] W. C. Elmore, "The transient response of damped linear networks with particular regard to wideband amplifiers," *Journal of Applied Physics*, Vol. 19, No. 1, January 1948, pp. 55–63.

[52] M. A. El-Moursy and E. G. Friedman, "Optimizing inductive interconnect for low power," *System-on-Chip for Real-Time Applications*, W. Badawy and G. A. Jullien (Eds.), Kluwer Academic Publishers, 2003, pp. 380–391.

[53] M. A. El-Moursy and E. G. Friedman, "Power characteristics of inductive interconnect," *IEEE Transactions on Very Large Scale Integration (VLSI) Systems*, Vol. 12, No. 12, December 2004, pp. 1295–1306.

[54] C. M. Lee and H. Soukup, "An algorithm for CMOS timing and area optimization," *IEEE Journal of Solid State Circuits*, Vol. 19, No. 5, October 1984, pp. 781–787.

[55] E. T. Lewis, "Optimization of device area and overall delay for CMOS VLSI designs," in *Proceedings of the IEEE*, Vol. 72, No. 5, June 1984, pp. 670–689.

[56] J. Yuan and C. Svensson, "Principle of CMOS circuit power-delay optimization with transistor sizing," in *Proceedings of the IEEE International Symposium of Circuits and Systems*, May 1996, pp. 637–640.

[57] M. Borah, R. M. Owens and M. J. Irwin, "Transistor sizing for low power CMOS circuits," *IEEE Transactions on Computer-Aided Design of Integrated Circuits and Systems*, Vol. 15, No. 6, June 1996, pp. 665–671.

[58] R. Rogenmoser and H. Kaeslin, "The impact of transistor sizing on power efficiency in submicron CMOS circuits," *IEEE Journal of Solid-State Circuits*, Vol. 32, No. 7, July 1997, pp. 1142–1145.

[59] J. P. Fishburn and S. Taneja, "Transistor sizing for high performance and low power," in *Proceedings of the IEEE Custom Integrated Circuits Conference*, May 1997, pp. 591–594.

[60] A. R. Conn et al., "Optimization of custom MOS circuits by transistor sizing," in *Proceedings of the IEEE/ACM International Conference on Computer-Aided Design*, November 1996, pp. 174, 180.

[61] T. Xiao and M. Marek-Sadowska, "Crosstalk reduction by transistor sizing," in *Proceedings of the Asia and Pacific Design Automation Conference*, January 1999, pp. 137–150.

[62] A. Vittal, L. H. Chen, M. Marek-Sadowska, K.-P. Wang and S. Yang, "Crosstalk in VLSI interconnection," *IEEE Transactions on Computer-Aided Design of Integrated Circuits and Systems*, Vol. 18, No. 12, December 1999, pp. 1817–1824.

[63] J. Cong, L. He, C.-K. Koh and P. H. Madden, "Performance optimization of VLSI interconnect layout," *Integration, The VLSI Journal*, Vol. 21, No. 1/2, November 1996, pp. 1–94.

[64] C. Tretz and C. Zukowski, "CMOS transistor sizing minimization of energy-delay product," in *Proceedings of the IEEE Great Lakes Symposium on VLSI*, March 1996, pp. 168–173.

[65] H. C. Lin and L. W. Linholm, "An optimized output stage for MOS integrated circuits," *IEEE Journal of Solid-State Circuits*, Vol. 10, No. 2, April 1975, pp. 106–109.

[66] R. C. Jaeger, "Comments on 'an optimized output stage for mos integrated circuits'," *IEEE Journal of Solid-State Circuits*, Vol. 10, No. 3, June 1975, pp. 185–186.

[67] A. Kanuma, "CMOS circuit optimization," *Solid-State Electronics*, Vol. 26, No. 1, January 1983, pp. 47–58.

[68] M. Nemes, "Driving large capacitances in MOS LSI systems," *IEEE Journal of Solid-State Circuits*, Vol. 19, No. 1, February 1984, pp. 159–161.

[69] T. Sakurai, "A unified theory for mixed CMOS/BiCMOS buffer optimization," *IEEE Journal of Solid-State Circuits*, Vol. 27, No. 7, July 1992, pp. 1014–1019.

[70] N. C. Li, G. L. Haviland and A. A. Tuszynski, "CMOS tapered buffer," *IEEE Journal of Solid-State Circuits*, Vol. 25, No. 4, August 1990, pp. 1005–1008.

[71] C. Prunty and L. Gal, "Optimum tapered buffer," *IEEE Journal of Solid-State Circuits*, Vol. 27, No. 1, January 1992, pp. 118–119.

[72] N. Hedenstiera and K. O. Jeppson, "CMOS circuit speed and buffer optimization," *IEEE Transactions on Computer-Aided Design of Integrated Circuits and Systems*, Vol. 6, No. 2, March 1987, pp. 270–281.

[73] B. S. Cherkauer and E. G. Friedman, "A unified design methodology for CMOS tapered buffers," *IEEE Transactions on Very Large Scale Integration (VLSI) Systems*, Vol. 3, No. 1, March 1995, pp. 99–111.

[74] J.-S. Choi and K. Lee, "Design of CMOS tapered buffer for minimum power-delay product," *IEEE Journal of Solid-State Circuits*, Vol. 29, No. 9, September 1994, pp. 1142–1145.

[75] S. R. Vemuru and A. R. Thorbjornsen, "Variable-taper CMOS buffer," *IEEE Journal of Solid-State Circuits*, Vol. 26, No. 9, September 1991, pp. 1265–1269.

[76] G. Chen and E. G. Friedman, "Low-power repeaters driving RC and RLC interconnects with delay and bandwidth constraints," *IEEE Transactions on Very Large Scale Integration (VLSI) Systems*, Vol. 14, No. 2, February 2006, pp. 161–172.

[77] C. Y. Wu and M. Shiau, "Delay models and speed improvement techniques for RC tree interconnections among small-geometry CMOS inverters," *IEEE Journal of Solid-State Circuits*, Vol. 25, No. 10, October 1990, pp. 1247–1256.

[78] M. Nekili and Y. Savaria, "Optimal methods of driving interconnections in VLSI circuits," in *Proceedings of the IEEE International Symposium on Circuits and Systems*, May 1992, pp. 21–23.

[79] M. Nekili and Y. Savaria, "Parallel regeneration of interconnections in VLSI & ULSI circuits," in *Proceedings of the IEEE International Symposium on Circuits and Systems*, May 1992, pp. 2023–2026.

[80] S. Dhar and M. A. Franklin, "Optimum buffer circuits for driving long uniform lines," *IEEE Journal of Solid-State Circuits*, Vol. 26, No. 1, January 1991, pp. 32–40.

[81] C. J. Alpert, "Wire segmenting for improved buffer insertion," in *Proceedings of the IEEE/ACM Design Automation Conference*, June 1997, pp. 588–593.

[82] V. Adler and E. G. Friedman, "Repeater design to reduce delay and power in resistive interconnect," *IEEE Transactions on Circuits and Systems II: Analog and Digital Signal Processing*, Vol. 45, No. 5, May 1998, pp. 607–616.

[83] H. B. Bakoglu and J. D. Meindl, "Optimal interconnection circuits for VLSI," *IEEE Transactions on Electron Devices*, Vol. 32, No. 5, May 1985, pp. 903–909.

[84] C. J. Alpert, A. Devgan, J. P. Fishburn and S. T. Quay, "Interconnect synthesis without wire tapering," *IEEE Transactions on Computer-Aided Design of Integrated Circuits and Systems*, Vol. 20, No. 1, January 2001, pp. 90–104.

[85] M. A. El-Moursy and E. G. Friedman, "Optimum wire sizing of RLC interconnect with repeaters," *Integration, the VLSI Journal*, Vol. 38, 2004, pp. 205–225.

[86] Y. I. Ismail and E. G. Friedman, "Effects of inductance on the propagation delay and repeater insertion in VLSI circuits," *IEEE Transactions on Very Large Scale Integration (VLSI) Systems*, Vol. 8, No. 2, April 2000, pp. 195–206.

[87] M.A. El-Moursy and E. G. Friedman, "Optimum wire sizing of RLC interconnect with repeaters," in *Proceedings of the IEEE Great Lakes Symposium on VLSI*, April 2003, pp. 27–32.

[88] K. T. Tang and E. G. Friedman, "Delay uncertainty due to on-chip simultaneous switching noise in high performance CMOS integrated circuits," in *Proceedings of the IEEE Workshop on Signal Processing Systems*, October 2000, pp. 633–642.

[89] K. T. Tang and E. G. Friedman, "Incorporating voltage fluctuations of the power distribution network into the transient analysis of CMOS logic gates," *Analog Integrated Circuits and Signal Processing*, Vol. 31, No. 3, June 2002, pp. 249–259.

[90] M. Saint-Laurent and M. Swaminathan, "Impact of power supply noise on timing in high-frequency microprocessors," in *Proceedings of the IEEE Topical Meeting on Electrical Performance of Electronic Packaging*, October 2002, pp. 261–264.

[91] A. Waizman and C.-Y. Chung, "Package capacitor impact on microprocessor maximum operating frequency," in *Proceedings of the IEEE Electronic Components and Technology Conference*, June 2001, pp. 118–122.

[92] J. J. Clement, "Electromigration reliability," *Design of High Performance Microprocessor Circuits*, A. Chandrakasan, W. Bowhill and F. Fox (Eds.), IEEE Press, New York, 2001, pp. 429–448, Chapter 20.

[93] I. A. Blech and H. Sello, *Mass Transport of Aluminum by Moment Exchange with Conducting Electrons*, Vol. 5, 1966, pp. 496–505. USAF-RADC Series

[94] J. R. Black, "Mass transport of aluminum by moment exchange with conducting electrons," in *Proceedings of the IEEE International Reliability Physics Symposium*, April 1967, pp. 148–159.

[95] E. G. Friedman, "Clock distribution networks in synchronous digital integrated circuits," in *Proceedings of the IEEE*, Vol. 89, No. 5, May 2001, pp. 665–690.

[96] E. G. Friedman and J. H. Mulligan Jr., "Clock frequency and latency in synchronous digital systems," *IEEE Transactions on Signal Processing*, Vol. 39, No. 4, April 1991, pp. 930–934.

[97] S. Y. Kung, *VLSI Array Processors*, Prentice-Hall, Englewood Cliffs, NJ, 1988.

[98] M. Nekili, Y. Savaria, G. Bois and M. Bennani, "Logic-based H-trees for large VLSI processor arrays: A novel skew modeling and high-speed clocking method," in *Proceedings of the International Conference on Microelectronics*, December 1993, pp. 1–4.

[99] H. B. Bakoglu, J. T. Walker and J. D. Meindl, "A symmetric clock-distribution tree and optimized high-speed interconnections for reduced clock skew in ULSI and WSI circuits," in *Proceedings of the IEEE International Conference on Computer Design*, October 1996, pp. 118–122.

[100] H. Kojima, S. Tanaka and K. Sasaki, "Half-swing clocking scheme for 75% power saving in clocking circuitry," in *Proceedings of the IEEE Symposium on VLSI Circuits*, June 1994, pp. 23–24.

[101] D. W. Dobberpuhl et al., "A 200-MHz 65-b dual-issue CMOS RISC microprocessor," *IEEE Journal on Solid-State Circuits*, Vol. 27, November 1992, pp. 1555–1565.

[102] J. L. Neves and E. G. Friedman, "Minimizing power dissipation in nonzero skew-based clock distribution networks," in *Proceedings of the IEEE International Symposium on Circuits and Systems*, May 1995, pp. 1576–1579.

[103] E. De Man and M. Schobinger, "Power dissipation in the clock system of highly pipelined ULSI CMOS circuits," in *Proceedings of the International Workshop on Low Power Design*, April 1994, pp. 133–138.

[104] M. Hatamian and G. L. Cash, "Parallel bit-level pipelined VLSI designs for high-speed signal processing," in *Proceedings of the IEEE*, Vol. 75, September 1987, pp. 1192–1202.

[105] M. Hatamian, L. A. Hornak, T. E. Little, S. T. Tewksbury and P. Franzon, "Fundamental interconnection issues," *AT&T Technical Journal*, Vol. 66, July/August 1987, pp. 13–30.

[106] J. P. Fishburn, "Clock skew optimization," *IEEE Transactions on Computation*, Vol. 39, July 1990, pp. 945–951.

[107] E. G. Friedman, "Performance limitations in synchronous digital systems," Ph.D. dissertation, University of California, Irvine, June 1989.

[108] M. Hatamian, "Understanding clock skew in synchronous systems," *Concurrent Computations (Algorithms, Architectures and Technology)*, S. K. Tewksbury, B. W. Dickinson and S. C. Schwartz (Eds.), Plenum, New York, 1988, pp. 87–96.

[109] J.W. Joyner et al., "Impact of three-dimensional architectures on interconnects in gigascale integration," *IEEE Transactions on Very Large Scale Integration (VLSI) Systems*, Vol. 9, No. 6, December 2001, pp. 922–928.

[110] J. W. Joyner and J. D. Meindl, "Opportunities for reduced power distribution using three-dimensional integration," in *Proceedings of the IEEE International Interconnect Technology Conference*, June 2002, pp. 148–150.

[111] W. E. Donath, "Placement and average interconnection lengths of computer logic," *IEEE Transactions on Circuits and Systems*, Vol. 26, No. 4, April 1999, pp. 272–277.

[112] D. Stroobandt, *A Priori Wire Length Estimates for Digital Design*, Kluwer Academic Publishers, Netherlands, 2001.

[113] B. Kim et al., "Factors affecting copper filling process within high aspect ratio deep vias for 3D chip stacking," in *Proceedings of the IEEE International Electronic Components and Technology Conference*, June 2006, pp. 838–843.

[114] M. W. Newman et al., "Fabrication and electrical characterization of 3D vertical interconnects," in *Proceedings of the IEEE International Electronic Components and Technology Conference*, June 2006, pp. 394–398.

[115] N. T. Nguyen et al., "Through-wafer copper electroplating for three-dimensional interconnects," *Journal of Micromechanics and Microengineering*, Vol. 12, No. 4, July 2002, pp. 395–399.

[116] C. S. Premachandran et al., "A vertical wafer level packaging using through hole filled via interconnect by lift-off polymer method for MEMS and 3D stacking applications," in *Proceedings of the IEEE International Electronic Components and Technology Conference*, June 2005, pp. 1094–1098.

[117] V. F. Pavlidis, *Interconnect-Based Design Methodologies for Three-Dimensional Integrated Circuits*, Ph.D. Dissertation, University of Rochester, Rochester, New York, June 2008.

CHAPTER

Networks-On-Chip

12

Over the past few years, the idea of using networks-on-chip (NoCs) as viable on-chip communication fabrics for future multiprocessor systems-on-chips (MPSoCs) has been gaining traction. NoCs are an attempt to scale down the concepts of large-scale networks, and apply them to the embedded system-on-chip (SoC) domain. Unlike traditional bus-based on-chip communication architectures, NoCs use packets to route data from the source to the destination component, via a network fabric that consists of switches (routers) and interconnection links (wires). An example of a NoC system is illustrated in Fig. 12.1. The figure shows an NoC interconnection architecture with a mesh type topology, consisting of several processing elements (*PEs*) connected together via routers and regular sized wires. A *PE* (also referred to as a *node*) in this case can be any component such as a microprocessor, application-specific integrated circuit (ASIC) block or memory, or a combination of components connected together. A *network interface* (*NI*) at the boundary of each *PE* is used to packetize any data generated by the *PE*. This NI is connected to a router, which has buffers at its input to accept data packets from a *PE* or from other routers connected to it. A *crossbar switch* inside the router is used to route the data packets from the input *buffers* to the appropriate output link, based on the address in the packet header. An *arbiter* component is used to determine which packets

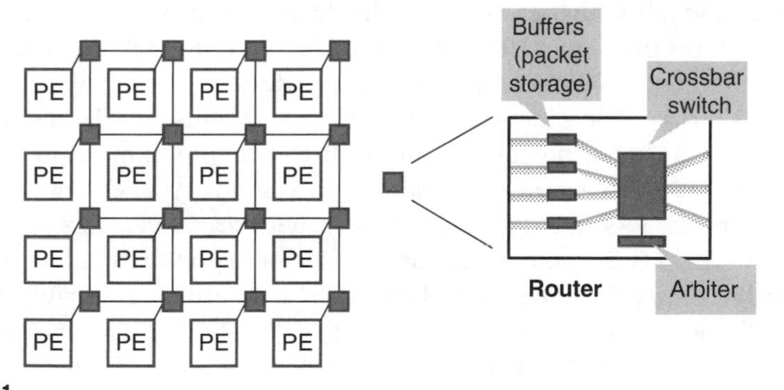

FIGURE 12.1

An example of a (mesh type) NoC interconnection fabric

get priority when multiple packets from different sources are vying for transfer on the same interconnection link. Thus packets traverse multiple links, and go through one or more routers in the NoC network as they move from the source to the destination. Note that this example is intended to show how NoCs are different from traditional bus-based architectures, but not all NoCs follow the structure shown in this example, as will be shown later. However, the idea of packet switched networks to interconnect components together as shown here is a good general representation of a typical NoC architecture.

There are two major factors that have motivated the growing interest in NoCs. First and foremost, it is an indisputable fact that electronic systems are becoming increasingly complex. In the 1990s and early in this decade, SoC designs consisted of a handful of components—typically a microprocessor or a digital signal processor (DSP), some memory, and interfaces to the external world. This has changed rapidly over the past few years, with yesterday's SoC designs evolving into multiprocessor SoC designs today, having tens to hundreds of components including several processing engines (e.g., IBM's cell chip [1]). It is envisioned that tomorrow's complex systems will have hundreds of components that will communicate on interconnects operating in the gigahertz (GHz) frequency range. In such a scenario, there is a need for a communication fabric that is *scalable* enough to handle the increasing performance requirements of such demanding systems. The communication architecture in such systems must also be able to support the *quality of service* (QoS) needs of heterogeneous systems that will require multiple modes of operations and with varying levels of real-time response requirements. NoCs have the potential to meet the scalability and QoS requirements of such complex systems. Secondly, integrated circuit technology has been advancing rapidly in the last few years. The shrinking of process technology into the deep submicron (DSM) domain has resulted in the emergence of key design problems that are challenging designers to seek novel solutions. A few of these key challenges affecting DSM technologies (i.e., below 90 nm) are the large undesirable wire delay on long interconnects due to inductive and capacitive coupling effects; greater probability of transmission errors due to factors such as crosstalk noise, electromagnetic interference (EMI), and synchronization failures; and non-determinism due to process/die/temperature variations. These DSM effects necessitate a communication fabric that can enable reliable and deterministic communication between the cores on a chip. The NoC fabric has the potential to overcome many of these effects, due to its inherent structure and design.

NoCs differ from wide-area networks because they must handle more stringent constraints, such as lower power consumption, lower communication latency, and a smaller area footprint for on-chip electronic systems. At the same time, NoCs must exhibit much less non-determinism than wide-area networks. Despite DSM variations in CMOS (complementary metal-oxide semiconductor) technologies, it is still possible to predict physical and electrical properties reasonably accurately for NoCs [2]. Figure 12.2 depicts a mapping of the NoC interconnection fabric onto a standard ISO/OSI network protocol stack model. We will give a high level overview of NoCs in the context of this mapping, starting from the bottom of the layer stack, and moving up.

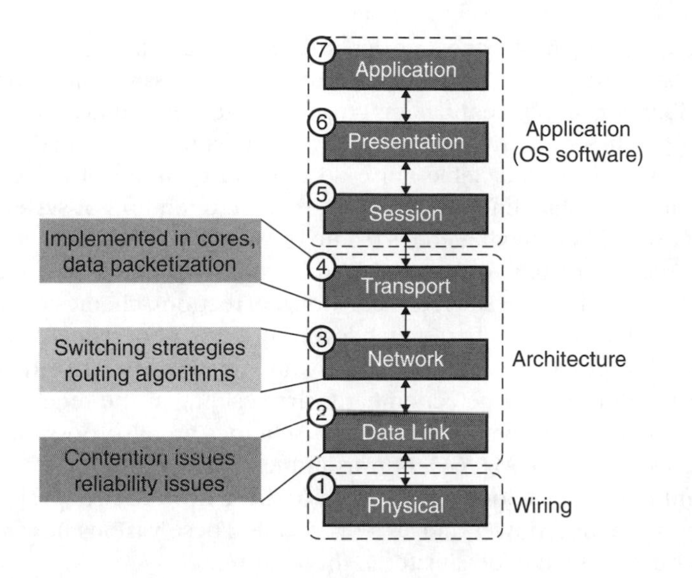

FIGURE 12.2

NoCs in the context of the 7 layer ISO/OSI protocol stack model

Physical layer: This is the lowest level of the protocol stack, and is concerned with the physical wiring that is the backbone of the NoC fabric. Bus-based architectures suffer from long global interconnect lengths that require several clock cycles to traverse in DSM technologies where wire delay is increasingly dominating gate delay. Pipelining or inserting buffers/registers can be used to overcome these large delays on long interconnects. Pipelining divides a long interconnect into shorter sized segments that are typically traversed in a single clock cycle by the data on a bus. However, resource utilization in such a case is usually inefficient because the whole bus is in a busy state (and vacant wire segments typically cannot be used by other masters) until a data transfer is completed. Pipelining for long interconnects is also normally done in an ad hoc manner much later in the design flow, usually during the place-and-route stage, and can be very time consuming. NoC fabrics address this problem of excessive wire delay on long interconnects by using shorter, regular sized wires (links) that are separated by switches (routers). Data packets typically traverse these links (also called switch-to-switch links) in a single cycle, and are then buffered in the routers, before being routed to another link in the subsequent cycle. Thus NoCs are designed from the very beginning to address wire delay in a more structured manner, which eases designer effort and reduces overall design time.

Data link layer: The data link layer can be divided into two sublayers—a media access control (MAC) sublayer that is closest to the physical channel, and the data link control (DLC) sublayer that sits above the MAC sublayer. The MAC is responsible for regulating access to the physical medium, and resolve contention issues using some form of arbitration. In bus-based architectures, arbitration for resolving contention on a bus is typically implemented in a centralized manner, within a separate arbiter module. However, in NoCs, arbitration is inherently distributed, occurring at each router. The router makes decisions about which packets to give priority to and transfer in the next clock cycle, when multiple data packets in the router contend for transfer in the same direction. The DLC is responsible for implementing an error-tolerant communication scheme, to ensure that glitches occurring due to factors such as process variations, manufacturing defects, and unpredictable noise do not cause failure of the chip. One of the most widely used techniques to ensure reliable operation of systems is to use error detecting and correcting codes (some of which are discussed in Chapter 7) which add redundant bits to the original data that help detect and rectify errors in the transmitted data. These error detection/correction schemes can be applied either at the link level (switch-to-switch level) by adding error detection hardware at each switch input, or in an end-to-end manner, where the error detection/correction occurs in the error detection hardware residing at the receiver end point. Hybrid techniques that combine these two schemes are also possible. Flow control protocols such as STALL/GO, T-Error, and ACK/NACK are used to implement fault tolerant communication in NoCs. Each of these flow control protocols entails different performance, power, and area overheads. These various flow control protocols will be discussed in detail later in this chapter.

Network and transport layers: The network layer is responsible for switching and routing of data packets through the NoC fabric. Switching techniques are used to establish the type of connection, whereas routing schemes determine the path followed by a message as it traverses the network from the source to its destination. Later in this chapter, we describe various packet switching techniques such as store and forward (SAF), virtual cut through (VCT), and wormhole (WH). Routing techniques such as *deterministic* (that define the same path between a source–destination pair) and *adaptive* (that use traffic and network condition information to dynamically create a path that avoids congestion) will also be discussed in detail. The main goal of the transport layer is to decompose messages into packets at the source and then assemble the packets back into messages at the destination. The granularity of packetization is an important design decision, since packet size can have a critical impact on the behavior of network control algorithms that affect the performance, power, and area footprint of the NoC fabric.

Session, presentation, and application layers: These "software" layers are made up of the *system software*, which provides an abstraction of the underlying hardware platform, and the *application software*, which leverages the hardware abstraction to effectively exploit the capability of the hardware. The system software is tightly coupled with the NIs that are responsible for connecting PEs to the NoC fabric. The NIs implement some or all of the functionality of the lower layers (e.g., data packetization, flow control, etc.) in the protocol stack.

The adoption of networks-on-chip (NoCs) is an evolutionary process, as shown in Fig. 12.3, the time frame for which will be determined by several factors such as the rate of increasing design complexity and how rapidly process technology will continue to advance, in the coming years. The advanced bus-based communication architectures being adopted for use today (such as the crossbar bus or bus matrix) are not far removed from some types of NoCs that have been proposed (such as the crossbar NoC), and consequently it is more accurate to consider NoCs as an evolutionary paradigm, rather than a revolutionary one. While

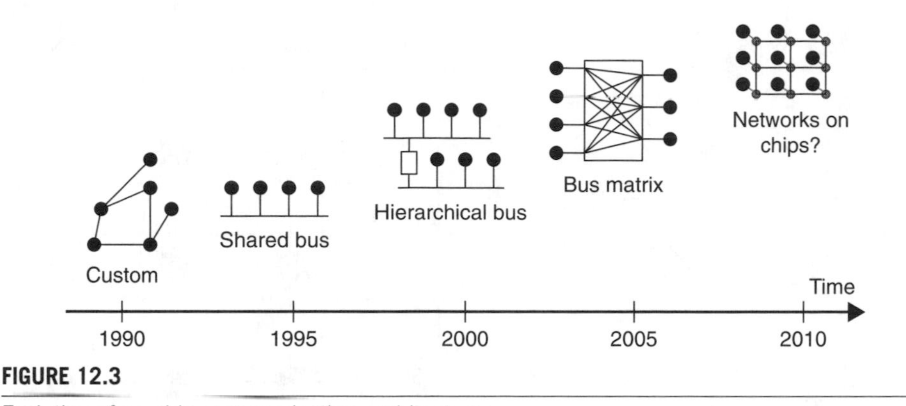

FIGURE 12.3

Evolution of on-chip communication architectures

the communication design space for NoC type architectures is much larger and more complex than for bus-based communication architectures, the design problems remain the same, requiring a trade-off between performance, power, cost, area, and reliability to create a communication fabric that can meet the requirements of a given application.

There are several variants of NoCs that have been proposed to date. An understanding of these proposed network fabrics requires an elaboration on the key characteristics that make NoCs unique. There are several characteristics that define an NoC, and allow a differentiation between the various NoC architectures that have been proposed. In the next few sections, we elaborate on these characteristics, which include the NoC topology (Section 12.1), switching strategies (Section 12.2), routing algorithms (Section 12.3), flow control schemes (Section 12.4), clocking schemes (Section 12.5), and QoS (Section 12.6). After describing these defining attributes of NoCs, a few examples of NoC architectures that have been proposed in literature are presented in Section 12.7. Finally, the status of current research into NoCs, and a discussion of a few of the open problems that still remain to be resolved are presented in Section 12.8.

12.1 NETWORK TOPOLOGY

The topology of an NoC specifies the physical organization of the interconnection network. It defines how nodes, switches, and links are connected to each other. Topologies for NoCs can be classified into three broad categories [3]—direct networks, indirect networks, and irregular networks. These are described below.

12.1.1 Direct Networks

In direct network topologies, each node has direct point-to-point links to a subset of other nodes in the system called *neighboring* nodes. The nodes consist of computational blocks and/or memories, as well as a NI block that acts as a router. This router is connected to the routers of the neighboring nodes through links

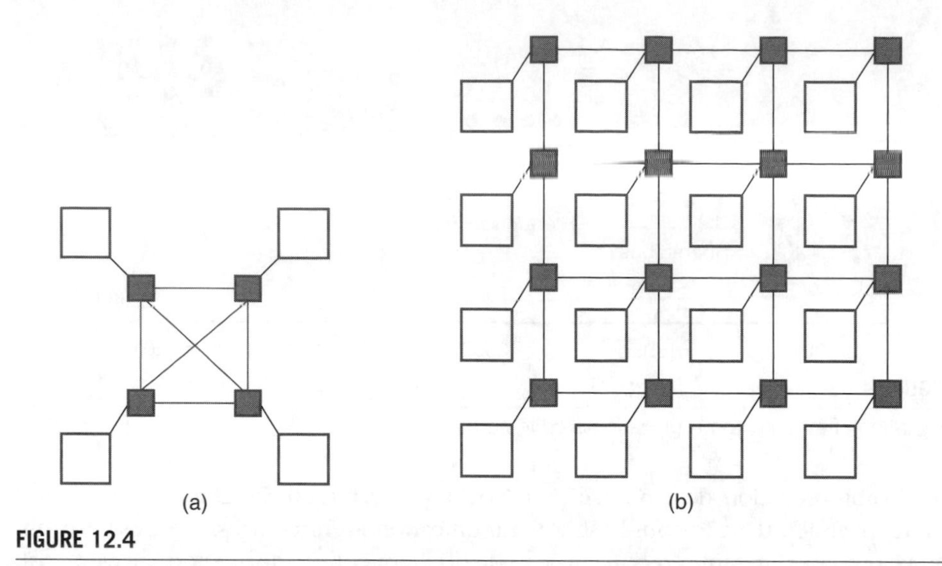

FIGURE 12.4

Examples of direct network topologies: (a) point-to-point. (b) mesh

(also referred to as channels). The direct network topology has the property that as the number of nodes in the system increases, the total available communication bandwidth also increases. A fundamental trade-off in direct network design is between connectivity and cost. Higher connectivity results in higher performance, but has greater energy and area costs for the router and link implementations. Thus a fully connected direct network topology, as shown in Fig. 12.4(a), where every node is directly connected to all the other nodes is quite prohibitive. Therefore, most practical implementations of direct networks require messages to traverse several intermediate nodes before reaching their destinations.

Most direct network topologies have an orthogonal implementation, where the nodes can be arranged in an *n*-dimensional orthogonal space, in such a way that every link produces a displacement in a single direction. Routing for such networks is fairly simple and can be implemented efficiently in hardware. Examples of popular orthogonal direct networks include the *n*-dimensional mesh, torus, folded torus, hypercube, and octagon topologies. The 2-D mesh (or grid) topology shown in Fig. 12.4(b) is one of the more popular NoC topologies because all links have the same length, which eases physical design. Every node in a 2-D mesh is connected to four neighboring nodes, except for the nodes at the edges. The area of a mesh grows linearly with the number of nodes. Meshes must also be designed in such a way as to avoid traffic accumulating in the center of the mesh, which reduces performance. The torus topology, also called a *k*-ary *n*-cube, is an *n*-dimensional grid with *k* nodes in each dimension. A *k*-ary 1-cube (1-D torus) is essentially a ring network with *k* nodes. Figure 12.4(c) shows an example of a 4-ary 1-cube, which is a ring network with $k = 4$ nodes. This is one of the simplest NoC topologies, but has limited scalability since performance decreases when more nodes are added. Figure 12.4(d) shows an example of the 4-ary 2-cube (i.e., 4×4, 2-D) torus. The 2-D torus topology is basically the same as a regular mesh with the only difference being that the nodes at the edges are connected to the switches at the opposite edge via

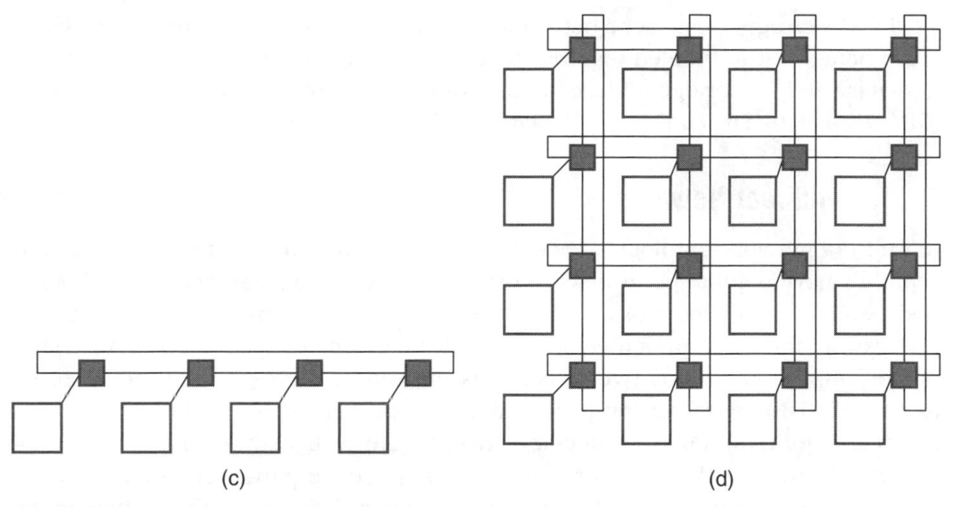

FIGURE 12.4

(c) Ring. (d) Torus

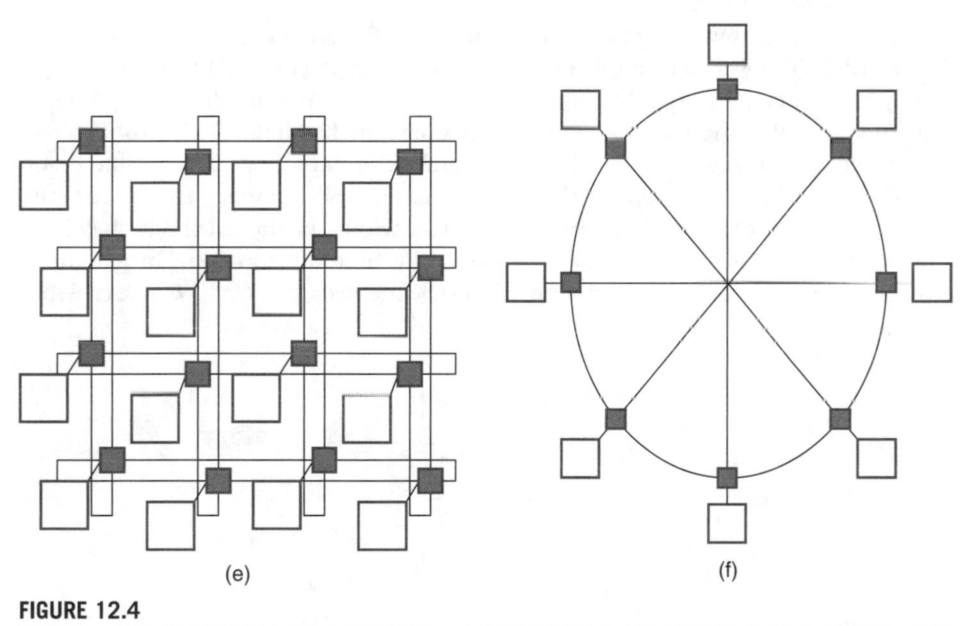

FIGURE 12.4

(e) Folded torus. (f) Octagor

wrap-around channels. Every node in this topology is connected to four neighboring nodes. The long end-around connections can, however, lead to excessive delays. This limitation is overcome by using a folding torus as shown in Fig. 12.4(e), where links have the same size. A k-ary n-cube with $k = 2$ is an n-dimensional cube, and is commonly referred to as a hypercube. Meshes and tori can be extended by adding bypass links to increase performance at the cost of higher area. The resulting topology is referred to as an express cube [4]. Finally, the octagon topology as shown in Fig. 12.4(f) is another example of a direct network, and has 8 nodes and

12 links. Messages being sent between any two nodes require at most two hops. More octagons can be tiled together to accommodate larger designs, if one of the nodes is used as a bridge node. Examples of direct networks include the Nostrum [5], SOCBUS [6], Proteo [7], and Octagon [8] NoC architectures.

12.1.2 Indirect Networks

In indirect network topologies, each node is connected to an external switch, and switches have point-to-point links to other switches. The NI associated with each node connects to a port of a switch. Switches do not perform any information processing, and correspondingly nodes do not perform any packet switching. One of the simplest indirect networks is a crossbar, where each PE node is connected to any other PE by traversing just a single switch. Similar to the fully connected direct network, the crossbar does not really scale well with an increasing number of PEs. Some techniques for realizing cost effective partial crossbars that are much smaller and consume less energy as compared to full crossbars have been proposed [9]. Multi-stage indirect networks can be realized by cascading many smaller crossbars together.

Figure 12.5 shows examples of popular multi-stage indirect networks [10]. Figure 12.5(a) shows an example of the fat-tree topology [11, 12]. In this tree topology, nodes are connected only to the leaves of the tree. For a simple-tree topology, it was observed that the root and its neighbors have higher traffic. This problem can be alleviated by using fat trees, where links among adjacent switches are increased as they get closer to the root of the tree. Increasing the number of links near the root of the tree essentially allocates more bandwidth on the channels that have higher traffic. Figure 12.5(b) shows a 2-ary, 3-fly butterfly topology. In general, a k-ary, n-fly butterfly network consists of k^n nodes, and n stages of k^{n-1} $k \times k$ crossbar

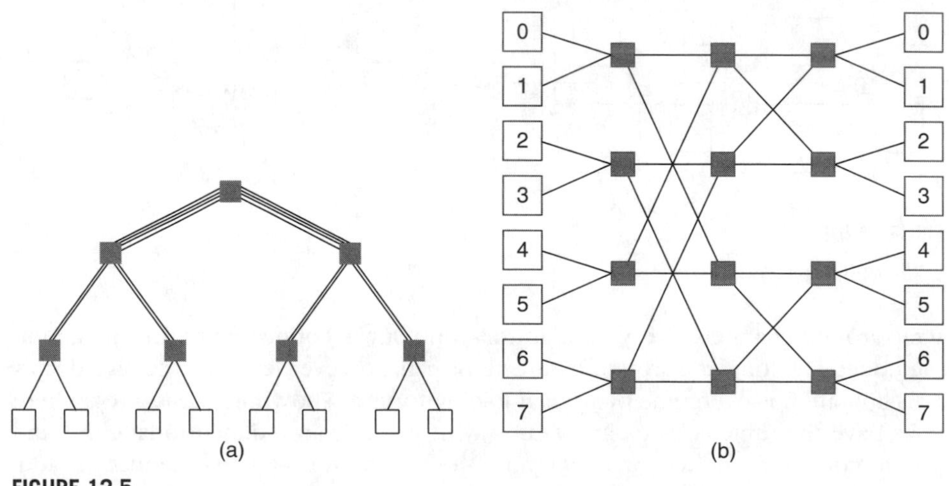

FIGURE 12.5

Examples of indirect network topologies: (a) fat tree; (b) butterfly switches. Note that the source and destination nodes, although shown separately for clarity, are the same (i.e., source node '1' is the same as destination node "1," etc.). Thus there are k^n nodes in a k-ary n-fly network. The butterfly network is a blocking multi-stage network, which implies that information may be temporarily blocked or dropped in the network if contention occurs. Figure 12.5(c) shows a (3, 3, 4) Clos topology. A Clos network topology consists of a three-stage network in which each stage is made up of a number of crossbar switches. A symmetric Clos is characterized by a triple (m, n, r), where m is the number of middle-stage switches, n is the number of input/output nodes on each input/output switch, and r is the number of input and output switches. The Clos network is an example of a non-blocking network, which is expensive because it consists of several full crossbars, but consequently also supports higher performance due to large available bandwidth. Figure 12.5(d) shows a (2, 2, 4) rearrangeable Clos network topology constructed using two (2, 2, 2) Clos networks with 4×4 middle switches. Such a Clos network topology composed of 2×2 switches is referred to as the Benes network topology. The Benes network is an example of a rearrangeable network in which paths may have to be rearranged to provide a connection, requiring an appropriate controller.

The SPIN [13] NoC architecture is an example of an indirect network.

12.1.3 Irregular Networks

Irregular or ad hoc network topologies are usually a mix of shared bus, direct, and indirect network topologies. The goal of these topologies is to increase available

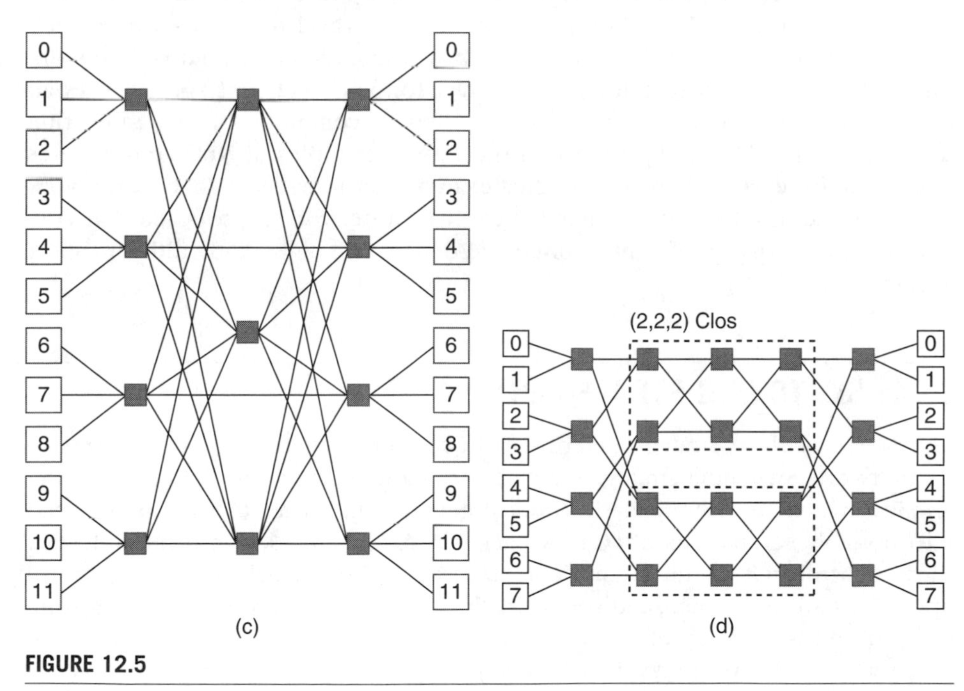

FIGURE 12.5

(c) Clos. (d) Benes

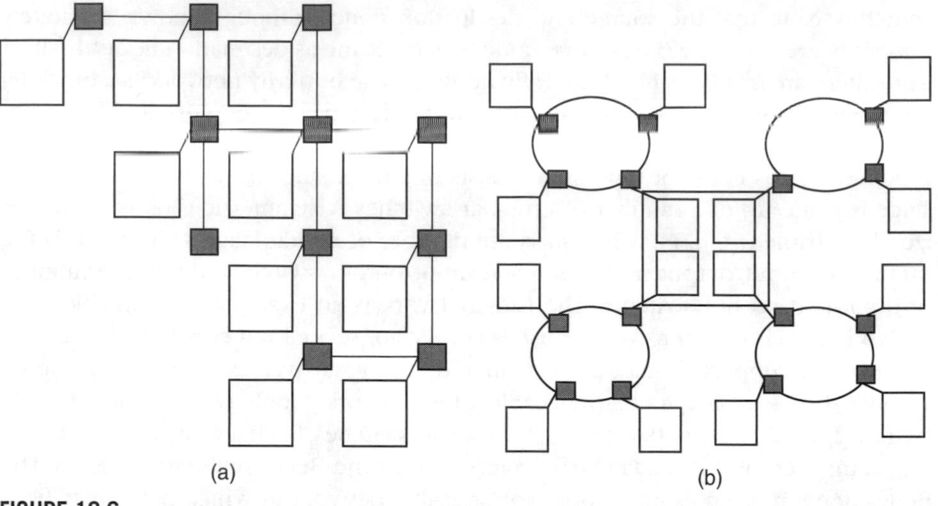

FIGURE 12.6

Examples of irregular network topologies: (a) optimized (reduced) mesh, (b) cluster-based hybrid (mesh + ring) topology

bandwidth as compared to traditional shared buses, and reduce the distance between nodes as compared to direct and indirect networks. Irregular topologies are typically customized for an application. An example of an irregular network topology is a reduced mesh in which unnecessary routers and links have been removed as shown in Fig. 12.6(a). A cluster-based hybrid network—where each cluster has any combination of a shared bus-based, direct, or indirect network topology—is another example of an irregular topology. Figure 12.6(b) shows an example of a cluster-based hybrid topology which combines a mesh and a ring topology. Xpipes [14] and Æthereal [15] are two examples of NoC architectures that allow irregular topologies. Another example of an irregular/hybrid network can be found in [70], where a hierarchical hybrid network is proposed that consists of several clusters of tightly connected nodes with high bandwidth and short intra-cluster links.

12.2 SWITCHING STRATEGIES

The NoC switching strategy determines how data flows through the routers in the network. Switching strategies define the granularity of data transfer and the applied switching technique. PEs (nodes) generate messages that are partitioned into possibly several data *packets*. A packet is further divided into multiple *flits* (flow control unit). A flit is an elementary packet on which link flow control operations are performed, and is essentially a synchronization unit between routers. Each flit is made up of one or more *phits* (physical units). A phit is a unit of data that is transferred on a link in a single cycle. The size of a phit is typically the width, in bits, of the communication link. Figure 12.7 shows the structure of

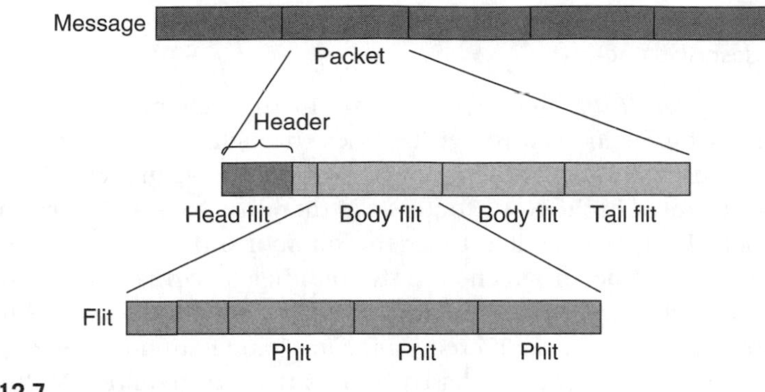

FIGURE 12.7

Structure of messages, packets, flits, and phits

phits, flits, packets, and messages. Different NoC architectures use different phit, flit, packet, and message sizes. The choice of size can have a significant impact on cost, performance, and power for NoC fabrics.

The two main modes of transporting flits in an NoC are circuit switching and packet switching. These are described below

12.2.1 Circuit Switching

In circuit switching, a physical path between the source and the destination is reserved prior to the transmission of data. The physical path is made up of a series of links and routers, and the messages from the sender are sent in their entirety to the receiver once a path (circuit) is reserved. A message header flit traverses the network from the source to the destination, reserving links along the way. If the header flit reaches the destination without any conflicts, all links in the path are available and an acknowledgement is sent back to the sender. Data transfer then commences for the sender upon receipt of the acknowledgement. If a link has been reserved by another circuit, however, a negative acknowledgement is sent back to the sender. The path is held until all the data has been transmitted, after which the path is torn down as part of the tail flit. The advantage of this approach is that the full link bandwidth is available to the circuit once it has been set up, which results in low latency transfers. Pure circuit switching is implemented in the SOCBUS [6] NoC architecture. However, pure circuit switching does not scale well as the size of the NoC grows because several links are occupied for the duration of the transmitted data, even when no data is being transmitted, for instance in the setup and tear down phases.

It is possible to multiplex multiple virtual links on a single physical link using virtual circuit switching. Circuit switching reserves physical links between routers. Virtual circuit switching creates virtual circuits that are multiplexed on links. The number of virtual links (also called virtual channels (VCs)) that can be supported by a physical link depends on the buffers allocated to the link. The two popular buffering schemes used for virtual circuit switching in NoCs involve either allocating one buffer per virtual link, or allocating one buffer per link. Both schemes are described below:

(i) *Allocating one buffer per virtual link*: In this scheme, a virtual circuit requires a buffer in each router it passes through. The number of buffers required in each router is determined by how the virtual circuits are spatially distributed in the NoC. Routers can therefore have a different number of buffers. It is also possible to fix the number of buffers in every router to a constant value, in which case the number of virtual circuits that can exist on a link is limited by the number of buffers allocated to a link in its routers. A virtual circuit buffered router implementation can be expensive due to the large number of shared buffers that are required. Multiplexing of virtual circuits on a single link also requires scheduling at each router and link (end-to-end schedule), and conflicts between different schedules can make it difficult to achieve bandwidth and latency guarantees. The MANGO [16] NoC architecture uses a variant of this scheme.

(ii) *Allocating one buffer per link*: In this scheme, virtual circuits are time multiplexed with a single buffer per link. This is achieved by using time division multiplexing (TDM) to statically schedule the usage of links among virtual circuits. Flits are typically buffered at the NIs and sent into the NoC according to the TDM schedule. Such a global scheduling with TDM makes it easier to achieve end-to-end bandwidth and latency guarantees. It also results in a less expensive router implementation, with fewer buffers. Nostrum [5] and Æthereal [15] are examples of NoC architectures that use this scheme.

12.2.2 Packet Switching

In packet switching, instead of establishing a path before sending any data as is done in circuit switching, the packets are transmitted from the source and make their way independently to the receiver, possibly along different routes and with different delays. While in circuit switching there is a start up waiting time, followed by a fixed minimal latency in the routers, packet switching involves a zero start up time, followed by a variable delay due to contention in routers along a packet's path. Without link reservation, multiple packets from different sources can arrive at a router and attempt to use a link at the same time. Since in the case of such a contention all but one of the packets must wait until the link becomes available again, QoS guarantees are harder to make in packet switching than in circuit switching. There are three popular packet switching schemes: (i) store and forward (SAF), (ii) virtual cut through (VCT), and (iii) wormhole (WH) switching. These are described below:

(i) *SAF switching*: In this simple switching technique, a packet is sent from one router to the next only if the receiving router has buffer space for the entire packet. Routers forward a packet only when it has been received in its entirety. The buffer size in the router is at least equal to the size of a packet. Because of large buffer size requirements for this technique, it is not commonly used in NoCs. However, at least one NoC architecture, the Nostrum [5] architecture, makes use of SAF switching (along with deflective routing).

(ii) *VCT switching*: This technique reduces the router latency over SAF switching by forwarding the first flit of a packet as soon as space for the entire packet is available in the next router (instead of first waiting for the entire packet to be received and then ensuring that sufficient buffer space is available in the next router before initiating packet transfer). The other flits follow the first flit without any delay. However, if no space is available in the receiving buffer, no flits are sent, and the entire packet is buffered. Since the buffering requirements of this scheme are the same as that for SAF, it is also not frequently used in NoCs.

(iii) *WH switching*: In this technique, buffer requirements are reduced to one flit, instead of an entire packet. A flit from a packet is forwarded to the receiving router if space for that flit is available in the router. If there is insufficient space in the next router to store the entire packet, parts of the packet are distributed among two or more routers. Such a distribution of packets among multiple routers can result in blocking of links, which leads to higher congestion than in SAF and VCT. WH switching is also more susceptible to deadlocks than both SAF and VCT switching, due to usage dependencies between links. Nonetheless, most NoC architectures use WH switching (e.g., SPIN [13]) or a combination of WH and virtual circuit switching (e.g., MANGO [16], Æthereal [15]).

12.3 ROUTING ALGORITHMS

Routing algorithms are responsible for correctly and efficiently routing packets or circuits from the source to the destination. The choice of a routing algorithm depends on trade-offs between several potentially conflicting metrics such as minimizing power required for routing, minimizing logic and routing tables to achieve a lower area footprint, increasing performance by reducing delay and maximizing traffic utilization of the network, and improving robustness to better adapt to changing traffic needs. Routing schemes can broadly be classified into several categories such as static or dynamic routing, distributed or source routing, and minimal or non-minimal routing. These are described below:

(i) *Static and dynamic routing*: Routing decisions in an NoC router can be either static (also called deterministic or oblivious) or dynamic (also called adaptive). In static routing, fixed paths are used to transfer data between a particular source and destination. This routing scheme does not take into account the current state of the network, and is unaware of the load on the routers and links when making routing decisions. One of the many advantages of static routing is that it is easy to implement, since very little additional router logic is required. Static routing also permits packets to be split among multiple paths between a source and destination, in a predetermined manner. If only a single path is used, static routing usually guarantees in-order delivery of data packets. This eliminates the need for adding bits to packets at the NI, in order to correctly identify and reorder them at the destination. In dynamic routing, routing decisions are made according to the current state of the network, considering factors such as availability and load on links. As such it is possible that the path between the source and destination changes over time, as traffic conditions and requirements of the application change. This adaptive behavior, however, comes at the cost of additional resources that continuously monitor the state of the network and dynamically change routing paths. Nonetheless, dynamic routing is able to better distribute traffic in a network, and is able to utilize alternate paths when certain directions become congested. This also allows support for more traffic on the same NoC topology. Typically, static routing is used for cases where traffic requirements are steady and known ahead of time, whereas dynamic routing is more desirable when traffic conditions are more irregular and unpredictable. Examples of static (or oblivious) routing algorithms include dimension order routing (DOR) [10], XY [17], pseudo-adaptive XY [17], surrounding XY [18], turn model (west-first, north-last, negative-first) [19], Valiant's random [10], ALOAS [20], topology adaptive [21], probabilistic flood [22], source [23], destination tag [10], directed flood [22], and random walk routing [22]. Examples of dynamic (or adaptive) routing algorithms include minimal adaptive [10], fully adaptive [10], congestion look-ahead [24], turnaround-turnback [13], turnback when possible [19], IVAL [19], 2TURN [19], Q [25], odd-even [26], slack time aware [27], and hot potato routing [28].

(ii) *Distributed and source routing*: Both static and dynamic routing schemes can be further classified depending on where the routing information is stored, and where routing decisions are made. In distributed routing, each packet carries the destination address (e.g., *XY* co-ordinates or number identifying the destination node or router), and routing decisions are made in each router by looking up the destination addresses in a routing table or by executing a hardware function. Thus every network router can be considered to implement a function that takes the destination address of a packet as an input and generates a routing decision as an output. When a packet arrives at the input port of a router, the routing table is consulted (or routing logic is executed) to determine the packet's output port based on its destination address. In source routing, pre-computed routing tables are stored at a node's (or PE's) NI. When a source node transmits a data packet, the routing information is looked up at the source router (or NI) based on the destination address, and this information is added to the header of the packet. Each packet thus carries the routing choices in its header for each hop in its path. When a packet arrives at a router, the routing information is extracted from the routing field in the packet header. Unlike distributed routing, source routing does not require a destination address in a packet, any intermediate routing tables, or functions needed to calculate the route. However, source routing requires additional routing information in a packet header, and the number of bits increases for longer paths. Additional routing tables may also be needed with specific entries for each source.

(iii) *Minimal and non-minimal routing*: Another way to distinguish between routing schemes is to classify them as minimal or non-minimal distance routing. A routing is minimal if the length of the routing path from the source to the destination is the shortest possible length between the two nodes. For instance, in a mesh NoC topology (where each node can be identified by its *XY* co-ordinates in the grid) if the source node is at (0, 0) and the destination node is at (i, j), then the minimal path length is $|i| + |j|$. In minimal routing, the source does not start sending a packet if a minimal path is not available. In contrast, a non-minimal routing scheme does not have such constraints, and can use longer paths if a minimal path is not available. By allowing non-minimal paths, the number of alternative paths is increased, which can be useful for avoiding congestion. Non-minimal routing can, however, have an undesirable overhead of additional power consumption in NoCs.

A routing algorithm is typically responsible for ensuring freedom from deadlocks in packet switched NoCs. A deadlock occurs when one or more packets in a network become blocked, and remain blocked for an indefinite amount of time waiting for an event that cannot occur. When a flit is transferred between neighboring routers, it releases a buffer in the first router and occupies the buffer in the second router. If the buffer in the second router is occupied, the flit is held until the buffer is available. Consider the scenario shown in Fig. 12.8, where four packets (labeled 1, 2, 3, and 4) are being routed in a circular manner in a ring (or square mesh) network topology (with the four links/channels labeled as *c1*, *c2*, *c3*, and *c4*, and the four routers labeled as *a*, *b*, *c*, and *d*). Channel *c1* is occupied by packet 1, with its flits distributed among the buffers in routers *a* and *b*. Packet 1 is waiting for packet 2 to release buffer space in router *b*. Packet 2 is occupying channel *c2*, with

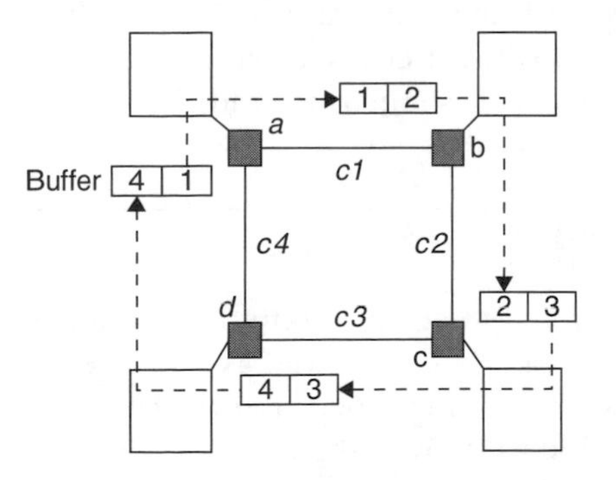

FIGURE 12.8

Deadlock scenario with a cyclic dependency between packets waiting for links its flits distributed among the buffers in routers *b* and *c*. Packet 2 in turn is waiting for packet 3 to release buffer space in router *c*. Packet 3 is occupying channel *c3*, with its flits distributed among the buffers in routers *c* and *d*. Packet 3 is waiting for packet 4 to release buffer space in router *d*. Packet 4 is occupying channel *c4*, with its flits distributed among the buffers in routers *d* and *a*, and it is waiting for packet 1 to release buffer space in router *a*. However, in this case, packet 1 cannot release buffer space in router *a*, and the packets will not be able to advance due to the cyclic dependency induced deadlock. Such a deadlock scenario is not uncommon in NoCs that make use of WH switching because of the fewer buffering resources used in the scheme. Deadlock avoidance strategies are used to analyze if a routing algorithm selected for a particular NoC topology creates potential deadlocks, and if so, then either additional hardware resources are added or restrictions are imposed on the routing to ensure freedom from deadlocks. Usually, a dependency graph of the shared network resources is built and analyzed either statically or dynamically to determine routing restrictions if deadlock conditions exist.

In a similar manner, livelocks (similar to deadlocks, except that the states of the resources involved constantly change with regard to one another, without making any progress) are also possible in NoCs especially when dynamic (adaptive) routing is used. One of the simplest dynamic routing techniques is the deflection (also called hot potato) routing. In this technique, when a packet enters a router, it is sent to the preferred output port, as determined by the routing table or routing function. However, if for some reason the preferred output port is busy (blocked or occupied by another packet), another port is selected as the output port. Since the routers in this scheme do not have additional buffers to store packets before they are moved, each packet is bounced around like a "hot potato" from one router to another (sometimes moving further away from its destination) until the packet reaches its destination. A livelock can occur in such a deflective routing case if a packet is bounced around over and over again between routers and never reaches its destination. Such a livelock scenario can usually be avoided by using simple priority rules.

Finally, in addition to deadlocks and livelocks, starvation is another problem in NoCs. It is possible that under scenarios where certain packets are prioritized during routing, some of the low priority packets never reach their intended destination. This happens when routing algorithms allow packets with higher priorities to reserve the resources without check. Starvation can be avoided by using a fair routing algorithm, or reserving some bandwidth for low priority data packets [2].

12.4 FLOW CONTROL

The goal of flow control is to allocate network resources for packets traversing an NoC. Alternatively, it can also be viewed as a problem of resolving contention during packet traversal [2, 10]. In particular, at the data link-layer level, when transmission errors occur, recovery from the error depends on the support provided by the flow control mechanism. For instance, if a corrupted packet needs to be retransmitted, the flow of packets from the sender must be stopped, and request signaling must be performed to reallocate buffer and bandwidth resources. Most flow control techniques can manage link congestion, but not all schemes can (by themselves) reallocate all the resources required for retransmission when errors occur. To guarantee reliable communication, either error correction or a scheme to handle reliable transfers at a higher layer in the communication protocol stack needs to be implemented.

12.4.1 Data Link-Layer Flow Control

There are several commonly used flow control schemes at the data link layer in NoCs, such as STALL/GO, T-Error [73], and ACK/NACK. Each of these schemes offers different fault tolerance features with different power, performance, and area overheads. These are described below.

12.4.1.1 *STALL/GO*

STALL/GO is a low overhead scheme that requires only two control wires: one going forward and signaling data availability, and the other going backward and signaling either a condition of buffers filled (STALL) or of buffers free (GO). STALL/GO can be implemented with distributed buffering (i.e., pipelining) along a link, as shown in Fig. 12.9(a), with every buffer stage designed as a two-stage first-in-first-out (FIFO). The sender only needs two buffers to cope with stalls in the first link segment. The scheme has low power overhead since there are no unnecessary transitions on the link wires when congestion occurs. The performance of this scheme is also good since the maximum throughput of one flit/cycle is easily achieved in the absence of congestion, and when congestion does occur, recovery from it is instantaneous with stalled/queued flits ready for flow immediately on resumption. STALL/GO, however, does not have any provision for fault handling. Higher level protocols are needed to handle cases where a flit is corrupted.

12.4.1.2 *T-Error*

T-Error is a more aggressive scheme that can detect faults by making use of a second delayed clock at every buffer stage. The delayed clock resamples the input data to detect any inconsistencies and then emits a *VALID* control signal. A resynchronization stage is added between the end of the link and the receiving switch,

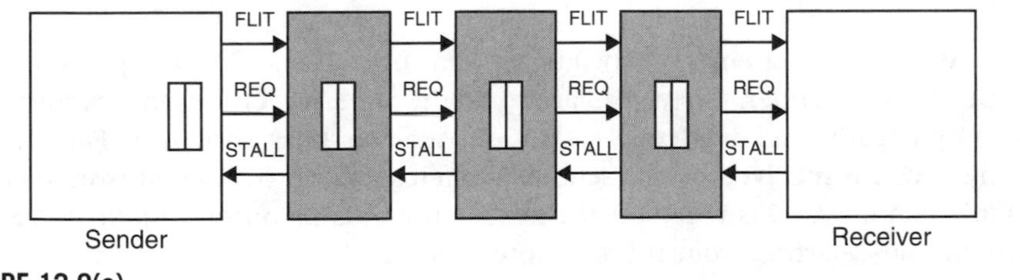

FIGURE 12.9(a)

Data link layer flow control schemes: (a) STALL/GO

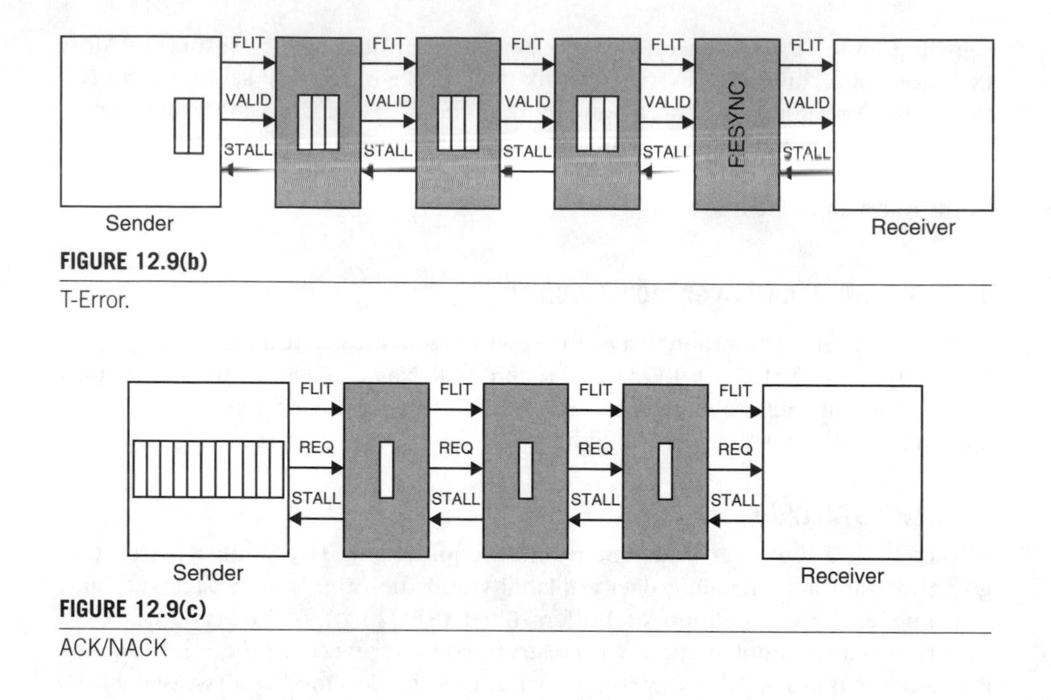

FIGURE 12.9(b)

T-Error.

FIGURE 12.9(c)

ACK/NACK

as shown in Fig. 12.9(b), to handle the offset between the original and delayed clocks. This stage realigns the timing of the *DATA* and *VALID* wires, which incurs a one cycle latency penalty. Three registers are required at each buffer stage in a link. Two registers are also needed at the sender end, along with substantial control logic, which drives up the area overhead of the scheme. Power consumption due to unnecessary flit transmission during congestion is avoided, similar to the STALL/GO scheme, but there is a power overhead due to the control logic power consumption. The timing budget provided by the T-Error scheme can be used to provide greater system reliability by configuring the links with the appropriate spacing and frequency. However, T-Error does not provide a thorough fault handling mechanism (e.g., errors with larger time constants are not detectable) and higher level fault tolerance schemes may be needed for noisy environments or time-critical systems. More details on this scheme can be found in [73].

12.4.1.3 ACK/NACK

In the ACK/NACK scheme, when flits are sent on a link, a local copy is kept in a buffer by the sender. When flits arrive at the receiver, either an acknowledge (ACK) or negative acknowledge (NACK) is sent back, as shown in Fig. 12.9(c). When an ACK is received by the sender, it deletes its copy of the flit from its local buffer. When a NACK is received, the sender rewinds its output queue and starts resending flits, starting from the corrupted one. ACK/NACK can be implemented either end-to-end between the sender and receiver, or at the switch-to-switch level. Each repeater on a link requires only a single register. However, the sender needs to have a buffer of size $2N + k$, where N is the number of buffers encountered between the source and destination. $2N$ buffers are needed at the sender to ensure maximum throughput, since the ACK/NACK feedback at the sender is only sampled after a round trip delay from the time a flit is transmitted. The value of k depends on the latency of the logic at the sender and receiver. Thus, overall a minimum of $3N + k$ buffers are required to ensure that no bandwidth penalties occur in a NACK-free environment. The extensive fault handling support with ACK/NACK comes at the cost of greater power and area overhead. Performance penalties are also noticeable in the presence of heavy congestion in an NoC or for long pipelined links.

12.4.2 Network and Transport-Layer Flow Control

Flow control techniques are also implemented at the higher network and transport layers, where flows between the sender and receiver are handled. In particular, flow control ensures that a sender does not send more data packets to the NoC than can be accepted by the receiver and its NI. Techniques for flow control at the network and transport layers can be classified according to whether they require resource reservations or not.

12.4.2.1 *Flow Control without Resource Reservation*

There are three main flow control techniques that do not require resource reservation. The first and also the simplest form of flow control is to ensure that packets always find space in the receiver NI buffers. In the case that the buffer is full, packets are simply dropped based on different policies (e.g., drop old packets first, or drop new packets first). However, such packet dropping tends to increase congestion in the long term, even though it might appear to solve contention and congestion problems in the short term. Another technique is to return the packets that do not fit into the buffers at the receiver end to the sender. To avoid deadlock scenarios, the rejected packets must be accepted by the sender. A more commonly used technique is to use deflection routing (e.g., in SPIN [13]) for end-to-end flow control. In this scheme, when a packet does not fit in the receiver NI buffers, it is temporarily sent back into the network, with the goal of accepting it at a later point of time when buffer space is available. The packet does not go back to the sender, but instead hops from one router to another, until it is accepted by the receiver.

12.4.2.2 *Flow Control with Resource Reservation*

Credit-based end-to-end flow control with resource reservation is one of the more popular flow control schemes. In this scheme, packets are allowed to leave the sender only if there is space for them at the receiver end. A credit counter at the sender interface tracks the free space available in the receiver NI buffers. The counter is initialized to the capacity of the receiver interface buffer and is decremented whenever a packet is sent. When the receiver removes a packet from its interface buffers, it sends a credit (in a packet) back to the sender, which then adds it to its credit counter. Separate credit-based flow control mechanisms exist for request (i.e., sender to receiver) and response (i.e., receiver to sender) flows between the sender and receiver. Since credit traffic can consume a large chunk of the NoC bandwidth, the requested credit packets can be combined with response packets, and vice versa. This is known as piggy backing [29], and is a common technique to save bandwidth. The Æthereal [15] and QNoC [30] architectures make use of end-to-end credit-based flow control. Nostrum [5] has also been extended to use the same flow control scheme. This credit-based flow control scheme can be applied at the link-to-link level as well. Typically, link-to-link credit-based flow control is used with virtual circuit switching, where a chain of link level flow controls is used to create an end-to-end flow control. MANGO [16] makes use of this approach.

12.5 CLOCKING SCHEMES

Clock distribution is an important component of modern synchronous digital system design because clock trees can consume a significant amount of the total power of a chip (over 30% in some cases) [31, 32]. The large contribution of the total power consumed by the clock network is due to the fact that a clock signal needs to connect to every single flip-flop and latch in the system, for the system to function properly. In NoCs, several different clocking schemes are possible, such as *synchronous*, *mesochronous*, *pleisochronous*, and *asynchronous*. In the fully *synchronous* case, a single global clock is distributed to synchronize the entire chip. The clock signal arrives simultaneously at the local flip-flops of routers, nodes, and buffered links all over the chip. This is, however, hard to achieve in practice, due to process variations and small differences in the length and loading of the clock network causing a small spread in the clock arrival time, referred to as *clock skew*. To overcome this problem, multiple clock domains are used. In the *mesochronous* case, local clocks are derived from a global clock that has been distributed all across the chip. All synchronous modules in a mesochronous system use the same clock source, but the phase between clock signals in different modules may differ due to an unbalanced global clock network. The clocking in this case is not sensitive to clock skew. If an NoC has regular placement (e.g., mesh), the phase difference with this scheme is deterministic. For irregular topologies, however, the phase difference can be non-deterministic, and synchronizers need to be used between clock domains. In the *pleisochronous* case, clock signals are produced locally. The local clock is almost at the same frequency as clocks produced elsewhere, causing a small frequency drift. Since clock frequencies among network components are not exactly the same, the receiver requires a mechanism to resolve the variable phase margin. The need for such a mechanism can, however, create a large overhead in NoC implementations. In the *asynchronous* case, clocks do not have to be present at all. Signals arrive arbitrarily, and asynchronous protocols such as two- or four-phase handshaking are used for synchronization between communicating components. A globally asynchronous locally synchronous (GALS) scheme [33] is an extension of this paradigm, where asynchronous communication protocols are used to communicate between locally clocked regions. This effectively eliminates the clock skew problem and exploits the advantages of both synchronous and asynchronous systems. However, using an asynchronous protocol (such as handshaking on global wires) can degrade performance because a signal must make a round trip for every signal transaction. One solution is to use asynchronous pipelining [34] which can improve throughput on long links.

12.6 QUALITY OF SERVICE

QoS in NoCs refers to the level of commitment for packet delivery. Such a commitment can be in the form of correctness of the transfer, completion of the transaction, or bounds on performance [32]. In most cases, however, QoS actually refers to bounds on performance (bandwidth, delay, and jitter) since correctness and completion are often the basic requirements of on-chip packet transfers. Correctness is concerned with packet integrity (corruption-less) and in-order transfer of packets from the source to the intended destination. This is achieved by using different strategies at different levels of the protocol stack, such as error correction at the data link layer or retransmission at the upper layers to guarantee packet integrity, and network or transport-layer protocols to ensure in-order packet delivery (which requires packet reordering). Completion requires that packets are not dropped or lost when being transferred from the source to the intended destination. Completion also ensures that no deadlocks or livelocks occur. Flow control mechanisms are used to ensure that the flow of packets from the source to the destination does not result in packet loss and freedom from deadlock or livelock conditions.

In terms of bounds on performance, QoS requirements can be classified into three basic categories: *best effort (BE)*, *guaranteed service (GS)*, and *differentiated service*. In BE, only the correctness and completion of communication is guaranteed and no other commitments can be made. Packets are delivered as quickly as possible over a connectionless (i.e., packet switched) network, but worst case times cannot be guaranteed, and can be an order of magnitude worse than the average case. A GS, such as guaranteed throughput (GT), makes a tangible guarantee on performance, in addition to the basic guarantees of correctness and completion for communication. GS is typically implemented using connection-oriented switching (i.e., virtual circuit switching). A differentiated service prioritizes communication according to different categories, and the NoC switches employ priority-based scheduling and allocation policies. For instance, Bolotin et al. [30] define four traffic classes—real-time traffic, signaling communication, read-write data transfer, and block data transfer. The signaling class is given the highest priority and the block data transfer the lowest priority. Unlike GS, such priority-based approaches can enable higher resource utilization, but do not provide strong guarantees.

12.7 NoC ARCHITECTURES

An NoC architecture defines a topology and a set of protocols that typically determine schemes for switching, routing, interfacing, clock distribution, and flow control. Several NoC architectures such as Æthereal [15], Butterfly Fat Tree (BFT) [35],

CHAIN [36], CLICHÉ [37], 2D Torus [68], HERMES [38], MANGO [16], Nostrum [5], Octagon [8], Proteo [7], QNoC [30], SPIN [13], SOCBUS [6], and Xpipes [14] have been proposed in literature that can support BE, GS, or hybrid BE/GS QoS for applications. Surveys of some of these architectures have been presented in literature [39–41]. Eventually, the choice of NoC architecture is determined by one or more design criteria, such as requirements for performance (latency and throughput), power consumption, QoS, reliability, scalability, and implementation cost. We describe a few of these NoC architectures below.

12.7.1 Æthereal

The Æthereal NoC [15, 42–47], developed by Philips, is a synchronous indirect network (but also supports irregular topologies) that uses WH switching and can provide GT as well as BE QoS. Virtual circuit switching is used to implement GT. Routers in Æthereal provide contention-free source routing based on TDM. Traffic is forwarded by routers based on slot allocation, and a sequence of consecutive slots in consecutive routers is used to implement a virtual circuit. In early router implementations, the next hop for a GT packet was determined by a local slot table in each router and thus no headers were required, but in recent versions the slot table has been removed to save area, thus requiring a GT packet header. Slots can be allocated statically at initialization phase, or dynamically at runtime. BE traffic makes use of non-reserved slots, and any unused reserved slots. BE packets are used to program the GT slots of the routers. A credit-based flow control scheme is used between the BE buffers to avoid loss of flits due to an overflowing of the buffers. Routers keep track of the buffer status of neighboring routers by exchanging credits as flits are forwarded. A range of standard NI sockets (such as OCP [48] and AXI [49]) are supported. In [50] and [51] an Æthereal implementation of a digital video receiver and a high end consumer TV system is compared to traditional interconnect solutions.

12.7.2 HERMES

Developed at the Faculdade de Informática PUCRS, Brazil, the HERMES NoC [38, 52] is a direct network with a 2-D mesh topology and employs WH switching with a minimal XY routing algorithm. Switches consist of control logic, five bidirectional ports (four connected to neighboring switches, and one to the local PE node), and buffers (at the input ports). The flit size is 8 bits, and the first two flits of a packet contain the header. The header consists of the target address and the number of flits in the packet. When a packet is granted access to an output port through its header flit, the connection between the input and output port is reserved for that packet. The number of flits is counted and compared to the flit number in the header until the entire packet has propagated through the switch. Once it has been determined that the entire packet has propagated through the switch, the connection closed. The connected ports inside the switch are stored in a table, and up to five connections can be active at any given time. Input queuing is used to reduce the number of switches affected by a blocked packet. The size of the input queue buffer is parameterizable, with a default size of eight flits. HERMES is connectionless, and as such cannot provide any form of bandwidth or latency GS.

12.7.3 MANGO

The MANGO network (Message-passing Asynchronous Network-on-chip providing GS over open core protocol (OCP) interfaces) [16, 53], developed at the Technical University of Denmark, is a clockless NoC that provides BE as well as GS services. NIs (or adapters) are used to convert between the synchronous OCP [48] domain and the asynchronous domain. To simplify design, routers allocate separate physical buffers for VCs. GS connections offer hard guarantees on bandwidth as well as latency, and are established by allocating a sequence of VCs through the NoC. A table in each router dictates how VCs are connected to form a virtual circuit. These tables are configured by writing to memory mapped addresses in the network. All configurations are carried out using OCP writes, which are converted into BE traffic if the configuration has to travel across the network to access a resource. At the local interface of the router three VCs can be accessed by the PE (node), in addition to the BE connections. BE connections are source routed, and the forward and return path can be different. This allows for great flexibility when routing packets, but the length of a BE path is limited to five hops because of the width of the flits. Links are implemented with delay insensitive signal encoding, which makes global timing robust since no timing assumptions are necessary. Scheduling of the links is done with an arbitration algorithm called ALG (asynchronous latency guarantee). A static scheduler gives link access to higher priority channels. An admission controller ensures that no VC is starved by monitoring the flits waiting to be scheduled, and only admits a higher priority channel if none of the lower priority channels have been stalled due to this higher priority channel. The admission controller takes a snapshot of the scheduler queues every time a flit leaves the scheduler. If a new flit arrives on that particular VC, it is stalled until all the flits that were in the schedulers queue at the time of the snapshot have been granted access to the link. This ensures that no VC waits for more than one flit at each high priority channel. Each VC has a control wire associated with it that handles flow control on the links. The ALG has eight channels, and the lowest priority channel is used by the BE router. The BE router uses credit-based buffers to handle flow control. Credits are exchanged as flits move between the routers ensuring that no buffer overflows.

12.7.4 Nostrum

The Nostrum NoC [5, 54, 55], developed at KTH in Stockholm, is a direct network with a 2-D mesh topology. Nostrum makes use of a hot potato (or deflective) routing scheme with SAF switching, and provides support for switch load distribution, guaranteed bandwidth (GB), and multicasting. GB is realized using what are called looped containers. These looped containers are implemented by virtual circuits, using a TDM mechanism referred to as temporally disjoint networks (TDNs). A container is a special type of packet which, once created, loops around on a predetermined path (i.e., the virtual circuit) and can either carry data or be marked as empty. To GB, insertion of GB data must be ensured and because the container is already in the network the insertion of data is guaranteed. In addition to the normal header of a Nostrum packet, the header of a container contains two additional bits—the first bit marks the packet as a container, while the second bit marks the container as empty or full. When a container arrives at the source input port of the virtual circuit, it will be forced to leave on the next clock cycle because of the deflective routing scheme and lack of buffers in the switch. The container will, however, not be deflected away on another path, but will always follow the path of the virtual circuit. The separation of the virtual circuits in each switch is done by TDN. Once the container has left the input port of the switch it will follow the path with precedence over the normal BE traffic to the destination. It is easy to introduce multicasting by simply having the container loop around on a virtual circuit consisting of all the recipients of such a multicast. Switch load distribution occurs in the following manner: Each switch indicates its current load by sending a stress value to its neighbors. Consequently, each switch has an idea of the load in its surroundings. Incoming packets are sorted by the number of switch cycles the packet has been traveling, and the packet with the highest priority chooses its output port first, followed by the remaining packets chosen in descending priority. Hot potato (or deflective) routing is employed for BE traffic. BE packets that find their desired output port occupied by a higher priority packet must choose another output port. Since routing decisions are made locally and on individual flits, there is no guarantee that flits, let alone packets, will arrive in order.

12.7.5 Octagon

The octagon NoC [8], developed by STMicroelectronics, is a direct network with an octagonal topology as the name suggests. It consists of 8 nodes and 12 bidirectional links, and was designed to meet requirements of network processor SoCs, such as Internet router implementations. A node in the octagon NoC is associated with a PE and a memory module, so only non-local memory requests generate octagon communication requests. Any node can reach any other node with a maximum of two hops. The octagon can operate in either packet switched or circuit switched mode. In circuit switched mode, a network arbiter allocates the path between the two nodes that must communicate with each other. The path is reserved for a number of clock cycles for that communication. The arbiter permits several pairs of nodes to communicate with each other at the same time, as long as the paths do not overlap. In packet switched mode, the nodes route the packet according to a destination field in each packet. The node calculates a relative address based on the destination address and the node's own address, and the packet is then routed either left, right, across, or into the node. The octagon NoC can be scaled if more than eight nodes are required, by connecting several basic octagon configurations together. The node that is common between the two octagon configurations acts as a bridge between them. The octagon architecture has been generalized to a polygon with diameters, and this extension has been named Spidergon [56].

12.7.6 QNoC

The Quality of Service NoC (QNoC) [30], developed at Technion in Israel, is a direct network with an irregular mesh topology. QNoC uses WH switching with a link-to-link credit-based flow control scheme, and an XY minimal routing scheme. A packet in QNoC consists of three parts: the target routing address, a command, and the payload. The command field identifies the payload format. The payload is of arbitrary length and can additionally contain operation-specific control information such as sender identification. Traffic is divided into four different service classes (or levels): signaling, real-time, read/write, and block-transfer, with signaling having the highest priority and block transfers the lowest priority. A packet is transmitted as multiple flits, with flit transfer over links being controlled by handshaking. The flit type (start packet, body, or end of packet) and the service level are indicated via separate out-of-band wires that are part of a link. When the first flit of a packet arrives at a router, the output port for that packet is determined. Subsequently, every flit arriving at that input port and with the same service level is forwarded to the output port. Every service level has its own small buffer (with a capacity of only a few flits) at the input of a switch. Unlike other WH-based switching schemes, packet forwarding is interleaved according to QoS rules, with high priority packets having the ability to preempt low priority packets. Hard guarantees are not possible due to the absence of circuit switching. Instead of providing a hard guarantee on service, QNoC provides a statistical guarantee.

12.7.7 SOCBUS

The SOCBUS NoC [6], developed at Linköping University, is a circuit switched, direct network with a 2-D mesh topology. A minimum path length routing scheme is used during the setup of a circuit to route a request from the source of the circuit to the end destination. When the request passes a switch, the route through the switch is locked. If a switch is unable to route the request closer to its destination, the switch returns a negative acknowledgement to the source, which releases all the locks as it traverses back along the route to the source. The source then retries by making a new request. When a request successfully reaches its destination, an acknowledgement is returned, which permanently locks the route. The source initiates the data transfer once it receives the acknowledgement, and ends the transfer with a cancel request that tears down the route by releasing all the locks along the route. The switches can then allocate the route to another circuit on a new request. This circuit switched scheme is deadlock free, requires simple routing hardware, has very little buffering (only for the request phase), and results in low latency (since each switch only needs a single latch). To accommodate wire delay and skew on the links between switches, mesochronous clocking with signal retiming is used. While SOCBUS does offer GS connections, hard guarantees are difficult to give because it takes a long time to set up a connection every time data needs to be sent. The setup phase takes a long time if the shortest path to the destination is blocked by one or more established paths.

12.7.8 SPIN

The scalable programmable integrated network (SPIN) [13, 57, 69] is a packet switched indirect network that implements a fat-tree topology, with two one way 32-bit link data paths. It makes use of WH switching, deflection routing, and packet reordering at the receiver NI. The virtual socket interface alliance (VSIA) virtual component interface (VCI) [58] protocol is used to interface between components in SPIN. Packets are sent as a sequence of flits, of size 4 bytes, using WH routing and no restrictions on packet size. The first flit contains the packet header, with 1 byte reserved for addressing (which limits the maximum number of nodes to 256), and the last byte of the packet contains the payload checksum. The additional bytes in the header can be used for packet tagging for special services, and for special routing options. There are three types of flits in SPIN: first, data, and last. Link level flow control is used to identify the flit type and act accordingly on its contents. Experiments with the architecture showed that although high bandwidth can be supported, random hiccups can be expected under high load. Thus GS are not supported, and only BE QoS is offered in SPIN.

12.7.9 Xpipes

Xpipes [14], along with the accompanying NetChip compiler (a combination of XpipesCompiler [59] and SUNMAP [60]) were developed by the University of Bologna and Stanford University. Xpipes uses source-based routing, WH switching, and supports the OCP standard for interfacing nodes with the NoC fabric. It consists of soft macros (a SystemC [61, 62] library) of switches and links which can be turned into instance-specific network components at instantiation time. Xpipes also supports the design of heterogeneous, customized (possibly irregular) network topologies. A go-back-*N* retransmission strategy is used as part of link level error control, which reduces switch complexity, though at the expense of considerable delay since each flit is not acknowledged until it has been transmitted across the destination switch. Errors are detected by a CRC (cycle redundancy check) block running concurrently with the switch operation. Once a floorplan for the chip is decided, the network architecture is fed into the XpipesCompiler. The XpipesCompiler is a tool that automatically instantiates an NoC communication infrastructure using Xpipe components. It can be used to tune parameters such as flit size, degree of redundancy of the CRC error detection, address space of cores, maximum number of hops between any two network nodes, flit size, etc. The NetChip compiler, which builds upon the xpipesCompiler, has been validated for different topologies such as mesh, torus, hypercube, Clos, and butterfly.

12.8 NoC STATUS AND OPEN PROBLEMS

As can be seen from the last section, there has been a great deal of interest in designing NoC architectures that offer varying levels of QoS, and are scalable to meet the needs of emerging high performance multiprocessor systems. The structure, regularity, and most importantly, predictability offered by NoC architectures allows designers to take a system level view of on-chip communication and make design choices that can withstand the variability that plagues DSM technologies. NoCs seem to be a promising communication fabric for massively parallel SoCs with tens to hundreds of components. Some semiconductor design houses such as STMicroelectronics [8, 63] and Philips [50, 51] have been experimenting with using NoCs in their designs. A proposal from Philips for an Æthereal NoC implementation of a digital video receiver and a high end consumer TV system was described in [50, 51], and showed potential for improvement over traditional bus-based schemes. STMicroelectronics has introduced the STNoC [63] NoC fabric that supports a family of topologies ranging from the Ring to the Octagon and Spidergon. Originally designed for the OC-768 network processor-based Internet router applications (octagon NoC) [8], its use is currently being evaluated for other application platforms. Arteris (an IP vendor) recently unveiled an IP (intellectual property) library of fundamental NoC components and tools to design and generate synthesizable RTL (register transfer level) for customized NoCs [64]. The Arteris NoC IP includes NIs to interface with current on-chip interface standards such as OCP 2.0 and AMBA AHB/AXI, user defined switches, and configurable synchronous or mesochronous 32- or 64-bit physical links. There have also been efforts in academia to create a complete NoC chip realization—for instance the BONE (basic on-chip network) project [2, 65] at KAIST (Korean Advanced Institute of Science and Technology, Daejeon, Korea).

In spite of the seemingly relentless push for adopting the NoC paradigm, there are still several key issues that need to be addressed to enable wide deployment of this approach. These challenges are described below:

(i) *Power*: NoC power consumption is several times greater than for current bus-based approaches, due to the need for several complex NI and switching/routing logic blocks. The choice of certain protocols (e.g., packet broadcasting) to ensure reliable data transmission can also lead to excessive power dissipation. Circuit and architecture innovations are needed to reduce this power gap.

(ii) *Latency*: Even though NoCs can offer superior bandwidth (i.e., throughput) compared to bus-based schemes, the latency response still lags behind what can be achieved with bus-based and dedicated wiring approaches today. An increase in latency occurs because of the additional delay to packetize/de-packetize data at the NIs, delays due to flow/congestion control and fault tolerance protocol overheads, and the delays at the numerous switching stages (having possible contention, buffering, and routing delays) encountered by packets as they traverse from the source to the destination. Even if circuit switching is used, there are overheads involved. For instance, in SOCBUS [6], setting up a circuit from source to destination usually requires a variable, non-deterministic delay (although latency is minimal once the circuit has been established). Innovative flow control strategies, lower diameter topologies, and native NoC support (especially in custom/new PEs) are needed to ensure that hard latency constraints are satisfied.

(iii) *Lack of tools and benchmarks*: The NoC design space is enormous, with numerous topologies and protocol/parameter choices for switching strategies, flow and congestion control schemes, buffer sizing, packet sizing, link sizing, etc. Design space exploration and implementation tools and flows are needed that can be integrated with standard tool flows used today to enable more widespread use of the NoC technology. There is also a need for open benchmarks [66] to compare the performance, cost, reliability, and other features of different NoC architecture alternatives, to select the one most suited for an application.

(iv) *Simulation speed*: Exploration and verification typically require simulation of a design in order to better understand, optimize, and verify the behavior of an application at execution time. Due to the large size of systems that will require NoCs (tens to hundreds of nodes), and the GHz frequency ranges that are possible on the NoC links, a cycle-by-cycle simulation on a workstation will take a prohibitively long time. Innovative strategies are needed that improve the simulation speed of these NoC-based designs by using techniques such as distributed simulation over multiple workstations and partial FPGA-prototyping [67].

12.9 SUMMARY

In this chapter we presented an overview of NoCs that are a promising new direction in on-chip communication design. We described the various design aspects that characterize NoCs, such as the topology, switching schemes, routing algorithms, flow control mechanisms, clocking strategies, and QoS support. A survey of several NoC architectures that have been proposed in literature was presented, and then finally the current status and open problems in the NoC domain were described. An important point that must be kept in mind is that the advent of NoCs will not result in a phasing out of bus-based communication architectures. In fact, the problem of designing bus-based fabrics will become complementary to the problem of NoC design, since large NoCs will encompass several subsystems that are interconnected using buses. Even in an era of multiprocessor systems with hundreds of cores in the future, the cost, area, performance, and power consumption needs of a given application will determine the choice of communication fabric based on trade-offs between the benefits of NoCs and bus-based schemes. Although the move to NoCs is inevitable, for applications with less aggressive requirements, bus-based architectures will still provide a low cost and attractive interconnection solution.

FURTHER READING

While the goal of this chapter was to provide an overview of the NoC paradigm, a more thorough treatment of the subject requires an entire book in itself. In recent years, there have been several books on NoCs that the interested reader can refer to for a more in-depth analysis of NoC design. A seminal book on the area of interconnection networks by Dally and Towles [10] explains several of the concepts and principles that have been adapted to NoCs in recent years. Jantsch and Tenhunen [32] compiled several papers on different areas within NoC design into a book that serves as a good introduction to the area. Micheli and Benini [2] describe some critical issues and research efforts focused at the different layers of the NoC protocol stack. Other books focus on specific topics within the NoC area. For instance, a book by Kogel et al. [71] focuses entirely on the system level modeling of NoCs. Another book by Grammatikakis et al. [72] describes the design and exploration of the STNoC architecture in great detail. All of these books are very useful resources for students, researchers, and designers interested in NoCs.

REFERENCES

[1] D. C. Pham et al., "Overview of the architecture, circuit design, and physical implementation of a first-generation cell processor," *IEEE Journal of Solid-State Circuits*, Vol. 41, No. 1, January 2006, pp. 179–196.

[2] G. De Micheli and L. Benini, "Networks on Chips," *Systems on Silicon Series*, 2006, Morgan Kaufmann.

[3] J. Duato, S. Yalamanchili and L. Ni, *Interconnection Networks—An Engineering Approach*, Morgan Kaufmann, 2002.

[4] W. J. Dally, "Express cubes: Improving the performance of k-ary n-cube interconnection networks," *IEEE Transactions on Computers*, Vol. 40, No. 9, September 1991, pp. 1016–1023.

[5] S. Kumar, A. Jantsch, J.-P. Soininen, M. Forsell, M. Millberg, J. Öberg, K. Tiensyrjä and A. Hemani, "A network on chip architecture and design methodology," in *Proceedings of IEEE Computer Society Annual Symposium on VLSI*, April 2002, pp. 105–112.

[6] P. T. Wolkotte, G. J. M. Smit, G. K. Rauwerda and L. T. Smit, "An energy-efficient reconfigurable circuit switched network-on-chip," in *Proceedings of the 19th IEEE International Parallel and Distributed Processing Symposium (IPDPS)*, 2005.

[7] D. Siguenza-Tortosa, T. Ahonen and J. Nurmi, "Issues in the development of a practical NoC: The Proteo concept," *Integration, the VLSI Journal*, Elsevier, 2004, pp. 95–105.

[8] F. Karim, A. Nguyen and S. Dey, "An interconnect architecture for networking systems on chips," *IEEE Micro*, Vol. 22, No. 5, September/October 2002, pp. 36–45.

[9] S. Pasricha, N. Dutt and M. Ben-Romdhane, "Constraint-driven bus matrix synthesis for MPSoC," in *Proceedings of Asia and South Pacific Design Automation Conference (ASPDAC)*, Yokohama, Japan, January 2006, pp. 30–35.

[10] W. J. Dally and B. Towles, *Principles and Practices of Interconnection Networks*, Morgan Kauffman, San Francisco, CA, 2004.

[11] P. Guerrier and A. Greiner, "A generic architecture for on-chip packet-switched interconnections," in *Proceedings of the Design, Automation and Test in Europe Conference and Exhibition (DATE)*, March 2000, pp. 250–256.

[12] C. E. Leiserson, "Fat-trees: Universal networks for hardware-efficient supercomputing," *IEEE Transactions on Computers*, Vol. 34, No. 10, October 1985, pp. 892–901.

[13] A. Andriahantenaina, H. Charlery, A. Greiner, L. Mortiez and C. Zeferino, "SPIN: A scalable, packet switched, on-chip micro - network," in *Design Automation and Test in Europe Conference and Exhibition (DATE)*, March 2003, pp. 70–73.

[14] M. D. Osso, G. Biccari, L. Giovannini, D. Bertozzi and L. Benini, "Xpipes: A latency insensitive parameterized network on chip architecture for multi-processor SoCs," in *Proceedings of 21st International Conference on Computer Design (ICCD)*, 2003, pp. 536–539.

[15] K. Goossens, J. Dielissen and A. Radulescu, "Æthereal network on chip: Concepts, architectures and implementations," *IEEE Design and Test of Computers*, Vol. 22, No. 5, 2005, pp. 414–421.

[16] T. Bjerregaard, "The MANGO clockless network-on-chip: Concepts and implementation," Ph.D. Thesis, Informatics and Mathematical Modelling, Technical University of Denmark, DTU, 2005.

[17] M. Dehyadgari, M. Nickray, A. Afzali-kusha and Z. Navabi, "Evaluation of pseudo adaptive XY routing using an object oriented model for NOC." *The 17th International Conference on Microelectronics*, December 13–15, 2005.

[18] C. Bobda, A. Ahmadinia, M. Majer, J. Teich, S. Fekete and J. van der Veen, "DyNoC: A dynamic infrastructure for communication in dynamically reconfigurable devices," *International Conference on Field Programmable Logic and Applications*, August 24–26, 2005, pp. 153–158.

[19] H. Kariniemi and J. Nurmi, "Arbitration and routing schemes for on-chip packet networks," *Interconnect-Centric Design for Advanced SoC and NoC*, Kluwer Academic Publishers, 2004, pp. 253–282.

[20] K. Kim, S. J. Lee, K. Lee and H. J. Yoo, "An arbitration look-ahead scheme for reducing end-to-end latency in networks on chip," *IEEE International Symposium on Circuits and Systems*, May 23–26, 2005, Vol. 3, pp. 2357–2360.

[21] T. A. Bartic, J.-Y. Mignolet, V. Nollet, T. Marescaux, D. Verkest, S. Vernalde and R. Lauwereins, "Topology adaptive network-on-chip design and implementation," *IEE Proceedings—Computers and Digital Techniques*, Vol. 152, No. 4, July 8, 2005, pp. 467–472.

[22] M. Pirretti, G. M. Link, R. R. Brooks, N. Vijaykrishnan, M. Kandemir and M. J. Irwin, "Fault tolerant algorithms for networks-on-chip interconnect," in *Proceedings, IEEE Computer Society Annual Symposium on VLSI*, February 19–20, 2004, pp. 46–51.

[23] W. J. Dally and B. Towles, "Route packets, not wires: On-chip interconnection networks," in *Proceedings, Design Automation Conference*, 2001, pp. 684–689.

[24] J. Kim, D. Park, T. Theocharides, N. Vijaykrishnan and C. R. Das, "A low latency router supporting adaptivity for on-chip interconnects," in *Proceedings, 42nd Design Automation Conference (DAC)*, June 13–17, 2005, pp. 559–564.

[25] M. Majer, C. Bobda, A. Ahmadinia and J. Teich, "Packet routing in dynamically changing networks on chip," in *Proceedings, 19th IEEE International Parallel and Distributed Processing Symposium*, April 4–8, 2005.

[26] J. Hu and R. Marculescu, "DyAD—Smart routing for networks-on-chip," in *Proceedings, 41st Design Automation Conference (DAC)*, 2004, pp. 260–263.

[27] D. Andreasson and S. Kumar, "Slack-time aware routing in NoC systems," *IEEE International Symposium on Circuits and Systems*, May 23–26, 2005, pp. 2353–2356.

[28] U. Feige and P. Raghavan, "Exact analysis of hot-potato routing," *33rd Annual Symposium on Foundations of Computer Science*, October 24–27, 1992, pp. 553–562.

[29] A. S. Tanenbaum, *Computer Networks*, Prentice-Hall, Upper Saddle River, NJ, 1996.

[30] E. Bolotin, I. Cidon, R. Ginosar and A. Kolodny, "QNoC: QoS architecture and design process for network on chip," *Journal of Systems Architecture, Special Issue on Network on Chip*, Vol. 50, February 2004, pp. 105–128.

[31] E. G. Friedman, "Clock distribution networks in synchronous digital integrated circuits," in *Proceedings of the IEEE*, Vol. 89, No. 5, May 2001, pp. 665–692.

[32] A. Jantsch and H. Tenhunen (Eds.), *Networks on Chip*, Kluwer Academic Publishers, 2003.

[33] A. Hemani, T. Meincke, S. Kumar, A. Postula, T. Olsson, P. Nilsson, J. Oberg, P. Ellervee and D. Lundqvist, "Lowering power consumption in clock by using globally asynchronous locally synchronous design style," in *Proceedings of the 36th ACM/IEEE Conference on Design Automation (DAC)*, 1999, pp. 873–878.

[34] T. Villiger, H. Käslin, F. K. Gürkaynak, S. Oetiker and W. Fichtner, "Self-timed ring for globally-asynchronous locally-synchronous systems," in *Proceedings of the 9th International Symposium on Asynchronous Circuits and Systems (ASYNC)*, 2003, pp. 1–10.

[35] P. P. Pande, C. Grecu, A. Ivanov and R. Saleh, "Design of a switch for network on chip applications," in *Proceedings of the International Symposium on Circuits and Systems (ISCAS)*, Vol. 5, May 2003, pp. 217–220.

[36] W. J. Bainbridge and S. B. Furber, "CHAIN: A delay insensitive chip area interconnect," *IEEE Micro Special Issue on Design and Test of System on Chip*, Vol. 142, No. 4, September 2002, pp. 16–23.

[37] S. Kumar, A. Jantsch, M. Millberg, J. Oberg, J.-P. Soininen, M. Forsell, K. Tiensyrjä and A. Hemani, "A network on chip architecture and design methodology," in *Proceedings International Symposium VLSI (ISVLSI)*, 2002, pp. 117–124.

[38] L. Ost, A. Mello, J. Palma, F. Moraes and N. Calazans, "MAIA—A framework for networks on chip generation and verification," in *Proceedings of the IEEE Asia and South Pacific Design Automation Conference (ASP-DAC)*, 2005, pp. 49–52.

[39] F. Moraes, N. Calazans, A. Mello, L. Möller and L. Ost, "HERMES: An infrastructure for low area overhead packet-switching networks on chip," in *Proceedings of the VLSI Integration*, 2004, pp. 69–93.

[40] T. Bjerregaard and S. Mahadevan, "A survey of research and practices of network-on-chip," *ACM Computing Surveys*, Vol. 38, No. 1, June 2006.

[41] C. P. Pederson, "Virtual circuits in network-on-chip," Master's Thesis, Technical University of Denmark, September 2006.

[42] E. Rijpkema, K. Goossens and P. Wielage, "A router architecture for networks on silicon," in *Proceedings of the 2nd Workshop on Embedded Systems*, 2001, pp. 181–188.

[43] E. Rijpkema, K. Goossens, A. Radulescu, J. Dielissen, J. V. Meerbergen, P. Wielage and E. Waterlander, "Trade offs in the design of a router with both guaranteed and best-effort services for networks-on-chip," in *Proceedings of the IEEE Design, Automation and Test in Europe Conference (DATE)*, 2003, pp. 350–355.

[44] A. Radulescu, J. Dielissen, K. Goossens, E. Rijpkema and P. Wielage, "An efficient on-chip network interface offering guaranteed services, shared-memory abstraction, and flexible network configuration," in *Proceedings of IEEE Design, Automation and Testing in Europe Conference (DATE)*, 2004, pp. 878–883.

[45] K. Goossens, J. V. Meerbergen, A. Peeters and P. Wielage, "Networks on silicon: Combining best-effort and guaranteed services," in *Proceedings of the IEEE Design, Automation and Test in Europe Conference (DATE)*, 2002, pp. 196–200.

[46] J. Dielissen, A. Radulescu, K. Goossens and E. Rijpkema, "Concepts and implementation of the phillips network-on-chip," in *Proceedings of the IFIP, IP based SOC (IPSOC)*, 2003, pp. 1–6.

[47] P. Wielage and K. Goossens, "Networks on silicon: Blessing or nightmare?" in *Proceedings of the IEEE Euromicro Symposium on Digital System Design (DSD)*, 2002, pp. 196–200.

[48] Open Core Protocol International Partnership (OCP-IP). OCP Datasheet, Release Version 1.0/2.0, http://www.ocpip.org.

[49] C. Bartels, J. Huisken, K. Goossens, P. Groeneveld and J. van Meerbergen, "Comparison of an Æthereal network on chip and a traditional interconnect for a multi-processor DVB-T system on chip," in *Proceedings of the IFIP International Conference on Very Large Scale Integration (VLSI-SoC)*, October 2006, pp. 80–85.

[50] F. Steenhof, H. Duque, B. Nilsson, K. Goossens and R. P. Llopis, "Networks on chips for high end consumer-electronics TV system architectures," in *Proceedings of Design, Automation and Test in Europe Conference and Exhibition (DATE)*, March 2006, pp. 148–153.

[51] F. Moraes, A. Mello, L. Möller, L. Ost and N. Calazans, "A low area overhead packet switched network on chip: Architecture and prototyping," in *IFIP Very Large Scale Integration (VLSI-SOC)*, 2003, pp. 318–323.

[52] ARM AMBA 3.0 AXI Specification, www.arm.com/armtech/AXI.

[53] T. Bjerregaard and J. Sparsø, "A router architecture for connection-oriented service guarantees in the MANGO clockless network-on-chip," in *Proceedings of Design, Automation and Testing in Europe Conference (DATE)*, 2005, pp. 1226–1231.

[54] H. Zimmer and A. Jantsch, "A fault tolerant notation and error-control scheme for switch-to-switch busses in a network-on-chip," in *Proceedings of Conference on Hardware/Software Codesign and System Synthesis Conference (CODES-ISSS)*, 2003, pp. 188–193.

[55] M. Millberg, E. Nilsson, R. Thid and A. Jantsch, "Guaranteed bandwidth using looped containers in temporally disjoint networks within the nostrum network-on-chip," in *Proceedings of IEEE Design, Automation and Testing in Europe Conference (DATE)*, 2004, pp. 890–895.

[56] M. Coppola, R. Locatelli, G. Maruccia, L. Pieralisi and A. Scandurra, "Spidergon: A novel on chip communication network," in *Proceedings of International Symposium on System on Chip*, Tampere, Finland, November 2004, p. 15.

[57] P. Guerrier and A. Greiner, "A generic architecture for on-chip packet-switched interconnections," in *Proceedings of IEEE Design, Automation and Test in Europe (DATE)*, March 2000, pp. 250–256.

[58] VSI Alliance™ On-Chip Bus Development Working Group, Virtual Component Interface Standard Version 2 (OCB 2 2.0), April 2001.

[59] A. Jalabert, S. Murali, L. Benini and G. D. Micheli, "XpipesCompiler: A tool for instantiating application specific networks-on-chip," in *Proceedings of Design, Automation and Testing in Europe Conference (DATE)*, 2004, pp. 884–889.

[60] S. Murali and G. D. Micheli, "SUNMAP: A tool for automatic topology selection and generation for NoCs," in *Proceedings of the 41st Design Automation Conference (DAC)*, IEEE, 2004, pp. 914–919.

[61] SystemC initiative, www.systemc.org.

[62] T. Grötker, S. Liao, G. Martin and S. Swan, *System Design with SystemC*, Kluwer Academic Publishers, 2002.

[63] M. D. Grammatikakis, M. Coppola and R. Locatelli, "Design of cost-efficient network-on-chip architectures: The STNoC," CRC Press, March 2008 (To be Published).

[64] http://www.arteris.com

[65] http://ssl.kaist.ac.kr/ocn/

[66] C. Grecu, A. Ivanov, P. Pande, A. Jantsch, E. Salminen, U. Ogras and R. Marculescu, "Towards open network-on-chip benchmarks," *First International Symposium on Networks-on-Chip (NOCS)*, 2007, pp. 205.

[67] P. T. Wolkotte, P. K. F. Holzenspies and G. J. M. Smit, "Fast, accurate and detailed NoC simulations," *First International Symposium on Networks-on-Chip (NOCS)*, 2007, pp. 323–332.

[68] W. J. Dally and B. Towles, "Route packets, not wires: On-chip interconnection networks," in *Proceedings of Design Automation Conference (DAC)*, 2001, pp. 683–689.

[69] A. Andriahantenaina and A. Greiner, "Micro-network for SoC: Implementation of a 32-port SPIN network," in *Design Automation and Test in Europe Conference and Exhibition (DATE)*, March 2003, pp. 1128–1129.

[70] S.-J. Lee, S.-J. Song, K. Lee, J.-H. Woo, S.-E. Kim, B.-G. Nam and H.-J. Yoo, "An 800 MHz star-connected on-chip network for application to systems on a chip," *IEEE International Solid-State Circuits Conference (ISSCC)*, 2003, pp. 468–469.

[71] T. Kogel, R. Leupers and H. Meyrs, *Integrated System-Level Modeling of Network-on-Chip Enabled Multi-Processor Platforms*, Springer, 2006.

[72] D. Grammatikakis, M. Coppola, R. Locatelli and G. Maruccia, *Design of Cost-Efficient Network-on-Chip Architectures: The STNoC*, CRC, 2008.

[73] R. Tamhankar, S. Murali and G. Micheli, "Performance driven reliable link for networks on chip," in *Proceedings of the Asian Pacific Conference on Design Automation (ASPDAC)*, 2005, pp. 794–754.

CHAPTER

Emerging On-Chip Interconnect Technologies 13

The previous chapters have reiterated the overall trend toward communication-centric design: while individual computation logic components have become considerably faster, the overall speed of system-on-chip (SoC) designs is limited by the communication between the different components on a chip. In deep submicron (DSM) VLSI technologies, it is becoming increasingly harder for a copper-based electrical interconnect (EI) to satisfy the design requirements of delay, power, bandwidth, and delay uncertainty [1]. Indeed, the situation is likely to become worse for future giga- and tera-scale electronic systems. The resistance of copper interconnects, in current and imminent technologies is increasing rapidly under the combined effects of enhanced grain boundary scattering, surface scattering, and the presence of a highly resistive diffusion barrier layer [2, 3]. Copper interconnects also constitute up to 70% of the total on-chip capacitance, and are major sources of power dissipation. The semiconductor industry has made major research and development investments in alternative (e.g., low *k* dielectric) materials in response to this urgent need. However, low *k* materials suffer from poor mechanical and thermal properties [4]. The steep rise in parasitic resistance and capacitance of copper interconnects poses serious challenges for interconnect delay (especially at the global level), power dissipation, and interconnect reliability [5]. According to the International Technology Roadmap for Semiconductors (ITRS) [6], interconnect innovation with optical, RF/wireless, and carbon nanotubes (CNTs) is the key to satisfying performance, reliability, and power requirements in the long term. These future interconnect technologies must support ultra-high data rates (e.g., greater than 100 Gbps/pin), be scalable enough to support tens to hundreds of concurrent communication streams, and involve fabrication techniques that are compatible with mainstream SoC and system-in-package (SiP) technologies. In this chapter, we present an overview of three such emerging interconnect technologies that promise to overcome the limitations of copper interconnects. Section 13.1 describes optical interconnects that make use of light and an on-chip optical medium to transfer data. Section 13.2 presents RF/wireless interconnects that forego the need for metal wires, and instead transfer data on a chip wirelessly, using transmitting and receiving antennas integrated on the same chip. Finally, Section 13.3 discusses CNTs which have been proposed as an evolutionary replacement for copper interconnects, just like copper previously replaced aluminum as the interconnect material of choice in electronic systems.

13.1 OPTICAL INTERCONNECTS

Optical interconnects (OIs) have the potential to overcome the communication bottleneck by replacing electrical wires with optical waveguides [7, 8]. Optical interconnects offer many advantages over traditional electrical (copper-based) interconnects: (i) they can support enormous intrinsic data bandwidths in the order of several Gbps using only simple on–off modulation schemes, (ii) they are relatively immune to electrical interference due to crosstalk, and parasitic capacitances and inductances, (iii) their power dissipation is completely independent of transmission distance at the chip level, and (iv) routing and placement is simplified since it is possible to physically intersect light beams with minimal crosstalk. Once a path is acquired, the transmission latency of the optical data is very small, depending only on the group velocity of light in a silicon waveguide: approximately 6.6×10^7 m/s, or 300 ps for a 2 cm path crossing a chip. After an optical path is established, data can be transmitted end to end without the need for repeating or buffering, which can lead to significant power savings.

While board-to-board and chip-to-chip OIs have been proposed and are actively under development [9–14], the feasibility of on-chip OIs is an open research problem that is garnering immense interest from academia and industry alike. Figure 13.1 shows a block diagram of an on-chip optical interconnect system, consisting of a laser light source, an optical transmitter, a waveguide optical link, and a receiver. Despite a lot of achievements in the area of optical gain in silicon over the last few years [10, 15, 16], a high speed, electrically driven, on-chip monolithic light source still remains to be realized. Consequently, for the subsequent discussion on optical on-chip transmission, it is assumed that the light source is an external (off-chip) laser.

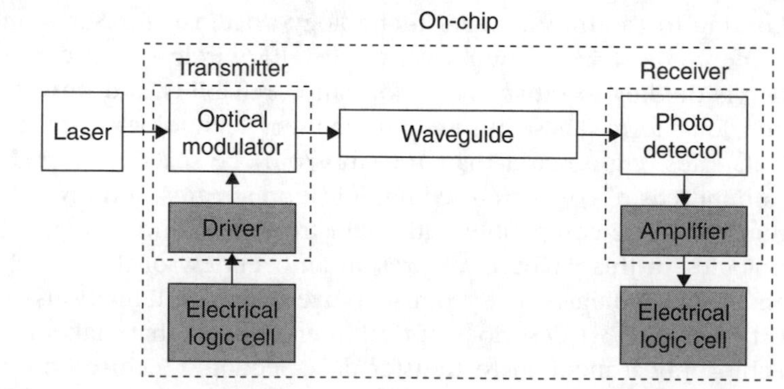

FIGURE 13.1

Block diagram of OI system [8]
© 2006 IEEE

On-chip optical transmission requires a laser source, an electro-optical modulator, and an electrical driver circuit (Fig. 13.2(a)). The laser source provides light to the modulator, which transduces electrical data supplied from the electrical driver into a modulated optical signal. Several high speed electro-optical modulators [17–20] have been proposed that can change the refractive index or the absorption coefficient of an optical path when an electrical signal is injected. The two most popular modulators in literature are the Mach-Zehnder interferometer-based silicon modulators [19] and microresonator-based P-I-N diode type modulators [20]. MOS capacitor structures such as the Mach-Zehnder interferometer-based silicon modulators have higher modulation speeds (several GHz) but a large power consumption and greater silicon footprint (around 10 mm). On the other hand, microresonator-based P-I-N diode type modulators are compact in size (10–30 μm) and have low power consumption, but possess low modulation speeds (several MHz). The performance of a modulator is dependent on the on-to-off light intensity ratio, referred to as the *extinction ratio*, which depends on the electrical input signal strength. A higher extinction ratio is desirable for proper signal detection, while a poor one may cause transmission errors. An extinction ratio greater than 10 dB has been recently reported with high input signal swing [21], which is high enough to enable proper signal detection without causing any transmission errors. Modulator size is also an important consideration for integrated applications, and significant efforts have been made to realize compact-sized modulators, such as the circular shaped 10 μm ring-modulators [21]. The modulator driver consists of a series of inverter stages that drive the modulator's capacitive load (refer to Fig. 13.2).

The waveguide is a path through which light is routed. The refractive index of the waveguide material has a significant impact on the bandwidth, latency, and area of an optical interconnect. Silicon and polymer are two of the most promising materials for on-chip waveguide realization. Table 13.1 compares some of the most relevant features of silicon and polymer waveguides. Silicon and polymer waveguides involve a trade-off in propagation speed and bandwidth. The smaller refractive index of polymer waveguides results in higher propagation speed. However, the polymer waveguide requires a larger pitch than silicon, which reduces bandwidth density (i.e., number of bits transmitted/unit surface area). Modulators for polymer waveguides [23] are also typically bulkier (requiring higher voltage drive and higher frequency operation) than modulators for silicon

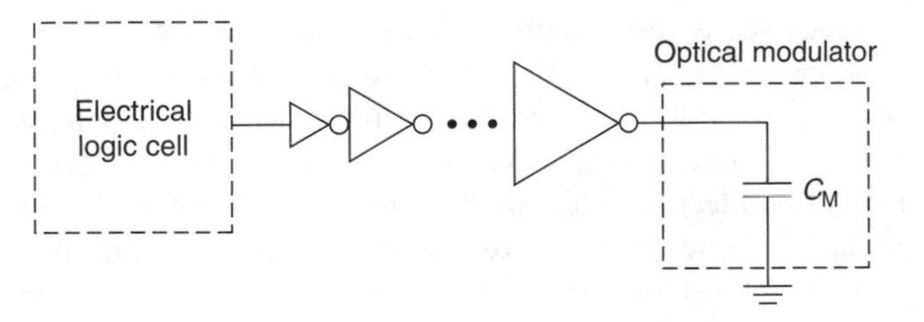

FIGURE 13.2(a)

Circuit schematic of modulator driver
© 2006 IEEE

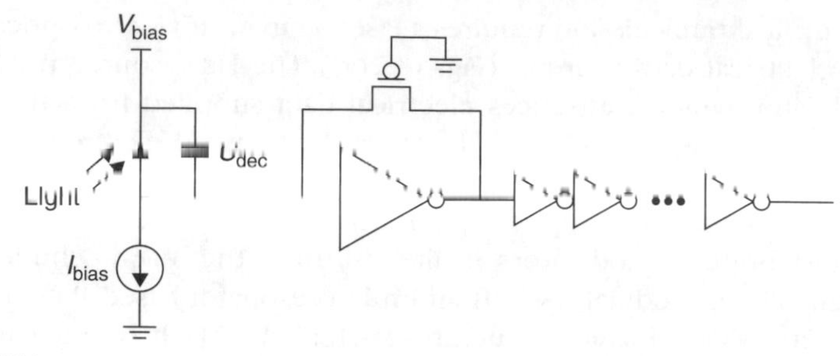

FIGURE 13.2(b)

Waveguide receiver (photo detector) and accompanying TIA [8]

Table 13.1 General characteristics of silicon and polymer on-chip waveguides [22]

Waveguide	Si	Polymer
Refractive index	3.5	1.5
Width (μm)	0.5	5
Separation (μm)	5	20
Pitch (μm)	5.5	25
Time of flight (ps/mm)	10.45	4.93
Loss (dB/cm)	1.3	1

waveguides. This limits the applicability of polymer waveguides to on-chip optical interconnects. Polymer waveguides are, however, feasible in a transmission system based on VCSELs (vertical cavity surface emitting laser) [24], where the modulator is not required. A VCSEL-based solution may, however, lead to an increase in on-chip power consumption as a result of the complex on-chip flip-bonded laser sources. Additionally, light is emitted vertically and needs to be transferred to the horizontal chip surface in this scenario, requiring integrated mirrors and sophisticated lithographic techniques.

The optical receiver is responsible for converting the optical signal to an electrical signal. It consists of a photo-detector and a trans-impedance amplifier (TIA) stage (Fig. 13.2(b)). In applications involving simultaneous transmission at different wavelengths per waveguide using wave division multiplexing (WDM), the receiver also requires a wave selective filter for each received wavelength. The P-I-N diode [25] is an example of one of the more popular photo-detectors in literature. The *quantum efficiency* of a photo-detector is an important figure of merit for the system. A high value for quantum efficiency means lower losses when converting optical information into an electrical form. One of the key trade-offs in the design of a photo-detector is between detector speed and quantum efficiency. Recently, interdigitated metal–semiconductor–metal (MSM) receivers have attracted attention

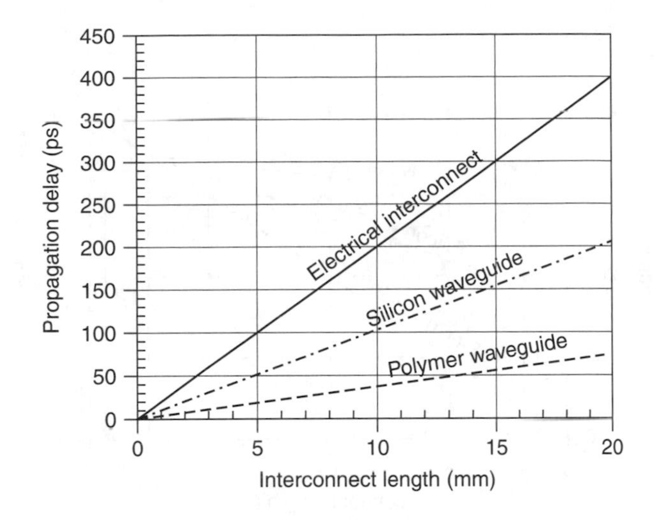

FIGURE 13.3

Propagation delay of silicon and polymer waveguides as compared to EIs [8]
© 2006 IEEE

due to their fast response and excellent quantum efficiency. High speed, low power inter-digitated MSM Ge and SiGe photo-detectors operating at telecommunication wavelengths were described in [26, 27]. Detector size is also an important criteria for both compactness and next stage capacitance [22]. Typically, the detector has large base capacitance, which poses a design challenge for the high speed gain stages following it. The TIA stage converts photo-detector current to a voltage which is thresholded by subsequent stages to digital levels [28]. To achieve high gain and high speed detection, an analog supply voltage higher than the digital supply voltage may be required, thereby dissipating higher power.

Modern on-chip EIs are realized using copper wires surrounded by a low k dielectric [29]. The on-chip EIs can be classified into two categories: (i) local interconnects, which are used for short distance communication and have a delay of less than a clock cycle and (ii) global interconnects, which are fewer in number, are used for long distance communication and have a delay spanning multiple cycles. While local interconnects are designed to have fast signal propagation speeds, long global EIs typically have high RC constants, which results in greater propagation delay, transition time, and crosstalk noise. Repeater insertion and increasing wire width can reduce the propagation delay somewhat, but the delay is still quite large in DSM technologies. In this respect, optical interconnects (OIs) possess an intrinsic advantage of low signal propagation delay in waveguides. Figure 13.3 shows a comparison of signal propagation delay in EI and the two common OI waveguides—silicon and polymer. Low refractive index polymer and high refractive index silicon waveguides are selected for the comparison, as they represent two opposite types of optical waveguides in terms of signal propagation delay and crosstalk. Regardless of the waveguide material, optical interconnects provide a lower propagation delay than EIs. This is because optical signal propagation is intrinsically faster than electrical signal propagation due to the absence of RLC impedances.

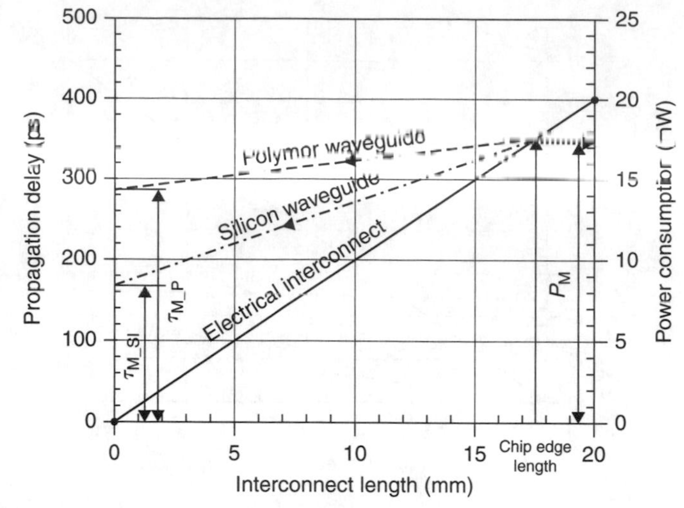

FIGURE 13.4

Maximum conversion cost for an on-chip OI [8]
© 2006 IEEE

In order to exploit the propagation delay advantage offered by optical waveguides, an electrical signal must first be converted into an optical signal and then back into an electrical signal. This conversion has a fixed delay associated with it, which is nearly independent of the interconnect length for a given technology. OIs will therefore have a delay advantage over EIs if the waveguide propagation delay dominates the overall delay. In addition to delay, power is another metric for which OIs need to have an advantage over EIs. If the average length of the global interconnects in a target architecture is known, it is possible to extract the delay and power requirements for OIs. Figure 13.4 shows the OI delay and power requirements for silicon and polymer waveguides, for an interconnect length equal to the ITRS projected chip edge length of 17.6 mm [8]. In the figure, the EI delay is plotted as a function of distance. The optical waveguide delay is then projected back from the 17.6-mm EI delay to the *y*-axis (as indicated by the arrows pointing to the left). The *y*-intercept of the optical waveguide delay curve indicates the maximum allowed conversion delay τ (for electrical to optical and optical to electrical conversion) in the chip edge length OI. From the figure, it can be estimated that the combined transmitter and receiver delay should be lower than 280–370 ps (τ_{M_P}) for polymer waveguides and 180–270 ps (τ_{M_Si}) for silicon waveguides, to have an advantage over EIs. Using the second *y*-axis in a similar way, the total power consumption for the chip-length OI should be less than 17–18 mW (P_M), to have an advantage over EI power consumption.

If OIs are to eventually replace EIs, then in addition to lower signal delay and lower power consumption, they must also possess superior bandwidth density compared to EIs [8]. Bandwidth density is a metric that characterizes information throughput through a unit cross section of an interconnect. Typically, the cross-section is defined as the pitch of the electrical wires in EIs or optical waveguides

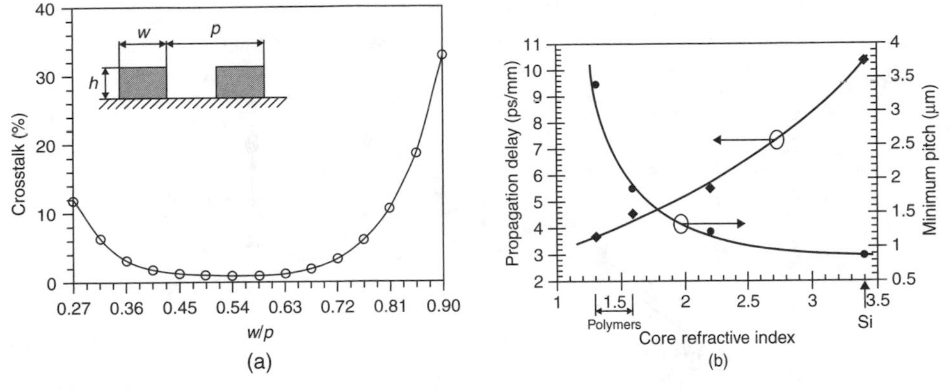

FIGURE 13.5

Optical waveguide analysis: (a) sketch of the modeled waveguides (inset) and the crosstalk as a function of the w/p ratio for a 10 mm polymer waveguide OI, (b) trade-off between waveguide density and propagation delay per unit length for a 10 mm interconnect and a maximum allowed crosstalk of 20% [8]
© 2006 IEEE

in OIs. The minimum pitch between two adjacent waveguides is determined by crosstalk considerations. Figure 13.5(a) shows the crosstalk in a polymer waveguide as a function of the waveguide width (w) and pitch (p). For a fixed pitch, if the waveguide is too wide, then the crosstalk is high due to the proximity between the sides of adjacent waveguides. If the waveguide is too narrow, the optical mode becomes less confined, causing a higher crosstalk due to a larger overlap between adjacent optical modes. As can be seen from Fig. 13.5(a), an optimum ratio exists between the waveguide width and pitch. In order to get the maximum bandwidth density for OIs, the minimum waveguide pitch is obtained by setting the crosstalk limit to 20%, for a 10 mm long interconnect. Figure 13.5(b) shows a plot of the optical signal propagation delay (curve with arrow pointing left) and minimum pitch (curve with arrow pointing right) vs. the refractive index of the waveguide core. It can be observed that a high index core offers a smaller waveguide pitch, while a low index core offers a lower propagation delay. To avoid significant crosstalk, optical waveguides should be spaced 0.5–3 μm from each other. In contrast, the pitch for delay-optimized global electrical wires (i.e., wires with optimal repeaters and wide pitch for low signal propagation delay) is around five to seven node sizes, providing a significant advantage in bandwidth density over OIs. Figure 13.6 shows a comparison of projected bandwidth density for delay-optimize EIs and optical waveguides (curves with arrows pointing to the left) over a number of technology nodes. The figure also shows projected increase in clock rate (curve with arrow pointing to right) over the technology nodes. It can be seen that with shrinking technology nodes, optical bandwidth density increases due solely to the higher bit rate through waveguides with a fixed pitch. EIs exploit more efficient repeaters, which leads to a higher growth in bandwidth density. Thus a single wavelength OI is inferior to a delay-optimized EI in terms of bandwidth density. So the question arises: why would anyone use OIs?

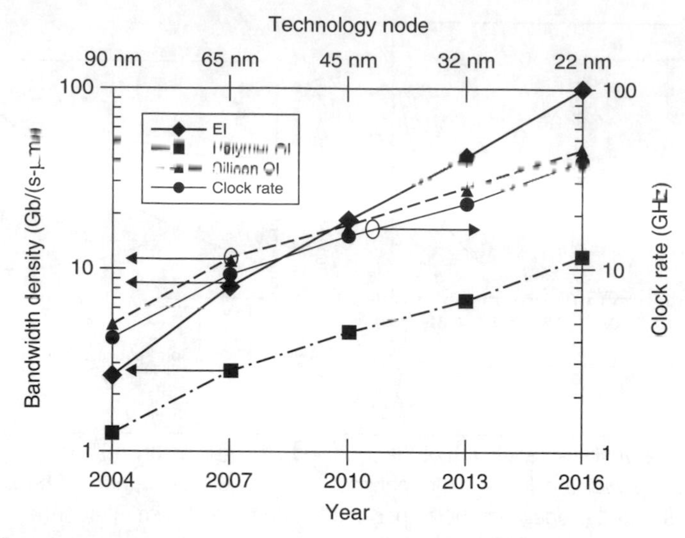

FIGURE 13.6

Comparison of bandwidth density of EIs and OIs as a function of year and technology node [8]
© 2006 IEEE

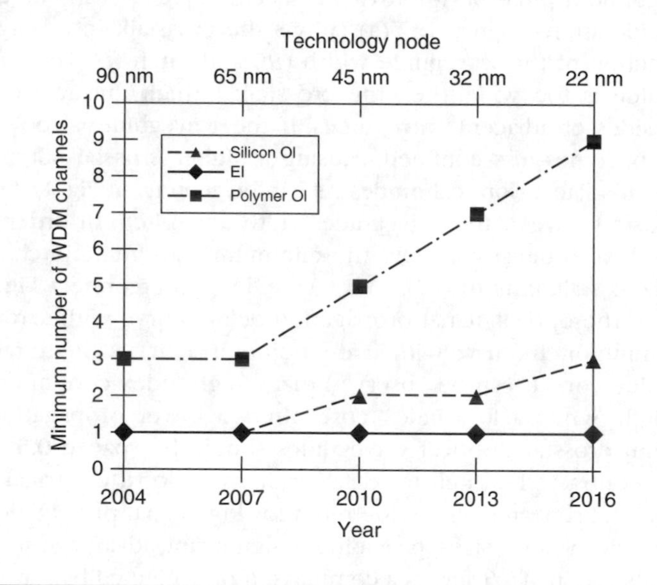

FIGURE 13.7

Number of OI WDM channels required to exceed the EI bandwidth density as a function of year [8]
© 2006 IEEE

To improve the bandwidth density in OIs, wavelength division multiplexing (WDM) [8, 30] can be used. Figure 13.7 shows the number of WDM channels required to match the EI bandwidth density for silicon and polymer waveguides over a number of technology nodes. Two types of trade-offs can be identified from this graph. Polymer core waveguides not only require a higher WDM to match the bandwidth density, but also allow for a larger conversion delay overhead. Silicon-core waveguides on the other hand permit lower WDM but require faster conversion (i.e., faster transmitters and receivers). Note, however, that each WDM channel has an area and delay penalty associated with it. The use of resonant structures for WDM architectures can help reduce WDM overhead [8].

13.1.1 Use of OIs for On-Chip Communication

On-chip optical interconnects have received significant attention in the past few years. It is claimed that optical interconnects will be suitable replacements for global on-chip interconnects, where they can be advantageous because of their inherent low propagation delay, low crosstalk, and a near constant power profile over long distances.

A few research efforts have studied how on-chip optical networks can be used for clock-tree networks [31–35], since clock networks are characterized by long interconnect lengths spanning the entire chip. Using optical links as a replacement for electrical clock networks can reduce clock distribution skew on the chip in the multi-GHz operating range. Since the clock source (i.e., the light source) can be external to the chip, it can remove energy consumption constraints and also alleviate the difficult problem of integrating III–V optoelectronic emitters on top of Si CMOS (complementary metal-oxide semiconductor) circuits [31]. A full CMOS-compatible process including Si-photo-detectors is feasible even if the quantum efficiency of CMOS compatible silicon photo-detectors is small [36–39]. Intel researchers have claimed [32] that WDM can enable optics to achieve high bandwidth and low latency for global signaling. However, they concluded that until efficient high speed and low capacitance CMOS-compatible modulators and detectors, and practical schemes for implementing WDM are realized, there is little power, jitter, or skew improvements from using optics in clock distribution. A similar claim was made by Chen et al. [33], where the authors argued that since most of the skew and power of clock signaling arises in local clock distribution, there is no significant skew and power advantages in using an optical solution. Ackland et al. [34] presented a study showing how an H-tree electrical clock network does not scale well, resulting in unacceptable levels of skew and jitter, compared to an optical tree. The authors claimed that while there are many technical challenges in implementing an optical tree, in principle the best solution might be a hybrid tree network, in which the front end is implemented optically, while the backend consists of a large number of small electrical trees.

A few recent works have proposed using optical links in an on-chip network-on-chip (ONoC) [22, 35, 40, 41]. O'Connor [35] and Briere et al. [40] gave a high level overview of a 4×4 multi-stage optical network-on-chip (NoC). Shacham et al. [41] presented an analysis of topology, routing algorithms, path-setup and teardown procedures, and deadlock avoidance for optical NoCs, and described simulation results for a 2-D folded torus optical NoC architecture, that can theoretically result in high bandwidth, low power intra-chip communication. An optical loop network was proposed by Kirman et al. [22]. Wavelengths were multiplexed in the silicon waveguide (that resides on a dedicated silicon layer) as part of WDM, either by assigning a wavelength to each address space, or assigning an equal number of wavelengths to the nodes in the system. A global optical waveguide connects electrical switch nodes, with each node being used to connect processors and memory. The authors modeled the speed, area, and power characteristics of electrical and recently developed optical components. They projected these models to the 32 nm technology, and determined that their proposed hybrid opto-electrical loop network yielded significant performance within reasonable power and area constraints. The performance improvements over purely EIs come mainly due to reduced contention for data transfers in the architecture. Extra wavelengths through WDM allowed virtually contention-less transfers over long distances in the network. However, the high power overhead of electrical routers and opto-electric/electro-optic conversion at the interface of each component, as well as a lack of availability of wideband photonic switching elements makes realizing these hybrid opto-NoC architectures a challenging proposition in the near future.

Recently, Pasricha and Dutt [105] proposed the ORB (optical ring bus) on-chip communication architecture for emerging multiprocessor SoC (MPSoC) designs. ORB is a novel opto-electric communication architecture that uses an ORB as a global interconnect between computation clusters (i.e., islands of multiple cores performing dedicated tasks) and traditional copper-based local interconnect within clusters. The ORB transfers data between clusters on a chip, while preserving the standard bus protocol interface (e.g., AMBA AXI) for inter- and intra-cluster communication. ORB consists of four major building blocks: (i) an *off-chip laser*, from which light is coupled onto the chip using optical fibers; (ii) *transmitters*, that convert electrical signals into optical waves and are made of hybrid Mach-Zehnder interferometer/microresonator-based P-I-N diode modulators, driven by a series of tapered inverters (i.e., drivers); (iii) an *optical waveguide*, made from a low refractive index polymer material (that has lower signal propagation delay than a silicon waveguide), having a ring shape (to avoid sharp turns that can lead to significant signal loss) and implemented on a dedicated layer, that is responsible for transporting data via light signals from the source modulator to the destination receiver; and (iv) *receivers*, consisting of a photo-detector to convert the light signal into an electrical signal, and TIA circuits to amplify the resulting analog electrical signal to a digital voltage level. To improve bandwidth density of the optical interconnect, wavelength division multiplexing (WDM) was used to transmit data on multiple wavelength channels on the same waveguide. The ORB communication architecture was shown to dissipate significantly lower power (more than a 10× reduction) and improve overall performance (more than 2×) compared to traditional pipelined, all-electrical global interconnects, across the 65–22 nm CMOS technology nodes, for several networking MPSoC design case studies.

From all the research to date in the area of on-chip optical interconnects, a strong case for their realization as global on-chip interconnects can be made, provided the following open problems are resolved in the coming years [8, 22]:

(i) *Efficient transmitter and receiver components*: High speed, low power, and small feature-size electro-optical modulators and photo-detector receivers need to be developed. The combined transmitter and receiver delays for these innovative solutions would have to be lower than 280–370 ps for polymer waveguides, and lower than 180–270 ps for silicon waveguides, for end-to-end chip global interconnects. Similarly, the total power consumption of the components should be reduced and at least comparable to that of EIs (approximately 18 mW) for chip-length global interconnects. Only then can optical interconnects be considered as advantageous over EIs, and replace them in future technologies. There is immense research interest in designing efficient electro-optical modulators and receivers, and some interesting results are emerging. For instance, recently, Mach-Zehnder electro-optic modulators with an ultra-compact length of 100–200 μm, having low power consumption and high modulation efficiency were presented in [114].

(ii) *Integrated on-chip light source*: The number of materials and processes available for optical interconnect fabrication is limited to those technologies that are compatible with microelectronics. Currently, this limitation results in the absence of a monolithic on-chip light source. While several exciting scientific achievements have been published in the area of optical gain in silicon [10, 15, 16], high speed, electrically driven monolithic light sources have remained elusive. Innovative solutions, such as the Indium Phosphide Hybrid Silicon Laser from Intel and UCSB [42] may, however, solve this problem in the future.

(iii) *Polymer waveguide*: Polymer waveguides provide a significant improvement in the waveguide refractive index over silicon waveguides, which translates into a compelling enough speed advantage to motivate research on it. However, a major drawback of polymer waveguides today is the lack of a suitable modulator. Polymer waveguide transmission systems based on VCSELs today have prohibitive manufacturing cost and complexity. Additionally, polymer waveguides have lower bandwidth density than silicon waveguides, requiring more aggressive levels of WDM than silicon for the same interconnect length. It may be possible in the future to design a hybrid system, consisting of fast, wider-channel polymer waveguides, and slower, but narrow channel silicon waveguide layers. The challenge then would be to architect the interconnect structure for an application to best exploit such a heterogeneous optical interconnect fabric.

(iv) *Temperature management*: Optical interconnect modules are very sensitive to temperature variations. Designers need to ensure that viable operating temperatures for components are maintained, or design new OI structures that are not so sensitive to temperature. Either an active or passive optical control method [43] will be required to maintain stable device operation.

13.2 RF/WIRELESS INTERCONNECTS

Another revolutionary approach to overcome the limitations of traditional EIs in future ultra-large scale integration (ULSI) technologies is to make use of *active* RF/

wireless interconnects. The main idea in such a system is to replace on-chip wires with integrated on-chip antennas communicating via electromagnetic waves. Data in such a scheme would be converted from baseband (i.e., digital) to RF/microwave signals and transmitted either through free space or guided mediums [44]. Free space signal broadcasting and reception is a common practice in modern wireless systems due to its low cost implementation and excellent channeling capabilities. However, free space transmission and reception of RF/microwave signals requires an antenna size that is comparable to its wavelength. This is a problem because even at near 100 GHz operating and cut-off frequencies in the future [6], the aperture size of the antenna for efficient transmission will be in the order of 1 mm^2, which is too large to implement in future ULSI designs. Microwave transmission in guided mediums such as microstrip transmission line (MTL) or coplanar waveguide (CPW) has a low attenuation up to at least 200 GHz [104], requires a smaller antenna size and is therefore a more viable alternative to free space transmission/reception for intra-chip communication. Simulation results [44] have shown that a 1 cm long CPW experiences extremely low loss (-1.6 dB at 100 GHz), and low dispersion (less than 2 dB for a frequency range 50-150 GHz). This is in contrast to a -60 and -115 dB loss per centimeter at 100 GHz, and a frequency dispersion of 30–40 dB across the same frequency range for conventional EIs. Thus microwave transmission over MTL or CPW has a clear advantage over conventional electrical wire-based transmission, especially for global interconnects in future ULSI designs running in the multi-GHz range.

The choice of transmission and receiving components for an RF/wireless interconnect system is an important one, in order to ensure compatibility with future VLSI designs [44]. Since the distance of global on-chip interconnects is relatively short (a few centimeters), the conventional *far field* antenna can be substituted for much smaller *near field* capacitive couplers [45]. The width of the center/top conductor of the CPW/MTL is typically 10–100 μm, depending on its characteristic impedance (25–100 Ω) and other signal transmission requirements [46, 47]. This size makes CPW or MTL more likely to be used as an off-chip but in-package transmission medium, which is shared by multiple ULSI I/Os. Based on these considerations, Fig. 13.8 shows the structure of a typical RF/wireless interconnect system. The system acts as a miniature wireless LAN (local area network), consisting of ULSI I/Os as users, capacitive couplers as near-field antennas, RF circuits

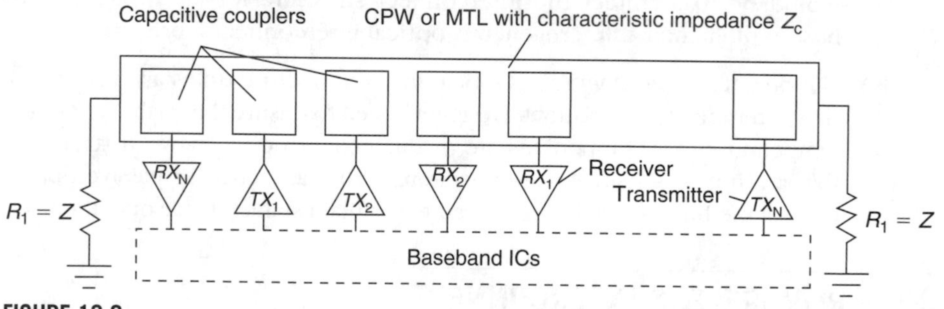

FIGURE 13.8

RF/wireless interconnect with multiple transmitters and receivers [44]
© 2001 IEEE

for transceivers, and a uniform and homogeneous MTL or CPW channel (with characteristic impedance Z_C) as a shared broadcasting medium. Output signals are *up-linked* to MTL or CPW via transmission capacitive couplers (TX_i), then down-linked via receiving capacitive couplers (RX_i) to input ports. Since the channel is bidirectional, both its ends are terminated with Z_C to avoid signal reflections. With orthogonal-coded and/or frequency-filtered RF transceivers, a passive MTL or CPW becomes suitable for relaying ultra-broadband signals up to 150 GHz [46, 48].

Frequency division multiple access (FDMA), code division multiple access (CDMA), or a combination of the two (FDMA/CDMA) can be used to achieve simultaneous communications in RF/wireless interconnects. In an FDMA interconnect, frequency bands of I/O channels can be allocated between 5–105 GHz with a bandwidth of approximately 5–20 GHz in each channel and sustaining a minimum data rate of 5–40 Gb/s depending on the modulation scheme. Whereas in a traditional EI, only the lowest frequency band (i.e., baseband) is occupied by the signals, RF modulated frequency bands can improve data bandwidth by transmitting data over multiple frequency bands. A high level block diagram of an FDMA interconnect is shown in Fig. 13.9. Each data stream at the transmitter is multiplied by a sinusoidal carrier, and the resulting signal is filtered through a band pass filter (BPF), and then eventually coupled into the CPW (or MTL). At the receiver, the received signals are boosted by a preamplifier and then demodulated. Subsequently, the signal is fed into a threshold comparator to recover the original data bits from the transmitter [49]. The FDMA transmitter and receiver architectures consist of I/O transceivers, frequency synthesizers, threshold comparators, frequency/code mixers, and BPFs. The receiver is made up of preamplifiers, mixers, and frequency synthesizers. Preamplifiers with 20–30 dB gains are required for input signal amplification. Balanced or double balanced active mixers, such as the Gilbert cell, may be used for modulation and demodulation functions. To simplify receiver implementation, a non-coherent detection scheme such as frequency shift keying (FSK) can be a suitable choice. Low loss and high selective BPFs that

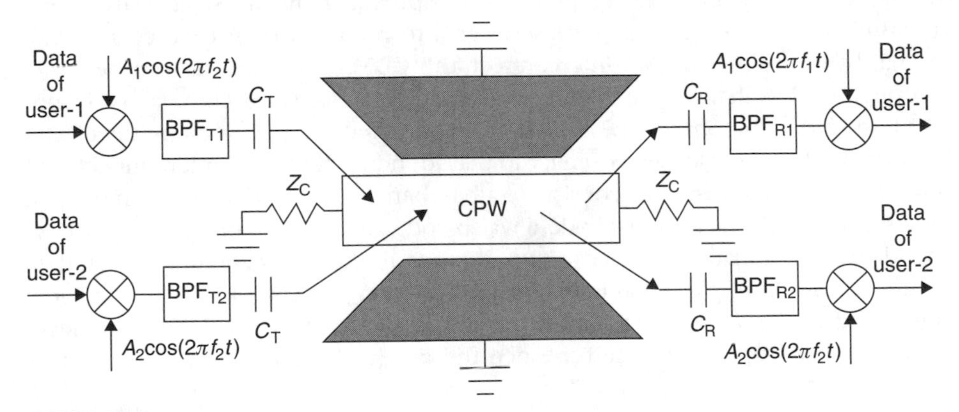

FIGURE 13.9

Schematic diagram of FDMA interconnect [44]
© 2001 IEEE

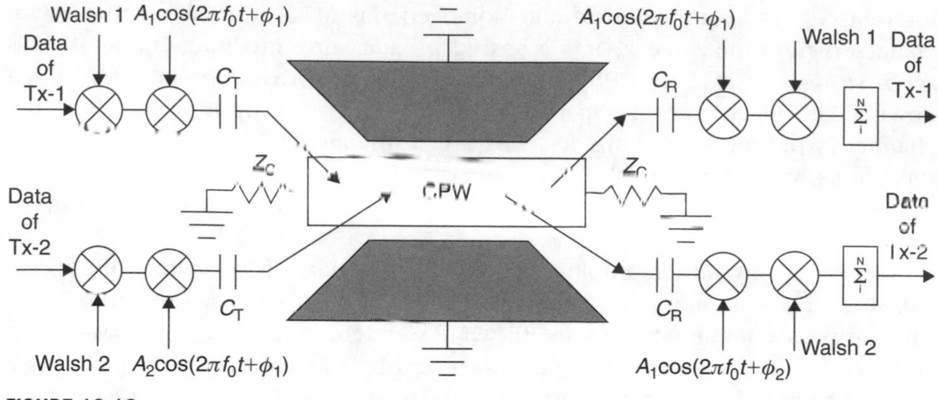

FIGURE 13.10

Schematic diagram of CDMA interconnect [44]
© 2001 IEEE

are needed in FDMA interconnects require tunable and high Q[1] inductors that are hard to realize due to their high energy loss to the conductive silicon substrate. This can be partially resolved by using a transformer type inductor design where the lost energy is recovered via a secondary inductor with delayed phase angles to attain high inductance and tunability [50]. Recent progress in MEMS (micro-electromechanical systems) have shown promise for high Q silicon resonators and filters in micrometer wave frequencies [51].

A high level schematic of a CDMA interconnect [44, 52–54], where baseband CDMA signals have been modulated with RF carriers, is depicted in Fig. 13.10. The data stream from a transmitter is first converted into a spread spectrum signal by orthogonal Walsh codes (or PN codes [55]) and then modulated with its RF sinusoidal carrier. Subsequently the resulting signals from the transmitters are capacitively coupled into a superposed multilevel signal on the shared CPW (or MTL) and transmitted to receivers. At the receiver end, coherent demodulation [49], sequence timing acquisition and tracking [56], and de-spreading by the same Walsh code are applied to the received superposed signal to recover the original data. Unlike the hardware-oriented FDMA interconnect, the CDMA interconnect can be easily reconfigured by changing spreading codes through software commands. Figure 13.11 shows the baseband CDMA transceiver architecture [53] in more detail. The RF portion of the transceiver is similar to that found in the FDMA interconnect, and is not shown. The transmitter consists of a baseband CDMA modulator to modulate the user data with the assigned code, a Walsh code generator to provide the orthogonal data "keys," a data combiner to put together all the user data into a serial data sequence with multiple signal levels for transmission, a phase-lock-loop clock generator, and finally a matching output buffer to transmit the signal. The multilevel signal is picked up at the receiver. Since the received signal has multiple values,

[1] The quality factor (or Q) of an inductor is the ratio of its inductive reactance to its resistance at a given frequency, and is a measure of its efficiency. The higher the Q factor of the inductor, the closer it approaches the behavior of an ideal, lossless inductor.

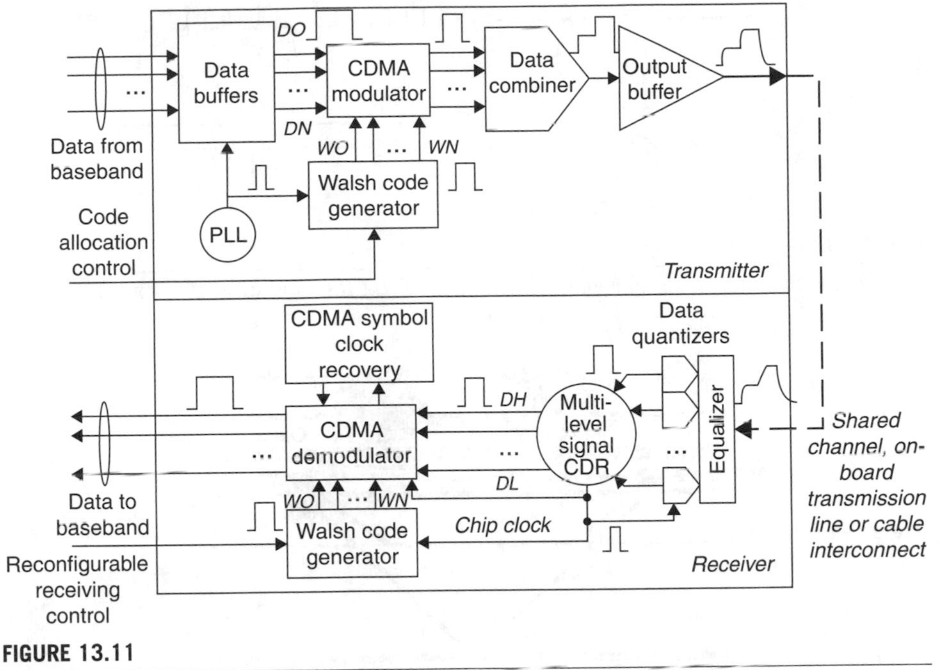

FIGURE 13.11

CDMA transceiver architecture [53]
© 2005 IEEE

a multilevel signal clock-data-recovery (CDR) circuit is needed to synchronize the clock and recover the baseband multilevel data. An Alexander type phase-detector CDR circuit [57] can be used for this purpose. It consists of two analog to digital (A/D) converters, which are used to quantize the input signal at the rising and falling edge of the clock. To achieve immunity to noise on the control line so as to satisfy jitter requirements, while using less silicon area, a differential control voltage controlled oscillator (VCO) using symmetrical metal-insulator-semiconductor (MIS) varactor pairs can be implemented [54]. After the clock is recovered from the multilevel signals, and the signal is quantized into a multi-bit data, the original data is recovered by demodulating the data with corresponding Walsh codes. An asynchronous error-detection correlator is used to obtain synchronization of the input data within one-symbol clock period. To improve data rate, a combined CDMA/FDMA [44, 53] access system (also referred to as multicarrier CDMA) is also a possibility, with frequency bands being divided by using different carriers while data is spread within each frequency band by using orthogonal codes.

The on-chip antennas fabricated on substrates are categorized as printed antennas [58–60], which include microstrip, dipole, and loop antennas. Figure 13.12 shows the popular linear dipole, and folded dipole antennas that have a compact implementation and are thus suitable as on-chip integrated antennas. The zig-zag dipole antenna emerged after Nakano and Yamauchi [61] showed that a halfwave dipole antenna whose arms were bent rectangularly at its central points maintained the basic characteristics of conventional linear dipole with the same length. The loop antenna is another suitable candidate for on-chip integrated antennas.

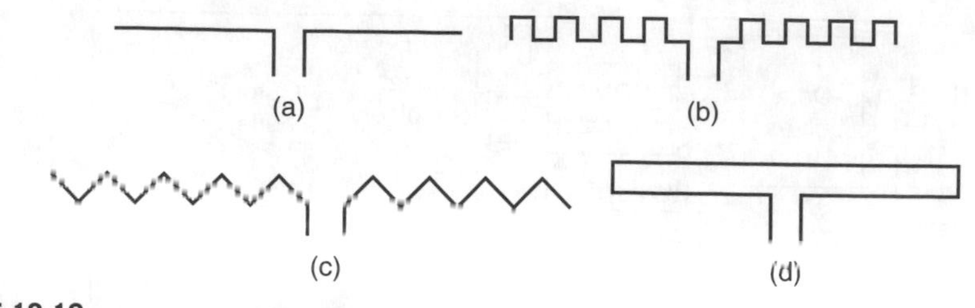

FIGURE 13.12

Types of on-chip antennas: (a) linear, (b) meander, (c) zig-zag, and (d) folded [58]
© 2006 IEEE

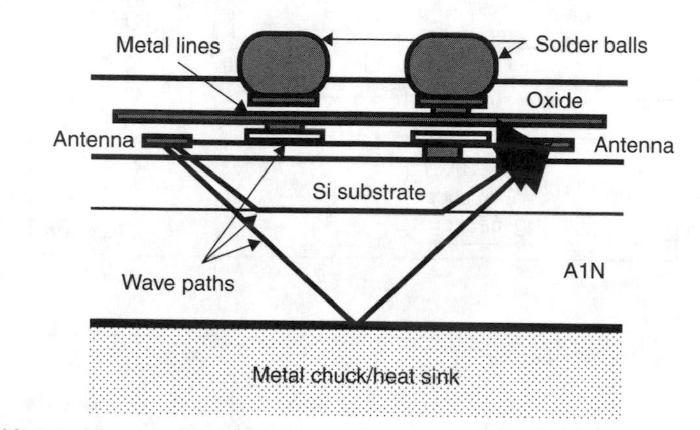

FIGURE 13.13

Antenna transmission paths [64]
© 2005 IEEE

It was shown to have an isotropic radiation pattern and good signal-to-noise ratio when the receiver is placed close to the loop antenna [62]. Wang et al. [58] showed that combining different antenna structures such as folded and meander can provide a higher power gain and a more compact on-chip antenna structure.

There has been some recent research on using the on-chip silicon substrate/dielectric layer as a transmission path, instead of using off-chip and in-package CPW or MTL guided mediums [63, 64]. At 24 GHz, the wavelength of electromagnetic waves in silicon is 3.7 mm. This implies that a quarter wave antenna needs to be only about 0.9 mm in silicon. This, in conjunction with increased chip sizes of 2 cm × 2 cm, has made the integration of antennas for wireless on-chip communication feasible. Figure 13.13 shows the possible paths for signal propagation between two on-chip integrated antennas. There is a direct path through air and paths through the silicon substrate. The paths through silicon substrate include a path formed by refraction through the SiO_2 layer and reflection at the interface between the silicon substrate and the underlying dielectric layer (A1N), and a path refracted through the SiO_2 and silicon layers and reflected by the metal chuck that emulates a heatsink in the back of a die. The signals propagating on these paths constructively and destructively interfere. There are also multiple reflected paths, but because of longer path lengths, the signals propagated on these paths suffer from greater attenuation and therefore these paths are not included. It was found in [64] that by increasing the A1N thickness, destructive signal interference is significantly reduced. Benech et al. [65] conducted a similar feasibility study and found that higher resistivity substrates are better suited for wireless communication. They showed that antennas on lower resistivity silicon substrate present gains of −30 dB, while antennas on SoI (silicon on insulator) substrates of higher resistivity present gains of −15 dB at frequencies around 30 GHz.

13.2.1 Use of RF/Wireless Interconnects for On-Chip Communication

RF/wireless interconnects have the potential to support high data bandwidths, effectively handle concurrent communication streams from multiple cores, and provide online firmware network reconfigurability (i.e., dynamic rewiring using software instructions). These wireless interconnects will not only reduce the wires in integrated circuits, but can also be used to replace I/O pins.

It has been proposed that RF/wireless interconnects can be used in clock networks to reduce signal skew [66]. Test chips have been created to demonstrate the feasibility of RF/wireless on-chip interconnects and their use in clock networks. For instance, on a wafer, a 15 GHz transmitted signal 2.2 cm away from a clock receiver with an on-chip antenna was shown to have been successfully picked up by the receiver and amplified to generate a digital output signal [67]. The receiver and transmitting antennas were fabricated using a 0.18 μm CMOS process [63]. Another demonstration of wireless clock was presented by Floyd et al. [68] at the frequency of 7.4 GHz and with a transmission gain of −49 dB between the transmitting and the receiving antenna placed at a distance of 3.3 mm from each other.

RF/wireless interconnect technology is still very much in its early stages of evolution. Much research is still needed to resolve the open problems in the area [44, 64]. Some of these issues are presented below:

(i) *Packaging and interference issues*: The most difficult problem anticipated for this technology is dealing with the packaging effects, which can add numerous metal structures that could potentially interfere with RF/wireless operation on a chip. Metal structures near antennas can change input impedances and phase of received signals. Design guidelines to exclude the interference structures which significantly change the input impedance and techniques to correct the phase changes are being developed [69, 70]. Another concern is the interference effects between the transmitted/received signal and switching noise of nearby circuits [71–74]. Some promising initial results were presented by Branch et al. [63], where a sine wave generated in a transmitter was transmitted through an on-chip antenna and the wave was picked up by a receiver on the same chip about 4 mm away, for an on-chip silicon substrate/dielectric transmission path. It was concluded that on-chip wireless interconnects can function correctly for more complex systems, provided the interference concerns are addressed.

(ii) *Ultra-high frequency requirements*: For antenna sizes (and related RF interconnect circuit sizes) to be feasible enough for on-chip fabrication,

RF circuits must operate in the ultra-high frequency domain (i.e., in the hundreds of GHz range). This makes them unsuitable for applications in the very near future that will not be able to achieve such high frequencies. However, due to technology scaling, CMOS circuits operating in the multi GHz and higher ranges are becoming gradually feasible. It is estimated [6, 64] that by the year 2015, it will become possible to implement RF circuits operating at 200–250 GHz, which will enable low cost wireless interconnect implementation.

(iii) *Power overhead*: If network flexibility is the highest concern (i.e., bidirectional communication and arbitrary transceiver distribution are top priorities) then each transceiver should have its own dedicated RF and CDR circuits. This of course will impose heavy circuit overhead as well as large power consumption to the RF/wireless interconnect implementation. To reduce the power consumption, multiple transmitters within a synchronous access range may need to share a common RF transmitter, while multiple receivers may end up sharing a common RF receiver and even a shared CDR without sacrificing significant channel reconfigurability.

(iv) *On-chip antennas*: While there has been a lot of past research on fabricating antennas on lossless or lower loss substrates such as polytetrafluoroethylene (PTFE), quartz, duroid, and GaAs in the millimeter wave range applications, there has not been sufficient research in the area of fabricating printed antennas on silicon substrate which is much more lossy than other types of substrates. The conductivity of silicon substrate will reduce the antenna efficiency [75], which necessitates choosing either low loss type of substrates (that may not be compatible with mass-scale fabrication techniques) or high transmission power to improve radiation efficiency [76].

(v) *Reference crystal oscillator*: Another issue for the adoption of RF/wireless interconnects is the reference crystal oscillator required for FDMA, which cannot be easily implemented on-chip and has a relatively large size compared with future ULSI. Consequently, this crystal oscillator will have to be implemented off-chip. It may, however, be possible to reuse a reference clock to minimize off-chip overhead.

(vi) *Security*: As with any wireless transmission system, RF/wireless interconnects are susceptible to hackers that intend to snoop on transmitted data and compromise the system. A lot of research efforts are needed to identify security issues with RF/wireless interconnects and, if needed, to develop new (or adapt previously developed) techniques that prevent wireless signal decryption and protect transmitted data from malicious entities.

13.3 CNT INTERCONNECTS

In order to alleviate the limitations associated with copper-based interconnects, changes in the materials used for on-chip interconnections are being actively explored. CNTs have been recently proposed as a replacement for metal interconnects in future technologies [3, 77, 78]. CNTs are sheets of graphite rolled into cylinders of diameters varying from 0.6 to about 3 nm. Depending on the direction in which they are rolled (called *chirality*), they can demonstrate either metallic or semiconducting properties. CNTs possess high mechanical and thermal stability, high thermal conductivity, and large current carrying capacity, making them promising candidates as on-chip interconnects in future technologies [79–81]. Due to their covalently bonded structure, they are highly resistant to electromigration and other sources of physical breakdown [106]. They can support very high current densities with very little performance degradation. For instance, it was shown in [107] that the current carrying capacity of CNTs did not degrade even after 350 hours at current densities of $\sim 10^{10}$ A/cm^2 at 250°C. CNTs possess high thermal conductivity in the range of 1700–3000 W/mK [108]. They also have much better conductivity properties than copper owing to longer electron mean free path (MFP) lengths in the micrometer range, compared to nanometer range MFP lengths for Cu [109].

It is predicted that isolated CNTs can replace copper at the local interconnect level because of their much lower lateral capacitance which improves latency for very short distances [110]. However, the high intrinsic resistance associated with an isolated CNT (greater than 6.45 KΩ) [82] necessitates the use of a bundle of CNTs conducting current in parallel to form longer on-chip interconnections [77, 78]. CNT bundle structures have recently been demonstrated and their metallic conducting properties reported in [83–85]. Due to the lack of control over chirality in current fabrication techniques, any bundle of CNTs consists of metallic nanotubes that contribute to current conduction, as well as semiconducting nanotubes that do not contribute to current conduction in an interconnect. CNTs are commonly classified into single-walled (SWCNT—shown in Fig. 13.14), and multi-walled (MWCNT—comprised of multiple concentric SWCNT shells) carbon nanotubes. Even though MWCNTs are primarily metallic, preliminary research

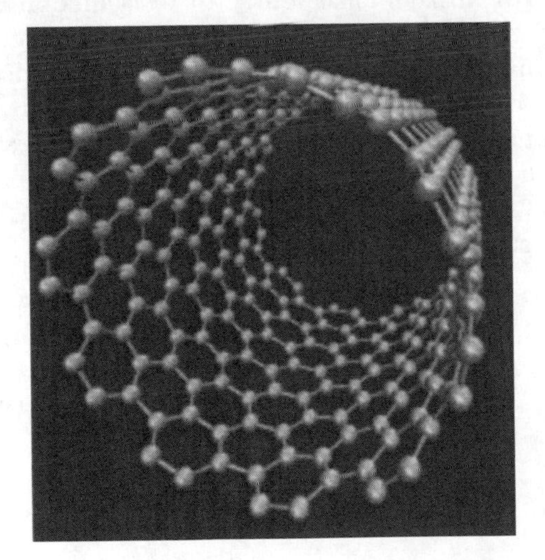

FIGURE 13.14

Single-walled carbon nanotube [88]
© 2004 IEEE

has shown that it is relatively more challenging to achieve transmission over long lengths with them [86, 87], compared to SWCNTs.

13.3.1 Circuit Parameters for Isolated SWCNTS

A detailed analysis of the circuit parameters of SWCNTs was presented in [3, 89]. Figure 13.15 shows the equivalent circuit model, for which the parameters are described below.

The conductance of a CNT can be evaluated using the two-terminal Landauer-Buttiker formula which states that, for a 1-D system with N channels in parallel, the conductance is given as $G = (Ne^2/h)T$, where e is the charge of an electron, h is Planck's constant, and T is the transmission coefficient for electrons through the sample [82]. As a result of spin degeneracy and sublattice degeneracy of electrons in graphite, each CNT has four conducting channels in parallel (i.e., $N = 4$). Hence the conductance of a SWCNT, assuming perfect contacts ($T = 1$), is given by $4e^2/h = 155\,\mu S$, which implies a resistance of $6.45\,K\Omega$ [82]. This is the fundamental resistance ($R_F = h/4e^2$) associated with a SWCNT that cannot be avoided [90]. Figure 13.15 shows R_F equally divided between the two contacts on either side of the nanotube. The MFP (L_0) of electrons is defined as the distance across which no scattering occurs. The MFP length is calculated as $L_0 = v_F d_t/\alpha T$, where α is the total scattering rate, v_F is the Fermi velocity of graphene, d_t is the nanotube diameter, and T is the temperature. Ideally, the MFP length (L_0) value is in the order of several μm. However, a rigorous analysis has shown that practically L_0 is in the order of 1 μm [82, 91, 92]. For lengths less than L_0, the resistance of a CNT is independent of length. However, for lengths larger than L_0, resistance increases with length, and this can be expressed as $R_{CNT} = R_F \cdot L/L_0$, where L is the length of the CNT [92]. In the circuit of Fig. 13.15, the additional scattering resistance would appear as a distributed resistance per unit length to account for resistive losses along the CNT length. An additional source of resistance in CNTs is due to the presence of imperfect metal-nanotube contacts. This resistance is often extremely high, and appears in series with R_F in Fig. 13.15, divided equally among the two end contacts. The observed total resistance associated with an isolated SWCNT (sum of fundamental, scattering, and imperfect contact resistances) has been found to be too high (~100 KΩ) for realizing on-chip interconnects. Hence, a bundle (or rope) of CNTs is needed that has much lower effective resistance (because of its larger cross-section area) to implement a viable on-chip interconnection.

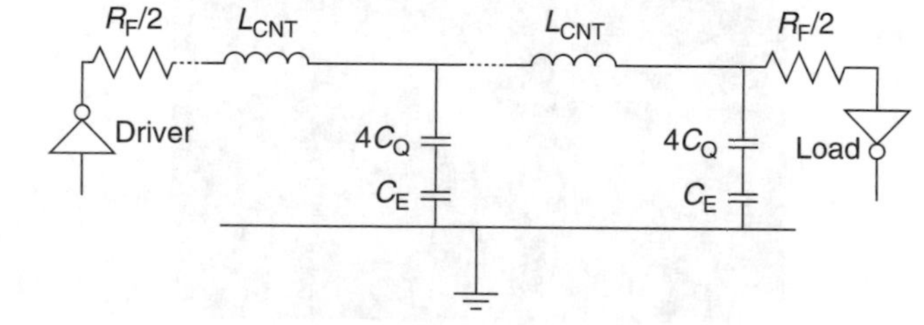

FIGURE 13.15

Equivalent circuit model for an isolated SWCNT [3, 89]
© 2005 IEEE

The capacitance of a CNT is comprised of its electrostatic capacitance (C_E) and its quantum capacitance (C_Q). The electrostatic capacitance C_E (also called intrinsic CNT capacitance) is calculated by treating the CNT as a thin wire with diameter d and placed a distance y away from a ground plane. For $y > 2d$, C_E is given as [89]:

$$C_E = \frac{2\pi\varepsilon}{\ln\left(\frac{y}{d}\right)}$$

The quantum capacitance (C_Q) accounts for the quantum electrostatic energy stored in the nanotube when it carries current [89]. Due to the Pauli exclusion principle, it is only possible to add electrons into the nanotube at an available quantum state above the Fermi energy level. By equating this energy to an effective capacitance, the expression for the quantum capacitance (per unit length) is obtained as:

$$C_Q = \frac{2e^2}{hv_F}$$

where h is the Planck's constant and v_F is the Fermi velocity. For a CNT ($v_F \approx 8 \times 10^5$ m/s), and $C_Q \approx 100$ aF/μm [89]. Since a CNT has four conducting channels (as described earlier), the effective quantum capacitance due to four parallel capacitances is given by $4C_Q$. The same effective charge resides on both these capacitances (C_E and $4C_Q$) when the CNT carries current, as is true for any two capacitances in series. Hence these capacitances appear in series in the effective circuit model shown in Fig. 13.15.

The inductance associated with an isolated SWCNT is comprised of the magnetic inductance (L_M) and kinetic inductance (L_K). L_M can be calculated from the magnetic field of an isolated current carrying wire some distance away from a ground plane, and is given as:

$$L_M = \frac{\mu}{2\pi} \ln\left(\frac{y}{d}\right)$$

The kinetic inductance is calculated in [89] by equating the kinetic energy stored in each conducting channel of the CNT to an effective inductance, and is expressed as:

$$L_K = \frac{h}{2e^2 v_F}$$

The four parallel conducting channels in a CNT give rise to an effective kinetic inductance of $L_K/4$. Typically, $L_K >> L_M$. Since large inductive effects expected due to L_K are not observed up to frequencies as high as 10 GHz, its effect on the interconnect delay can be ignored [93].

13.3.2 Circuit Parameters for a Bundle of SWCNTs

It is clear that an isolated SWCNT has very high intrinsic resistance, which makes it inferior to copper interconnects in terms of performance. Even using parallel SWCNTs (Fig. 13.16(a)) to reduce resistance does not offer much advantage over

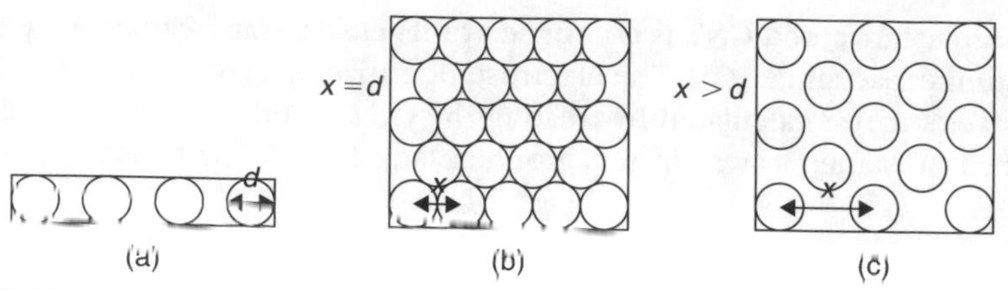

FIGURE 13.16

SWCNT bundle interconnects: (a) flat SWCNT array, (b) dense SWCNT bundle, and (c) sparse SWCNT bundle [3]

Cu interconnects [94, 95]. Bundled (or stacked) SWCNTs can, however, reduce CNT interconnect resistance significantly. A bundled SWCNT interconnect consists of hexagonally packed identical metallic SWCNTs [3]. Each SWCNT is surrounded by six immediate neighbors, with their centers separated from each other by a distance x. A densely packed structure with $x = d$ (SWCNT diameter), shown in Fig. 13.16(b), will lead to the best interconnect performance. In reality, not all SWCNTs of a bundle are metallic—there are non-metallic SWCNTs that do not contribute to current conduction and their presence is taken into account by considering a "sparsely" populated SWCNT bundle model, as shown in Fig. 13.16(c). A metallic density (P_m) parameter associated with an SWCNT bundle refers to the probability that an SWCNT in the bundle is metallic. The value of $P_m \approx 1/3$ with today's best fabrication techniques, which implies that only 1/3 of the SWCNTs in a bundle are metallic (i.e., conducting).

The resistance of a SWCNT bundle is given by $R_{bundle} = R_{isolated}/n_{CNT}$, where $R_{isolated}$ is the resistance of an isolated SWCNT, and n_{CNT} is the total number of SWCNTs forming the bundle. The value of n_{CNT} is given as:

$$
\begin{aligned}
n_w &= \frac{w - d_t}{x}, \qquad n_h = \frac{h - d_t}{(\sqrt{3}/2)x} + 1 \\
n_{CNT} &= n_w n_h - \frac{n_h}{2} \quad \text{if } n_h \text{ is even} \\
&= n_w n_h - \frac{n_h - 1}{2} \quad \text{if } n_h \text{ is odd}
\end{aligned}
$$

where the spacing between SWCNTs $x = d_t/\sqrt{P_m}$, and n_w and n_h are the number of SWCNTs along the width and the height of the bundle, respectively.

As far as bundle capacitance is concerned, it is observed [3] that the mutual capacitance between SWCNTs of the same bundle is not important as it does not place any additional load on the interconnect. This is because when the SWCNT bundle carries current, it is assumed that each SWCNT conducting current is held at the same potential as any other. The electrostatic capacitance of the bundle (C_E^{bundle}) is expected to arise mainly from the SWCNTs lying at the edges of the bundle that are capacitively coupled with the adjacent interconnects (left and right neighbors) as well as the substrate. The coupling to the adjacent interconnects is expected to be higher than that to the substrate due to higher proximity.

In the absence of actual data on the electrostatic capacitance of a SWCNT bundle of these dimensions, a coarse model was developed in [3] using the field solver FastCap [96]. It was observed that the SWCNTs completely surrounded by other nanotubes had a very small electrostatic coupling capacitance to ground compared to those CNTs along the edges of the bundle (three orders of magnitude smaller). SWCNTs that are nearly completely surrounded had capacitance (to ground) over an order of magnitude less than CNTs along the edges. Their contribution to the total electrostatic capacitance can thus be neglected. The electrostatic capacitance contributed by each edge CNT in a bundle, in terms of intrinsic plate capacitance (C_E) for an isolated CNT, is expressed as:

$$C_E^{bundle} = 2C_{En} + \frac{n_W - 2}{2} C_{Ef} + \frac{3(n_H - 2)}{5} C_{En}$$

where n_W and n_H are the number of CNTs along the width and height of the bundle, respectively, and C_{En} and C_{Ef} are the intrinsic plate capacitances. C_{En} is the value of C_E calculated for a separation distance of s from the "near" adjacent interconnect. C_{Ef} is the value of C_E calculated for a separation distance of $s + w$ from the "far" adjacent interconnect. For simplicity, coupling between different interconnect layers is not considered in the analysis. The quantum capacitances of all the CNTs forming a bundle appear in parallel. The effective quantum capacitance of the bundle is therefore the sum of the individual quantum capacitances, $C_Q^{bundle} = C_Q^{CNT} \cdot n_{CNT}$. The effective capacitance (C_{bundle}) of the series combination of the electrostatic and quantum capacitances is given by the expression:

$$\frac{1}{C_{bundle}} = \frac{1}{C_Q^{bundle}} + \frac{1}{C_E^{bundle}}$$

Finally, the kinetic inductance of an SWCNT bundle is the parallel combination of individual SWCNT kinetic inductances (L_k). The magnetic inductance (L_m) remains relatively constant with wire dimensions. The mutual inductance between SWCNTs in a bundle is accounted for using the partial element equivalent circuit (PEEC) model [111]. The total SWCNT bundle inductance per unit length is given by:

$$L_{SWCNT\ bundle} = \left(\frac{L_k}{4n_{CNT}} + L_m \right)$$

13.3.3 Comparison between Copper and SWCNT-Bundles

Srivastava and Banerjee [3] presented a performance comparison between copper (Cu) interconnect and a SWCNT bundle having the same dimensions as the Cu interconnect. Figure 13.17 shows a comparison of the resistance and capacitance of Cu and CNT bundle interconnects for local interconnect lengths (no more than a few microns in length and consequently less than MFP length L_0). Figure 13.17(a) shows that the resistance of CNT bundles with imperfect contacts (120 KΩ) remains higher than Cu interconnect across technology generations. Perfect contacts

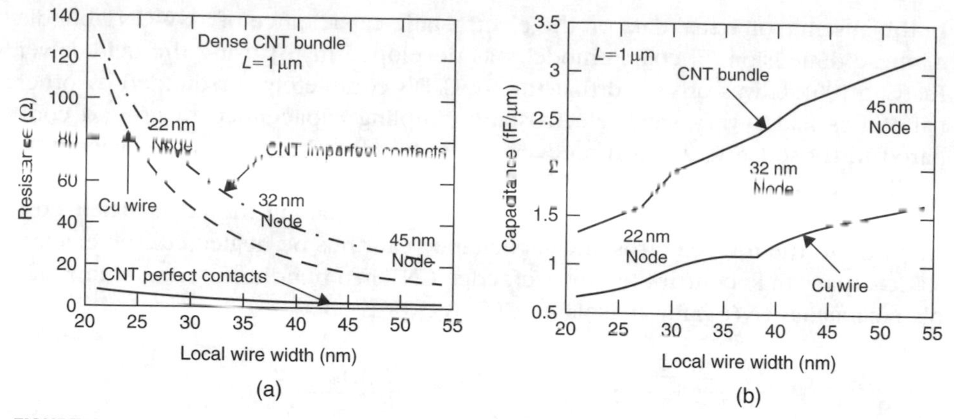

FIGURE 13.17

Comparison of (a) resistance, (b) capacitance between a Cu interconnect and CNT bundle interconnect [3]
© 2005 IEEE

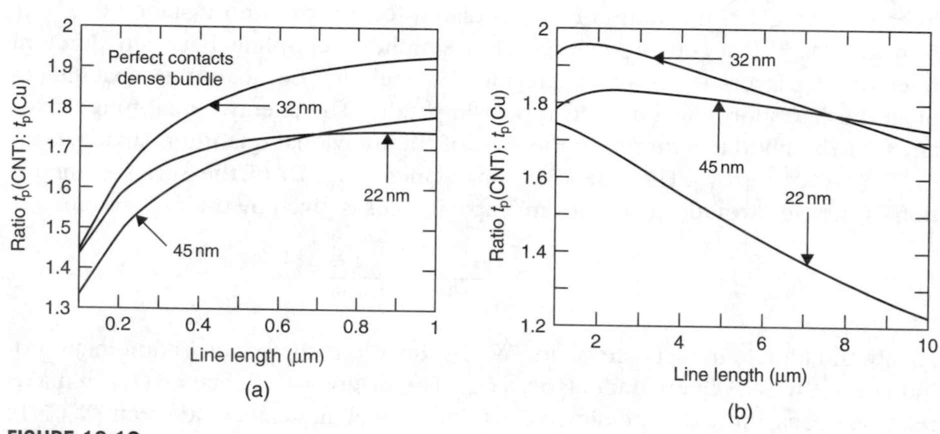

FIGURE 13.18

(a) Ratio of CNT bundle propagation delay with Cu wire propagation delay as a function of local interconnect length and (b) ratio for larger values of L, assuming MFP $L_0 > L$ [3]
© 2005 IEEE

can, however, make the resistance much lesser than Cu interconnect. The experiment assumes maximum densely packed CNT bundles. With the reduced densities that are realistically achievable, the resistance of CNT bundles will be higher. Capacitance of CNT bundles, on the other hand, remains higher than copper wires across all technology generations, as shown in Fig. 13.17(b).

Figure 13.18 shows that the propagation delay of local interconnects with densely packed CNT bundles is higher than that with Cu interconnects across all technology generations, even if contacts are perfect and an MFP (L_0) as large as 10μm can be achieved. This is because the higher capacitance of CNT bundles and the high resistance of minimum sized drivers at the local interconnect level overshadow the advantage from low CNT bundle resistance.

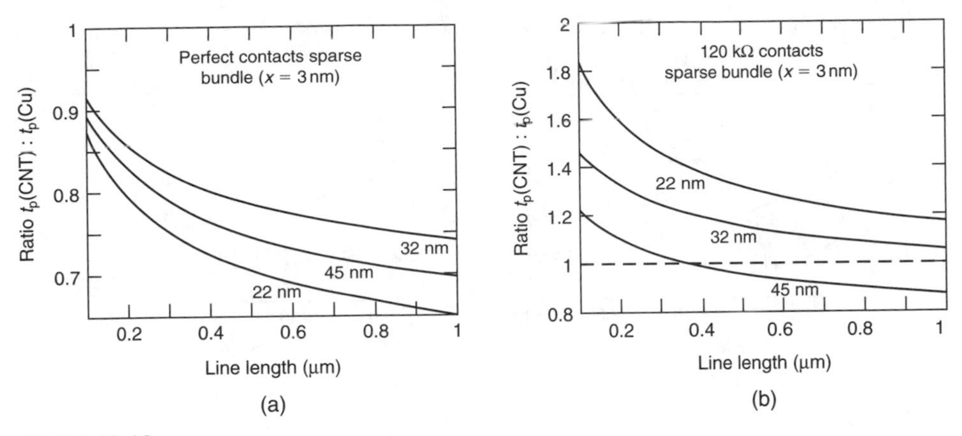

FIGURE 13.19

Ratio of local interconnect propagation delay with sparse CNT bundle interconnect having (a) perfect contacts, (b) imperfect (120 KΩ) contacts, to that with Cu interconnect as a function of interconnect length [3]

Figure 13.19(a) shows that the performance of CNT bundles with perfect contacts becomes better than Cu wires if the distance between adjacent metallic CNTs forming a bundle is increased. However, with realistic (i.e., imperfect) contacts (Fig. 13.19(b)), the delay is still higher than Cu interconnects. It is worth noting that it is very difficult to achieve the maximum packing density for CNTs [5], and these lower densities may be more easily achievable. However, the density can be reduced only up to a small extent beyond which the improvement in performance is lost due to increasing resistance of the bundle.

Figure 13.20 shows how the ratio of the propagation delay for densely packed CNT bundle interconnects t_p(CNT) with the propagation delay for Cu interconnects t_p(Cu) varies with wire line length at the intermediate level (assuming MFP $L_0 = 1\,\mu$m). A ratio value of greater than 1 indicates that the propagation delay for Cu interconnect is lower than that for CNT bundle interconnect, while a value of less than 1 indicates that propagation delay for the CNT bundle interconnect is lower than that for the Cu interconnect. It can be observed from the figures that beyond a certain minimum length (e.g., about 35 μm at the 45 nm technology node), the performance of CNT bundle interconnects at this level is better than Cu wires. As technology scales, this minimum length decreases while the improvement in performance increases. Figure 13.21(a) shows how propagation delay for densely packed CNT bundle interconnects depends on the length at the global level. It can be seen that global interconnects implemented with CNT bundles can achieve significantly better performance than copper. In the case of both intermediate and global interconnects, the improvement in performance with CNT bundles is larger for longer interconnects and saturates beyond a certain length. These observations hold if the MFP has an ideal value of 1 μm. Unfortunately, the presence of defects in a nanotube leads to the MFP being much less than the typical 1 μm [97]. The impact of reduced MFP lengths on global interconnect propagation delay with CNT

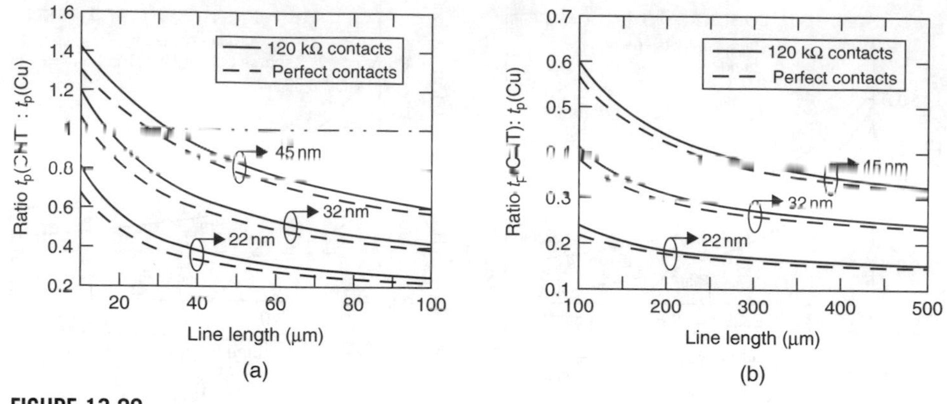

FIGURE 13.20

Ratio of propagation delay with dense CNT bundle interconnect to that with Cu interconnect as a function of intermediate level interconnect length ranging from (a) 0–100 μm and (b) 100–500 μm [3]

© 2005 IEEE

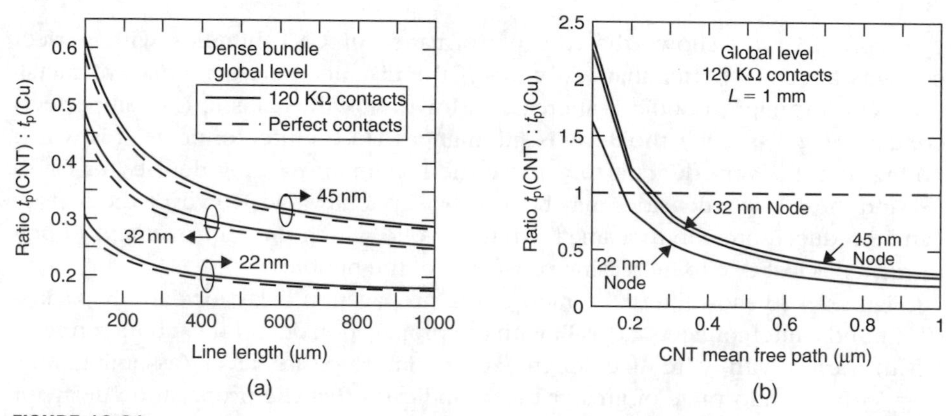

FIGURE 13.21

Ratio of global interconnect propagation delay with dense CNT bundle to that with Cu: (a) as a function of long interconnect length (L_0=1 μm), (b) as a function of L_0 for 1 mm long interconnect [3]

© 2005 IEEE

bundles is shown in Fig. 13.21(b). It can be seen that it is critical to maintain MFP lengths in the range of a micrometer by ensuring freedom from such defects. The improved performance of intermediate and global level CNT bundle interconnects is because of the much lower resistance compared to Cu global interconnects. While Cu interconnect resistance increases linearly with length, in the case of CNT bundle interconnects it is only the scattering resistance (proportional to $h/4e^2$) that increases linearly. The additional resistance arising from imperfect metal-nanotube contacts (which dominates the resistance in the case of short local interconnects) does not increase with length. Hence, Figures 13.20 and 13.21 show only a minor difference between the performance of CNT bundles with perfect contacts and with imperfect contacts. The performance improvement in intermediate and global interconnects as a result of using CNT bundles decreases when the CNT packing density is decreased. This is because when CNT packing density is decreased, the effective resistance increases and hence performance degrades. This is in contrast to short local interconnects for which the effect of interconnect capacitance dominates and a slightly lower CNT density (and hence lower interconnect capacitance) actually leads to improved performance.

13.3.4 Multi-Wall Carbon Nanotubes (MWCNT)

Multi-wall carbon nanotubes (MWCNTs) can be considered to be a coaxial assembly of SWCNT cylinders (shells), one within another. While SWCNTs have diameters in the few nanometer range, MWCNTs may have diameters in a wide range varying from a few to hundreds of nanometers. It has been recently shown that if properly connected to contacts, all the shells in a MWCNT can conduct [99–101]. The conductance of a (SWCNT) graphene shell in a MWCNT increases as the diameter of a shell becomes larger. Since the basic building blocks of an MWCNT are essentially SWCNTs of varying diameters, many of the properties of SWCNTs hold for MWCNTs. The number of shells (N_s) in a MWCNT is diameter dependent, and given by:

$$N_s = 1 + \frac{D_{outer} - D_{inner}}{2\delta}$$

where $\delta = 0.34$ nm (van der Waals distance) is the spacing between adjacent concentric shells, and D_{outer} and D_{inner} are the maximum and minimum shell diameters. The ratio of D_{outer}/D_{inner} has been observed to vary from 0.35 to 0.8 [100, 102]. Preliminary results shown in Fig. 13.22 [98] indicate that for long lengths (a few hundreds of micrometers) MWCNTs can have conductivities several times larger than that of copper or SWCNT bundles.

13.3.5 Using CNTs for On-Chip Communication

CNT interconnects offer a promising alternative to traditional Cu-based interconnects that are reaching their limits with scaling technology, especially at the global interconnect level [3, 4, 88, 94, 95, 98, 103]. While an isolated SWCNT is not a viable option to replace a Cu wire for on-chip communication [3, 94, 95], a CNT bundle consisting of stacked SWCNTs can be a viable replacement for Cu at the intermediate and global interconnect levels. MWCNTs also have the potential to be potential candidates for on-chip communication. Much work, however, still needs to be done to develop and improve fabrication techniques, and resolve the open problems in the area, some of which are enumerated below:

(i) *Inefficient metal-nanotube contacts*: It has been observed [97] that imperfect metal-nanotube contacts that are fabricated today give rise to an additional contact resistance which makes propagation delay on CNT interconnects higher than with Cu interconnects. Making a reliable contact to CNT is very challenging [97], but is a critical issue that needs to be addressed. A few studies [112, 113] have shown that state-of-the-art

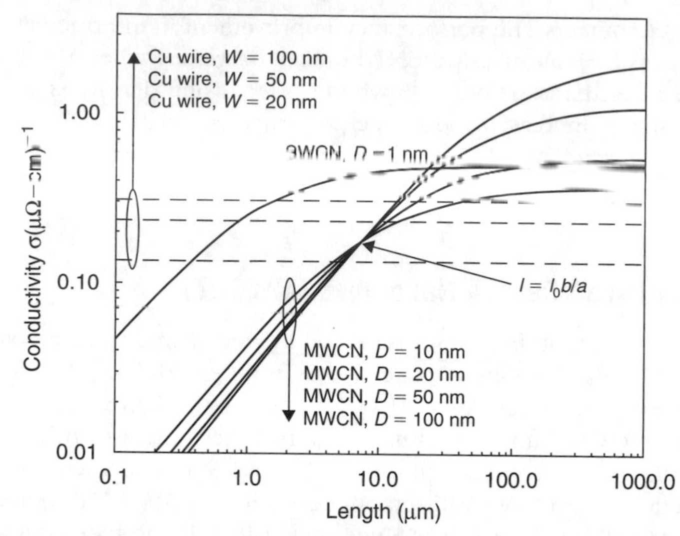

FIGURE 13.22

Conductivity of MWCNTs with various diameters, SWCNT bundle, and Cu interconnect vs. length [98]
© 2006 IEEE

fabrication techniques can make it possible to reduce contact resistance down to a very small value (a few hundred Ω).

(ii) *Small MFP length*: It is important to maintain MFP length for CNTs in the range of a micrometer. Due to limitations in nanotube fabrication technology today and the presence of defects in nanotubes, the MFP is much less than a micrometer. This leads to propagation delays in CNT bundles that are larger than for Cu interconnects in future technologies. Better fabrication techniques are needed to reduce these defects and keep the MFP length in the micrometer range.

(iii) *Density of nanotube bundles*: Densely packed nanotubes are needed for global CNT interconnects that perform better than Cu interconnects. However, due to the lack of control on chirality during nanotube fabrication, any bundle of CNTs consists of metallic as well as semiconducting nanotubes. The semiconducting nanotubes do not contribute to current conduction in CNT bundles, which results in a lower effective density for the CNT bundle (creating a sparse bundle). The sparse CNT bundle that is created has a higher resistance, and consequently poor propagation delay, compared to Cu interconnects. CNT bundles currently fabricated do not have a very high density of CNTs [5], but improvements are needed if CNT interconnects are to become viable replacements for Cu interconnects.

(iv) *Inductive effects at high frequencies*: Currently, inductive effects have been ignored while calculating propagation delay in CNTs. Inductive effects are expected to become significant at very high frequencies (greater than 10 GHz). At such frequencies, these inductive effects can negatively influence performance on CNTs, which needs to be addressed when comparing CNTs to Cu interconnects in future technologies.

13.4 SUMMARY

Traditional on-chip metallic interconnects are beginning to show their limitations due to a number of reasons: high resistivity (due to electron grain and grain boundary scatterings, leading to large propagation delays, and poor performance), low reliability, and high susceptibility to electromigration (at high current densities) with technology and interconnect scaling. Consequently, novel on-chip interconnection schemes need to be explored for future high frequency (giga- and tera-scale) electronic systems. In this chapter, we presented three of the most promising on-chip interconnection technologies that have the potential to mitigate the difficulties faced by current on-chip metallic interconnects, because of their desirable crosstalk and bandwidth properties. Optical interconnects convert data into light and transmit them over an optical waveguide on a chip. Wireless interconnects convert data in RF/wireless signals that can propagate between the transmitting and receiving antennas on the chip, without the need for any metal wires. Finally, CNTs (more specifically CNT bundles or MWCNTs) utilize rolled sheets of graphene to transmit current, with less crosstalk, and lower propagation delay than metallic interconnects. All three of these emerging technologies have several issues and open problems (such as the need for improvements in fabrication technology) that must be resolved before they can be adopted as part of on-chip interconnect fabrics. With the rapid advances in technology, it is only a matter of time before one or more of these technologies becomes feasible and advantageous to use in tomorrow's ULSI designs.

REFERENCES

[1] G. Chen, H. Chen, M. Haurylau, N. A. Nelson, D. H. Albonesi, P. M. Fauchet, and E. G. Friedman, "On-chip copper-based vs. optical interconnects: Delay uncertainty, latency, power, and bandwidth density comparative predictions," in *Proceedings of the IEEE International Interconnect Technology Conference*, June 2006, pp. 39–41.

[2] W. Steinhogl et al., "Size-dependent resistivity of metallic wires in the mesoscopic range," *Physical Review B*, 66, 2002, 075414.

[3] N. Srivastava and K. Banerjee, "Performance analysis of carbon nanotube interconnects for VLSI applications," in *Proceedings of IEEE/ACM International Conference on Computer Aided Design (ICCAD)*, 2005, pp. 383–390.

[4] A. Naeemi, R. Sarvari and J. D. Meindl, "On-chip interconnect networks at the end of the roadmap: limits and nanotechnology opportunities," *2006 International Interconnect Technology Conference (IITC)*, 2006, pp. 201–203.

[5] N. Srivastava and K. Banerjee, "A comparative scaling analysis of metallic and carbon nanotube interconnections for nanometer scale VLSI technologies," in *Proceedings of the 21st International VLSI Multilevel Interconnect Conference*, 2004, pp. 393–398.

[6] Semiconductor Industry Association, "International technology roadmap for semiconductors," 2003.

[7] J.W. Goodman, F. J. Leonberger, S.Y. Kung and R.A. Athale, "Optical interconnections for VLSI systems," in *Proceedings of IEEE*, Vol. 72, No. 7, July 1984, pp. 850–866.

[8] M. Haurylau, G. Chen, H. Chen, J. Zhang, N. A. Nelson, D. H. Albonesi, E. G. Friedman and P. M. Fauchet, "On-chip optical interconnect roadmap: Challenges and critical directions," *IEEE Journal of Selected Topics in Quantum Electronics*, Vol. 12, No. 6, November–December 2006, pp. 1699–1705.

[9] J. Fritts and R. D. Chamberlain, "Breaking the memory bottleneck with an optical data path," In *Proceedings of the 35th Annual Simulation Symposium (April 14–18)*, 2002, pp. 352–362.

[10] X. Han, G. Kim, G. J. Lipovski and R. T. Chen, "An optical centralized shared-bus architecture demonstrator for microprocessor-to-memory interconnects," *IEEE Journal of Selected Topics in Quantum Electronics*, Vol. 9, No. 2, March–April 2003, pp. 512–517.

[11] A. G. Kirk, D. V. Plant, M. H. Ayliffe, M. Chateauneuf and F. Lacroix, "Design rules for highly parallel free-space optical interconnects," *IEEE Journal of Selected Topics in Quantum Electronics*, Vol. 9, No. 2, March–April 2003, pp. 531–547.

[12] D.V. Plant, "System design of chip and board level optical interconnects," in *Proceedings of the 2004 Meeting Bipolar/BiCMOS Circuits and Technology*, September 13–14, 2004, pp. 72–78.

[13] N. Savage, "Linking with light," *IEEE Spectrum*, Vol. 39, No. 8, August 2002, pp. 32–36.

[14] H. Cho, P. Kapur and K. C. Saraswat, "Power comparison between high-speed electrical and optical interconnects for interchip communication," *Journal of Lightwave Technology*, Vol. 22, No. 9, September 2004, pp. 2021–2033.

[15] H. Rong, R. Jones, A. Liu, O. Cohen, D. Hak, A. Fang and M. Paniccia, "A continuous-wave Raman silicon laser," *Nature*, Vol. 433, February 2005, pp. 725–728.

[16] L. Pavesi, "Routes toward silicon-based laser," *Materials Today*, Vol. 8, No. 1, January 2005, pp. 18–25.

[17] C. A. Barrios, V. R. de Almeida and M. Lipson, "Low-power consumption short-length and high-modulation-depth silicon electrooptic modulator," *Journal of Lightwave Technology*, Vol. 21, No. 4, April 2003, pp. 1089–1098.

[18] L. Liao, D. Samara-Rubio, M. Morse, A. Liu, D. Hodge, D. Rubin, U. Keil and T. Franck, "High-speed silicon Mach-Zehnder modulator," *Optics Express*, Vol. 13, No. 8, April 2005, pp. 3129–3135.

[19] A. Liu, R. Jones, L. Liao, D. Samara-Rubio, D. Rubin, O. Cohen, R. Nicolaescu and M. Paniccia, "A high-speed silicon optical modulator based on a metal-oxide-semiconductor capacitor," *Nature*, Vol. 427, No. 6975, February 2004, pp. 615–618.

[20] Q. F. Xu, B. Schmidt, S. Pradhan and M. Lipson, "Micrometre-scale silicon electro-optic modulator," *Nature*, Vol. 435, No. 7040, May 2005, pp. 325–327.

[21] V. R. Almeida, C. A. Barrios, R. R. Panepucci, M. Lipson, M. A. Foster, D. G. Ouzounov and A. L. Gaeta, "All-optical switching on a silicon chip," *Optics Letters*, Vol. 29, No. 24, December 2004, p. 2867.

[22] N. Kirman, M. Kirman, R. K. Dokania, J. F. Martinez, A. B. Apsel, M. A. Watkins and D. H. Albonesi, "Leveraging optical technology in future bus-based chip multiprocessors," In *Proceedings of the 39th Annual IEEE/ACM International Symposium on Microarchitecture*, 2006, pp. 492–503.

[23] P. Rabiei, W. H. Steier, C. Zhang and L. R. Dalton, "Polymer micro-ring filters and modulators," *Journal of Lightwave Technology*, Vol. 20, No. 11, November 2002, pp. 1968–1975.

[24] J. Tatum, "VCSELs for 10 GB/s optical interconnects," In *IEEE Emerging Technologies Symposium on BroadBand Communications for the Internet Era*, Richardson, TX, September 2001, pp. 58–61.

[25] T. Yin, A. M. Pappu and A. B. Apsel, "Low-cost, high efficiency, and high-speed SiGe phototransistors in commercial BiCMOS," *IEEE Photonics Technology Letters*, Vol. 18, No. 1, January 2006, pp. 55–57.

[26] M. R. Reshotko, D. L. Kencke and B. Block, "High-speed CMOS compatible photodetectors for optical interconnects," in *Proceedings of SPIE,* Vol. 5564, October 2004, pp. 146–155.

[27] S. J. Koester, J. D. Schaub, G. Dehlinger, J. O. Chu, Q. C. Ouyang, and A. Grill, "High-efficiency, Ge-on-SOI lateral PIN photodiodes with 29 GHz bandwidth," in *Proceedings of the Device Research Conference*, Notre Dame, IN, 2004, pp. 175–176.

[28] A. M. Pappu and A. B. Apsel, "A low power, low delay TIA for on-chip applications," *Conference on Lasers and Electro-Optics*, May 2005, pp. 594–596.

[29] R. Ho, K. W. Mai and M. A. Horowitz, "The future of wires," in *Proceedings of IEEE*, Vol. 89, No. 4, April 2001, pp. 490–504.

[30] M. Kobrinsky, B. Block, J.-F. Zheng, B. Barnett, E. Mohammed, M. Reshotko, F. Robertson, S. List, I. Young and K. Cadien, "On-chip optical interconnects," *Intel Technology Journal*, Vol. 08, No. 02, May 2004, pp. 129–141.

[31] J. H. Collet, F. Caignet, F. Sellaye and D. Litaize, "Performance constraints for on chip optical interconnects," *IEEE Journal of Selected Topics in Quantum Electronics*, Vol. 9, No. 2, March–April 2003, pp. 425–432.

[32] M. J. Kobrinsky, B. A. Block, J.-F. Zheng, B. C. Barnett, E. Mohammed, M. Reshotko, F. Robertson, S. List, I. Young and K. Cadien, "On-chip optical interconnects," *Intel Technology Journal*, Vol. 8, No. 2, May 2004, pp. 129–141.

[33] K.-N. Chen, M. J. Kobrinsky, B. C. Barnett and R. Reif, "Comparisons of conventional, 3-D, optical, and RF interconnects for on-chip clock distribution," *IEEE Transactions on Electron Devices*, Vol. 51, No. 2, February 2004, pp. 233–239.

[34] B. Ackland, B. Razavi and L. West, "A comparison of electrical and optical clock networks in nanometer technologies," in *Proceedings of the IEEE 2005 Custom Integrated Circuits Conference*, 2005, September 18–21, 2005, pp. 779–782.

[35] I. O'Connor, "Optical solutions for system-level interconnect," in *Proceedings of the 2004 International Workshop on System Level interconnect Prediction (SLIP)*, 2004, pp. 79–88.

[36] J. Qi, C. L. Schow, L. D. Garrett and J. C. Campbell, "A silicon NMOS monolithically integrated optical receiver," *IEEE Photonics Technology Letters*, Vol. 9, May 1997, pp. 663–665.

[37] T. K. Woodward and A. V. Krishnamoorthy, "1Gb/s CMOS photoreceiver with integrated detector operating at 850 nm," *Electronics Letters*, Vol. 34, 1998, pp. 1252–1253.

[38] T. Heide, A. Ghazi, H. Zimmermann and P. Seegebrecht, "Monolithic CMOS photoreceivers for short-range optical data communications," *Electronics Letters*, Vol. 35, 1999, pp. 1655–1656.

[39] H. Zimmermann, T. Heide and A. Ghazi, "Monolithic high-speed CMOS-photoreceiver," *IEEE Photonics Technology Letters*, Vol. 11, February 1999, pp. 254–256.

[40] M. Briere, E. Drouard, F. Mieyeville, D. Navarro, I. O'Connor and F. Gaffiot, "Heterogeneous modelling of an optical network-on-chip with SystemC," *16th IEEE International Workshop on Rapid System Prototyping (RSP)*, 2005, pp. 10–16.

[41] A. Shacham, K. Bergman and L. P. Carloni, "On the design of a photonic network-on-chip," *First International Symposium on Networks-on-Chip*, 2007, pp. 53–64.

[42] http://www.intel.com/research/platform/sp/hybridlaser.htm.

[43] S. M. Weiss, M. Molinari and P. M. Fauchet, "Temperature stability for silicon-based photonic band-gap structures," *Applied Physics Letters*, Vol. 83, No. 10, September 2003, pp. 1980–1982.

[44] M. F. Chang, V. P. Roychowdhury, L. Zhang, H. Shin and Y. Qian, "RF/wireless interconnect for inter- and intra chip communications," in *Proceedings of the IEEE*, Vol. 89, No. 4, April 2001, pp. 456–466.

[45] H. Arai and T. Itoh, "Proximity coupled power combiner/divider using parasitic element," in *1997 Asia-Pacific Microwave Conference Digest*, 1997, pp. 1133–1136.

[46] K. C. Gupta et al., *Computer-Aided Design of Microwave Circuits*, Artech House, Norwell, MA, 1981, pp. 69.

[47] G. Hasnain et al., "Dispersion of picosecond pulses in coplanar transmission lines," *IEEE Transactions on Microwave Theory and Techniques*, Vol. 34, June 1989, pp. 738–741.

[48] M. Y. Frankel et al., "Terahertz attenuation and dispersion characteristics of coplanar transmission lines," *IEEE Transactions on Microwave Theory and Techniques*, Vol. 39, June 1991, pp. 910–916.

[49] K. Feher, *Digital Communications*, Prentice-Hall, Englewood Cliffs, NJ, 1981.

[50] D. R. Pehlke, A. Burstein and M. F. Chang, "Extremely high-Q tunable inductor for Si-based RF integrated circuit applications," in *International Electron Devices Meeting, Technical Digest 1997*, Washington, DC, 1997, pp. 63–66.

[51] L. P. B. Katehi and K. Herrick, "Si-micromachining in MM-wave circuits," in *Proceedings of the 1997 Topical Symposium on Millimeter Waves*, New York, 1997, pp. 85–88.

[52] R. T. Chang, N. Talwalkar, C. P. Yue and S. S. Wong, "Near speed-of-light signaling over on-chip electrical interconnects," *IEEE Journal of Solid-State Circuits*, Vol. 38, No. 5, May 2003, pp. 834–838.

[53] M. F. Chang, I. Verbauwhede, C. Chien, Z. Xu, J. Kim, J. Ko, Q. Gu and B. Lai, "Advanced RF/baseband interconnect schemes for inter- and intra-ULSI communications," *IEEE Transactions on Electron Devices*, Vol. 52, No. 7, July 2005, pp. 1271–1285.

[54] Z. Xu, H. Shin, J. Kim, M. F. Chang and C. Chien, "Giga bit/s CDMA interconnect transceiver chip-set with multilevel signal data recovery for re-configurable VLSI system," in *Proceedings of the IEEE International Solid-State Circuits Conference Technical Digest*, San Francisco, CA, February 2003, pp. 322–323.

[55] J. S. Lee and L. E. Miller, *CDMA Systems Engineering Handbook*, Artech House, Norwell, MA, 1998.

[56] A. J. Viterbi, "CDMA principles of spread spectrum communication," *Communication Series*, Addison-Wesley, Reading, MA, 1995.

[57] R. Farjad-Rad et al., "A 0.3-μm CMOS 8-Gb/s 4-PAM serial link transceiver," *Symposium on VLSI Circuits*, May 1999, pp. 41–44.

[58] Y. Wang, D. Makadia and M. Margala, "On-chip integrated antennas—The first challenge for reliable on-chip wireless interconnects," *Canadian Conference on Electrical and Computer Engineering*, May 2006, pp. 2322–2325.

[59] K. Carver and J. W. Mink, "Microstrip antenna technology," *IEEE Transactions on Antennas and Propagation*, Vol. AP-29, No. 1, January 1981, pp. 2–24.

[60] David M. Pozar, "Considerations for millimeter wave printed antennas," *IEEE Transactions on Antennas and Propagation*, Vol. AP-31, No. 5, September 1983, pp. 740–747.

[61] H. Nakano and J. Yamauchi, "A consideration of curved halfwave dipoles," *Transactions of IECE Japan*, Vol. J61-B, 1978, pp. 210–213.

[62] D. M. Pozar, *Microwave and RF Design of Wireless Systems*, John Wiley & Sons, New York, 2000, p. 137.

[63] J. Branch, X. Guo, L. Gao, A. Sugavanam, J.-J. Lin and K. K. O, "Wireless communication in a flip-chip package using integrated antennas on silicon substrates," *Electron Device Letters, IEEE*, Vol. 26, No. 2, February 2005, pp. 115–117.

[64] K. K. O, K. Kim, B. Floyd, J. Mehta, H. Yoon, C.-M. Hung, D. Bravo, T. Dickson, X. Guo, R. Li, N. Trichy, J. Caserta, W. Bomstad, J. Branch, D.-J. Yang, J. Bohorquez, E.-Y. Seok, L. Gao, A. Sugavanam, J.-J. Lin, S. Yu, C. Cao, M.-W. Hwang, Y.-P. Ding, S.-H. Hwang, H. Wu, N. Zhang and J. E. Brewer, "The feasibility of on-chip interconnection using antennas," *IEEE International Conference on Computer Aided Design*, San Jose, CA, November 2005, pp. 976–981.

[65] P. Benech, F. Ndagijimana, A. Triantafyllou, A. Farcy and J. Torres, "Design and performance of integrated antennas for wireless intra chip interconnections," *IEEE 32nd Annual Conference on Industrial Electronics (IECON)*, November 2006, pp. 2953–2957.

[66] K. K. O, K. Kim, B. A. Floyd, J. Mehta and H. Yoon, "Inter and intra-chip wireless clock signal distribution using microwaves: A status of an initial feasibility study," *Government Microcircuit Applications Conference*, Monterey, CA, March 1999, pp. 306–309.

[67] X. Guo, J. Caserta, R. Li, B. Floyd and K. K. O, "Propagation layers for intra-chip wireless interconnection compatible with packaging and heat removal," *2002 Symposium on VLSI Technology*, June 2002, pp. 36–37.

[68] B. A. Floyd et al., "Wireless interconnection in a CMOS IC with integrated antennas," *Proceeding of ISSCC*, 2000, pp. 328–329.

[69] X. Guo, R. Li and K. K. O, "Design guidelines for reducing the impact of metal interference structures on the performance on-chip antennas," *IEEE AP-S International Symposium and USNC/URSI National Radio Science Meeting*, Vol. 1, June 2003, pp. 606–609.

[70] E.-Y. Seok and K. K. O, "Design rules for improving predictability of on-chip antenna characteristics in the presence of other metal structures," *2005 International Interconnect Technology Conference*, San Francisco, CA, June 2005, pp. 120–123.

[71] J. Mehta and K. K. O, "Switching noise of integrated circuits (IC's) picked up by a planar dipole antenna mounted near the IC's," *IEEE Transactions on Electro-Magnetic Compatibility*, Vol. 44, No. 5, May 2002, pp. 282–290.

[72] D. Bravo, H. Yoon, K. Kim, B. Floyd and K. K. O, "Estimation of the signal-to-noise ratio for on-chip wireless clock signal," in *Proceedings of International Interconnect Technology Conference*, June 2000, pp. 9–11.

[73] T. O. Dickson, D. Bravo and K. K. O, "Noise coupling to on-chip antennas," *2002 IEEE International Symposium on EMC*, Vol. 1, 2000, pp. 340–344.

[74] T. O. Dickson, B. Floyd and K. K. O, "Jitter in a wireless clock distribution system," *International Interconnect Technology Conference*, June 2002, pp. 154–156.

[75] K. Kim, "Design and characterization of RF components for inter and intra-chip wireless communications," PhD Dissertation, University of Florida, 2000.

[76] K. K. O et al., "On-chip antennas in silicon ICs and their application," *IEEE Transactions on Electron Devices*, Vol. 52, No. 7, July 2005, pp. 1312–1323.

[77] F. Kreupl et al., "Carbon nanotubes in interconnect applications," *Microelectronic Engineering*, 2002, pp. 399–408.

[78] J. Li et al., "Bottom-up approach for carbon nanotube interconnects," *Applied Physics Letters*, Vol. 82, No. 15, April 2003, pp. 2491–2493.

[79] B. Q. Wei et al., "Reliability and current carrying capacity of carbon nanotubes," *Applied Physics Letters*, Vol. 79, No. 8, 2001, pp. 1172–1174.

[80] P. G. Collins et al., "Current saturation and electrical breakdown in multiwalled carbon nanotubes," *Physical Review Letters*, Vol. 86, No. 14, 2001, pp. 3128–3131.

[81] S. Berber et al., "Unusually high thermal conductivity of carbon nanotubes," *Physical Review Letters*, Vol. 84, No. 20, 2000, pp. 4613–4616.

[82] P. L. McEuen et al., "Single-walled carbon nanotube electronics," *IEEE Transactions on Nanotechnology*, Vol. 1, No. 1, 2002, pp. 78–85.

[83] A. Thess et al., "Crystalline ropes of metallic carbon nanotubes," *Science*, Vol. 273, No. 5274, 1996, pp. 483–487.

[84] J. Li et al., "Carbon nanotube interconnects: A process solution," *IEEE International Interconnect Technology Conference*, 2003, pp. 271–272.

[85] M. Nihei et al., "Carbon nanotube vias for future LSI interconnects," *IEEE International Interconnect Technology Conference*, 2004, pp. 251–253.

[86] C. Schonenberger et al., "Interference and interaction in multi-wall carbon nanotubes," *Applied Physics A*, Vol. 69, 1999, pp. 283–295.

[87] A. Bachtold et al., "Scanned probe microscopy of electronic transport in carbon nanotubes," *Physical Review Letters*, Vol. 84, No. 26, 2000, pp. 6082–6085.

[88] T. Raja, V. D. Agrawal and M. L. Bushnell, "A tutorial on the emerging nanotechnology devices," *17th International Conference on VLSI Design*, 2004, pp. 343–360.

[89] P. J. Burke, "Luttinger liquid theory as a model of the gigahertz electrical properties of carbon nanotubes," *IEEE Transactions on Nanotechnology*, Vol. 1, No. 3, 2002, pp. 129–144.

[90] S. Datta, "Electrical resistance: An atomistic view," *Nanotechnology*, Vol. 15, 2004, pp. S433–S451.

[91] J. Kong, E. Yenilmez, T. W. Tombler, W. Kim, H. Dai, R. B. Laughlin, L. Liu, C. S. Jayanthi and S. Y. Wu, "Quantum interference and ballistic transmission in nanotube electron waveguides," *Physical Review Letters*, Vol. 87, No. 10, 2001, pp. 106801-1–106801-4.

[92] J.-Y. Park, S. Rosenblatt, Y. Yaish, V. Sazonova, H. Ustunel, S. Braig, T. A. Arias, P. Brouwer and P. L. McEuen, "Electron–phonon scattering in metallic single-walled carbon nanotubes," *Nano Letters*, Vol. 4, No. 3, 2004, pp. 517–520.

[93] Z. Yu and P. J. Burke, "Microwave transport in metallic single-walled carbon nanotubes," *Nano Letters*, Vol. 5, No. 7, 2005, pp. 1403–1406.

[94] A. Raychowdhury and K. Roy, "Modeling of metallic carbon nanotube interconnects for circuit simulation and a comparison with Cu interconnects for scaled technologies," *IEEE Transactions on Computer-Aided Design of IC's*, January 2006, pp. 58–65.

[95] A. Raychowdhury and K. Roy, "A circuit model for carbon nanotube interconnects: Comparative study with Cu interconnects for scaled technologies," in *Proceedings of the 2004 IEEE/ACM International Conference on Computer-Aided Design (ICCAD)*, 2004, pp. 237–240.

[96] K. Nabors and J. White, "FastCap: A multipole accelerated 3-D capacitance extraction program," *IEEE Transactions on Computer-Aided Design of Integrated Circuits and Systems*, Vol. 10, No. 11, 1991, pp. 1447–1459.

[97] Th. Hunger, B. Lengeler and J. Appenzeller, "Transport in ropes of carbon nanotubes: Contact barriers and Luttinger liquid theory," *Physical Review B*, Vol. 69, 2004, pp. 195406-1–195406-4.

[98] A. Naeemi and J. D. Meindl, "Compact physical models for multiwall carbon-nanotube interconnects," *Electron Device Letters, IEEE*, Vol. 27, 2006, pp. 338–340.

[99] H. J. Li, W. G. Lu, J. J. Li, X. D. Bai and C. Z. Gu, "Multichannel ballistic transport in multiwall carbon nanotubes," *Physical Review Letters*, Vol. 95, No. 8, August 2005, pp. 086601-1–086601-4.

[100] M. Nihei, D. Kondo, A. Kawabata, S. Sato, H. Shioya, M. Sakaue, T. Iwai, M. Ohfuti and Y. Awano, "Low-resistance multi-walled carbon nanotube vias with parallel channel conduction of inner shells," in *Proceedings of IEEE International Interconnect Technology Conference*, June 2005, pp. 234–236.

[101] Q. Yan, J. Wu, G. Zhou, W. Duan and B. Gu, "Ab initio study of transport properties of multiwalled carbon nanotubes," *Physical Review B, Condensed Matter*, Vol. 72, No. 15, October 2005, pp. 155425-1–155425-5.

[102] L. Forró and C. Schönenberger, "Physical properties of multi-wall nanotubes," *Topics in Applied Physics, Carbon Nanotubes: Synthesis, Structure, Properties and Applications*, M. S. Dresselhaus, G. Dresselhaus and P. Avouris (Eds.), Springer-Verlag, Berlin, Germany, 2000.

[103] A. Naeemi and J. D. Meindl, "Monolayer metallic nanotube interconnects: Promising candidates for short local interconnects," *Electron Device Letters, IEEE*, Vol. 26, 2005, pp. 544–546.

[104] M. Y. Frankel et al., "Terahertz attenuation and dispersion characteristics of coplanar transmission lines," *IEEE Transactions on Microwave Theory and Techniques*, Vol. 39, June 1991.

[105] S. Pasricha and N. Dutt, "ORB: An on-chip optical ring bus communication architecture for multi-processor systems-on-chip," *IEEE Asia and South Pacific Design Automation Conference (ASPDAC 2008)*, Seoul, Korea, January 2008.

[106] M. S. Dresselhaus, G. Dresselhaus and P. Avouris, *Carbon Nanotubes: Synthesis, Structure, Properties, and Applications*, Springer, 2001.

[107] B. Q. Wei, R. Vajtai and P. M. Ajayan, "Reliability and current carrying capacity of carbon nanotubes," *Applied Physics Letters*, 2001, pp. 1172–1174.

[108] J. Hone, M. Whitney, C. Piskoti and A. Zettl, "Thermal conductivity of single-walled carbon nanotubes," *Physical Review, B*, Vol. 59, No. 4, 1999, pp. R2514–R2516.

[109] F. Kreupl, A. P. Graham, M. Liebau, G. S. Duesberg, R. Seidel and E. Unger, "Carbon nanotubes for interconnect applications," *IEDM Technical Digest*, December 2004.

[110] A. Naeemi and J. D. Meindel, "Design and performance modeling for single-walled carbon nanotubes as local, semiglobal, and global interconnects in gigascale integrated systems," *Transactions on EDL*, January 2007.

[111] M. W. Beattie and L. T. Pileggi, "Inductance 101: Modeling and extraction," in *Proceedings of the Design Automation Conference (DAC)*, 2001, pp. 323–328.

[112] S. Sato, M. Nihei, A. Mimura, A. Kawabata, D. Kondo, H. Shioya, T. Iwai, M. Mishima, M. Ohfuti and Y. Awano, "Novel approach to fabricate carbon nanotube via interconnects using size-controlled catalyst nanoparticles," in *Proceedings of International Interconnect Techno logy Conference (IITC)*, 2006, pp. 230–232.

[113] O. Hjortstam, P. Isberg, S. Söderholm and H. Dai, "Can we achieve ultra-low resistivity in carbon nanotube-based metal composites?," *Applied Physics A, Materials Science and Processing*, 2004, pp. 1175–1179.

[114] W. M. Green, M. J. Rooks, L. Sekaric and Y. A. Vlasov, "Ultra-compact, low RF power, 10 Gb/s silicon Mach-Zehnder modulator," *Optics Express*, Vol. 15, No. 25, December 2007, pp. 17106–17113.

Index

B

C

D

E

I

J

K

L

Q

R

S

T

U

V

W

X

Z